I0047450

James Harkness, Frank Morley

A treatise on the theory of functions

James Harkness, Frank Morley

A treatise on the theory of functions

ISBN/EAN: 9783742892454

Manufactured in Europe, USA, Canada, Australia, Japa

Cover: Foto ©berggeist007 / pixelio.de

Manufactured and distributed by brebook publishing software
(www.brebook.com)

James Harkness, Frank Morley

A treatise on the theory of functions

ON THE

THEORY OF FUNCTIONS

BY

JAMES HARKNESS, M.A.

ASSOCIATE PROFESSOR OF MATHEMATICS IN BRYN MAWR COLLEGE, PA.
LATE SCHOLAR OF TRINITY COLLEGE, CAMBRIDGE

AND

FRANK MORLEY, M.A.

PROFESSOR OF PURE MATHEMATICS IN HAVERFORD COLLEGE, PA.
LATE SCHOLAR OF KING'S COLLEGE, CAMBRIDGE

New York
MACMILLAN AND CO.
AND LONDON
1893

All rights reserved

PREFACE.

In this book we have sought to give an account of a department of mathematics which is now generally regarded as fundamental. A list of the men to whom the successive advances of the subject are due, includes, with few exceptions, the names of the greatest French and German mathematicians of the century, from Cauchy and Gauss onward. And in line with these advances lie the chief fields of mathematical activity at the present day.

The most legitimate extensions of elementary analysis lead so directly into the Theory of Functions, that recent writers on Algebra, Trigonometry, the Calculus, etc., give theories which are indispensable parts of our subject. But since these theories are not found in many current text-books, it appears most convenient for the generality of readers to make the earlier chapters complete in themselves. Thus an account is given in ch. i. of the geometric representation of elementary operations; and in ch. iii., before the introduction of Weierstrass's theory of the analytic function, the theory of convergence is discussed at some length.

We have aimed at a full presentation of the standard parts of the subject, with certain exceptions. Of these exceptions, three must be stated. In ch. ii., the theory of real functions of a real variable is given only so far as seems necessary as a basis for what follows. In the account of Abelian integrals (ch. x.), our object is to induct the reader as simply and rapidly as possible into what is itself a suitable theme for more than one large volume. And we have entirely passed over the automorphic functions, since it was not possible to give even an introductory sketch within the space at our disposal. However, an account of some of Kronecker's work, which is necessary for the study of Klein's recent developments of the theory of Abelian functions, is included in ch. vi.; and ch. viii. is devoted to a somewhat condensed treatment of double theta-functions, which goes further than is necessary for our immediate purpose, for the reason that the subject is not very accessible in the English language.

Progress is intentionally slow in some places, where what is required is a formation of certain new concepts, rather than an enlargement of ideas that pre-exist.

As to the place of the Theory of Functions in the order of those mathematical studies which appear in all curricula, its more elementary parts can be attacked with advantage so soon as a sound knowledge of the Integral Calculus is gained. It will be found that, though collateral subjects such as the theory of Algebraic Equations and the analytic theory of Plane Curves are freely referred to, a previous knowledge of them is rarely necessary for the understanding of what follows. It may be added that an acquaintance with the present subject is requisite for the study of the modern theory of Differential Equations. It is presupposed, for example, in Dr. Craig's treatise.

The many writers to whom we are indebted for theories or for elucidations are, of course, referred to in the text. There has appeared quite recently the very important treatise of Dr. Forsyth, unfortunately too late to be included in these references.

A Glossary is added, which gives the principal technical terms employed by German and French writers, with the adopted equivalents. The page on which these equivalents are defined will be found by consulting the Index.

To our respective colleagues, Professor C. A. Scott and Professor E. W. Brown, our hearty thanks are due for valuable assistance with the proof-sheets.

May, 1893.

CONTENTS.

The reader to whom the subject is new may omit chapter ii., chapter vi. from § 184 to the end, chapter viii. from § 233 to the end, and chapter ix.

In the following table the numbers refer to the pages.

CHAPTER I.

Geometric Introduction.

CHAPTER II.

Real Functions of a Real Variable.

CHAPTER III.

The Theory of Infinite Series.

CHAPTER IV.

Algebraic Functions.

CHAPTER V.

INTEGRATION.

CHAPTER VI.

RIEMANN SURFACES.

CHAPTER VII.

ELLIPTIC FUNCTIONS.

CHAPTER VIII.

Double Theta-Functions.

CHAPTER IX.

Dirichlet's Problem.

CHAPTER X.

Abelian Integrals.

CHAPTER I.

Geometric Introduction.

§ 1. A real number x can be represented by a point on a straight line, which lies at a distance x units of length from a fixed point o of the line; and conversely to every point of the line corresponds a definite number. This representation presupposes a precise definition of irrational real numbers (see Chapter II.); also it raises the question of the continuity of the system of real numbers.

To represent the complex number $x+iy$, take two lines ox, oy at right angles to each other, and measure off distances om along ox, and mp parallel to oy, such that om, $mp = x$, y units of length. The number $x+iy$ is considered as attached to the point p. In this representation, neighbouring numbers are attached to neighbouring points, and there is a (1, 1) correspondence between the numbers of the complete number system and the points of the plane.

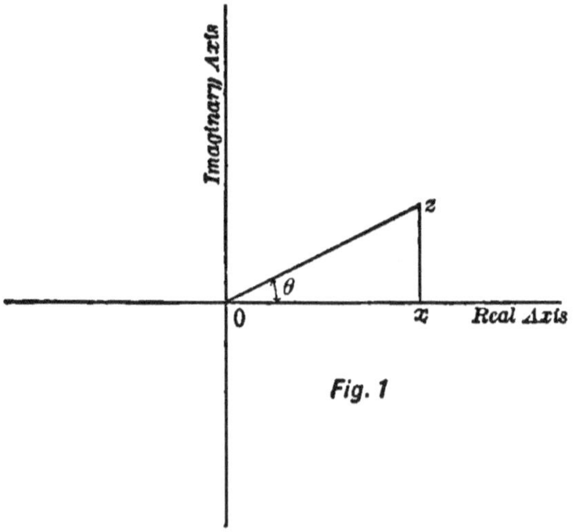

Fig. 1

From this point of view the *complex number* $x+iy$ is attached to the *point* whose rectangular Cartesian co-ordinates are x, y.

The number $x+iy$ can be replaced by a single symbol z, called the *affix* of the point. We shall name the point by its affix, so that by the point z we shall mean the point to which z or $x+iy$ is attached. It is understood throughout that x and y are real.

The complex number is constructed out of two elements, each of which is capable of assuming a singly infinite series of values; this is the reason that it needs for its representation a doubly infinite series of points. The plane is chosen as the field on which these points are to lie; and the figure formed by representing points in this way is referred to as Argand's diagram. Other representations are possible; as, for instance, on a sphere. We shall have occasion to establish a (1, 1) correspondence between the points of a sphere and the doubly infinite series of values of z, and thus to use the sphere as the field on which z is represented.

§ 2. *Absolute value and amplitude.*

Let ρ, θ_0 be the polar co-ordinates of the point z, where $0 \leqq \theta_0 < 2\pi$. The quantity ρ, or $+\sqrt{x^2+y^2}$, is called the *absolute value* or modulus of z, and is denoted by $|z|$. Thus the absolute value is represented by the distance of the point from the origin taken positively. The *amplitude* of a point z is the angle $\theta_0 + 2m\pi$, where m is any integer, positive or negative. It differs from the absolute value in not being one-valued. It may be denoted by $am(z)$.*

It is clear that as z describes a continuous curve in the plane of z (which will henceforth be called the z-plane), $|z|$ changes continuously. When an amplitude of z has been selected, that of any point on the curve, which is infinitely near to z, is understood to be that one which is infinitely near to the selected amplitude. In this way the amplitude is determined throughout the curve. There is one exception to the continuity of the amplitude; when z passes through the origin, there is an abrupt change of π in the value of its amplitude. There is an obvious advantage in choosing for the amplitude, when possible, the least positive turn which will bring the line $0x$ into coincidence with the line $0z$; this turn, which we have called θ_0, may be called the chief value of the amplitude. This plan will be generally adopted, and will not need explicit statement. The context will show when it is inapplicable.

§ 3. *The representation of the point $z_1 + z_2$.*

We are to construct the point whose affix is the sum of the affixes of two given points.

* The amplitude is sometimes called the argument.

Let $$z_1 = x_1 + iy_1,\ z_2 = x_2 + iy_2;$$
then $$z_1 + z_2 = x_1 + x_2 + i(y_1 + y_2).$$

The co-ordinates of the fourth corner z_3 of the parallelogram $0z_1z_3z_2$ are $x_1 + x_2$, $y_1 + y_2$; hence, z_3 is the point which represents $z_1 + z_2$.

The representation of the point $z_1 + z_2 + z_3$.

Let z_1, z_2, z_3 be any three points in the plane. The point $z_1 + z_2 + z_3$ can be found by the following construction: draw lines $\overline{z_1p_2}$, $\overline{p_2p_3}$ parallel and equal to $\overline{0z_2}$, $\overline{0z_3}$. The point p_3 has for its affix $z_1 + z_2 + z_3$. And so on for the sum of any number of z's.

§ 4. *Subtraction.* If $z_1 + z_2 = z_3$, then of course $z_3 - z_1 = z_2$. The construction is as follows: join z_1 to z_3, and draw $\overline{0z_2}$ parallel and equal to $\overline{z_1z_3}$. The point z_2 represents the difference $z_3 - z_1$.

§ 5. *Multiplication.* To construct the point z_1z_2, use polar co-ordinates, and write $z_1 = \rho_1(\cos\theta_1 + i\sin\theta_1)$, $z_2 = \rho_2(\cos\theta_2 + i\sin\theta_2)$. By ordinary multiplication

$$z_1z_2 = \rho_1\rho_2\{\cos(\theta_1 + \theta_2) + i\sin(\theta_1 + \theta_2)\},$$

so that the point z_1z_2 has polar co-ordinates $\rho_1\rho_2$, $\theta_1 + \theta_2$.

We see that the absolute value of the product is equal to the product of the absolute values, and the amplitude of the product is equal to the sum of the amplitudes; this has been proved for two factors, and can be proved at once for any number.

§ 6. *Division.* $$z_1/z_2 = \rho_1(\cos\theta_1 + i\sin\theta_1)/\rho_2(\cos\theta_2 + i\sin\theta_2)$$
$$= \rho_1/\rho_2 \cdot \{\cos(\theta_1 - \theta_2) + i\sin(\theta_1 - \theta_2)\}.$$

Thus the absolute value of a quotient is the quotient of the absolute values, and the amplitude is the difference of the amplitudes of the numerator and denominator.

§ 7. So far we have treated z as a quantity attached to a point without assigning any meaning to z itself. But a meaning can be assigned to z which is very useful for many applications. To this we now proceed.

Strokes. When the magnitude and direction, but not the position, of a quantity are taken into account, the directed quantity is called a *stroke* or vector. The magnitude is called its absolute value or modulus; the direction of the stroke (*i.e.* its inclination to some fixed line) is called its amplitude or argument. A displacement of a point from the position a to the position b is a stroke, which we denote by $\overline{ab}$. If a rigid body be displaced from one position to another, without rotation, so that a, a_1 are displaced to b, b_1, the definition implies that the strokes $\overline{ab}$ and $\overline{a_1b_1}$ are equal. The strokes $\overline{ab}$, $\overline{b_1a_1}$ are not equal, but opposite. Equality of strokes implies two equalities: equality of absolute values and equality of amplitudes.

In the comparison of strokes it is convenient, in general, to refer them to the same initial or zero-point, which we shall call the origin, and denote by 0. From 0 we draw a directed straight line $0x$, which serves as zero of direction. On this line we take points a, b equidistant from 0. Then $\overline{b0} = \overline{0a}$. Now by the expression $\overline{b0} + \overline{0b}$ is meant a displacement from b to 0, followed by a displacement from 0 to b. The resultant displacement is evidently zero, or $\overline{b0} + \overline{0b} = 0$. Thus $\overline{b0} = -\overline{0b}$, and $\overline{0a} = \overline{b0}$; therefore $\overline{0b} = (-1)\overline{0a}$. In this equation we may regard -1 as an operator which turns $\overline{0a}$, in the plane of the paper, through the angle π. Mark off in the line $0y$ perpendicular to $0x$ points c and d, whose distances from 0 are equal to that of a or b. Let i be the operator which changes the stroke $\overline{0a}$ into the stroke $\overline{0c}$. Then $\overline{0c} = i \cdot \overline{0a}$, and we may regard i as a turn through $\pi/2$, counter-clockwise, in the plane of the paper. Now $\overline{0b}$ arises from $\overline{0c}$ by such a turn, therefore

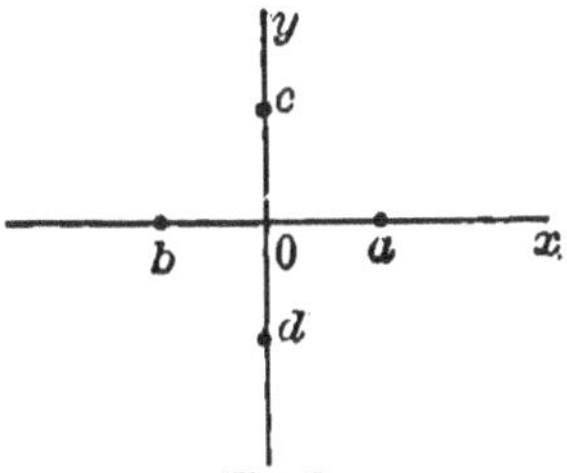

Fig. 2

$$\overline{0b} = i \cdot \overline{0c} = i^2 \cdot \overline{0a}.$$

But $$\overline{0b} = (-1)\,\overline{0a};$$

hence $$i^2 \cdot \overline{0a} = (-1) \cdot \overline{0a}.$$

Since the operators -1 and i^2 have the same effect on $\overline{0a}$, we are led to represent the operator i by the symbol $\sqrt{-1}$, and to interpret the symbol i by a turn through a right angle. If, in Fig. 2, $\overline{0a}$ be of length 1, we denote it merely by 1, and generally by any real positive quantity x we mean the stroke $\overline{0x}$; $\overline{0c}$ is $i \cdot 1$, or simply i, it being understood that the subject of operation, when not mentioned, is the stroke unity. Further, $\overline{0b} = i^2 = -1$, $\overline{0d} = i^3 = -i$, $\overline{0a} = i^4 = 1$. The stroke iy is the line which joins 0 to a point on the axis of y, distant y units from the origin, the point being above or below the line $0x$ according as y is positive or negative.

To interpret the sum of two real numbers, say $2 + 3$, we suppose the displacement 3 to follow on the displacement 2, so that the beginning of 3 is the end of 2. Thus in a sum only the first stroke is drawn from the origin. We define the addition of any two strokes in the same way. Thus, $2+i3$ shall mean two steps to the right and 3 upwards; the sum is represented by the single stroke from 0 to the point so obtained. If we write $z=x+iy$, we mean that the stroke from the origin to the point whose co-ordinates are x, y is called z.

We may, then, regard z as a stroke drawn from the origin instead of (as at first) merely the point which marks the end of the stroke. And though in general we shall represent z by a point, it is sometimes convenient to fall back on the underlying stroke idea. We now show that the method of strokes gives interpretations of the fundamental operations which agree with those already given.

§ 8. *The addition of strokes.*

The sum of two strokes z_1, z_2 is the diagonal of the parallelogram whose sides are z_1, z_2, and z_1+z_2 means that z_1 begins at o and z_2 at z_1 (see Fig. 3). To add several strokes, we add the third to the sum of

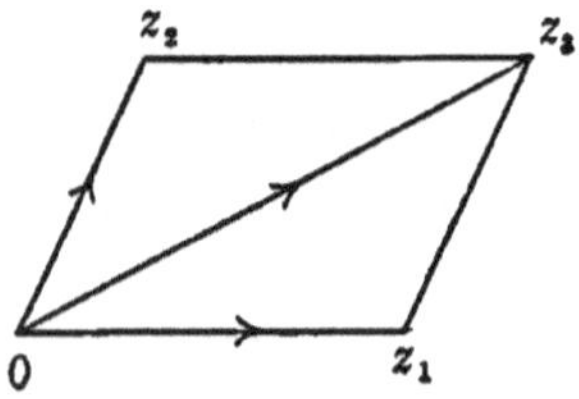

Fig. 3

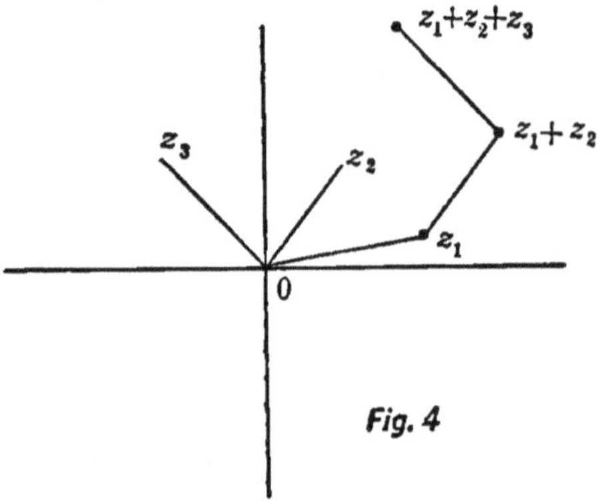

Fig. 4

the first two, giving a new stroke; the fourth to this, and so on. Since the length of the resultant stroke is not greater than the sum of the lengths of the component strokes, we have the theorem that the absolute value of a sum is not greater than the sum of the absolute values of the terms.

§ 9. *Subtraction of two strokes.*

Let $z_1+z_2=z_3$, then $z_3-z_2=z_1$. Hence, to represent the difference of two strokes z_3, z_2, let both start from the origin and join the point z_2 to the point z_3. The stroke $\overline{z_2z_3}$ (Fig. 3) represents z_3-z_2.

§ 10. *Multiplication.*

Using polar co-ordinates, $x+iy=\rho(\cos\theta+i\sin\theta)$; we may regard this expression as an *operator* on the stroke 1, which is understood. Then this stroke 1 is to be stretched in the ratio $\rho/1$, and turned through an angle θ.

To multiply z_2 by z_1, the *operator* z_1 is to affect the stroke z_2 in the same way as it affects unity in order to obtain the *stroke* z_1. Let ρ_2, θ_2 be the absolute value and amplitude of z_2; ρ_1, θ_1 those of z_1. Then we must stretch ρ_2 till it becomes $\rho_1\rho_2$, and increase the amplitude θ_2 by θ_1. Thus we obtain the same rule as before.*

Geometrically, if the triangle $0z_2z_3$ be similar to the triangle $01z_1$, then z_3 represents the product z_1z_2. When we say that abc and $a'b'c'$ are similar, it is to be understood that the triangles are congruent, or of like sign; abc is positive when, in describing the perimeter from a to c by way of b, the area lies on the left. It is further to be understood that corresponding angles are mentioned in their proper order, *i.e.* that the angles at a and a' or at b and b' or at c and c' are equal.

§ 11. *Division.* Let $z_3/z_2 = z_1$; therefore $z_3 = z_1z_2 = z_2z_1$.

Hence, to determine z_1, we construct a triangle $01z_1$ similar to the triangle $0z_2z_3$.

§ 12. The interpretation of $\cos\theta + i\sin\theta$ as a turn through an angle θ affords an immediate proof of De Moivre's theorem. For a succession of turns $\theta_1, \theta_2, \cdots, \theta_k$ is effected by the operator

$$(\cos\theta_k + i\sin\theta_k)(\cos\theta_{k-1} + i\sin\theta_{k-1})\cdots(\cos\theta_1 + i\sin\theta_1),$$

and the single turn $\theta_1 + \theta_2 + \cdots + \theta_k$ by

$$\cos(\theta_1 + \theta_2 + \cdots + \theta_k) + i\sin(\theta_1 + \theta_2 + \cdots + \theta_k).$$

But the single resultant turn produces the same effect as the succession (in any order) of component turns, therefore

$$(\cos\theta_1 + i\sin\theta_1)(\cos\theta_2 + i\sin\theta_2)\cdots(\cos\theta_k + i\sin\theta_k)$$
$$= \cos(\theta_1 + \theta_2 + \cdots + \theta_k) + i\sin(\theta_1 + \theta_2 + \cdots + \theta_k).$$

An important special case is when $\theta_1 = \theta_2 = \cdots = \theta_k = 2\lambda\pi/k$. The formula becomes

$$(\cos 2\lambda\pi/k + i\sin 2\lambda\pi/k)^k = \cos 2\lambda\pi + i\sin 2\lambda\pi = 1.$$

Hence $\cos 2\lambda\pi/k + i\sin 2\lambda\pi/k$ is a kth root of unity, and the k distinct roots are obtained by making $\lambda = 0, 1, \cdots, k-1$.

---

* It is to be observed that z_1z_2 is not the product of two vectors in the Quaternion sense, but the effect of an operator denoted by z_1 on a vector denoted by z_2. For this reason there is an advantage in replacing the word *vector* by *stroke*. [See Clifford's Common Sense of the Exact Sciences, p. 199.]

§ 13. *Powers of z.* The positive integral power is included under the multiplication law. Let the absolute value of z be ρ, the amplitude θ. Then z^n has, when n is a positive integer, an absolute value ρ^n and an amplitude $n\theta$. To interpret $z^{p/q}$, where p and q are integers, call its absolute value ρ', its amplitude ϕ. By $z^{p/q}$ is to be understood a quantity which, when multiplied by itself $q-1$ times, gives z^p. Therefore $\rho'^q = \rho^p$, $q\phi = p\theta + 2\lambda\pi$, where λ is any integer. The absolute value ρ being positive, there is a real positive value of $\rho^{p/q}$, and this we choose for ρ'. The amplitude ϕ has q distinct values, got by giving to λ the values $0, 1, 2, \cdots, q-1$. No two of these are congruent with regard to 2π, but any other value of the integer λ gives an angle congruent to one of these q amplitudes. The absolute values of the q roots are equal to ρ', and the amplitudes form an arithmetic progression of q terms, for which the common difference is $2\pi/q$. If $\lambda = 0$ and $\theta = \theta_0$, then $\phi_0 = p\theta_0/q$ gives the *chief* root. The representative points form a regular q-gon.

A negative power is the quotient of 1 by a positive power.

§ 14. The Arithmetic Mean of n points z_r $(r = 1, 2, \cdots, n)$ is the point defined by the equation

$$z = 1/n \cdot \sum_1^n z_r.$$

Its co-ordinates are $1/n \cdot \sum_1^n x_r$, $1/n \cdot \sum_1^n y_r$; accordingly the point is the centroid of the n points.

A Geometric Mean of n points z_r $(r = 1, 2, \cdots, n)$ is a point defined by the equation

$$z^n = z_1 z_2 \cdots z_n = \prod_1^n z_r.$$

Thus the geometric means are n in number, and form a regular polygon.

In particular, the geometric means of z_1, z_2 are the points z, z', which lie on the bisectors of the angles $z_1 0 z_2$, $z_2 0 z_1$, at a distance from 0 which is the positive square root of $|z_1|\,|z_2|$. These points are more properly the geometric means

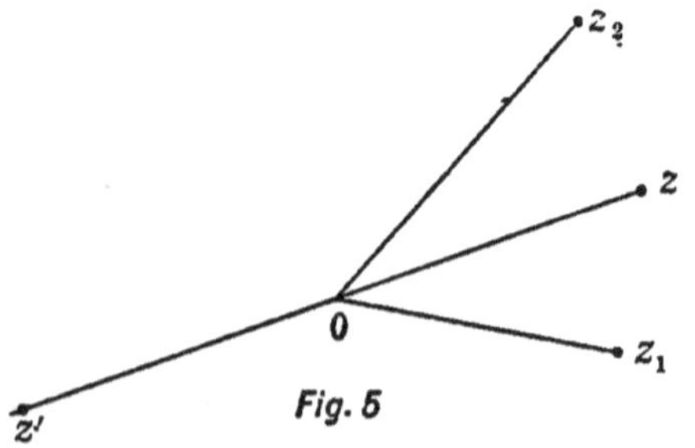

Fig. 5

with regard to the origin. A geometric mean with regard to any point z_0 is defined by the equation

$$(z - z_0)^n = \prod_1^n (z_r - z_0).$$

The Harmonic Mean of n points (with regard to the origin) is the point defined by

$$n/z = \sum_1^n 1/z_r.$$

The harmonic mean with regard to any point z_0 is defined by

$$n/(z - z_0) = \sum_1^n 1/(z_r - z_0).$$

The arithmetic mean is peculiar in that it is independent of the origin.

§ 15. An extension of the Argand representation to points in space is not consistent with the maintenance of the rules of ordinary algebra.

If such an extension be possible, a point (ξ, η, ζ) will have for its affix $\xi + i\eta + j\zeta$. The product of two such expressions, $(a + ib + jc)(\alpha + i\beta + j\gamma)$, must be of the same form. Hence i^2, ij, j^2 must be linear functions of i, j.

Let $i^2 = p_1 + q_1 i + r_1 j$, $ij = ji = p_2 + q_2 i + r_2 j$, $j^2 = p_3 + q_3 i + r_3 j$, and let the product in question be zero. This requires that the constant term and the coefficients of i, j shall vanish separately. Let the resulting equations be

$$D_r\alpha + E_r\beta + F_r\gamma = 0 \quad (r = 1, 2, 3).$$

These equations can be satisfied by values of α, β, γ other than $\alpha=\beta=\gamma=0$, provided $\Delta = 0$, where

$$\Delta = \begin{vmatrix} D_1 & E_1 & F_1 \\ D_2 & E_2 & F_2 \\ D_3 & E_3 & F_3 \end{vmatrix}.$$

Now Δ is a cubic expression in a, b, c. If real values be given to b, c, the equation $\Delta = 0$ must have one real root; choose this for a. Thus the product $(a + ib + jc)(\alpha + i\beta + j\gamma)$ can be made to vanish in cases when neither of the factors vanishes, a result contrary to the laws of ordinary algebra. [See Königsberger, Elliptische Functionen, p. 10; Stolz, Allgemeine Arithmetik, t. ii. ch. 1.]

§ 16. *Continuity of one-valued functions.*

When to each value of a complex variable z another complex variable w is assigned, so that when z is given, w can be constructed uniquely, w is a *one-valued function of z*. When, for a given z, n values of w exist, w is an *n-valued function of z*.*

A one-valued function $f(z)$ is continuous at a point $z = a$ of a region if to every positive quantity ϵ, however small, a positive quantity δ can be assigned, such that for all points z of the region for which $|z - a| < \delta$,

$$|f(z) - f(a)| < \epsilon.$$

* The idea of a function will be more carefully considered in Chapters II. and III.

For example, if m be a positive integer, z^m is a continuous one-valued function of z at every finite point a. For let $z-a=h$; then

$$z^m - a^m = ma^{m-1}h + \frac{m(m-1)}{1\cdot 2}a^{m-2}h^2 + \cdots,$$

therefore (§ 8)

$$|z^m - a^m| \leqq m\,|a|^{m-1}|h| + \frac{m(m-1)}{1\cdot 2}|a|^{m-2}|h|^2 + \cdots.$$

By taking $|h|$ small enough, the series on the right can be made $<$ any given positive quantity ϵ; let δ be such a value of $|h|$; then, when $|z-a|<\delta$, $|z^m-a^m|<\epsilon$, and therefore z^m is continuous at $z=a$.

If m be a negative integer, z^m is discontinuous at $z=0$.

A function is continuous in a region if it be continuous at all points of that region.

The sum of two functions, which are continuous at $z=a$, is continuous at $z=a$. For if, given ϵ, we can assign δ_1 and δ_2 so that

when $\qquad |z-a|<\delta_1,\ |f_1(z)-f_1(a)|<\epsilon,$

and when $\qquad |z-a|<\delta_2,\ |f_2(z)-f_2(a)|<\epsilon,$

then, when $|z-a|<$ the smaller of the two δ_1 and δ_2,

$$|f_1(z)+f_2(z)-f_1(a)-f_2(a)| \leqq |f_1(z)-f_1(a)| + |f_2(z)-f_2(a)|$$
$$< 2\epsilon.$$

Corollary.—The same theorem holds for the sum of a finite number of functions. In particular

$$w = c_m z^m + c_{m-1}z^{m-1} + \cdots + c_0,$$

where m is a positive integer, is a continuous function at every finite point. Such a function is called an integral rational function, or shortly, an *integral function.*

The quotient of two integral functions is called a *rational function.* It may be left to the reader to prove that a rational function is continuous at a finite point a, unless the denominator is zero at a.

The integral and rational functions are special cases of the *algebraic function,* which is defined by the algebraic equation $F(w^n, z^m)=0$, where $F(w^n, z^m)$ is an integral function of both z and w, of degrees m in z and n in w.*

---

* Definitions similar to those of this article are required in the case of several independent variables. In particular, a one-valued function $f(z_1, z_2, \ldots, z_r)$ is continuous at $a_1, a_2, \ldots, a_r$ if to an arbitrary positive ϵ positive quantities δ_κ can be assigned such that

$$|f(z_1, z_2, \ldots, z_r) - f(a_1, a_2, \ldots, a_r)| < \epsilon \text{ when } |z_\kappa - a_\kappa| < \delta_\kappa, \text{ where } \kappa = 1, 2, \ldots, r.$$

§ 17. In the Cartesian geometry of two dimensions we consider an integral algebraic equation between two variables, $F(x, y)=0$. For graphical purposes x and y must be real, but to maintain the generality of the algebraic processes employed, imaginary points are introduced.

(i.) The values x, y, when both real, are the affixes of real points on two straight lines ox, oy, which are both axes of real numbers; but besides these real points, there may be imaginary points on the axes, to which complex numbers are attached.

(ii.) The curve does not represent the values of either x or y separately, but merely shows to the eye the relation between the real variable x and the corresponding y, when real.

In the Theory of Functions $F(w, z)=0$ is an equation in which both w and z are complex, and in which some or all of the constants may be complex. To represent z a complete plane is needed. The values of w which correspond to a given z are represented by as many points in the w-plane, which may or may not be the same as the z-plane. When we wish to show to the eye the relation of w to z, we seek to draw the paths and mark the regions in the w-plane, which correspond to assigned paths and regions in the z-plane.

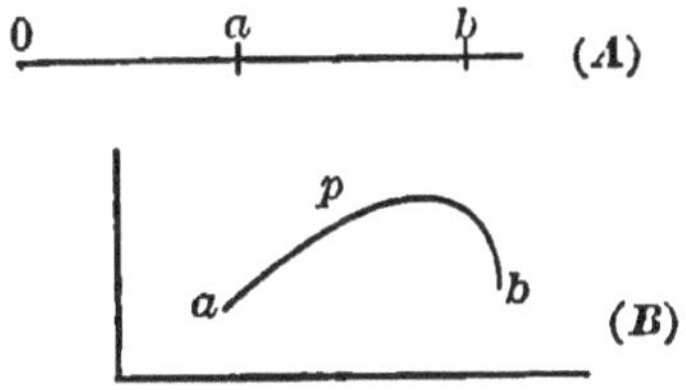

Fig. 6

The origin in the w-plane will be denoted by $0'$.

A most important difference between the theories of real and complex variables lies in the fact that in the former the routes from a point a to a point b, are restricted to the straight line ab (Fig 6, A); whereas, in the latter, infinitely many paths apb connect a and b (Fig. 6, B).

§ 18. *The fundamental theorem of algebra.*

One of the earliest uses made of the geometric representation of the complex variable was in the proof that the equation

$$w=f(z)=\sum_{0}^{n}c_\kappa z^\kappa=0$$

has n roots. While z ranges over the whole plane, w ranges over part, if not all, of the w-plane. Let us assume that w never reaches $0'$, then there must be some point or points in the z-plane for which w is at a minimum distance ρ' from $0'$. Let the circle whose centre is $0'$ and radius ρ' be called Γ. By hypothesis it is impossible for w to move into the interior of Γ, whatever be the z-path. We shall disprove this hypothesis by showing that when $z+h$ describes a small circle once round z, $w'=f(z+h)$ turns at least once round w. By the theory of equations,

$$w'-w=\sum_1^n h^r f^r(z)/r!.$$

(i.) Let $f'(z)\neq 0$, and let $w'-w=h\{f'(z)+\zeta\}$. When $|h|$ is sufficiently small, $|\zeta|$ becomes less than an arbitrarily assigned small positive quantity ϵ, by § 8; and $am\,(f'(z)+\zeta)$ tends to the limit $am\,f'(z)$. Hence, approximately, $w'-w=hf'(z)$. Now $am\,(h)$ takes all values from 0 to 2π and $am\,f'(z)$ remains constant, while $z+h$ describes the small circle. Hence $am\,(w'-w)$ increases by 2π. This shows that w' describes a small closed path once round w, and penetrates into Γ, contrary to hypothesis.

(ii.) Let $f'(z)=0$. All the coefficients cannot vanish, since the last coefficient is a constant. Let $f^r(z)/r!$ be the first of the coefficients which does not vanish. The approximate equation $w'-w=h^r f^r(z)/r!$ shows that w' turns r times round w, while z describes its circle. As before, this leads to a result which is inconsistent with the original hypothesis.

It results from this argument that there is a value, say z_1, for which $f(z)=(z-z_1)f_1(z)$. By the same argument $f_1(z)=(z-z_2)f_2(z)$, and so on. Thus $f(z)$ has as many zeros as dimensions. [See Argand, Sur une manière de représenter les quantités imaginaires, Reprint of 1874; Clifford's Collected Works, p. 528; Chrystal's Algebra, t. i., p. 244.]

A result of Cauchy's follows at once. We have

$$f(z)=a\,(z-z_1)\,(z-z_2)\cdots(z-z_n).$$

Therefore the amplitude of $f(z)$ is equal to the sum of the amplitudes of $a, z-z_1, z-z_2, \cdots, z-z_n$. Now let z describe a closed contour returning to its starting point. If a root z_r be inside the contour, the amplitude of $z-z_r$ increases continuously, and the final amplitude is greater than the initial amplitude by 2π (Fig. 7, *A*). If z_r be outside the contour, the amplitude returns to the old value (Fig. 7, *B*). Hence if s be the number of roots inside the contour, the change in the amplitude of z is $2\pi s$.

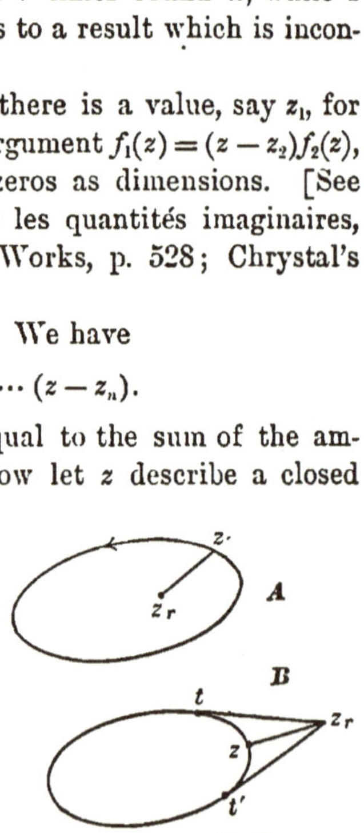

Fig. 7

§ 19. When a function has more than one value for a given z, each value is called a *branch* of the function. Each branch is a one-valued and continuous function at z, certain critical points being excluded. Its analytic expression is, in general, only possible by the aid of infinite series. For example, an algebraic function, given by $F(w^n, z^m)=0$ has n branches (§ 18); the critical points are those at which a branch is infinite or equal to another branch. Except at these points, each branch is evidently one-valued, and will be proved to be continuous.* (See Chapter IV.)

§ 20. Since, when z is given, x and y are also given, any function w of the real variables x and y was termed by Cauchy a function of z. Let $w=u+iv$, where u, v are real functions of x, y; also let $u,\ v,\ \frac{\delta u}{\delta x},\ \frac{\delta u}{\delta y},\ \frac{\delta v}{\delta x},\ \frac{\delta v}{\delta y}$ be continuous functions of x, y, at the point z. We propose to find the necessary conditions in order that $(du+idv)/(dx+idy)$ may be independent of dy/dx. Let x, y, z receive increments Δx, Δy, Δz $(=\Delta x+i\Delta y)$; then

$$u(x+\Delta x,\ y+\Delta y)-u(x,\ y)=\Delta x\left(\frac{\delta u}{\delta x}+\xi_1\right)+\Delta y\left(\frac{\delta u}{\delta y}+\eta_1\right),$$

$$v(x+\Delta x,\ y+\Delta y)-v(x,\ y)=\Delta x\left(\frac{\delta v}{\delta x}+\xi_2\right)+\Delta y\left(\frac{\delta v}{\delta y}+\eta_2\right),$$

where ξ_1, ξ_2, η_1, η_2 tend to zero simultaneously with Δx, Δy. Hence,

$$\frac{\Delta u+i\Delta v}{\Delta x+i\Delta y}=\frac{\left(\frac{\delta u}{\delta x}+\xi_1\right)+\frac{\Delta y}{\Delta x}\left(\frac{\delta u}{\delta y}+\eta_1\right)+i\left(\frac{\delta v}{\delta x}+\xi_2\right)+i\frac{\Delta y}{\Delta x}\left(\frac{\delta v}{\delta y}+\eta_2\right)}{1+i\frac{\Delta y}{\Delta x}}.$$

The limit of the expression on the right is

$$\frac{\frac{\delta u}{\delta x}+i\frac{\delta v}{\delta x}+\frac{dy}{dx}\left(\frac{\delta u}{\delta y}+i\frac{\delta v}{\delta y}\right)}{1+i\frac{dy}{dx}}.$$

The necessary and sufficient condition that this limit be independent of dy/dx, *i.e.* independent of the mode of approach to z, is

$$\frac{\delta u}{\delta y}+i\frac{\delta v}{\delta y}=i\left(\frac{\delta u}{\delta x}+i\frac{\delta v}{\delta x}\right),$$

and this condition is equivalent to the two

$$\frac{\delta u}{\delta x}=\frac{\delta v}{\delta y},\quad \frac{\delta u}{\delta y}=-\frac{\delta v}{\delta x} \quad \text{. . . . . (i.)}$$

* The definition of continuity assigns no meaning to the phrase 'continuous function' except when for the region in question the function is one-valued. When the phrase occurs, the attention is fixed on a continuous branch of the function; and the points where there is confusion of branches are to be excluded, for the present, from the region considered.

Analytic expressions, which involve x, y only in the combination $x+iy$, satisfy these relations. Cauchy used for such analytic expressions the term *monogenic* function. As we shall be concerned almost exclusively with functions of this kind it is convenient to omit the adjective. Functions such as $x-iy$, x^2-y^2, will not be regarded as functions of z, inasmuch as they are non-monogenic.

The equations (i.) are equivalent to

$$\frac{\delta w}{\delta x}=\frac{1}{i}\frac{\delta w}{\delta y} \quad \text{. . . . .} \quad \text{(ii.)},$$

and sufficient conditions in order that a function of x and y may possess a determinate differential quotient which is independent of dy/dx, are that $\delta w/\delta x$, $\delta w/\delta y$ are to be continuous and satisfy the equation (ii.) throughout a region which encloses z. [See Harnack's Diff. Calc., trans. by G. L. Cathcart, p. 141.] It will be seen subsequently that the initial assumptions as to the continuity of u, v, and of their first derivatives, imply the existence and continuity of the remaining derivatives. Hence, by ordinary differentiation, it follows from (i.) that u, v satisfy

$$\frac{\delta^2 u}{\delta x^2}+\frac{\delta^2 u}{\delta y^2}=0, \quad \frac{\delta^2 v}{\delta x^2}+\frac{\delta^2 v}{\delta y^2}=0 \quad \text{. . . . . . .} \quad \text{(iii.)}$$

The equations (iii.) show that u, v are solutions of Laplace's equations for two dimensions. It is evident from (i.) that the solutions u and v are related. They are, in fact, the well-known conjugate functions of Physics. [See Clerk Maxwell's Electricity aud Magnetism, 3d ed., t. i., p. 285; Minchin's Uniplanar Kinematics, p. 226.]

§ 21. A theorem of great importance follows from the fact that dw/dz does not depend on dz.* Let z, z_1, z_2 be three near points in the z-plane and w, w_1, w_2 the corresponding points of a branch w; then

$$\lim\,(w_1-w)/(z_1-z)=\lim\,(w_2-w)/(z_2-z).$$

Hence, if dw/dz be neither zero nor infinite,

$$\lim\,(w_2-w)/(w_1-w)=\lim\,(z_2-z)/(z_1-z).$$

Now the absolute value of $(z_2-z)/(z_1-z)$ is the ratio of the two sides of the triangle z_1zz_2, and the amplitude is the included angle. Therefore the triangles w_1ww_2, z_1zz_2 are ultimately similar. In other words, corresponding figures in the two planes are similar in their infinitesimal parts, except at points which make $dw/dz=0$, or ∞. The scale dw/dz depends upon the part selected.

---

* Riemann (Werke, p. 5) defined a function of z as a variable whose differential quotient is independent of dz.

The significance of the equations $dw/dz = 0$, $dw/dz = \infty$ can be readily discovered in the case of the algebraic function. It will appear from the theory of series (Chapters III., IV.), which affords a precise basis for a knowledge of functions, that for a finite pair of values w, z the condition $dw/dz = \infty$ requires that two or more values of w become equal at z, so that w cannot be regarded as one-valued. Similarly it will appear that $dw/dz = 0$ requires that two or more values of z become equal at w. Accordingly the theorem holds when the correspondence between w, z is (1, 1) at and near the points considered. It is not implied that near points at which dw/dz is neither zero nor infinite, w must be one-valued. There may be values of w and z which afford more than one value of dw/dz. (See Chapter IV.)

If, for a given pair w, z, n values of w and m values of z become equal, the theory of series will show that an angle in the w-plane is to an angle in the z-plane in the ratio m/n. As a simple example consider $w^2 = z^3$. Here three points z correspond to two points w, and when any one of the three z's describes a circle, whose centre is 0, with angular velocity ω, each w describes a circle whose centre is $0'$, with angular velocity $3\omega/2$.

Transformations which conserve angles are known as *isogonal*, or *orthomorphic;* and the one figure is called the *conform representation* of the other.

If $w = u + iv$ be a one-valued monogenic function of $x + iy$, the systems of orthogonal straight lines $x = a$, $y = b$ transform into systems of orthogonal curves in the w-plane. There may be exception at points given by $dw/dz = 0$.

We proceed to a brief discussion of some simple cases of orthomorphic transformation.

§ 22. I. $w = az + b$. Let z describe any curve.

The w-curve is obtained from the z-curve by (1) changing the length of each vector from the origin in the ratio $|a| : 1$, (2) turning the curve about the origin through an angle $=$ am (a), (3) giving the curve a displacement b. The nature of the curve is evidently unaltered by these processes.

§ 23. II. $$w = z^2.$$

Let $$z = \rho\,(\cos\theta + i\sin\theta),\ w = \rho'\,(\cos\theta' + i\sin\theta').$$

Then $$\rho' = \rho^2,\ \theta' = 2\,\theta.$$

When z describes the line $\rho\cos(\theta - \alpha) = \kappa$, w describes the curve

$$\rho'\cos^2(\theta'/2 - \alpha) = \kappa^2,$$

a parabola with focus at the origin $0'$. When the line passes through 0, $\kappa = 0$; θ and $\theta + \pi$ give congruent values of θ'; and the w-curve is a line from $0'$ to infinity, described twice. A semicircle with centre at 0 gives a circle with centre at $0'$.

If z describe a circle, with centre a (which may without loss of generality be taken on the real axis) and radius c, then

$$\rho^2 - 2\,a\rho\cos\theta = c^2 - a^2,$$

and the w-curve is $$\rho' - 2\,a\,\sqrt{\rho'}\cos\theta'/2 = c^2 - a^2,$$

a Cartesian. Since there is a node at $\theta' = \pi$, $\rho' = c^2 - a^2$, the curve must be a limaçon with a focus at the origin. [See Salmon, Higher Plane Curves, 3d edition, p. 252.]

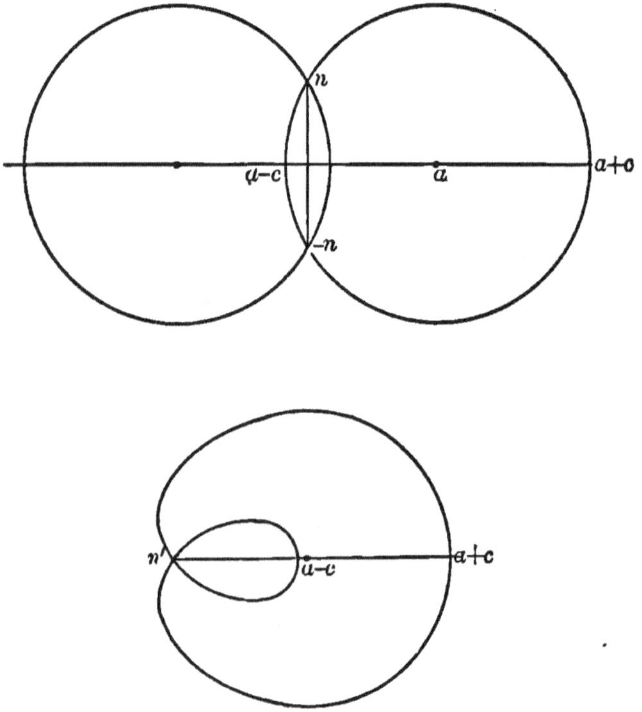

Fig. 8

To study the relations between the z-path and the w-path, we observe that, since both z and $-z$ give the same point w, the limaçon corresponds not only to the circle already considered, but equally to another circle, the two forming a figure symmetric with regard to the origin (Fig. 8). The points of intersection of these circles give the same w-point, say n'. If we start from n and describe the right-hand circle positively, when z is at $-n$, w has returned to n', so that the arc n, $-n$ of the circle gives a loop of the w-path. The other arc $-n$, n gives another loop of the w-path. The reader will find it interesting to contrast the manners in which the nodes arise in this paragraph and in § 47.

§ 24. III. $w = z^n.$

Here $\rho' = \rho^n$, $\theta' = n\theta$. The line $\rho \cos\theta = \kappa$ gives the curve

$$\rho'^{1/n} \cos\theta'/n = \kappa.$$

We then get a well-known group of curves. A fundamental property of these curves follows from the principle of isogonality. Let ψ be the angle which the z-line makes with the stroke z; then ψ is also the angle which the w-curve makes with the stroke w. Since $\theta' = n\theta$, and $\theta + \psi = \pi/2$, therefore $\psi = \pi/2 - \theta'/n$, the property in question.

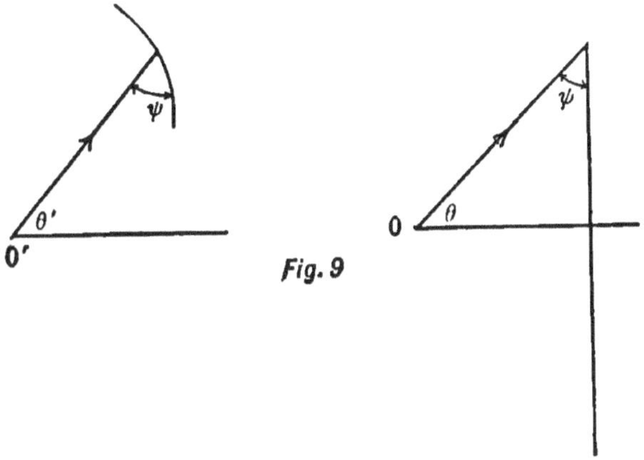

Fig. 9

The half-line $\theta' \equiv \beta \pmod{2\pi}$ represents the n equiangular half-lines $\theta = (\beta + 2m\pi)/n$ $(m = 0, 1, \cdots, n-1)$, while the whole line $2\theta' \equiv 2\beta$ represents the $2n$ equiangular half-lines $\theta = (\beta + m\pi)/n$ $(m = 0, 1, \cdots, 2n-1)$, which make up n whole lines.

The circle $\rho^2 - 2a\rho\cos\theta + a^2 = c^2,$

gives the curve $\rho'^{2/n} - 2a\rho'^{1/n}\cos\theta'/n + a^2 = c^2.$

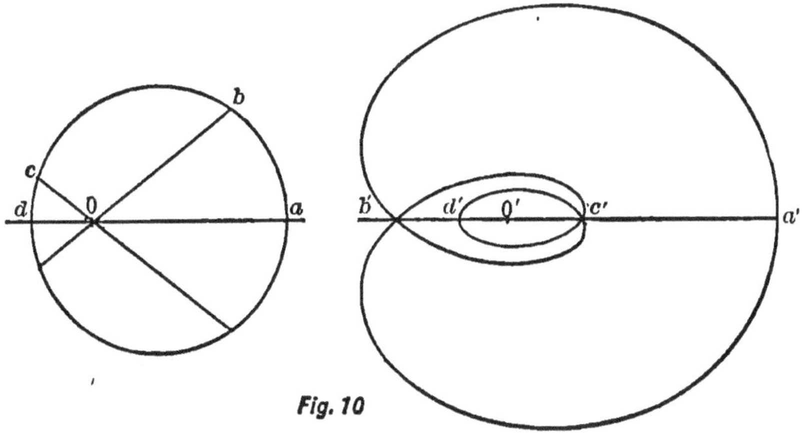

Fig. 10

This curve is easily sketched directly from the relation $w = z^n$. Let us, for instance, take $n = 3$ and suppose that the origin o is inside the circle and that the axis of x is a diameter. The points whose amplitudes are $\pm \pi/3$ give the same value of w, *i.e.* give a node; the points whose amplitudes are $\pm 2\pi/3$ give another node. As am (z) increases from 0 to π, $|z|$ (Fig. 10) decreases, and therefore, as am(w) increases from 0 to 3π, $|w|$ decreases.

§ 25. IV. $w = az^p + bz^q$ where a, b are real.

Let z describe a unit circle with centre 0; let the angular velocity be unity. The motion of w is the composition of two rotations: (1) a point w_1 rotates with angular velocity p in a circle of radius a, centre $0'$; (2) w rotates with angular velocity q in a circle of radius b, centre w_1. The motion is thus epicyclic. The reduction to trochoidal motion can be effected in two ways.

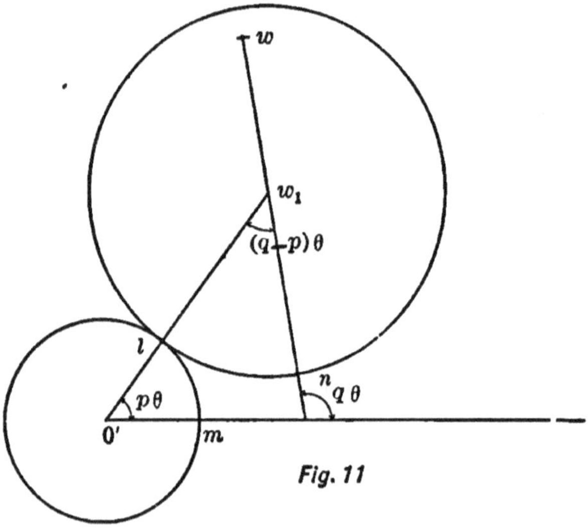

Fig. 11

(1) If we divide the stroke w_1 at l, and draw circles as in Fig. 11, choosing l so that the arcs lm, ln are equal, then w is a point fixed in the circle (w_1) which rolls on the circle $(0')$. This happens if the radii of the circles be in the ratio $(q-p)/p$, so that the radius of the fixed circle is $(q-p)a/q$, that of the moving circle pa/q.

(2) We may write our original relation in the form

$$w = bz^q + az^p;$$

interchanging a, b and p, q, the same curve is produced by the rolling of a circle of radius qb/p on a fixed circle of radius $(p-q)\,b/p$, the tracing-point being at a distance a from the centre of the rolling circle.

The w-path becomes cycloidal when the tracing-point is on the rolling circle, *i.e.* when $pa = qb$. A convenient pair of equations for the trochoid is obtained by writing $\bar{w}$ for the complex quantity conjugate to w. Then, since $z\bar{z} = 1$,

$$w = az^p + bz^q,\ \bar{w} = az^{-p} + bz^{-q}.$$

§ 26. V. In III. let $n=-1$; then $\rho'=1/\rho$, $\theta'=-\theta$. If w be represented on the z-plane, the point w has polar co-ordinates $1/\rho$ and $-\theta$, and is therefore the reflexion in the real axis of the geometric inverse of z with respect to a circle whose radius is unity and centre 0. This combination of reflexion and inversion is called by Professor Cayley quasi-inversion.

Since $wz=1$, to each value of z, other than zero, corresponds one value w, but when $z=0$, $w=\infty$. This leads us to consider infinity as consisting of a single point, not of infinitely many. In the Theory of Functions, all points at an infinite distance from 0 are supposed to coalesce in a single point $z=\infty$.

§ 27. To make this supposition a natural one Neumann, following out one of Riemann's ideas, chose as the field of the complex variable a sphere instead of a plane. Let 0, 0′ be opposite points on a sphere of diameter 1. Take the tangent plane at o as z-plane. Join each point z to 0′, and let the joining line cut the sphere at p. Then the value z is to be attached to p, and we may treat the sphere as the field on which z assumes all its values. Each point at a

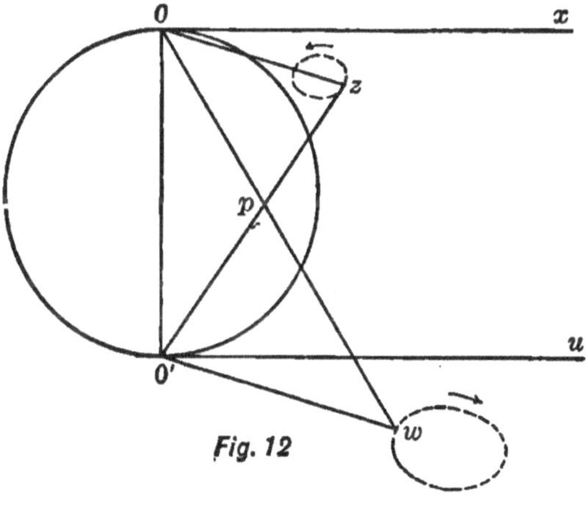

Fig. 12

finite distance in the plane is replaced by a single point on the sphere, but all the infinitely distant points of the plane are replaced by the single point 0′. This process of replacing the plane by the sphere may be regarded as the converse of stereographic projection; or, again, as the inversion of the z-plane with regard to the external point 0′.

Let the tangent plane at 0′ be the w-plane, with 0′ as origin, and the line $0'u$ in which $x00'$ meets this plane as real axis. Let angles be measured counter-clockwise for observers standing on the sphere at 0 and 0′, and let the line $0p$ meet the w-plane at w. Then the amplitude of $w = -$(the amplitude of z). Again, since the triangles $00'w$, $z00'$ are each similar to $0p0'$,

$$|z|\,|w| = 1.$$

Hence $wz = 1$, or the curve traced out by w is the representation, for $wz = 1$, of the z-curve. When z describes positively a curve which includes 0, w describes negatively a curve which includes 0′; but otherwise a positive description of the z-path is accompanied by a positive description of the w-path.

This transformation $w = 1/z$ is of great use, for by its help the consideration of the behaviour of a function of w at $w = 0$ is substituted for that of a function of z at $z = \infty$.

§ 28. The transformation $wz = k$ gives $|w|\,|z| = |k|$, and $\operatorname{am} w + \operatorname{am} z = \operatorname{am} k$. It is (1) an inversion with respect to a circle, centre 0 and radius $|k^{1/2}|$, (2) a reflexion in a line which bisects the angle $x0k$. That is, it is a quasi-inversion.

Now any circle remains a circle after inversion and also after reflexion. Hence, any circle in the z-plane becomes a circle in the w-plane. The following are the special cases:—

(1) A circle through the origin in the z-plane becomes a line in the w-plane (*i.e.* a circle of infinite radius);

(2) A line in the z-plane becomes a circle through 0′;

(3) A line through 0 becomes a line through 0′.

When w and z are represented in the same plane, all the pairs of points which satisfy $wz = k$ are said to form an involution. The points $\pm\sqrt{k}$, at which w and z coincide, are the double points; the points of a pair are said to be conjugate; and the point $z = 0$, whose conjugate is $z = \infty$, is the centre of the involution.*

§ 29. The most general bilinear transformation is

$$w = (az + b)/(cz + d).$$

It contains four constants, but only three ratios. We may therefore assume any relation between a, b, c, d. The most suitable one is $ad - bc = 1$.

This transformation can be built up out of simpler ones, for writing $w - a/c = w'$, $z + d/c = z'$, so as to change both origins, we get

$$w' = -1/c^2 z',$$

* Generally, if U and V be integral functions of z, of degree n, the points given by $U + \lambda V = 0$, where λ is arbitrary, form an n-ic involution. When $n = 2$, this gives a relation between two points of the form $z_1 z_2 - \mu(z_1 + z_2) + \nu = 0$; when $z_2 = \infty$, $z_1 = \mu$, and taking μ as a new origin we have a relation of the form in question.

which is the quasi-inversion. Accordingly, circles change into circles with the following exceptions:—

(1) A line through the z'-origin, that is, a line through $-d/c$ in the z-plane, becomes a line through a/c in the w-plane;
(2) A circle through $-d/c$ becomes a line through a/c;
(3) A line in the z-plane becomes a circle through a/c.

It is to be noticed that when $z=-d/c, w=\infty$, and when $z=\infty, w=a/c$. Hence if we regard a straight line as a circle of infinite radius, the bilinear transformation transforms circles into circles, without exception. We shall now show that the transformation is orthomorphic throughout the plane. We have

$$dw/dz=1/(cz+d)^2.$$

The doubtful points are $z=-d/c, w=\infty$, and $z=\infty, w=a/c$.

To study what happens in the neighbourhood of $z=\infty$, write $z=1/z'$, and examine the behaviour of w in the region of the new origin $z'=0$. We have

$$w=(a+bz')/(c+dz'),$$

$$dw/dz'=(bc-ad)/(c+dz')^2.$$

Hence when $z'=0$, $w=a/c$, $dw/dz'=-1/c^2$, and the transformation is orthomorphic at $z=\infty$. If the equation

$$w=(az+b)/(cz+d)$$

be solved for z and w be put $=1/w'$, we find in the same way that dz/dw' is finite and not zero, when $w'=0$. Thus the orthomorphism exists throughout the plane.

Take four points z_1, z_2, z_3, z_4, and let the corresponding w's be w_1, w_2, w_3, w_4. Then

$$w_r=(az_r+b)/(cz_r+d),\quad w_s=(az_s+b)/(cz_s+d),$$

$$w_r-w_s=(ad-bc)(z_r-z_s)/(cz_r+d)(cz_s+d).$$

Hence

$$(w_1-w_2)(w_3-w_4)/(w_1-w_3)(w_2-w_4)=(z_1-z_2)(z_3-z_4)/(z_1-z_3)(z_2-z_4).$$

Each of these is called an anharmonic ratio of the four points considered, and we have the theorem that an anharmonic ratio is unchanged by a bilinear transformation. [This theorem was given by Möbius: Die Theorie der Kreisverwandtschaft in rein geometrischer Darstellung. Abh. d. Kgl. Sächs. Ges. d. W., t. ii.; or Ges. Werke, t. ii.]

Since the relation between w and z involves three arbitrary ratios, any three points in the w-plane can be made to correspond to any three points in the z-plane. Möbius's theorem then gives the point in the w-plane, which corresponds to any fourth point in the z-plane.

§ 30. *The anharmonic ratios of four points.*

Let z_1, z_2, z_3, z_4 be any four points, and let λ, μ, ν stand for $(z_2-z_3)(z_1-z_4)$, $(z_3-z_1)(z_2-z_4)$, $(z_1-z_2)(z_3-z_4)$, so that

$$\lambda+\mu+\nu=0.$$

Any one of the six ratios

$$-\mu/\nu,\quad -\nu/\lambda,\quad -\lambda/\mu,$$
$$-\nu/\mu,\quad -\lambda/\nu,\quad -\mu/\lambda,$$

is an anharmonic ratio of the four points. Let σ be any one of these, say $-\mu/\nu$. Then $-\lambda/\nu=1-\sigma$, and the six ratios are

$$\sigma,\ 1/(1-\sigma),\ (\sigma-1)/\sigma,$$
$$1/\sigma,\ 1-\sigma,\ \sigma/(\sigma-1).$$

The number of substitutions which can be formed out of the four suffixes 1, 2, 3, 4, is twenty-four. Six of these leave 4 unchanged; they are

$$1;\ (123);\ (132);\ (23);\ (31);\ (12)\quad .\ .\ .\ .\quad (A).$$

These change λ, μ, ν into

$$\lambda, \mu, \nu;\ \mu, \nu, \lambda;\ \nu, \lambda, \mu;\ -\lambda, -\nu, -\mu;\ -\nu, -\mu, -\lambda;\ -\mu, -\lambda, -\nu;$$

and therefore change σ into

$$\sigma;\ 1/(1-\sigma);\ (\sigma-1)/\sigma;\ 1/\sigma;\ \sigma/(\sigma-1);\ 1-\sigma.$$

Any substitution of the following set,

$$1,\ (23)(14),\ (31)(24),\ (12)(34)\quad .\ .\ .\ .\ .\ .\quad (B),$$

leaves each of the six ratios unaltered. All the 24 substitutions of four letters can be compounded out of (A) and (B).

Since the six anharmonic ratios are bilinearly related, it follows that if we denote by σ a point in a plane, when σ describes a circle, so do the points which denote the other ratios; and therefore while σ describes a region bounded by circular arcs, the other points describe regions bounded by circular arcs. It is possible to choose the region (σ) in such a way that the six regions fill the σ-plane without overlapping. With centres 0 and 1, describe two circles of radius 1. These intersect at the points $-v^2, -v$, where $v = \frac{-1+i\sqrt{3}}{2}$. When σ is confined to the unshaded triangle whose vertices are $0, 1/2, -v^2$, $1/\sigma$ is confined to a triangle which arises from this by inversion with regard to the circle whose centre is 0, and reflexion in the real axis. The point 1/2 becomes 2, 0 becomes ∞, and $-v^2$ becomes $-v$; the region $(1/\sigma)$ is bounded by the arc of the second circle from 2 to $-v$, the part of the imaginary axis from $-v$ to $-\infty$, and the part of the real axis from $+\infty$ to 2. Again, the points σ and $1-\sigma$ are symmetric with regard to 1/2, and therefore their regions are symmetric with regard to 1/2. The region of $1/(1-\sigma)$ is the inverse of the region of $1-\sigma$, and therefore has

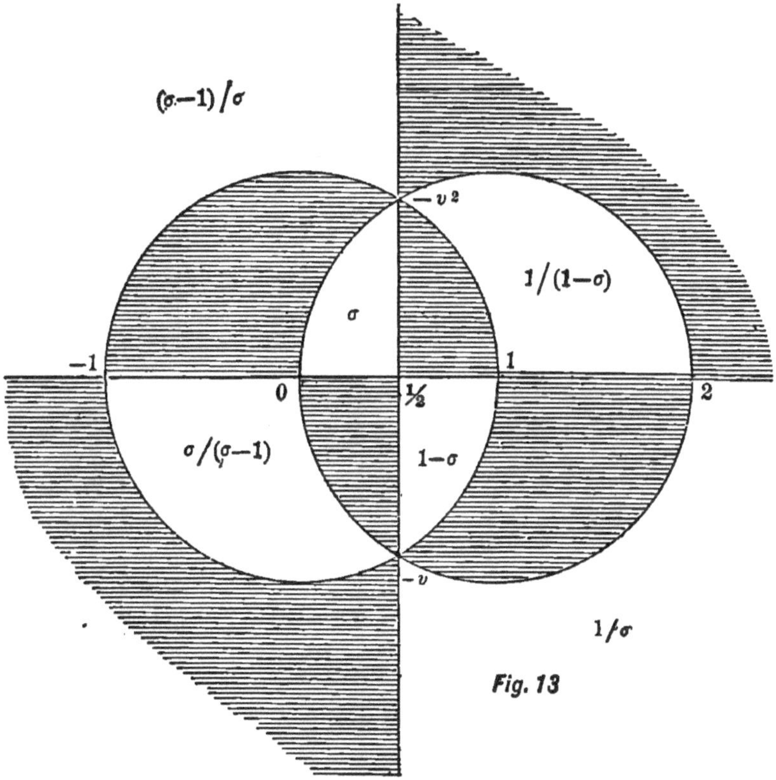

Fig. 13

the vertices $1, 2, -v^2$. The region of $\sigma/(\sigma-1)$, and that of $1/(1-\sigma)$ are symmetric with regard to 1/2; as also are those of $1/\sigma$ and $(\sigma-1)/\sigma$. In this way the marking in the figure is seen to be correct. When σ describes a shaded region adjacent to (σ), the other ratios describe the remaining shaded regions.

At the point 1/2 draw the normal to the σ-plane, and take a point on the normal at a distance $\sqrt{3}/2$ from the point 1/2. Invert the plane with this point as origin; the points $\infty, -1, 0, 1/2, 1, 2$ become the vertices of a regular hexagon on a sphere, lying in a great circle which we will call the equator. The points $-\upsilon^2$ and $-\upsilon$ become the poles of the equator, since they subtend a right angle at the centre of inversion. The shaded and unshaded regions of the plane pass into 12 equal regions of the sphere.

The Special Anharmonic Ratios.

Among the six ratios equalities may arise in two ways:—

(1) $\mu/\nu = \nu/\mu = \pm 1$. The lower sign gives $\lambda = 0$, and involves coincidence of points. Excluding this, we have $\mu = \nu$. When two of the quantities λ, μ, ν become equal, the four points are said to be *harmonic.*

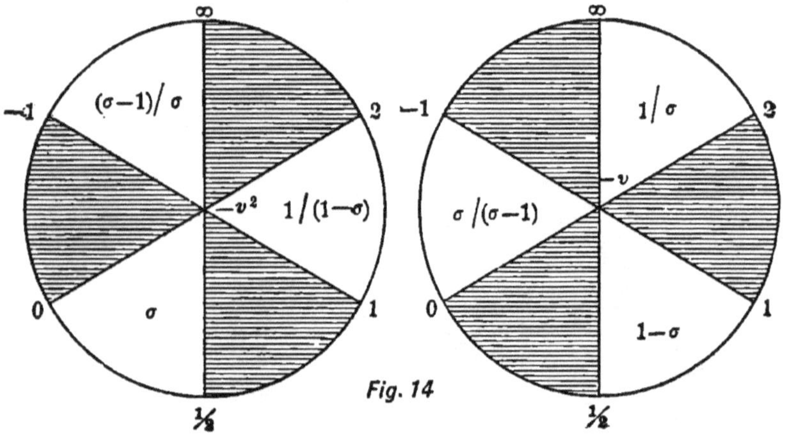

Fig. 14

(2) $\mu/\nu = \nu/\lambda = (\mu+\nu)/(\nu+\lambda) = \lambda/\mu = 1^{1/3}$. The real root gives $\lambda = \mu = \nu$ and involves coincidences. Excluding this, $\mu/\nu = \nu/\lambda = \lambda/\mu = \upsilon$ or υ^2. In either case the points are equianharmonic. We shall now study these special arrangements of four points.

§ 31. *The Harmonic Quadrangle.*

We have said that four points are harmonic when $\mu = \nu$, or

$$(z_3 - z_1)(z_2 - z_4) = (z_1 - z_2)(z_3 - z_4) \quad . \quad . \quad . \quad \text{(i.)},$$

or

$$2(z_2 z_3 + z_1 z_4) = (z_2 + z_3)(z_1 + z_4) \quad . \quad . \quad . \quad . \quad . \quad . \quad \text{(ii.)}$$

From the symmetry of this last equation it appears that the points have paired, z_2 with z_3, and z_1 with z_4. When four points are harmonic, two points which pair in this way are said to be conjugate; and either pair is said to be harmonic with the other.

To obtain the geometric relations of the four points, let us take the origin at the mean of z_2 and z_3. Then $z_2+z_3=0$, and $z_1z_4=z_2^2=z_3^2$. Hence, given z_1, z_2, z_3, we construct z_4 in the following way: let the

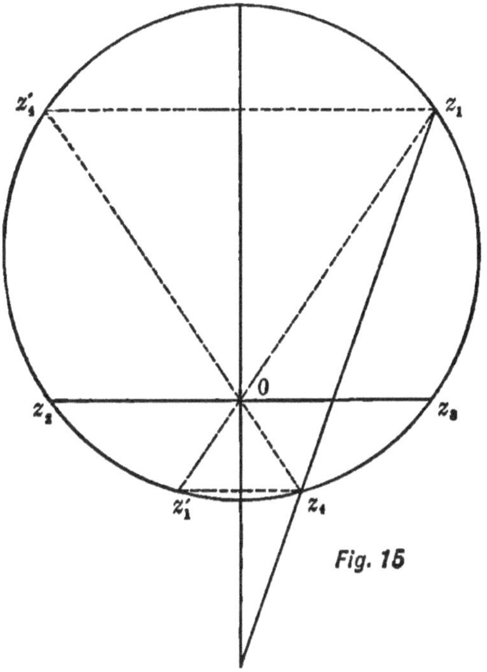

Fig. 15

line z_10 meet the circle $z_1z_2z_3$ at z_1'. The image of z_1' in that diameter of the circle which passes through 0 is z_4. Hence the four points are concyclic.

Using for an instant the methods of projective geometry, we see that the lines joining z_1' to the four points z_2, z_3, and z_1, z_4 cut the line z_20z_3 harmonically, so that the pencil whose vertex is any point of the circle and whose rays pass through the four points is harmonic. Taking the vertex first at z_2 and secondly at z_3, we see that the tangents at z_2, z_3 meet on the line z_1z_4; in other words, the chords z_2z_3 and z_1z_4 are conjugate chords of the circle.

From equation (i.) we see, by equating absolute values, that the rectangles of opposite sides are equal, and, by equating amplitudes, that

$$\angle z_2z_1z_3+\angle z_3z_4z_2=\pi,$$

whence again the points are concyclic.

Given three points z_1, z_2, z_3, we have, evidently, three points each of which forms with the given three a harmonic system. Let these points be j_1, j_2, j_3; z_1, j_1 being harmonic with z_2, z_3, etc.

We can prove that the lines $z_r j_r$ $(r = 1, 2, 3)$ meet at a point k by Brianchon's Theorem, or in the following manner. Let us pay attention not only to the magnitude and direction of a stroke, but also to the point of application. For instance, regard z as a force applied at the origin; $z_2 - z_1$ as a force applied at z_1, etc. The expression $1/\bar{z}$ will symbolize a force of magnitude $1/\rho$ and inclination θ, applied at the origin. Now equation (i.) may be written

$$2/(z_4 - z_1) = 1/(z_2 - z_1) + 1/(z_3 - z_1),$$

or, writing j_1 for z_4, and changing the sign of i,

$$2/(\bar{j}_1 - \bar{z}_1) = 1/(\bar{z}_2 - \bar{z}_1) + 1/(\bar{z}_3 - \bar{z}_1),$$

so that the resultant of forces along $z_2 - z_1$, $z_3 - z_1$ and inversely proportional to the lengths of these strokes, lies along $j_1 - z_1$ and is inversely proportional to half its length. Similarly for the other vertices; but, by addition,

$$\Sigma\, 1/(\bar{j}_r - \bar{z}_r) = 0;$$

hence the three resultants are in equilibrium and therefore meet at a point k. This point is called the symmedian point of the triangle $z_1 z_2 z_3$.

§ 32. *The Equianharmonic Quadrangle.*

We have said that four points z_1, z_2, z_3, z_4 are equianharmonic when $\mu/\nu = \upsilon$ or υ^2. Taking first the case $\mu/\nu = \upsilon^2$, the relation which connects the points is

$$(z_1 - z_3)(z_2 - z_4) = -\upsilon^2 (z_1 - z_2)(z_3 - z_4) \quad \cdots \quad \text{(i.)},$$

which may be written

$$z_2 z_3 + z_1 z_4 + \upsilon(z_3 z_1 + z_2 z_4) + \upsilon^2 (z_1 z_2 + z_3 z_4) = 0 \quad \cdots \quad \text{(ii.)},$$

and also

$$1/(z_1 - z_4) + \upsilon/(z_2 - z_4) + \upsilon^2/(z_3 - z_4) = 0 \quad \cdots \quad \text{(iii.)}$$

Other forms of the relations (i.) and (iii.) are obtained by the use of the substitutions,

$$(z_2 z_3)(z_1 z_4),\ (z_3 z_1)(z_2 z_4),\ (z_1 z_2)(z_3 z_4).$$

From (i.) we have, by equating absolute values,

$$|z_2 - z_3|\,|z_1 - z_4| = |z_3 - z_1|\,|z_2 - z_4| = |z_1 - z_2|\,|z_3 - z_4| \quad \cdots \quad \text{(iv.)},$$

i.e. the rectangles of the three pairs of opposite sides are equal. Again, equating amplitudes,

$$\left.\begin{aligned} \angle z_2 z_4 z_3 &= \angle z_2 z_1 z_3 - \pi/3 \\ \angle z_3 z_4 z_1 &= \angle z_3 z_2 z_1 - \pi/3 \\ \angle z_1 z_4 z_2 &= \angle z_1 z_3 z_2 - \pi/3 \end{aligned}\right\} \quad \cdots \quad \text{(v.)}$$

Now regard z_1, z_2, z_3 as a given triangle. From (iv.) z_4 is given as an intersection of three circles, each of which passes through a vertex and through the points where the bisectors of the angles at that vertex meet the opposite side. Either of the two intersections of these circles forms with z_1, z_2, z_3 an equianharmonic quadrangle. Call these points h_+, h_-, then from (ii.)

$$\left.\begin{aligned} h_+ &= -(z_2z_3 + vz_3z_1 + v^2z_1z_2)/(z_1 + vz_2 + v^2z_3) \\ &\text{and, interchanging } v \text{ and } v^2, \\ h_- &= -(z_2z_3 + v^2z_3z_1 + vz_1z_2)/(z_1 + v^2z_2 + vz_3) \end{aligned}\right\} \quad . \quad . \quad . \quad \text{(vi.)}$$

The equations i., ii., iii., v., refer to h_+. When $z_4 = h_-$, we must change the sign of $\pi/3$ in the equations v.

§ 33. *The Covariants of the Cubic.*

Let z_1, z_2, z_3 be the roots of $z^3 + 3a_1z^2 + 3a_2z + a_3 = 0$. We have proved (§ 29) that a bilinear transformation does not change an anharmonic ratio. The points h_+, h_- are defined by special anharmonic ratios and are therefore covariant. Hence if we form the quadratic whose zeros are h_+, h_- and notice that its coefficients involve z_1, z_2, z_3 symmetrically, we see that it is a covariant of order 2, *i.e.* it is the Hessian of the given cubic U.

With regard to the cubic whose zeros are j_1, j_2, j_3, which is equally a covariant, we know that the cubic, each of whose zeros forms with those of the given cubic a harmonic system, is the cubicovariant or Jacobian. [See Salmon's Higher Algebra, 4th ed. p. 183; Clifford's paper, On Jacobians and Polar Opposites, Collected Works, p. 27.]

§ 34. *The canonical form of the cubic.*

The analytic condition that a triangle be equilateral is, if z_1', z_2', z_3' be the points in positive order,

$$(z_1' - z_3')/(z_1' - z_2') = \cos \pi/3 + i \sin \pi/3 = -v^2,$$

or

$$z_1' + vz_2' + v^2z_3' = 0 \quad . \quad . \quad . \quad . \quad . \quad . \quad . \quad . \quad \text{(vii.)}$$

But from (iii.) we have, when h_+ is origin,

$$1/z_1 + v/z_2 + v^2/z_3 = 0.$$

Thus the quasi-inverse of z_1, z_2, z_3 as to h_+ is positively equilateral, and therefore the inverse triangle itself is negatively equilateral.

The triangle is now reduced to its canonical or equilateral form. The inversion has sent one Hessian point to infinity. If we now invert with regard to the other, the triangle must remain equilateral. Hence this other Hessian point must be the centre of the triangle. Taking this point as origin, we may write for the cubic U

$$z^3 = a.$$

In the canonical form, j_1 and z_1 are harmonic with z_2 and z_3. Hence, $j_1j_2j_3$ is the counter-triangle, and the figure $z_1j_3z_2j_1z_3j_2$ is a regular hexagon.

§ 35. After the Jacobian and Hessian points we consider the polars or emanants of the triangle. We shall consider, in the first place, the general case in which there are n points z_r given by an equation $f(z)=0$, of the nth order in z.

Introduce the unit y by writing z/y for z, and multiply by y^n so that the equation takes the homogeneous form $f(z, y)=0$. Take any points z/y and z'/y', and let

$$z_r=(z+\lambda_r z')/(1+\lambda_r)=(z+\lambda_r z')/(y+\lambda_r y');$$

the n quantities λ_r are roots of

$$f(z+\lambda z',\ y+\lambda y')=0,$$

which, if we write

$$\left.\begin{aligned}\Delta &= z'\frac{\delta}{\delta z}+y'\frac{\delta}{\delta y},\\ \Delta' &= z\frac{\delta}{\delta z'}+y\frac{\delta}{\delta y'},\end{aligned}\right\}$$

is either

$$f(z, y)+\lambda\Delta f(z, y)+\frac{\lambda^2}{2!}\Delta^2 f(z, y)+\cdots+\frac{\lambda^n}{n!}\Delta^n f(z, y)=0$$

or $$\frac{1}{n!}\Delta'^n f(z', y')+\frac{\lambda}{n-1!}\Delta'^{n-1} f(z', y')+\cdots+\lambda^n f(z', y')=0.$$

The coefficients of the powers of λ are well-known covariants. We call the $n-r$ points, given by $\Delta^r f(z, y)=0$, the rth polar of the pole z' with regard to the given n points.

The geometric meaning is easily seen. When $\Delta^s f(z, y)=0$, the sum of the products of the λ's, $n-s$ at a time, is zero. But

$$\lambda_r=(z-z_r)/(z_r-z').$$

Hence the sums of the products of $(z_r-z)/(z_r-z')$, $n-s$ at a time, or of $(z_r-z')/(z_r-z)$, s at a time, is zero. For the first polar

$$\Sigma\,(z_r-z')/(z_r-z)=0$$

or $$\Sigma\left(1-\frac{z'-z}{z_r-z}\right)=0,$$

or $$n/(z'-z)=\Sigma\,1/(z_r-z),$$

so that z' is the harmonic mean of the n points z_r with regard to any first polar point of z'.

Comparing the two equations for λ, we see that $\Delta^s f(z, y)$ and $\Delta'^{n-s} f(z', y')$ differ only by a numerical factor. Hence if z be an sth polar point of z', z' is an $(n-s)$th polar point of z.

§ 36. By means of the polars of a group of points it is possible to frame a geometric definition of the Hessian and Jacobian. Let us consider the points whose first polars have two coincident points, or (say) a double point. We have

$$z'\delta f/\delta z + y'\delta f/\delta y = 0,$$

and if z be a double point,

$$z'\delta^2 f/\delta z^2 + y'\delta^2 f/\delta y\delta z = 0,$$
$$z'\delta^2 f/\delta y\delta z + y'\delta^2 f/\delta y^2 = 0.$$

Hence such double points are given by the equation

$$n^2(n-1)^2 H \equiv \begin{vmatrix} \dfrac{\delta^2 f}{\delta z^2} & \dfrac{\delta^2 f}{\delta y\delta z} \\ \dfrac{\delta^2 f}{\delta y\delta z} & \dfrac{\delta^2 f}{\delta y^2} \end{vmatrix} = 0.$$

The expression H is the Hessian of the given function f.

Again, consider the points whose first polars with regard to f and H have a common point. We have

$$z'\delta f/\delta z + y'\delta f/\delta y = 0,$$
$$z'\delta H/\delta z + y'\delta H/\delta y = 0,$$

whence such common points are given by the equation

$$n(n-2)\, J \equiv \begin{vmatrix} \dfrac{\delta f}{\delta z} & \dfrac{\delta f}{\delta y} \\ \dfrac{\delta H}{\delta z} & \dfrac{\delta H}{\delta y} \end{vmatrix} = 0.$$

The expression J is the Jacobian of the given function.

For a full discussion of the theory of the polars of a binary form, see Clebsch, Geometrie, t. i., ch. 3.

In the case of two points, say $z^2 = a$, the polar of z' is given by $zz' = a$, so that the polar point is the conjugate of z' with regard to the given points.

For the triangle, we use the canonical form

$$z^3 = a.$$

The polar pair of z' is given by

$$z'z^2 = a.$$

Hence the polar pair of any point is harmonic with the Hessian points 0 and ∞; and, in particular, the points which form the polar pair of a Hessian point coincide with the other Hessian point.

§ 37. We shall now transfer the figure from a plane to a sphere. This can be done by geometric inversion with regard to an *external* point s'. Let the constant of inversion be equal to the distance from s' to the plane, so that the plane and the sphere, derived from the plane by inversion, touch at a point t.

The fundamental facts of ordinary geometric inversion hold good, — that a circle becomes a circle and that the magnitudes of an angle and of an anharmonic ratio are unaffected. Hence a harmonic quadrangle becomes a harmonic quadrangle; an equianharmonic quadrangle becomes a tetrahedron in which the rectangles formed by opposite edges are equal, while the inverse of such a tetrahedron is another of the same kind. In particular, the inverse as to a vertex is an equilateral triangle and the point at infinity in its plane.

We wish to choose s' so that the fundamental triangle $z_1z_2z_3$ shall become an equilateral triangle on the sphere. This will happen when the tetrahedron $s'z_1z_2z_3$ is equianharmonic; that is when, if λ_1, λ_2, λ_3 denote the lengths of the sides, and δ_1, δ_2, δ_3 the distances from s' to the vertices, $\delta_1\lambda_1 = \delta_2\lambda_2 = \delta_3\lambda_3$.

Each of these equations giving a sphere, we have for the locus of s' a circle in a plane perpendicular to the given plane, and meeting that plane at the Hessian points h_+, h_-.

Choosing any point s' on this circle, let us denote the inverse of any point z in the plane by z'; then the triangle $z_1z_2z_3$ becomes an equilateral triangle $z_1'z_2'z_3'$ on the sphere, and the points h_+, h_-, which are equianharmonic with z_1, z_2, z_3, become the ends h_+', h_-' of the axis of the small circle $z_1'z_2'z_3'$. If we choose s' so that this circle is a great circle, we have a simple dihedral configuration. [See Klein's Ikosaeder.]

Consider the set of great circles through h'_+, h'_-, and the orthogonal set of small circles. These become in the z-plane two sets of coaxial circles; h_+, h_- are the common points of the first set and the limiting points of the second set.

We have now to consider the plane $z_1'z_2'z_3'$ as the projection of the z-plane from the vertex s'. Let us denote the projection of z by ζ, of a by α, etc. It is evident that whenever a circle is projected into a circle, the symmedian point of an in-triangle becomes the symmedian point of the projected triangle. For the symmedian point was defined by means of conjugate chords of the circumcircle; and conjugate chords project into conjugate chords. In an equilateral triangle the symmedian point is the centre; hence the projection of k is κ, the centre of the circle $z_1'z_2'z_3'$.

Next let c be the circumcentre of the original triangle. It is the pole of the line at infinity with regard to the circumcircle; hence its projection γ is the pole of the intersection of the ζ-plane and the tangent plane at s', with regard to the circle $\zeta_1\zeta_2\zeta_3$. Hence the chords $c's'$ and $\alpha\beta$ (Fig. 16) of the meridian circle

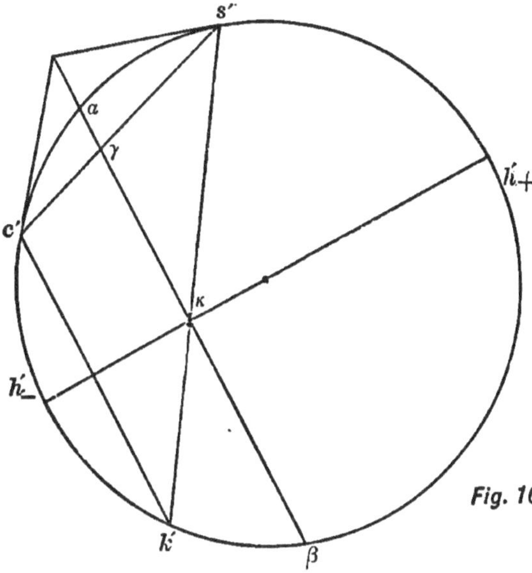

Fig. 16

$s'h'_+h'_-$ are conjugate, and therefore the chords $c'k'$ and $\alpha\beta$ are parallel (see Fig. 15); this shows that the points c', k' are harmonic with h'_+, h'_-. Hence c, k are harmonic with h_+, h_-, and further the four points are collinear.

§ 38. *The three quadratic involutions determined by four points.*

Let z_1, z_2, z_3, z_4 be the zeros of a quartic. The points can be paired off in three ways:

$$(23)(14); \quad (31)(24); \quad (12)(34).$$

Now two pairs of points define a quadratic involution. Let e_1, e_2, e_3 be the centres of the three involutions obtained from the three divisions of the four points into pairs. We have, to determine e_1,

$$(z_2 - e_1)(z_3 - e_1) = (z_1 - e_1)(z_4 - e_1),$$

whence $$e_1 = (z_2z_3 - z_1z_4)/(z_2 + z_3 - z_1 - z_4).$$

The focus of the parabola which touches the lines joining z_2, z_3 to z_1, z_4 is e_1. For let the line joining z_l and z_m touch the parabola at the point z_{lm}, and let f be the focus. Then by the properties of the geometric mean

$$z_1-f=\pm\sqrt{(z_{31}-f)(z_{12}-f)},$$

$$z_2-f=\pm\sqrt{(z_{12}-f)(z_{24}-f)},$$

$$z_3-f=\pm\sqrt{(z_{31}-f)(z_{34}-f)},$$

$$z_4-f=\pm\sqrt{(z_{24}-f)(z_{34}-f)}.$$

The amplitudes of points on the parabola, when referred to the focus and axis, vary continuously from 0 to 2π. Let am $(+\sqrt{z-f})$ be $\theta/2$; then

$$\text{am}\,(+\sqrt{(z_{lm}-f)(z_{ln}-f)})=(\theta_{lm}+\theta_{ln})/2.$$

But geometric considerations show that am $(z_l-f)=$ the same quantity. Hence the $+$ sign must be chosen throughout, and

$$(z_2-f)(z_3-f)=(z_1-f)(z_4-f),$$

whence f and e_1 coincide.

Accordingly, to construct the point e_1, we observe that it lies on the circumcircle of the triangle formed by any three of the four lines

$$z_1z_2,\ z_1z_3,\ z_2z_4,\ z_3z_4.$$

[See R. Russell on the Geometry of the Quartic, Lond. Math. Soc., t. xix., p. 56.]

Let c_1^2, c_2^2, c_3^2 be the constants of the three involutions, and let us eliminate the z's from the six equations,

$$\left.\begin{array}{l}(z_2-e_1)(z_3-e_1)=(z_1-e_1)(z_4-e_1)=c_1^2\\ (z_3-e_2)(z_1-e_2)=(z_2-e_2)(z_4-e_2)=c_2^2\\ (z_1-e_3)(z_2-e_3)=(z_3-e_3)(z_4-e_3)=c_3^2\end{array}\right\}\quad\cdots\cdots\quad\text{(i.)}$$

We have
$$z_3-e_1=c_2^2/(z_1-e_2)+e_2-e_1,$$

$$z_2-e_1=c_3^2/(z_1-e_3)+e_3-e_1.$$

Hence

$$c_1^2=\{c_2^2/(z_1-e_2)+e_2-e_1\}\{c_3^2/(z_1-e_3)+e_3-e_1\}\quad\cdot\ \cdot\quad\text{(ii.)}$$

And, by proceeding similarly with the other pairs of quantities, we have precisely the same equation for z_2, z_3, z_4. Hence every coefficient of the quadratic (ii.) is zero. Now (ii.) is

$$\{c_1^2-(e_2-e_1)(e_3-e_1)\}(z_1-e_2)(z_1-e_3)-c_2^2c_3^2-c_2^2(e_3-e_1)(z_1-e_3)$$
$$-c_3^2(e_2-e_1)(z_1-e_2)=0.$$

Hence $$c_1^2 = (e_2 - e_1)(e_3 - e_1),$$
and similarly $$c_2^2 = (e_3 - e_2)(e_1 - e_2),$$
$$c_3^2 = (e_1 - e_3)(e_2 - e_3).$$

If we regard the points e_1, e_2, e_3 as given, the constants c_1^2, c_2^2, c_3^2 are also given. We have only three independent equations to determine the four z's, say

$$\left.\begin{aligned}(z_1 - e_1)(z_4 - e_1) &= (e_2 - e_1)(e_3 - e_1)\\ (z_2 - e_2)(z_4 - e_2) &= (e_3 - e_2)(e_1 - e_2)\\ (z_3 - e_3)(z_4 - e_3) &= (e_1 - e_3)(e_2 - e_3)\end{aligned}\right\} \quad . \quad . \quad . \quad . \quad . \quad \text{(iii.)}$$

From these follow three equations of the form

$$(z_2 - e_1)(z_3 - e_1) = (e_2 - e_1)(e_3 - e_1),$$

a result which can be verified directly.

We can state this in words as follows:

Three involutions are determined by any four points; the conjugates of any point in two of these involutions are themselves conjugates in the third involution.

From (iii.) we see that one of the points, z_4, is arbitrary; and when it is chosen, the rest are at once constructed from the preceding equations. Corresponding to the values $z_4 = \infty$, e_1, e_2, e_3 we have

$$z_1 = e_1,\ \infty,\ e_3,\ e_2;\quad z_2 = e_2,\ e_3,\ \infty,\ e_1;\quad z_3 = e_3,\ e_2,\ e_1,\ \infty.$$

§ 39. *The Jacobian of the quartic.*

The double points of the preceding involution are given by

$$(z - e_\lambda)^2 = c_\lambda^2 \qquad (\lambda = 1, 2, 3).$$

Let

$$j_\lambda = e_\lambda + c_\lambda = e_\lambda + \sqrt{(e_\lambda - e_\mu)(e_\lambda - e_\nu)} \quad (\lambda, \mu, \nu = 1, 2, 3 \text{ in any order}),$$

$$j_\lambda' = e_\lambda - c_\lambda = e_\lambda - \sqrt{(e_\lambda - e_\mu)(e_\lambda - e_\nu)}.$$

Now when we invert the points of an involution we obtain a new involution, and the old double points become the new double points. This is clear because any pair of points is harmonic with the double points. Accordingly the six points j are covariant. They are the zeros of a covariant of the sixth degree, the Jacobian of the quartic.

§ 40. *Any two pairs of Jacobian points are harmonic.*

For $$(j_\mu - e_\nu)(j_\mu' - e_\nu) = (e_\mu - e_\nu + c_\mu)(e_\mu - e_\nu - c_\mu)$$
$$= (e_\mu - e_\nu)^2 - c_\mu^2.$$

But $$c_\mu^2 + c_\nu^2 = (e_\mu - e_\nu)^2.$$

Hence $$(j_\mu - e_\nu)(j_\mu' - e_\nu) = c_\nu^2,$$

and j_μ, j_μ' are harmonic with j_ν, j_ν'.

§ 41. The number of quartics which have a common Jacobian is singly infinite, for we have seen that one of the points, z_4, is arbitrary. Each of these quartics is determined conveniently by means of its centroid m, as follows:

We have $$4m = z_1 + z_2 + z_3 + z_4,$$

$$e_1 = (z_2z_3 - z_1z_4)/(z_2 + z_3 - z_1 - z_4);$$

therefore $$4(e_1 - m) = \{(z_1 - z_4)^2 - (z_2 - z_3)^2\}/(z_2 + z_3 - z_1 - z_4),$$

and $$16(e_2 - m)(e_3 - m) = (z_2 + z_3 - z_1 - z_4)^2.$$

Now $(z_2 + z_3 - z_1 - z_4)/4$ is the stroke from m to the centroid of z_2 and z_3. Hence the centroids of the sides are determined, and the quartic is at once constructed.

We shall denote the quartic of the system, whose centroid is m, by U_m.

Let us write the cubic whose zeros are e_1, e_2, e_3,

$$4z^3 - g_2z - g_3,$$

so that the origin is the centroid of e_1, e_2, e_3, and therefore also of the Jacobian.

Regarding this as a quartic with an infinite root, let us calculate the Hessian H_∞. We have

$$U_\infty = 4z^3y - g_2zy^3 - g_3y^4,$$

$$H_\infty = \frac{1}{144}\begin{vmatrix} 24zy, & 12z^2 - 3g_2y^2 \\ 12z^2 - 3g_2y^2, & -6g_2zy - 12g_3y^2 \end{vmatrix}$$

$$= -(z^4 + \tfrac{1}{2}g_2z^2 + 2g_3z + \tfrac{1}{16}g_2^2);$$

thus the centroid of this quartic is at the origin, so that $H_\infty = U_0$.

Any quartic U_m is now

$$mU_\infty + H_\infty,$$

for in this last expression the coefficients of z^4 and z^3 are -1 and $4m$, so that the centroid is m. It is proved in works on Covariants (see Burnside and Panton, 2nd ed., § 170; or Clebsch, Geometrie, t. i., ch. iii.) that the Hessian of $\kappa U + \lambda H$ is

$$\tfrac{1}{8}\left(H\frac{\delta\Omega}{\delta\kappa} - U\frac{\delta\Omega}{\delta\lambda}\right),$$

where $4\Omega = 4\kappa^3 - g_2\kappa\lambda^2 - g_3\lambda^3$; so that if the Hessian be $\kappa' U + \lambda' H$, the ratio of κ' to λ' is given by $\kappa'\frac{\delta\Omega}{\delta\kappa} + \lambda'\frac{\delta\Omega}{\delta\lambda} = 0$; and the geometric statement of this is that the centroid of the Hessian is the second polar of the centroid of the quartic with regard to the points e_1, e_2, e_3.

Thus the centroid of the Hessian is determined, and the Hessian can be constructed.

If the Hessian be given, the first polar of its centroid, with regard to e_1, e_2, e_3, consists of two points, which are the centroids of the two quartics which have the given Hessian.

§ 42. *The anharmonic ratios of the four points.*

If the four points z_1, z_2, z_3, z_4 of § 30 be replaced by e_1, e_2, e_3, ∞, we have

$$\mu/\nu = (e_3 - e_1)/(e_1 - e_2),$$

or $\lambda/(e_2 - e_3) = \mu/(e_3 - e_1) = \nu/(e_1 - e_2) = (\mu - \nu)/(e_2 + e_3 - 2e_1)$.

Now $\quad e_1 + e_2 + e_3 = 0,\ e_2e_3 + e_3e_1 + e_1e_2 = -g_2/4,\ e_1e_2e_3 = g_3/4,$

$$(e_2 - e_3)^2(e_3 - e_1)^2(e_1 - e_2)^2 = (g_2^3 - 27g_3^2)/16 \equiv \Delta/16.$$

Therefore $\quad \mu - \nu = ke_1,\ \nu - \lambda = ke_2,\ \lambda - \mu = ke_3,$

$$(\mu - \nu)(\nu - \lambda)(\lambda - \mu) = k^3 g_3/4,$$

$$\Sigma\mu\nu - \Sigma\lambda^2 = -k^2 g_2/4;$$

or, since $\quad \Sigma\lambda = 0,$

$$3\Sigma\mu\nu = -k^2 g_2/4;$$

and since $\quad 3\lambda = k(e_3 - e_2),$

$$3^6\lambda^2\mu^2\nu^2 = k^6\Delta/16.$$

Eliminating k,

$$-\frac{4(\Sigma\mu\nu)^3}{g_2^3} = \frac{(\mu-\nu)^2(\nu-\lambda)^2(\lambda-\mu)^2}{27g_3^2} = \frac{27\lambda^2\mu^2\nu^2}{\Delta}.$$

The meaning of the vanishing of an invariant is evident from these equations. If $g_2 = 0$, then $\Sigma\mu\nu = 0$, and this, combined with $\Sigma\lambda = 0$, gives either

$$\lambda = \upsilon\mu = \upsilon^2\nu, \quad \text{or} \quad \lambda = \upsilon^2\mu = \upsilon\nu;$$

that is, the points are equianharmonic. If $g_3 = 0$, then two of the quantities λ, μ, ν are equal and the points are harmonic.

§ 43. *The invariants of* $mU_\infty + H_\infty$.

In works on the theory of covariants it is shown that the invariants of this form are

$$-3H'/4, \quad J'/16,$$

where H' is the Hessian and J' the Jacobian of the cubic

$$4m^3 - g_2m - g_3.$$

Hence the quartic is equianharmonic when m is a Hessian point of the triangle $e_1e_2e_3$, and harmonic when m is a Jacobian point. In other words, we have in the quartic involution $mU_\infty + H_\infty$ two equianharmonic and three harmonic quadrangles; the polar points of z_4, with regard to them, are respectively the Hessian points and Jacobian points of $z_1z_2z_3$.

§ 44. *Canonical form of the quartic.*

Let one Jacobian point j_λ be sent to infinity by inversion and let j_λ' be chosen as origin. Since two pairs of points are harmonic with j_λ, j_λ', the quadrangle becomes a parallelogram, and the quartic takes the form

$$(1-z^2)(1-k^2z^2).$$

The other pairs of Jacobian points, being harmonic with j_λ, j_λ' and with one another, form a square.

§ 45. *The representation of the quartic on the sphere.*

Let the regular octahedron be constructed, of which the abovementioned square contains four vertices, let one of the new vertices be selected as origin, and let the z-plane be inverted with regard to this origin. The six Jacobian points become the vertices of a regular octahedron in a sphere. Let ξ, η, ζ be the co-ordinates of any point on the sphere, referred to the three diameters of the octahedron as rectangular axes. The other points which, with this point, make up a quartic with the given Jacobian, have co-ordinates

$$\xi, -\eta, -\zeta; \quad -\xi, \eta, -\zeta; \quad -\xi, -\eta, \zeta.$$

For evidently any two of the four points are harmonic with regard to the points where an axis meets the sphere.

The extremities of those four diameters of the sphere, which are perpendicular to the faces of the octahedron, are the vertices of a cube inscribed in the sphere; the co-ordinates of these vertices are, if the radius of the sphere be 1,

$$\pm 1/\sqrt{3}, \quad \pm 1/\sqrt{3}, \quad \pm 1/\sqrt{3},$$

for we have $\qquad \xi^2 = \eta^2 = \zeta^2$, and $\xi^2 + \eta^2 + \zeta^2 = 1.$

Choosing from the eight points those whose co-ordinates have an even number of minus signs, we obtain a regular tetrahedron. The other four, whose co-ordinates have an odd number of minus signs, form another regular tetrahedron, which we call the counter-tetrahedron of the former one. These tetrahedrons represent equianharmonic quartics of the involution.

Again, draw the six diameters which are perpendicular to the edges of the octahedron, and, therefore, also to the edges of the cube. We thus get twelve points on the sphere whose co-ordinates are given by the scheme

$$0, \ \pm 1/\sqrt{2}, \ \pm 1/\sqrt{2},$$

$$\pm 1/\sqrt{2}, \ 0, \ \pm 1/\sqrt{2},$$

$$\pm 1/\sqrt{2}, \ \pm 1/\sqrt{2}, \ 0.$$

The four which lie in any co-ordinate plane form a square, which from the form of the co-ordinates belongs to the involution. Hence the twelve points represent the three harmonic quartics of the involution.

[For further information on these subjects the student is referred to Klein's Ikosaeder, to a memoir by Wedekind published in Math. Ann., t. ix., and to Beltrami's paper, Ricerche sulla geometria delle forme binarie cubiche, Accademia di Bologna, 1870.]

§ 46. *Examples of many-valued functions.*

Example 1. If w be defined by the equation $w^2 = z$, there are for a given z two values of w. If $z = \rho(\cos\theta + i\sin\theta)$, these values are

$$\left.\begin{aligned} w_1 &= \sqrt{\rho}\,(\cos\theta/2 + i\sin\theta/2) \\ w_2 &= \sqrt{\rho}\,(\cos(\theta+2\pi)/2 + i\sin(\theta+2\pi)/2) \end{aligned}\right\} \quad . \quad . \quad . \quad . \quad \text{(i.)}$$

We have always $w_1 + w_2 = 0$, and when $z = 0$ or ∞ the two roots w_1, w_2 become equal. When z describes, in the z-plane, a path which does not pass near the origin, there corresponds to a small change in z, a definite small change in either root; in this way it can be seen that the two points w_1, w_2 in the w-plane trace out continuous paths as z moves continuously in the z-plane from one position to another. Moreover, assigning to one of the initial values of w the name w_1, the final value of w_1 is completely determinate.

If z describe a closed curve round the origin, starting from a point ρ, θ, the final values of ρ and θ are ρ, $\theta + 2\pi$, and the final values of w_1, w_2 are

$$w_1 = \sqrt{\rho}\{\cos(\theta/2 + \pi) + i\sin(\theta/2 + \pi)\},$$
$$w_2 = \sqrt{\rho}\{\cos(\theta/2 + 2\pi) + i\sin(\theta/2 + 2\pi)\}.$$

Comparing these with equations (i.), we see that w_1, w_2 have changed into w_2, w_1.

If z describe a contour which does not include the origin, the final values of ρ, θ are the same as the initial ones, and the final values of w_1, w_2 are w_1, w_2.

Example 2. $\quad w^2 = (z - a_1)(z - a_2)\cdots(z - a_n)$.

Let C be a contour which contains a_1, a_2, $\cdots a_\kappa$, but not $a_{\kappa+1}$, $a_{\kappa+2}$, $\cdots a_n$. When z, starting from z', describes C, each of the amplitudes θ_1, θ_2, $\cdots$, θ_κ of $z - a_1$, $z - a_2$, $\cdots$, $z - a_\kappa$ increases by 2π, while those of $z - a_{\kappa+1}$, $z - a_{\kappa+2}$, $\cdots$, $z - a_n$ return to their initial values. Let the initial value of the branch w_1 be

$$\sqrt{\rho_1\rho_2\cdots\rho_n}\{\cos(\theta_1 + \theta_2 + \cdots + \theta_n)/2 + i\sin(\theta_1 + \theta_2 + \cdots + \theta_n)/2\},$$

where $z - a_\lambda = \rho_\lambda(\cos\theta_\lambda + i\sin\theta_\lambda)$. The final value is

$$\sqrt{\rho_1\rho_2\cdots\rho_n}\{\cos(\theta_1+\theta_2+\cdots+\theta_n+2\kappa\pi)/2+i\sin(\theta_1+\theta_2+\cdots+\theta_n+2\kappa\pi)/2\}.$$

This is the same as the initial value of w_1 or w_2, according as κ is even or odd.

§ 47. *Example* 3. $\quad a_0w^2 + 2a_1w + a_2 = z$.

By a linear transformation of w the equation becomes

$$w^2 - 1 = z.$$

Let the branches be w_1 and w_2. We have always

$$w_1 + w_2 = 0;$$

when z describes the circle $|z| = \kappa$, the path of either w-point is given by

$$|w - 1||w + 1| = \kappa,$$

or

$$\rho_1\rho_2 = \kappa,$$

ρ_1 and ρ_2 being the distances of the point from ± 1. This is the equation of a Cassinian.

(1) Let κ be small. Then ρ_1 or ρ_2 must be small, so that the curve consists of two small ovals round the points $+1$, -1; w_1 must remain on the one oval, w_2 on the other.

(2) Let κ grow; the ovals grow, but remain distinct until $\kappa = 1$, when the circle passes through the point $z = -1$, and the ovals join at $w = 0$.

If z, instead of passing through -1, describe a small semicircle round it positively, then (Fig. 17) q_2 is consecutive to p_1, and q_1 to p_2. But if z describe a semicircle negatively, q_1 is consecutive to p_1, and q_2 to p_2. The lines p_1p_2 and q_1q_2 are of course at right angles.

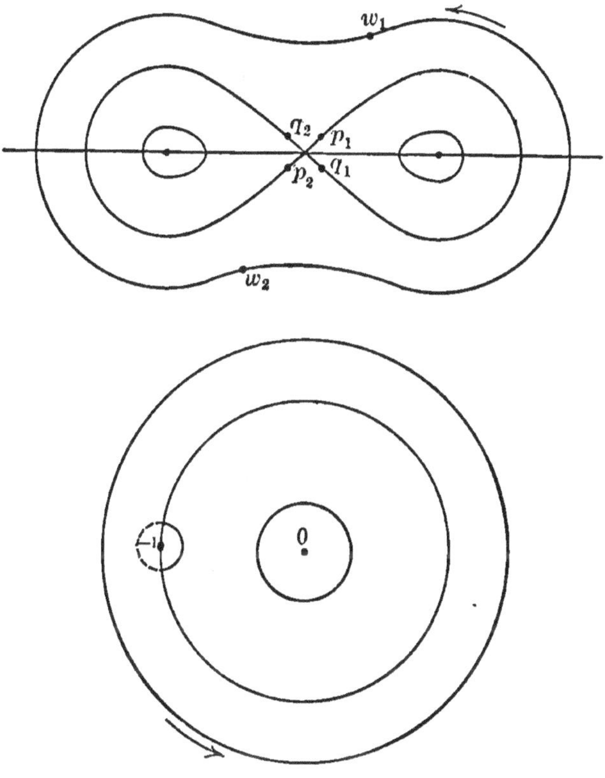

Fig. 17

(3) When $\kappa > 1$, the curve is at first an indented oval, and afterwards an ordinary oval. As z describes its circle, w_1 and w_2 together describe the whole oval, and the final position of either is the initial position of the other.

§ 48. *The relation* $a_0w^3 + 3a_1w^2 + 3a_2w + a_3 = z$.

By linear transformation this reduces to

$$w^3 - 3w = z \quad . \quad . \quad . \quad . \quad . \quad . \quad . \quad . \quad . \quad . \quad . \quad . \quad . \quad (1).$$

As in Cardan's solution of the cubic, put

$$w = t + 1/t \quad \text{(2)},$$

then

$$z = t^3 + 1/t^3 \quad \text{(3)}.$$

If z be given, we have six values of t, namely,

$$t,\ \nu t,\ \nu^2 t,\ 1/t,\ \nu^2/t,\ \nu/t,$$

and three values of w, namely,

$$w_1 = t + 1/t,\ w_2 = \nu t + \nu^2/t,\ w_3 = \nu^2 t + \nu/t.$$

Let t describe the circle $|t| = \kappa$; the point z will describe an ellipse whose foci are $z = \pm 2$ and whose semi-axes are $\alpha = \kappa^3 + 1/\kappa^3$, $\beta = |\kappa^3 - 1/\kappa^3|$, and the three w's will describe an ellipse whose foci are ± 2 and semi-axes $\alpha_1 = \kappa + 1/\kappa$, $\beta_1 = |\kappa - 1/\kappa|$, whence $(\alpha_1 + \beta_1)/2 = \sqrt[3]{(\alpha + \beta)/2}$, $(\alpha_1 - \beta_1)/2 = \sqrt[3]{(\alpha - \beta)/2}$. From either of these equations, and from $\alpha_1^2 - \beta_1^2 = \alpha^2 - \beta^2$, the w-ellipse is determined when z is given.

If θ be the amplitude of t, the eccentric angle of z is 3θ, and those of w_1, w_2, w_3 are θ, $\theta + 2\pi/3$, $\theta + 4\pi/3$; so that the points w form a maximum triangle in the w-ellipse. Therefore, when z is given, the points w are determined. In this way, if z describe any curve, say a circle with centre z', it is easy to construct the w-curve. If w_1', w_2', w_3', be the points which correspond to z', the equation of the w-curve is

$$|w - w_1'|\,|w - w_2'|\,|w - w_3'| = \text{a constant},$$

as is evident from the relations

$$(w - w_1')(w - w_2')(w - w_3') = w^3 - 3w - z' = z - z'.$$

Two of the family of curves should be specially noticed. The points at which values of w become equal are given by $w^2 - 1 = 0$, and are $z = \pm 2$. Thus to the circles with centre z' through $z = \pm 2$ correspond w-curves with nodes at $w = \pm 1$. Since for these values of z two values of w become equal, we have, approximately, near these points

$$(w \pm 1)^2 = c(z \pm 2),$$

so that the branches of the w-curve cut at right angles at these points. Fig. 18 shows these two special w-curves.

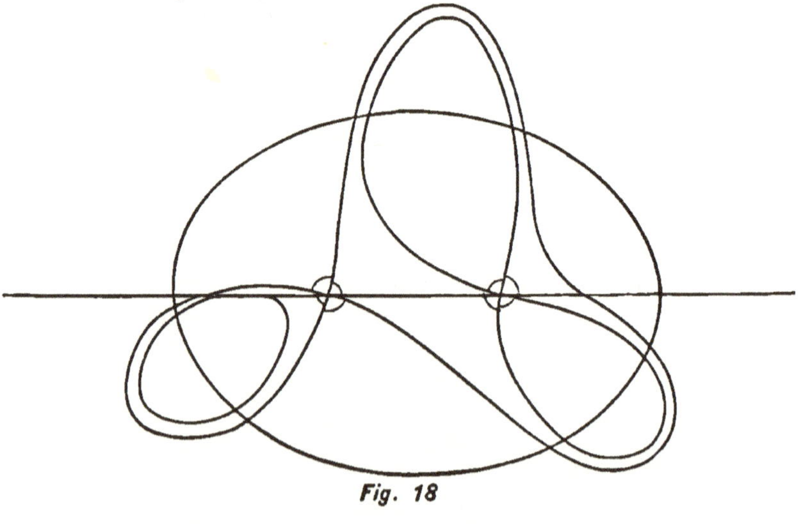

Fig. 18

The orthogonal family of curves correspond to z-lines through z'. Their equations are of the form

$$\theta_1 + \theta_2 + \theta_3 = \text{a constant},$$

where θ_r is the inclination, measured from the real axis, of the line from the fixed point w_r' to w. Since z' is any point on the z-line, the fixed points w_r' can be chosen in infinitely many ways; they may be conveniently taken to correspond to the point where the z-line meets the real axis. If the z-line pass through a point at which two branches become equal, the equation takes the form

$$2\,\theta_1 + \theta_2 = \text{a constant},$$

a nodal cubic, which can be very readily constructed.

CHAPTER II.

Real Functions of a Real Variable.

§ 49. One of the most noteworthy features in the history of mathematical activity during the nineteenth century is the careful revision to which the fundamental processes and concepts of analysis have been subjected. Cauchy, Gauss, and Abel were the first to insist upon the importance of basing the theories of their day on securer foundations, and they have been followed by a long line of brilliant successors. B. Bolzano of Prague (1781–1848) worked on the same lines, but his writings failed to attract the attention they deserved, until a comparatively recent date. The discovery that continuous functions exist which do not possess differential quotients and that a discontinuous function may possess an integral, the new light thrown upon discontinuous functions by Dirichlet's well-known memoir on Fourier's series and by Riemann's masterly researches on the same subject, have all tended to reveal how little is known of the nature of functions in general, and to emphasize the need for rigour in the elements of Analysis.

In the course of this revision it became necessary to place the number-concepts of Algebra upon a basis independent of, but consistent with, Geometry. If a zero-point be selected on a straight line and also a fixed length, measured on this line, be chosen as the unit of length, any real number a can be represented by a point on this line at a distance from the zero-point equal to a units of length. Conversely, each point on the line is at a distance from the origin equal to a units of length, where a is a real number. This theorem may seem evident, but a little reflexion will show that it cannot be true unless the word *number* is so defined as to make the number-system continuous instead of discrete.

Definitions which will fulfil this requirement have been given by Weierstrass, G. Cantor, Heine, Dedekind, and others. Much of the theory of the higher Arithmetic is due to Weierstrass and was communicated to the world in University lectures at Berlin. His starting-point is the positive integer, and the number-system is extended so as to include, successively, positive fractions, negative, irrational, and complex numbers. One noticeable element in his theory is the exclusion of geometric concepts. Certain laws of calculation can be stated, in general terms, only after the introduction of the new numbers, and it is for this reason, and not because of any geometric representation by extensive magnitudes, that these new numbers, namely the fractional, the negative, the irrational, and the complex numbers, are introduced.

§ 50. *The irrational numbers.*

Three different definitions have been employed, due to Weierstrass, Dedekind, and G. Cantor. The following account of Cantor's method is based on a memoir by Heine (Crelle, t. lxxiv.).

Cantor's method. The starting-point is a sequence $a_1, a_2, a_3, \cdots$ of rational numbers. This sequence is determinate when the law of construction of the general term a_n is known. A *regular* sequence is one for which a finite number μ can be found, such that for all values of n $(>\mu)$ the absolute value of $a_{n+p}-a_n$ $(p=1, 2, 3, \cdots)$ is less than an arbitrarily small positive rational number ϵ. [Throughout this chapter and the next, ϵ will be used, always, to denote an arbitrarily small positive number.] When the sequence $a_1, a_2, a_3, \cdots$ is regular and when a rational number α exists which is such that a finite integer μ can be found which makes $|\alpha-a_{\mu+p}|<\epsilon$ $(p=1, 2, \cdots)$, the rational number α is said to be the *limit* of the sequence. This statement can be expressed shortly in the form

$$\lim_{n=\infty} a_n = \alpha,$$

or, using a convenient notation, $La_n=\alpha$. It is not always possible to find a rational number α. For example, there is no rational limit of the sequence,

$$a_1=1,\ a_2=1-1/2,\ a_3=1-1/2+1/3,\ a_4=1-1/2+1/3-1/4, \text{ etc.}$$

To the regular sequence $a_1, a_2, a_3, \cdots$ is attached a symbol A, (a), or $(a_1, a_2, a_3, \cdots)$, which is associated with the *totality* of numbers contained in the sequence. If, after any term a_κ (κ finite), all the numbers of the regular sequence be equal to a rational number C, *let C be the symbol attached to the sequence.* In this way all rational numbers provide symbols. We use the word *symbol* advisedly, because we have not as yet proved that a quantity which possesses the properties of a number can be defined by a regular sequence. When the terms *equality* and *greater* or *less inequality* have been defined in respect to the symbols, it will be found possible to replace the term *symbol* by *number*.

In the regular sequence $a_1, a_2, \cdots$ three cases occur:—

(1) A term a_n can be found such that the values of $a_n, a_{n+1}, a_{n+2}, \cdots$ all lie, for a sufficiently great value of n, between the arbitrarily small rational numbers $-\epsilon, \epsilon$.

(2) A term a_n can be found such that the values of $a_n, a_{n+1}, a_{n+2}, \cdots$ are all greater than some definite positive rational number g.

(3) A term a_n can be found such that the values of $a_n, a_{n+1}, a_{n+2}, \cdots$ are all less than some definite negative rational number $-k$.

In case (1) the symbol attached to the sequence is 0; in cases (2), (3) the symbols attached to the sequences are said to be positive and negative.

§ 51. *Definitions.* It is easy to frame definitions of the sum, difference, product, and quotient of symbols attached to regular sequences. If A and B be attached to the regular sequences $a_1, a_2, a_3, \cdots, b_1, b_2, b_3, \cdots$, $A \pm B$ and AB are attached to the sequences $(a_n \pm b_n)$, $(a_n b_n)$, or writing the sequences at full length, to

$$a_1 \pm b_1,\ a_2 \pm b_2, \cdots,\ a_n \pm b_n, \cdots,$$

and

$$a_1 b_1,\quad a_2 b_2, \cdots\cdots,\ a_n b_n, \cdots.$$

If B be different from 0, A/B is attached to the sequence $a_1/b_1, a_2/b_2, a_3/b_3, \cdots, a_n/b_n, \cdots$. These definitions are consistent with those of the four fundamental operations in the case of rational numbers. For example, let A, B be two rational numbers. The symbol $A = (A, A, A, \cdots)$, the symbol $B = (B, B, B, \cdots)$, the symbol $A + B$ = the symbol A + the symbol B, and the number $A + B$ = the number A + the number B. As we wish all these symbols to be attached to regular sequences, we must show that the sequences

$$(a_n \pm b_n),\ (a_n b_n),\ (a_n/b_n),$$

are regular.

I. The sequence $(a_n \pm b_n)$ is regular; for a number μ can be found such that, when $n > \mu$,

$$|a_{n+p} - a_n| < \epsilon/2,\ |b_{n+p} - b_n| < \epsilon/2\ (p = 1, 2, \cdots);$$

and from these two inequalities it is an immediate deduction that

$$|(a_{n+p} \pm b_{n+p}) - (a_n \pm b_n)| < \epsilon\ (p = 1, 2, \cdots).$$

II. The sequence $a_1b_1, a_2b_2, a_3b_3, \cdots$ is regular; for

$$a_{n+p}\, b_{n+p} - a_n b_n = a_{n+p}\,(b_{n+p} - b_n) + b_n(a_{n+p} - a_n),$$

a quantity which is numerically less than $(a_{n+p}+b_n)\epsilon$, which, in turn, can be made less than any positive rational number, however small, by a proper choice of ϵ.

III. The sequence $a_1/b_1, a_2/b_2, a_3/b_3 \cdots$ is regular, if B be not 0.

For $$\frac{a_{n+p}}{b_{n+p}}-\frac{a_n}{b_n}=\frac{(a_{n+p}-a_n)b_n-a_n(b_{n+p}-b_n)}{b_nb_{n+p}},$$

a fraction which can be made less than any positive rational number, however small, by a proper choice of ϵ.

§ 52. *Definitions of the terms equality and inequality.*

If the symbols A, B be attached to the sequences $a_1, a_2, a_3, \cdots$, $b_1, b_2, b_3, \cdots$, and if the elements of the sequence a_1-b_1, a_2-b_2, a_3-b_3, $\cdots$, from a definite term onwards, be always greater than some positive rational number, A is said to be greater than B. If the elements after that term be always less than some negative rational number, B is said to be greater than A.

If a number μ can be found such that $|a_n-b_n|<\epsilon$, $(n>\mu)$, $A-B=0$, and A is said to be equal to B. This definition agrees with the facts in the case of rational numbers. For, if it can be proved that the difference of two rational numbers is less than ϵ, the numbers must be equal. The two sequences

$$a_1, a_2, a_3, \cdots a_k, a_{k+1}, \cdots;\quad a_1', a_2', \cdots, a_k', a_{k+1}, a_{k+2}, \cdots$$

define the same symbol, for a_1-a_1', a_2-a_2', $\cdots$, $a_{k+1}-a_{k+1}, \cdots$ has for its symbol 0. It is to be noticed that the definitions of equality and inequality for A, B do not overlap.

Let the sequence $a_1, a_2, \cdots, a_n, \cdots$ have the rational limit α; the symbol attached to $a_1-\alpha$, $a_2-\alpha$, $a_3-\alpha$, $\cdots$ is 0, since by hypothesis μ can be so determined that $|\alpha-a_n|<\epsilon$, when $n>\mu$.

Hence, $$(a_1, a_2, a_3, \cdots)-(\alpha, \alpha, \alpha, \cdots)=0;$$

i.e. the symbol attached to the sequence $a_1, a_2, a_3, \cdots$ is α, or La_n. Thus, when a rational limit is known to exist, it must be the symbol attached to the sequence.*

Example. If $a_n=.333\cdots$, where the figure 3 occurs n times, $L(\frac{1}{3}-a_n)=0$. Hence $\frac{1}{3}$ is the symbol attached to .3, .33, .333, $\cdots$.

---

* When each number of the sequence is a decimal fraction, differing from the preceding number only by having one additional digit assigned, the necessary and sufficient condition for a rational limit is that there must be, sooner or later, a periodic recurrence of the digits. See Stolz, Allgemeine Arithmetik, t. i., p. 100.

§ 53. *Definitions.* (i.) The absolute value of (a) is $(|a|)$; (ii.) the symbols $A_1, A_2, A_3, \cdots, A_n$ are said to decrease below every assignable value, when for every symbol E, except 0, a number μ can be found such that, when $n > \mu$, $A_n < E$ in absolute value.

Consider the regular sequence $a_1 - a_n, a_2 - a_n, a_3 - a_n, \cdots$; we have

$$A - a_n = (a_1 - a_n, a_2 - a_n, \cdots) = (a_{n+1} - a_n, a_{n+2} - a_n, \cdots).$$

When n increases beyond any assignable rational number, the elements of the sequence decrease below any assignable rational number, by the definition of a regular sequence. The absolute value of the symbol attached to the second sequence decreases below any assignable absolute value; therefore $A - a_n$ decreases below any assignable value, as n increases. Hence we have the theorem that a limit exists for a_n, namely the symbol A. When La_n is not a rational number, we call it an *irrational* number.

When the limit of a sequence is a rational number, the symbol attached to the sequence can be regarded as that rational number. By the definitions of § 52, *all* symbols can be arranged in order of magnitude. Hence, when the limit is not a rational number, the corresponding symbol can be regarded as of like nature with the rational numbers, and is called an irrational number. From this point onwards the word *symbol* can be replaced by *number.*

The symbols 1/3, 1 attached to the sequences .3, .33, .333, ⋯; 1, 1, 1, ⋯ are now the ordinary numbers 1/3, 1, and such an inequality as

$$(1, 1, 1, \cdots) < (1, 1.4, 1.41, 1.414, \cdots) < (1.5, 1.5, 1.5, \cdots),$$

where 1, 1.4, 1.41, 1.414, ⋯ are numbers which occur in the process of extracting the square root of 2, shows that the symbol (1, 1.4, 1.41, 1.414, ⋯) can be regarded as a number intermediate in value between 1 and 1.5.

Cantor (Math. Ann., t. xxi., p. 568) calls attention to the importance of perceiving that

$$A = La_n$$

is a theorem and not a definition. The number A was not defined as the limit of a_n $(n = \infty)$, for this would presuppose the existence of the limit. A different course was adopted. First, numbers were defined by regular sequences; next, a meaning was given to the elementary operations as applied to the new numbers; thirdly, a definition was given of equality and of greater or less inequality;

finally, there followed from these definitions and laws of construction the theorem

$$A = La_n.$$

To see that no new numbers are obtained by using a regular sequence of *irrational* numbers $a_1, a_2, a_3, \cdots$, consider a regular sequence $b_1, b_2, b_3, \cdots$ of rational numbers defined in the following way:

Let a_1 and b_1 have the same integral part, a_2 and b_2 agree to the first decimal place, a_3 and b_3 to the second decimal place, and so on. Form the sequence $a_1 - b_1, a_2 - b_2, a_3 - b_3, \cdots$. The elements become ultimately $< \epsilon$, and the sequence defines 0. That is, the numbers $(a_1, a_2, a_3, \cdots)$, $(b_1, b_2, b_3, \cdots)$ are equal. Rational and irrational numbers constitute the system of *real* numbers.

Deductions from the theorem $La_n = (a_1, a_2, a_3, \cdots)$. By definition

$$AB = (a_1b_1, \cdots, a_nb_n, \cdots) = La_nb_n.$$

Hence $$La_n \cdot Lb_n = La_nb_n,$$

similarly $$La_n / Lb_n = La_n/b_n, \text{ where } Lb_n \neq 0.$$

We leave to the reader, as an instructive exercise, the proof that

$$(La_n)^k = La_n^k,$$

$$(La_n)^{1/k} = La_n^{1/k},$$

where k is a positive integer, and also the proof that $(La_n)^{Lb_n} = L(a_n^{b_n})$, when the a's and b's form regular sequences. If b be irrational and a rational, a^b is the number defined by the series

$$a^{b_1}, \ a^{b_2}, \ a^{b_3}, \cdots.$$

In this way it is possible to arrive at a definition of such expressions as $a^{\sqrt{2}}$. The simplest mode of treatment is obtained from the exponential theorem.

§ 54. *The unicursal, or one-one, correspondence between the totality of real numbers and the points of a straight line.*

I. To every point on a straight line corresponds a distance from the zero-point, whose ratio to the unit line is a real number.

Let the line which extends from the zero-point to the point in question be named B, and let the unit line be A. To fix ideas, let the line $B(= ob)$ be longer than the line $A(= oa)$. Let lines oc_1, c_1c_2, $c_2c_3, \cdots$ be measured along the straight line so that oc_1 is the last multiple of A contained within ob, c_1c_2 is the last multiple of $A/10$ contained within c_1b, c_2c_3 is the last multiple of $A/10^2$ contained within c_2b, and so on. Either the point b is reached after a finite number of divisions, or it is not. In the former case the ratio of B to A is evidently a rational number. In the latter the points $c_1, c_2, c_3, \cdots$ approach b. The sum of the lines $oc_1, c_1c_2, c_2c_3, \cdots$ falls short of B, but the amount by which it falls short can be made less than an arbitrarily assigned fractional portion of B. If the ratios of oc_1, $c_1c_2, \cdots$ to A, $A/10, \cdots = m_1, m_2 \cdots$, where $m_1, m_2, \cdots$ are positive integers of the set $0, 1, 2, \cdots, 9$, the ratio of B to $A = m_1 + m_2/10 + m_3/10^2 + \cdots =$ the number defined by the regular sequence

$$m_1,\ m_1 + m_2/10,\ m_1 + m_2/10 + m_3/10^2, \cdots.$$

II. To every real number corresponds a point of the line, as soon as a unit line has been selected.

There is no difficulty in proving this for rational numbers. If the number be irrational, let it be defined by a regular sequence $a_1, a_2, a_3, \cdots$ of rational numbers, in *ascending* order of magnitude. Let p, q be two points on the line such that the numbers which correspond to op, oq are respectively less than and greater than (a). As the point moves from p to q it must come to a position r such that for the first time the line or is not less than any of the lines oa_1, oa_2, $oa_3, \cdots$ which represent the rational numbers $a_1, a_2, a_3, \cdots$. This position r possesses the property that every line, which starts from o and does not extend to r, is less than some number a_n of the sequence. The number represented by the line or is equal to that represented by the number-sequence $(a_1, a_2, \cdots)$, for it is greater than every a_n, but, when diminished by any number however small, is less than *some* member of the sequence.

The preceding theorem is capable of the following extension. The complex number, $a + ib$, is built up out of the units 1, i. There is a unicursal correspondence between the system of complex numbers and the points of a plane, each system forming a doubly extended manifoldness. It is, therefore, permissible to use the word point instead of number. [See Fine's Number-System of Algebra, §§ 40–42.]

§ 55. *Variable quantities.* Prior to a consideration of functions of a real variable, it is advantageous to examine the range of values of the independent variable x. In some questions x cannot pass through all the rational and irrational real numbers from $-\infty$ to ∞. Whether it passes through a finite or infinite number of values will depend upon circumstances, but in any case there will be an upper limit for the values, which may be either finite or infinite. If x take only a finite number of values, the greatest of these will be the limit and, moreover, this upper limit will be attained. It is not to be inferred that this will always be the case when x takes an infinite number of values. For example, if $x = 1 - 1/n$, $(n = 1, 2, 3, \cdots)$, the upper limit 1 is unattainable.

Theorem. If all the values of x be less than a finite number A, there is one and only one number G, which is such that no value of x is greater than G, while at least one value of x is greater than $G-\epsilon$. G is called by Weierstrass the *upper limit* of the variable.*

To prove this theorem let $a_1, a_2, a_3, \cdots$ be the values which x can assume. If a_1 be greater than every succeeding term, it is itself the upper limit; but if there should be one or more terms of the sequence $a_2, a_3, \cdots$ which are $> a_1$, let b_2 be the first of these. Either b_2 is greater than all the succeeding a's or it is not. In the former case b_2 is the upper limit, but, in the latter, some subsequent number b_3 of the series is greater than b_2. A continuation of this process leads to the sequence of ascending numbers $a_1, b_2, b_3, b_4, \cdots$; if the sequence contain only a finite number of terms, the last of these is the upper limit, whereas, if it contain an infinite number of terms, it is regular and defines a number $< A$. The number so defined is G. When no number A can be found with the assigned properties, the upper limit is said to be ∞. There is a like theorem with respect to the existence of a *lower limit* K.

A variable x is said to be *continuous* in the interval $x_0 \leqq x \leqq x_1$, when it passes through all the rational and irrational numbers of the interval (x_0 and x_1 inclusive).

Example. Let the values of x be

$$1-1/2,\ -1+1/3,\ 1-1/4,\ -1+1/5, \cdots$$
$$\cdots,\ 1-1/2n,\ -1+1/(2n+1), \cdots \text{ (Thomæ).}$$

The upper and lower limits, which are *unattainable*, are $+1$ and -1. The variable is not continuous in the interval.

§ 56. *Limiting numbers.* In the neighbourhood of a limiting number there is an infinite accumulation of points. To make the meaning more precise, let us consider an infinite sequence $a_1, a_2, a_3, \cdots, a_n, \cdots$ whose members are finite and distinct. There must be in the neighbourhood of some member H, ($K \leqq H \leqq G$), an infinite accumulation of points. To prove this, divide the finite interval (K to G) into p parts, where p is some integer; in at least one of these parts, say (K_1 to G_1), there must lie infinitely many points of the sequence. Divide (K_1 to G_1) into p parts. In at least one of these parts, say (K_2 to G_2), there must lie infinitely many points. By continuing this process we arrive at the two sequences

$$\left.\begin{matrix} K_1,\ K_2,\ K_3 \cdots \\ G_1,\ G_2,\ G_3 \cdots \end{matrix}\right\}$$

* Bolzano was the first mathematician to draw attention to the existence of upper and lower limits of a variable. See Stolz, B. Bolzano's Bedeutung in der Geschichte der Infinitesimalrechnung, Math. Ann. t. xviii. Their introduction into modern analysis is due to Weierstrass.

which are in ascending and descending order. Each member of the second sequence is greater than every member of the first, and $G_n - K_n$ can be made less than any assignable number, if n be taken sufficiently great. The two sequences define one and the same number H. By the mode of construction of H it follows that, however small δ may be, infinitely many values of x exist within the interval ($H-\delta$ to $H+\delta$). This is what is meant by saying that there is an infinite accumulation of numbers at H, and that H is a *limiting number*. H need not be itself a member of the system $a_1, a_2, a_3, \cdots$. A distinction must be drawn between "the upper limit" and a "limiting number" of a sequence. For example let $a_1 \leqq a_2 \leqq a_3 \cdots < A$. There will be, in all cases, an upper limit for this sequence, but there cannot be a limiting number when the series contains only a finite number of elements. Again there can be more than one limiting number, but only one upper limit.

§ 57. Many of the more serious difficulties in the study of functions are caused by infinite accumulations of values x in the neighbourhoods of particular points. Consider, for instance, Riemann's statement of the necessary and sufficient conditions in order that a function $f(x)$ may be integrable between the limits $x = a$, $x = b$. The function $f(x)$ is said to be *continuous* at the point c of the interval, if a field ($c - h$ to $c + h$) can be found such that, for all points of this field, $|f(x) - f(c)| < \epsilon$. Otherwise the function is said to be *discontinuous* at c.

Interpolate between a and b the values $x_1, x_2, \cdots, x_{n-1}$, and let $x_1 - a = \delta_1$, $x_2 - x_1 = \delta_2$, $x_3 - x_2 = \delta_3, \cdots$. If $f(x)$ be a discontinuous function of x, which is always finite between a and b, its values within an interval δ will have an upper and a lower limit. The difference between these upper and lower limits is named the oscillation of the function. Riemann's theorem is the following: —

If the sum of the intervals in which the oscillations of $f(x)$ are greater than a given finite number λ become, always, infinitely small with the intervals δ, the series

$$\delta_1 f(a + \theta_1\delta_1) + \delta_2 f(x_1 + \theta_2\delta_2) + \cdots + \delta_n f(x_{n-1} + \theta_n\delta_n),$$

in which $\theta_1, \theta_2, \cdots, \theta_n$ are positive proper fractions, will converge to a definite sum, when every δ is indefinitely diminished. This sum is called $\int_a^b f(x)dx$.* This definition does not require that the function should be continuous, and thus there arises the possibility of finite changes in the value of the function for infinitely small changes of x. Evidently it is important to know where these changes take place.

§ 58. This illustration has been introduced to show the need for a theory of point-distribution. Two theories of this kind have been elaborated. One of these is due to Hankel (Math. Ann. t. xx.), while the other has been developed by Cantor in a remarkable series

* For a proof of the convergence of this sum, see Riemann's memoir Ueber die Darstellbarkeit einer Function durch eine trigonometrische Reihe (Werke, pp. 213–253); also a paper by H. J. S. Smith, On the Integration of Discontinuous Functions, Proc. Lond. Math. Soc. t. vi.

of papers published in the Mathematische Annalen and the Acta Mathematica.

To Hankel we owe the idea of a *discrete mass of points*. Let the range from $x - \delta$ to $x + \delta$, where δ is finite though arbitrarily small, be called the *neighbourhood* of x. When a finite or infinite number of points, lying in an interval (a to b), can all be included in neighbourhoods whose sum $< \epsilon$, the points are said to form a *discrete mass*. When such a choice of neighbourhoods is not possible for the interval (a to b), or for any arbitrary part of it, the points are said to form a *linear mass* throughout (a to b). This distinction was adopted by Hankel, with the object of making Riemann's theory of integration more precise.*

Cantor's division of masses into species constitutes a farther reaching theory than that of Hankel, and has led to important generalizations in the hands of recent writers upon the subject of essential singularities in the Theory of Functions. [See, for example, Mittag-Leffler's paper in the Acta Mathematica, t. iv.]

Let P be a mass of different points arranged along a line. These points will, in general, be accumulated in infinitely great numbers at certain points of the line. The limiting points of these infinite accumulations form a *derived* mass of points, which can be represented by P_1. If P_1 contain infinitely many points, it will, in turn, give rise to a new derived mass of points P_2, and so on indefinitely, unless at some stage of the process a derived mass P_n is reached, in which there is nowhere an infinite accumulation. The process then stops.

Points of P_1 need not be points of P, for it has been pointed out that a limiting number of P does not necessarily belong to P. All the points of P_2 must be points of P_1. If possible, let a point c of P_2 be not included in P_1. At c there is an infinite accumulation of P_1-points. Hence at c there is an infinite accumulation of P-points. This latter statement is tantamount to saying that c is a limiting point of P, or that c belongs to P_1. Hence in the process of derivation new points can be introduced only at the first stage.

Examples. (1) $P \equiv (1, 1/2, 1/2^2, 1/2^3 \cdots)$; we have $P_1 \equiv (0)$, the process stopping with P_1.

$$(2) \left\{ \begin{array}{l} P \equiv (1,\ 1/2,\ 1/2^2,\ 1/2 + 1/2^2,\ 1/2^3,\ 1/2 + 1/2^3,\ 1/2^2 + 1/2^3, \\ \quad 1/2^4,\ 1/2 + 1/2^4, \cdots), \end{array} \right.$$

a series of numbers formed by taking 1, $1/2$, $1/2^2$, $1/2^3$, ⋯ one at a time, and $1/2$, $1/2^2$, $1/2^3$, ⋯ two at a time; we have

$$P_1 \equiv (0,\ 1/2,\ 1/2^2,\ 1/2^3, \cdots);\quad P_2 \equiv (0),$$

the process stopping with P_2.

* The nomenclature adopted, and also the examples of this paragraph, are taken from Harnack (Bulletin des Sc. Math., Sér. 2, t. vi., p. 242; Calculus, § 144).

(3) $P \equiv$ the rational numbers between 0 and 1, these two numbers exclusive; we have $P_1 \equiv$ the rational and irrational numbers between 0 and 1, these two numbers inclusive. The remaining masses, $P_2, P_3, \cdots$, contain the same points, and the process never terminates.

Example (2) illustrates the difference between Hankel's discrete mass and Cantor's derived masses. The mass P is discrete, despite the circumstance that it possesses infinitely many places of accumulation. For, drawing from 0 towards 1 a small interval 0 to δ, the remaining interval δ to 1 contains only a finite number of the points of accumulation, and the sum of the corresponding fields can be made arbitrarily small, however small δ may be.

§ 59. *Meaning attached to the word Function.*

As a clear understanding of what is meant by a function of a variable x is essential for the comprehension of modern researches in the Theory of Functions, we shall give here a brief sketch of the evolution of the modern view of functionality. We shall have occasion to speak, from time to time, of the arithmetic expression of a function. By this we shall mean an expression which arises from certain quantities by the four fundamental operations of addition, subtraction, multiplication, and division. As Weierstrass has pointed out,* there are no other known elementary operations. No new elementary operations are needed for powers, roots, and logarithms, for if the relationship between numbers can be put, in any manner, into an analytic form, in which only the four elementary operations are used, there is no occasion to regard the indicated operation as elementary. No example is known of an analytic connexion, which is not reducible to a combination of these fundamental operations.

The three secondary operations arose in the following manner. The power is the result of multiplying equal factors. From the equation $a^m = b$, the two other operations can be found; namely, they are involved in the problems:—

(i.) To find a, given m and b;
(ii.) To find m, given a and b.

Prior to the introduction of the Infinitesimal Calculus, the functions used in Analysis were of so simple a type that no special name was needed. The word *function* was used by Leibnitz (Acta Eruditorum, 1692).

First definition (Bernoulli-Euler). The number y was called a function of x, when it could be constructed from x by prescribed combinations of the four fundamental and the three secondary operations. If y can be called a function of x only when these conditions

---

* See Kossak, Die Elemente der Arithmetik, p. 29.

are fulfilled, x cannot logically be called a function of y, until a proof has been given that x can be derived from y by a combination of the seven operations. Cases were known in which this could not be effected by a finite number of operations, and no examination was made as to whether, in such a case, a valid representation could be afforded by infinitely many of these operations.*

Second definition. Bernoulli also gave this definition: If two variable magnitudes be so related, the one to the other, that to every value of the one corresponds a certain number of values of the other, the first variable is said to be a function of the second, and the second of the first.

Since the magnitudes considered can always be represented by numbers, this definition includes the former one. This definition was intended to apply solely to continuous curves, and to express the connexion between the ordinate and abscissa of an arbitrarily drawn continuous curve which had everywhere a tangent. It was assumed that a continuous function must be expressible by such a curve; that is, that a continuous function must possess a differential quotient. (See § 65.)

Curves drawn arbitrarily on a plane may present peculiarities which mark them off, at once, as non-algebraic. These curves were known as 'curvæ discontinuæ seu mixtæ seu irregulares' (Euler, Introductio in Analysin infinitorum, t. ii., p. 6), and were regarded as incapable of arithmetic expression.

A controversy between D'Alembert, Daniel Bernoulli, Euler, and Lagrange led to a discussion of the question whether every function represented by a continuous curve (in the then accepted sense of the term) could be expressed as a trigonometric series. In 1807 Fourier greatly enlarged the concept of a function by proving that certain curves which obey different laws in different parts can be represented by a single arithmetic expression. Fourier's discovery showed that a portion of a curve does not fix the form of the complete curve and that some, at least, of the curves hitherto excluded resulted from arithmetic expressions. For information on Fourier's series we refer the student to Fourier's collected works, and to memoirs by Dirichlet (Crelle. t. iv.), Riemann (Ges. Werke.,

---

* John Bernoulli, in a memoir dated 1718, gave this definition: "On appelle . . . Fonction d'une grandeur variable, une quantité composée de quelque manière que ce soit de cette grandeur variable et de constantes." Mém. de l'Acad. Roy. des Sciences de Paris, 1718, p. 100. This definition was adopted by Euler in 1748: "Functio quantitatis variabilis est expressio analytica quomodocunque composita ex illa quantitate variabili et numeris seu quantitatibus constantibus."

p. 213), Sachse (Bulletin des Sciences Mathématiques, 2e série., t. iv., 1880).

Dirichlet's definition. Dirichlet defined y as a function of x in the following manner:—

Let x take certain values in an interval (x_0 to x_1); if y possess a definite value or definite values for each of these, y is said to be a function of x. It is not necessary that y should be related to x by any law or arithmetic expression. Moreover, the function may be defined for certain values only of x within the interval (x_0 to x_1), *e.g.* for all rational numbers; but usually it is defined for all the values within the interval. [Repertorium der Physik, her. v. Dove, t. i., p. 152.]

According to this definition the value of y, when $x = a$, may be entirely unrelated to the value of y for any other value of x, $x = b$. This definition, in contrast to those used before Dirichlet's time, errs on the side of excessive generality; for it does not of itself confer properties on the function. The functions so defined must be subjected to restrictive conditions before they can be used in analysis. Nevertheless, this definition forms and must continue to form the basis for researches upon discontinuous functions of a real variable.

In Dirichlet's sense $f(x)$ is a function of x throughout an interval when, to every value of x within the interval, belongs a *definite* value of $f(x)$. A value of x which makes the function ∞ is excluded.

Ex. 1. Sin $1/x$ has no *definite* value at $x = 0$.

Ex. 2. Sin $x/1 - \sin 2x/2 + \sin 3x/3 - \cdots$ is, in spite of sudden discontinuities, a function which is definite for all finite values of x.

§ 60. *The limit of a function.*

Let y be defined for an infinite mass of points, continuous or discontinuous, of which a is a limiting point. If, for every ϵ, a positive number δ can be assigned such that, when $a < x < a + \delta$, $|y - A| < \epsilon$, we call A the *limit* of y when x tends to a from the right, and write $\lim\limits_{x=a+0} y = A$. Similarly if, when $a - \delta < x < a$, $|y - B| < \epsilon$, we call B the limit of y when x tends to a from the left, and write $\lim\limits_{x=a-0} y = B$.

If, when $a<x<a+\delta$, $y>1/\epsilon$ (or $<-1/\epsilon$), $\lim\limits_{x=a+0} y = +\infty$ (or $-\infty$).

If, when $x > \delta$, $|y - A| < \epsilon$, $\lim\limits_{x=+\infty} y = A$. And so in the remaining cases. It is not implied that y exists when $x = a$.

Each of the following five theorems refers to a selected limiting point and a selected side of that point.

(1) A function cannot have two distinct limits.

(2) $\text{Lim}(y_1 + y_2) = \lim y_1 + \lim y_2$, provided that, if one of these two limits be $+\infty$, the other is not $-\infty$.

(3) $\text{Lim}(y_1 - y_2) = \lim y_1 - \lim y_2$, provided that these two limits, if both infinite, have not the same sign.

(4) $\text{Lim}(y_1 y_2) = \lim y_1 \cdot \lim y_2$, provided that the last expression is not $0 \cdot \infty$ or $\infty \cdot 0$.

(5) $\text{Lim}(y_1/y_2) = \lim y_1/\lim y_2$, provided that the last expression is not $0/0$ or ∞/∞.

These are re-statements of familiar theorems of the Calculus. To deal with the excluded cases, restrictions must be imposed on the functions. The reader is referred to Stolz, Allgemeine Arithmetik, t. i., section ix.

§ 61. *Continuous functions.* The definition of the continuity of a function at a point was given in § 57. A function is said to be continuous over an interval (x_0 to x_1), when it is continuous at every point of the interval, and when $f(x_0 + h)$, $f(x_1 - h)$ have the limits $f(x_0)$, $f(x_1)$, as h tends to zero. A function may be continuous on one side of $x = a$ without being continuous on the other.

It is to be remarked that when a function is continuous at a, the limit when x tends to a either from the right or left is the value of the function at a. This may be conveniently expressed by writing $f(a+0) = f(a-0) = f(a)$.

If $f(x)$ be continuous, and $= b$, at $x = a$, and if $\phi(y)$ be continuous at $y = b$, then $\phi\{f(x)\}$ is continuous at $x = a$. This theorem follows immediately from the definition.

§ 62. *Upper and lower limits of a function.*

Theorem. Let $f(x)$ be a function which remains finite for *all* values of x between $x = x_0$ and $x = x_1$. The values taken by the function will have an upper limit G and a lower limit K. For no value of x in the interval can the function be $> G$ or $< K$; but further, however small ϵ may be, there must be at least one value of x for which $f(x)$ is greater than $G - \epsilon$, and at least one value of x for which $f(x)$ is less than $K + \epsilon$.

The finiteness of the values of the function shows that there must be two finite integers A, B, between which all these values must lie. Divide the interval (A, B) into the parts

$$(A, A+1),\ (A+1, A+2),\ \cdots,\ (B-1, B).$$

Let A' be the last of these numbers which is not greater than all the values of the function, and let $A'+1=B'$. Divide the interval (A', B') into 10 parts

$$(A', A'+1/10), (A'+1/10, A'+2/10), \cdots, (A'+9/10, B').$$

Let A'' be the last value which is not greater than all the functional values, and let $B''=A''+1/10$. A continuation of this process shows that there can be found a value $A^{(n)}$, such that

(i.) $B^{(n)}-A^{(n)}<\epsilon$;

(ii.) There exists a value of $f(x)$ which is greater than $A^{(n)}$, and no value so great as $B^{(n)}$.

The ascending sequence $A, A', A'', \cdots$, and the descending sequence $B, B', B'', \cdots$, define a number G. This proves that a value of x can be found for which the value of the function differs infinitely little from G, but does not prove that there exists a value which is exactly G. Precisely similar results hold with respect to the lower limit K.

An example will clear up the preceding reasoning: let $y=x$ for all values of x in the interval 0 to 1, 1 exclusive, and let $y=0$ when $x=1$. The upper limit of the discontinuous function y, namely 1, is never attained, but a value of y exists which is $>1-\epsilon$.

When a function is said to be infinite for $x=a$, it is to be understood that this is only a short mode of expressing the fact that no finite upper limit exists. The value $x=a$ must then be excluded from the list of possible values of x. When a function is continuous throughout an interval, it is implied that there is no place of this kind in the interval.

Example 1. $y=1/x$ is a law which may define values of y for all values of x in the interval $(-a$ to $a)$, $x=0$ exclusive. As x approaches 0 from either side, y tends to infinity; but if $x=0$ be included in the range of values, y is not defined by this law, but must be otherwise assigned.

Example 2. The function $y=(x^2-a^2)/(x-a)$ has a determinate value, provided $x\neq a$. It is natural to assign to y the value $2a$ when $x=a$; but that this is an arbitrary assumption has been pointed out by Darboux. Let $\phi(x)$ be a function, for example a definite integral, which $=1$ when $x\neq a$, and $=A$ when $x=a$. The product $\phi(x)\cdot(x+a)=(x^2-a^2)/(x-a)$ when $x\neq a$; but when $x=a$, the left-hand side $=2Aa$.

§ 63. *Two theorems which relate to continuous functions.*

Theorem I. If the function $f(x)$ be continuous in the interval (x_0 to x_1) of the real variable, and if it take opposite signs when $x=x_0$ and $x=x_1$, $f(x)$ must equal zero for one or more values contained between x_0 and x_1.

Divide the interval (x_0, x_1) into 10 equal parts. Suppose $f(x_0)$ negative, $f(x_1)$ positive. The sequence

$$f(x_0),\ f(x_0+(x_1-x_0)/10),\ f(x_0+2(x_1-x_0)/10), \cdots,$$
$$f(x_0+9(x_1-x_0)/10),\ f(x_1),$$

will start with a negative term. If any one of the interpolated values equal zero, the theorem is established. But if the terms be all different from zero, let $f(x_0')$ be the last which is negative, $f(x_1')$ the first which is positive. Divide the interval (x_0', x_1') into 10 new intervals, and pick out of the sequence of values

$$f(x_0'),\ f(x_0'+(x_1'-x_0')/10),\ \cdots,\ f(x_0'+9(x_1'-x_0')/10),\ f(x_1'),$$

the last one which is negative, namely $f(x_0'')$, and the first which is positive, namely $f(x_1'')$. The process may stop owing to the occurrence of a zero-value for $f(x)$, a result which would at once establish the theorem. If it do not stop, the two infinite sequences

$$f(x_0),\ f(x_0'),\ f(x_0''),\ \cdots,$$
$$f(x_1),\ f(x_1'),\ f(x_1''),\ \cdots,$$

have a common limit $f(X)$, where X is the number defined by the two sequences

$$(x_0, x_0', x_0'', \cdots) \text{ and } (x_1, x_1', x_1'', \cdots),$$

where

$$x_1-x_0=10(x_1'-x_0')=10^2(x_1''-x_0'')=10^3(x_1'''-x_0''')=\cdots;$$

and $f(X)$ must $=0$, for every member of the sequence

$$f(x_0),\ f(x_0'),\ f(x_0''),\ \cdots$$

is negative, and every member of the sequence

$$f(x_1),\ f(x_1'),\ f(x_1''),\ \cdots$$

positive. The first rigorous proof of this theorem was given by Cauchy in his Cours d'Analyse.

Corollary. A function $\phi(x)$ which is continuous between two values x_0, x_1, must pass through every value A intermediate to $\phi(x_0)$, $\phi(x_1)$. This can be proved by putting $\phi(x)-A=f(x)$, and applying the above theorem. Darboux has established the existence of discontinuous functions which possess the same property. [Annales de l'Éc. Norm. Sup., Sér. 2, t. iv. pp. 51–112; t. viii., pp. 195–202.]

Theorem II. If $f(x)$ be a function which is finite and continuous in the interval (x_0 to x_1), the function will possess maximum and minimum values which are equal to the upper and lower limits G, K, and consequently it will pass through all values between G and K.

The function takes values $> G - \epsilon$, and values $< K + \epsilon$. Does it actually attain to G and K? Divide (x_0, x_1) into two equal intervals. In one at least of these, the upper limit is G. In fact, if the two upper limits were G', G'', quantities $< G$, the function would not take, in the interval (x_0, x_1), the values comprised between G and the greater of the two numbers G', G''. Hence one, at least, of the limits G', $G'' = G$. Divide that half interval, in which the upper limit is G, into two equal parts, and continue the process.

The result is a sequence of sub-intervals, one comprised within the other, which tend toward zero as limit. For each of these intervals the upper limit is G. The further extremities of the sub-intervals are constant in position or else approach x_0; the nearer extremities of the sub-intervals are constant in position or else move away from x_0. Let a be the point defined by these sub-intervals. Within the neighbourhood ($a-h$ to $a+h$) the upper limit of the function is G, *however small h may be*. That is, there is a value of $x (= a \mp \theta h)$ comprised within this interval for which $G - f(a \mp \theta h) < \epsilon$. On the other hand we can take h small enough to make $|f(a \mp \theta h) - f(a)| < \epsilon$, because the function is continuous at a. Hence the absolute value of $G - f(a) < 2\epsilon$; therefore $G = f(a)$. Similarly for K. If a be one of the points x_0, x_1 a slight modification of the proof is required.

§ 64. If $f(x)$ be continuous at every point of the interval (x_0 to x_1), we know by the definition that, at a point x_κ, it is possible to find a value Δx_κ such that $|f(x_\kappa \pm \theta \Delta x_\kappa) - f(x_\kappa)|$ shall be less than ϵ. If ϵ be fixed and x_κ allowed to vary, two cases may conceivably arise: (i.) it is possible to choose for Δx_κ a value h which is independent of x_κ throughout the interval (x_0 to x_1); (ii.) it is impossible to find such a value h. In Case (i.) the continuity is termed uniform, while in Case (ii.) the continuity is non-uniform. The following theorem shows that Case (ii.) does not arise:—

Divide the interval into 2 equal parts, each of these into 2 equal parts, and so on.

(i.) If, after the operation has been performed a finite number of times, the oscillation of the function within each of the divisions of length δ be $< \epsilon/2$, δ is the required value of h. To confirm this statement, let ξ be any one of the points of division, $\xi - \lambda$, $\xi + \mu$ two points within an interval of length δ; then

$$|f(\xi) - f(\xi - \lambda)| < \epsilon/2, \quad |f(\xi) - f(\xi + \mu)| < \epsilon/2,$$

and therefore
$$|f(\xi + \mu) - f(\xi - \lambda)| < \epsilon.$$

The difference $(\xi+\mu)-(\xi-\lambda)$ is any real number whose absolute value $<\delta$. Thus δ is the required value of h.

(ii.) If possible, let the oscillation within one of the divisions be $>\epsilon$, however often the operation of division may have been performed. The oscillations within *some* interval $(x_0^{(m)}, x_1^{(m)})$, which decreases indefinitely as m increases, are always $\geqq \epsilon/2$. Now the sequences

$$x_1, x_1^{(1)}, x_1^{(2)}, x_1^{(3)}, \dots x_1^{(m)}, \dots$$
$$x_0, x_0^{(1)}, x_0^{(2)}, x_0^{(3)}, \dots x_0^{(m)}, \dots$$

define a number X. Hence, since the function $f(x)$ is continuous at the point X, it must be possible to find a number h, such that

$$|f(X \mp \theta h)-f(X)|<\epsilon/4,$$

for all values of θ from 0 to 1, inclusive. Let X_1, X_2 be any two of the values of $\pm\theta h$. We can deduce, at once, from the two inequalities

$$|f(X-X_1)-f(X)|<\epsilon/4,$$
$$|f(X+X_2)-f(X)|<\epsilon/4,$$

the further inequality

$$|f(X+X_2)-f(X-X_1)|<\epsilon/2,$$

which implies that the oscillation within the interval ($X-X_1$ to $X+X_2$) $<\epsilon/2$, contrary to supposition. Hence a function of one variable, which is continuous at every point of an interval, is uniformly continuous. Cantor and Heine were the first to enunciate the above theorem.*

§ 65. The continuous function plays a much more important part in analysis than the discontinuous, but the study of the latter throws light on the problems suggested by Fourier's series and similar questions. Enough has been said to show the necessity for precise definitions and to emphasize our present ignorance of the discontinuities presented by functions of a highly transcendental character.

The properties which are most commonly associated with a continuous function are the possession of a differential quotient and an integral, and expansibility by Taylor's theorem. It is not our intention to examine these cases for the real variable, but we shall give an example of Weierstrass's which relates to the existence of a differential quotient.

Weierstrass's example of a continuous function which has nowhere a differential quotient.

The function $f(x)=\sum_0^\infty b^n \cos(a^n x\pi)$, in which x is real, a an odd

* Heine, Crelle., t. lxxiv., p. 188: Thomæ, Theorie der analytischen Functionen.

positive integer, b a positive constant < 1, is a continuous function, which has nowhere a determinate differential quotient if $ab > 1 + 3\pi/2$. We reproduce Weierstrass's proof:—

Let x_0 be a definite value of x, and m an arbitrarily chosen positive integer. There is a determinate integer α_m, for which

$$-1/2 < a^m x_0 - \alpha_m \leqq 1/2.$$

Write $\qquad x_{m+1}$ for $a^m x_0 - \alpha_m$,

and put $\qquad x' = (\alpha_m - 1)/a^m, \quad x'' = (\alpha_m + 1)/a^m.$

Thus $\quad x' - x_0 = -(1 + x_{m+1})/a^m, \quad x'' - x_0 = (1 - x_{m+1})/a^m,$

and $x' < x_0 < x''$. The integer m can be chosen large enough to insure that x', x'' shall differ from x_0 by as small a quantity as we please. We have

$$\{f(x') - f(x_0)\}/(x' - x_0) = \sum_{n=0}^{\infty} \left\{ b^n \frac{\cos(a^n x' \pi) - \cos(a^n x_0 \pi)}{x' - x_0} \right\}$$

$$= \sum_{n=0}^{m-1} \left\{ (ab)^n \frac{\cos(a^n x' \pi) - \cos(a^n x_0 \pi)}{a^n (x' - x_0)} \right\}$$

$$+ \sum_{n=0}^{\infty} \left\{ b^{m+n} \frac{\cos(a^{m+n} x' \pi) - \cos(a^{m+n} x_0 \pi)}{x' - x_0} \right\}.$$

Since

$$\frac{\cos(a^n x' \pi) - \cos(a^n x_0 \pi)}{a^n (x' - x_0)} = -\pi \sin\left(a^n \frac{x' + x_0}{2} \pi\right) \cdot \frac{\sin\left(a^n \frac{x' - x_0}{2} \pi\right)}{a^n \frac{x' - x_0}{2} \pi},$$

and since the value of $\dfrac{\sin\left(a^n \frac{x' - x_0}{2} \pi\right)}{a^n \frac{x' - x_0}{2} \pi}$ always lies between -1 and $+1$, the absolute value of the first part of the expression is less than

$$\pi \sum_0^{m-1} (ab)^n = \pi \frac{(ab)^m - 1}{ab - 1},$$

and therefore $< \pi (ab)^m/(ab - 1)$, if $ab > 1$.

Further, because a is an odd number,

$$\cos(a^{m+n} x' \pi) = \cos(a^n (\alpha_m - 1)\pi) = -(-1)^{\alpha_m},$$

$$\cos(a^{m+n} x_0 \pi) = \cos(a^n \alpha_m \pi + a^n x_{m+1} \pi) = (-1)^{\alpha_m} \cos(a^n x_{m+1} \pi),$$

therefore

$$\sum_{n=0}^{\infty} b^{m+n} \frac{\cos(a^{m+n}x'\pi) - \cos(a^{m+n}x_0\pi)}{x' - x_0}$$
$$= (-1)^{\alpha_m}(ab)^m \sum_{n=0}^{\infty} \frac{1 + \cos(a^n x_{m+1}\pi)}{1 + x_{m+1}} b^n.$$

All terms of the sum

$$\sum_{n=0}^{\infty} b^n \{1 + \cos(a^n x_{m+1}\pi)\}/(1 + x_{m+1})$$

are positive, and the first term $\not< 2/3$, since $\cos(x_{m+1}\pi)$ is not negative and $1 + x_{m+1}$ lies between 1/2 and 3/2. Accordingly

$$\{f(x') - f(x_0)\}/(x' - x_0) = (-1)^{\alpha_m}(ab)^m \eta(2/3 + \pi\eta_1/(ab - 1)),$$

where η denotes a positive quantity >1, while η_1 lies between -1 and $+1$.

Similarly

$$\{f(x'') - f(x_0)\}/(x'' - x_0) = -(-1)^{\alpha_m}(ab)^m \eta'(2/3 + \pi\eta_1'/(ab - 1)),$$

where η' is positive and >1, while η_1' lies between -1 and $+1$.

If a, b be so chosen as to make

$$ab > 1 + 3\pi/2,$$

that is,

$$2/3 > \pi/(ab - 1),$$

the two expressions

$$\{f(x') - f(x_0)\}/(x' - x_0), \quad \{f(x'') - f(x_0)\}/(x'' - x_0),$$

have always opposite signs, and are both infinitely great when m increases without limit. Hence $f(x)$ possesses neither a determinate finite nor a determinate infinite differential quotient. [Weierstrass, Abh. a. d. Functionenlehre, p. 97.]

A real function is represented graphically by drawing ordinates equal to the values of the function and marking the terminal points. The greater the number of these points, the more closely will the polygon through them resemble a continuous curve which admits tangents. This polygon will, in the limit, appear to the eye indistinguishable from the curve, but the polygon may have, in its smallest parts, infinitely many re-entrant angles. This is what actually happens in the case of functions which are continuous, without admitting differential quotients.*

---

* For further information on the representability of functions by curves we refer the reader to a paper by Klein, Ueber den allgemeinen Functionsbegriff, Math. Ann., t. xxii., p. 249, to two memoirs by Köpcke in the same journal, tt. xxix., xxxiv., and to Pasch's Einleitung in die Differential und Integralrechnung.

§ 66. *Functions of two real variables.*

Let x_1, y_1 be the centre of a rectangle, whose sides are $2h$, $2k$. Any point within this rectangle has for its co-ordinates $x_1 \mp \theta_1 h$, $y_1 \mp \theta_2 k$, where θ_1, θ_2 are proper fractions. Allow θ_1, θ_2 to take, independently of each other, all values from 0 to 1. The one-valued function $f(x, y)$ is said to be continuous at (x_1, y_1), when finite values h, k exist, for which $|f(x_1 \mp \theta_1 h, y_1 \mp \theta_2 k) - f(x_1, y_1)| < \epsilon$, for every combination (θ_1, θ_2).

Erroneous statements are sometimes made with regard to the continuity of a function of two variables.* Let $f(x, y)$ be a continuous function of y when x is put $= x_1$, and a continuous function of x when y is put $= y_1$. It is an illegitimate inference that $f(x, y)$ is continuous at (x_1, y_1). For example, let $f(x, y) = xy/(x^2+y^2)$. When $x = 0$, this is a continuous function of y, and, when $y = 0$, a continuous function of x; but the function is not a continuous function of x, y, conjointly, at (0, 0), as is easily seen by writing $y = mx$. Cauchy fell into this error in his Cours d'Analyse. As an example of the discontinuities which occur in the case of functions of two variables we may instance $\sin(\tan^{-1} y/x)$, which is continuous at every point off the axis of x and discontinuous on that axis. That the discussion of such discontinuities is of more than purely theoretic importance will be evident from these examples: —

Example 1. Let $f(x, y)$ be a function of x, y, which equals $x \sin(4 \tan^{-1} y/x)$ when $x \neq 0$, and equals 0 when $x = 0$, whatever be the value of y.

When $x \neq 0$, $\delta^2 f/\delta x \delta y = \delta^2 f/\delta y \delta x$;

but when $x = 0$, $y = 0$,

$$\delta^2 f/\delta x \delta y = \pm \infty, \text{ according to the sign of } \Delta x,$$

and $$\delta^2 f/\delta y \delta x = 0,$$

so that $$\delta^2 f/\delta x \delta y \neq \delta^2 f/\delta y \delta x \text{ at } (0, 0).$$

Example 2. Let $f(x, y) = y^2 \sin x/y$ when $y \neq 0$, and $= 0$ when $x = 0$, whatever be the value of y. Then, at (0, 0),

$$\delta^2 f/\delta x \delta y = 1, \quad \delta^2 f/\delta y \delta x = 0.$$

[See F. D'Arcais, Corso di Calcolo Infinitesimale, t. i. Padua, 1891.]

Theorems strictly analogous to those already proved for functions of one variable exist with regard to upper and lower limits, limiting values, etc.

---

* On this subject see a memoir by Schwarz, Abhandlungen, t. ii., p. 275.

Dirichlet's definition can be at once extended to functions of two real variables x and y. The restriction that the function takes only real values may be removed. Such a definition therefore includes both monogenic and non-monogenic functions of $x+iy$. The Theory of Functions, in the modern sense, discards Dirichlet's definition as too general and treats only of the monogenic function.

References. Cantor's researches are to be found in memoirs published in the Mathematische Annalen and the Acta Mathematica. For those of Weierstrass, see Pincherle, Saggio di una introduzione alla Teorica delle funzione analitiche secondo i principii del Prof. C. Weierstrass, Giornale di Matematiche, t. xviii., 1880; and Biermann's Analytische Functionen. An important memoir on Cantor's theory of irrational numbers was published by Heine in Crelle, t. lxxiv., Die Elemente der Functionenlehre. The student who wishes to find a thorough discussion of many of the more difficult questions connected with the theory of the real variable should consult Dini's standard work, Fondamenti per la Teorica delle Funzioni di Variabili Reali. Pisa, 1878. Also Stolz, Allgemeine Arithmetik; Tannery, Introduction à la Théorie des Fonctions d'une Variable; Fine, The Number-System of Algebra; Thomae, Elementare Theorie der analytischen Functionen einer complexen Veränderlichen. Many references to original memoirs are given at the end of the German translation of Dini's work, Leipzig, 1892. The student is also referred to the appendix in t. iii., of Jordan's Cours d'Analyse; to Dedekind's pamphlets (i.) Stetigkeit und irrationale Zahlen. Brunswick, 1872 (ii.), Was sind und was sollen die Zahlen. Brunswick, 1888; to Cathcart's translation of Harnack's Introduction to the Calculus; and to Du Bois-Reymond's Die Allgemeine Functionen-theorie. Erster Theil. Tübingen, 1882.

CHAPTER III.

The Theory of Infinite Series.

§ 67. We shall begin this chapter with a sketch, in outline, of the chief properties of infinite series of real terms, referring the student for a fuller discussion to treatises on Algebra.

Series with Real Terms.

An infinite series $\sum_1^\infty a_n$ contains infinitely many terms, defined by a law which permits the calculation of the general term. Let

$$s_n = a_1 + a_2 + \cdots + a_n;$$

if s_n tend to a finite limit s, the series is said to converge, and s is called its sum. In all other cases s_n either tends to ∞, as in $1+1+1+\cdots$, or oscillates, as in $1-1+1-1+\cdots$. Strictly speaking, an oscillating series is distinct from a divergent series, but it is usual to speak of non-convergent series as divergent.

The necessary and sufficient condition for the convergence of the series $\sum_1^\infty a_n$ is that a number μ* can be found such that

$$|a_{n+1} + a_{n+2} + \cdots + a_{n+p}| < \epsilon \quad (n > \mu,\ p = 1, 2, 3, \cdots).$$

The condition is necessary, for $\underset{n=\infty}{\mathrm{L}}\, s_{n+p} = Ls_n = s$, and therefore

$$L(s_{n+p} - s_n) = L(a_{n+1} + a_{n+2} + \cdots + a_{n+p}) = 0.$$

To see that it is sufficient it will be enough to observe that when the condition holds good, a number μ can be found such that

$$|s_{n+p} - s_n| \text{ and } |s_{n+q} - s_n| < \epsilon,$$

when $n > \mu$, whatever positive integral values are assigned to p, q, and that these relations give $\underset{n=\infty}{\mathrm{L}}\, s_{n+p} = \underset{n=\infty}{\mathrm{L}}\, s_{n+q}$.

Corollary. The condition $|a_{n+1}| < \epsilon\ (n > \mu)$, is necessary, but not sufficient.

---

* Here, as elsewhere, ϵ is given in advance, and μ depends upon ϵ.

§ 68. Many of the tests of convergence are derived from a comparison of the series in question with a few standard series. Such a series is

$$1+x+x^2+x^3+\cdots\begin{pmatrix}x>-1\\x<1\end{pmatrix}.$$

By comparing with it a series of positive terms $a_1+a_2+a_3+\cdots$, we can prove that the latter series converges when $a_{n+1}/a_n<\kappa$ for all values of n greater than some finite positive integer μ, provided κ be a proper fraction. The series also converges if, from some term a_μ onwards, $a_n<\kappa^n\ (n>\mu)$; that is, if $\sqrt[n]{a_n}<1$. The cases

$$La_{n+1}/a_n=1,\quad L\sqrt[n]{a_n}=1,$$

require special treatment. A third test is to compare a series $\sum\limits_1^\infty a_n$ with another series $\sum\limits_1^\infty b_n$, where $Lb_n/a_n=1$. When the latter series is convergent, the former is also convergent.

Convergent series of positive terms have the properties of finite series, for they are subject to the associative, commutative, and distributive laws. The same is always the case with series in which the negative terms have finite suffixes; but series whose terms are ultimately both positive and negative present some new features.

Theorem. If the series formed from a given series by the change of all the negative signs into positive be convergent, the original series must be convergent.

Let s_n be the sum to n terms of the original series and let $s_n', -t_n'$ be the sums of the positive and negative terms of s_n. If $L(s_n'+t_n')$ be finite, Ls_n' and Lt_n' must be finite, and therefore also $L(s_n'-t_n')$.

A series is said to converge *absolutely* when it still converges after all negative signs have been changed into positive. A series which converges, but does not converge absolutely, is called *semi-convergent.*

Theorem. If $a_1, a_2, a_3, \cdots$ be positive numbers arranged in descending order of magnitude and if $La_n=0$, the series

$$a_1-a_2+a_3-a_4+\cdots \text{ converges.}$$

For $\quad a_{n+1}-a_{n+2}+a_{n+3}-\cdots+(-1)^{p-1}a_{n+p}$

$$=a_{n+1}-(a_{n+2}-a_{n+3})-(a_{n+4}-a_{n+5})-\cdots<a_{n+1};$$

hence $\quad |(-1)^n a_{n+1}|>|a_{n+1}-a_{n+2}+\cdots+(-1)^{p-1}a_{n+p}|.$

Now a_{n+1} can be made less than any assigned positive number ϵ, when n is taken sufficiently large; hence the series

$$a_1 - a_2 + a_3 - a_4 + \cdots \text{ is convergent.}$$

Example. $1 - 1/2 + 1/3 - 1/4 + \cdots$ is semi-convergent. The series

$$a_1 - a_2 + a_3 - \cdots$$

will not converge, if $La_n \neq 0$. It may diverge if $a_1, a_2, a_3, \cdots$ be not in descending order, even when $La_n = 0$.

§ 69. It is natural to ask under what circumstances the component terms of a convergent series are subject to the associative and commutative laws.

(i.) The terms of a convergent series can be united into groups, when this does not affect the arrangement of the terms of the original series.

Let s_n be the sum of n terms of the original series, s_m' the sum of m terms of the series of groups.

Whatever n may be, m can always be taken large enough to insure that s_m' shall contain all the terms of s_n. When this has been done,

$$s_m' - s_n = a_{n+1} + a_{n+2} + \cdots + a_{n+p},$$

supposing that s_m' contains $n+p$ terms. But, by the condition of convergence,

$$|a_{n+1} + a_{n+2} + \cdots + a_{n+p}| < \epsilon \quad (n > \mu);$$

hence $$L(s_m' - s_n) = 0, \text{ or } Ls_m' = Ls_n.$$

(ii.) In a convergent series such as

$$(1/2 - 2/3) + (2/3 - 3/4) + (3/4 - 4/5) + \cdots,$$

it is not permissible to remove the brackets, without investigation. In the particular example selected, the series oscillates after the removal of the brackets.

In connexion with the commutative law it is convenient to define unconditional and conditional convergence.

A convergent series which is subject to the commutative law is said to be *unconditionally* convergent; otherwise it is said to be *conditionally* convergent.

A semi-convergent series is conditionally convergent. To Riemann is due the theorem that it is possible, by suitable derangements, to make a semi-convergent series have for its sum any assigned real number A. To fix ideas suppose A positive.

Let the positive terms, unchanged in order, be $a_1+a_2+a_3+\cdots$, and let the negative terms, unchanged in order, be $-(b_1+b_2+b_3+\cdots)$. If the first n terms of the original series consist of q positive and r negative terms, $s_n=s_q'-s_r''$, where

$$s_q'=a_1+a_2+\cdots+a_q, \quad s_r''=b_1+b_2+\cdots+b_r.$$

When n becomes infinite, q and r also become infinite, and s_q', s_r'', tend to the values s', s'', where

$$s'=\sum_1^\infty a_\kappa, \quad s''=\sum_1^\infty b_\kappa.$$

If the original series be semi-convergent, both the series $\sum_1^\infty a_\kappa$, $\sum_1^\infty b_\kappa$ must be divergent; for, if $\sum_1^\infty a_\kappa$, $\sum_1^\infty b_\kappa$ have finite sums, the series $\sum_1^\infty a_\kappa+\sum_1^\infty b_\kappa$ has a finite sum, and the original series is absolutely convergent; and if only one be divergent, the original series $=s'-s''=\infty$, and is divergent. The terms of the semi-convergent series can be rearranged in the following way: we write down the positive terms in order, and stop at the first term which makes the sum $>A$; next we write down the negative terms, and stop at the first term which makes the sum $<A$; we continue the process by adding on further positive terms, stopping at the first term which makes the sum $>A$, and so on. The values of the sums thus obtained oscillate about A, and the range through which the oscillation takes place diminishes continually, tending to a limit 0. Hence Riemann's theorem is proved. Since the value of A is arbitrary, a semi-convergent series can be made divergent by suitable derangements of the terms. [See Dirichlet, Abh. d. Berl. Akad., 1837; Riemann, Werke. p. 221.]

Absolute convergence implies unconditional convergence. Let $a_1+a_2+a_3+\cdots$ be an absolutely convergent series, and let $a_1'+a_2'+a_3'+\cdots$ be a series derived from it by derangements such that no term a_n with a finite suffix is displaced to infinity. If s_n, s_m' be the sums to n and m terms of these two series, it is always possible to choose such a value of m as will make s_m' contain all the terms $a_1, a_2, \cdots, a_n$, and others besides. If s_μ' be the first partial sum of this nature, the sole limitation on m is that it must be greater than μ. Suppose that $s_m'-s_n=$ the sum of a certain number of terms which lie between a_n and $a_{n+\lambda+1}$. This sum is certainly less than $|a_{n+1}|+|a_{n+2}|+\cdots+|a_{n+\lambda}|$, a quantity which can be made $<\epsilon$ by choosing n sufficiently great. Hence

$$Ls_m'=Ls_n.$$

The preceding reasoning breaks down for a semi-convergent series $\sum_1^\infty a_\kappa$, since $|a_{n+1}|+|a_{n+2}|+\cdots+|a_{n+\lambda}|$ is not necessarily less than ϵ.

Series with Complex Terms.

§ 70. Let $u_m=a_m+i\beta_m$; then $s_n=\sum_1^n a_m+i\sum_1^n\beta_m$. If s_n tend to a limit s when n is arbitrarily great, the series is said to be convergent and its sum is s. Clearly, the necessary and sufficient conditions are that $\sum_1^\infty a_m$ and $\sum_1^\infty\beta_m$ both converge. In the limit both $|a_m|$ and $|\beta_m|=0$; but $|a_m|+|\beta_m|>|a_m+i\beta_m|$, hence $L|u_m|=0$. A series $\sum_1^\infty(a_m+i\beta_m)$ is said to converge absolutely, when $\sum_1^\infty|a_m+i\beta_m|$ converges; this definition presupposes that the former series converges simultaneously with the latter, but this is evidently a legitimate assumption, for $|a_m|$ and $|\beta_m|$ are separately less than $|a_m+i\beta_m|$. Unconditional convergence has the same meaning as before; it is evident that it can exist only when $\sum_1^\infty a_m$ and $\sum_1^\infty\beta_m$ are unconditionally convergent.

Absolute convergence implies unconditional convergence. Let $a_m+i\beta_m=\rho_m(\cos\theta_m+i\sin\theta_m)$. The terms $\rho_m\cos\theta_m$, $\rho_m\sin\theta_m$ are less than ρ_m in absolute value, and therefore, if $\sum_1^\infty\rho_m$ be finite, the two real series $\sum_1^\infty\rho_m\cos\theta_m$, $\sum_1^\infty\rho_m\sin\theta_m$ are absolutely convergent. But absolute convergence, in the case of real series, has been shown to imply unconditional convergence. Therefore

$$\sum_1^\infty\rho_m(\cos\theta_m+i\sin\theta_m)$$

is unconditionally convergent.

Semi-convergence implies conditional convergence. Because $\sum_1^\infty|u_m|$ diverges, $\sum_1^\infty\{|a_m|+|\beta_m|\}$ must diverge. This shows that one, at least, of the series $\sum_1^\infty|a_m|$, $\sum_1^\infty|\beta_m|$, diverges. Let the former diverge; then $\sum_1^\infty a_m$ is semi-convergent, and therefore conditionally convergent. But, if the series $\sum_1^\infty u_m$ be unconditionally convergent, $\sum a_m$ must be unconditionally convergent. Hence the series $\sum u_m$ is conditionally convergent.

Unconditional convergence implies absolute convergence. This is proved indirectly. We have seen that semi-convergence implies conditional convergence; hence a semi-convergent series can never be unconditionally convergent. Therefore an unconditionally convergent series can never be semi-convergent; that is, it must be absolutely convergent. The terms *unconditional* and *absolute* are, as applied to convergence, co-extensive.

§ 71. The *sum*, or *difference*, of two series $\sum_1^\infty u_n$, $\sum_1^\infty v_n$ is the series $\sum_1^\infty (u_n+v_n)$, or $\sum_1^\infty (u_n - v_n)$.

The product of two absolutely convergent series. If the absolutely convergent series $\sum_1^\infty u_n$, $\sum_1^\infty v_n$ converge to sums s, s', the series

$$u_1v_1 + (u_1v_2 + u_2v_1) + (u_1v_3 + u_2v_2 + u_3v_1) + \cdots$$
$$+ (u_1v_n + u_2v_{n-1} + \cdots + u_nv_1) + \cdots$$

will converge to ss'.

Let $\quad A_n = |u_1| + |u_2| + \cdots + |u_n|,$

$$B_n = |v_1| + |v_2| + \cdots + |v_n|,$$

then

$$\begin{aligned} A_nB_n = &|u_1||v_1| + |u_2||v_1| + \cdots + |u_n||v_1| \\ &+ |u_1||v_2| + |u_2||v_2| + \cdots + |u_n||v_2| \\ &\cdots\cdots\cdots\cdots\cdots\cdots \\ &\cdots\cdots\cdots\cdots\cdots\cdots \\ &+ |u_1||v_n| + |u_2||v_n| + \cdots + |u_n||v_n|. \end{aligned}$$

If

$$\begin{aligned} C_n = &|u_1||v_1| + \{|u_1||v_2| + |u_2||v_1|\} + \cdots \\ &+ \{|u_1||v_n| + |u_2||v_{n-1}| + \cdots + |u_n||v_1|\}, \end{aligned}$$

it is evident that $A_nB_n > C_n$, and that

$$A_nB_n < C_{2n} < C_{2n+1}.$$

Therefore, whether n be of the form $2\kappa + 1$, or of the form 2κ, the inequalities

$$A_\kappa B_\kappa < C_n < A_nB_n$$

must hold good, and $\quad LC_n = LA_n \cdot LB_n.$

Now consider the series

$$\sum_1^\infty u_m,\ \sum_1^\infty v_m,\ \sum_1^\infty (u_1v_m + u_2v_{m-1} + \cdots + u_mv_1),$$

and let the sums to n terms be s_n, s_n', t_n. The difference

$$s_ns_n' - t_n = \{u_2v_n + u_3v_{n-1} + \cdots + u_nv_2\} + \cdots + u_nv_n;$$

and the absolute value of this difference is less than

$$|u_2||v_n|+|u_3||v_{n-1}|+\cdots+|u_n||v_n|,$$

an expression which is itself $=A_nB_n-C_n$. When n increases indefinitely, $A_nB_n-C_n$ tends to the limit 0; hence

$$L|s_ns_n'-t_n|=0,$$

or

$$Lt_n=ss'.$$

Uniform Convergence.

§ 72. Hitherto the term u_n of the infinite series $\sum\limits_1^\infty u_n$ has been regarded as dependent merely on n. We now allow u_n to be a function of z, as well as of n, and suppose that there is a region of the z-plane at all points of which u_1, u_2, $\cdots$ are one-valued and continuous.

The sum of a finite number of rational algebraic functions is itself a rational algebraic function; it is only when the number of these functions is infinite that new properties present themselves. In fact the passage from the algebraic to the transcendental function is effected by infinite series, whose terms are algebraic functions of z.

Let $f_1(z), f_2(z), f_3(z), \cdots$ be algebraic or transcendental functions of z, which are one-valued and continuous within a region A of the z-plane, and let the series $\sum\limits_1^\infty f_n(z)$ be convergent at every point of A. For a given value of z, the necessary and sufficient condition for convergence is that a number μ can be found such that, when $n>\mu$,

$$|f_{n+1}(z)+f_{n+2}(z)+\cdots+f_{n+p}(z)|<\epsilon \quad (p=1, 2, \cdots).$$

When this condition is satisfied, we call the sum of the series $F(z)$. The complete system of points, for which the series converges, covers a region of the z-plane which is known as the region of convergence. This region may consist of one or more separate pieces, and the boundary of a piece may consist of discrete masses of points, or of linear masses of points. For such functions as we shall consider, each piece will consist of a continuum of points covering a doubly extended region.

The series $\sum\limits_1^\infty f_n(z)$ is said to converge *uniformly* in a part of the region of convergence, when a positive integer μ can be found such that when $n>\mu$,

$$|f_{n+1}(z)+f_{n+2}(z)+\cdots+f_{n+p}(z)|<\epsilon \ (p=1, 2, \cdots),$$

whatever be the position of z in that part. Attention is specially called to the fact that μ depends on the arbitrarily selected ϵ, but can be made independent of z. If, when z approaches a point z_0 of the region of convergence, the value of μ tend to ∞, the series is said to be *non-uniformly* convergent in any region which contains z_0, and the convergence is said to be *infinitely slow* in the neighbourhood of z_0.

Example 1 (*Du Bois-Reymond*). Let

$$f_n(x)=\frac{x}{(nx+1)(nx-x+1)}=\frac{nx}{nx+1}-\frac{(n-1)x}{(n-1)x+1}.$$

The sum of n terms of the series, when x is real, is $\frac{nx}{nx+1}$, and the sum to infinity, when $x \neq 0$, is 1. The remainder after n terms is $\frac{1}{nx+1}$. Let x move in an interval such that $x >$ a positive number a; the remainder $< \epsilon$, if $nx+1 > 1/\epsilon$, and therefore, *a fortiori*, if $na+1>1/\epsilon$. Let μ be a number which satisfies the inequality $\mu a+1>1/\epsilon$; then, for all values of $n>\mu$,

$$|f_{n+1}(x)+f_{n+2}(x)+\cdots \text{ to } \infty|<\epsilon;$$

but μ is independent of x, hence the convergence is uniform. If x be allowed to approach the value 0, a becomes smaller and smaller and μ tends to ∞. The convergence, when x is infinitely small, is infinitely slow. When $x \neq 0$,

$$\underset{n=\infty}{\mathrm{L}}\ nx/(nx+1)=1,$$

and when $x=0$,

$$\underset{n=\infty}{\mathrm{L}}\ nx/(nx+1)=0;$$

hence the non-uniformly convergent series is discontinuous at the point $x=0$ of the interval of convergence.

Example 2. If x be a real positive number <1, the series

$$\sum_1^\infty (1-x)x^{n-1}$$

has for its sum $(1-x)/(1-x)=1$. This is true no matter how nearly x may approach the value 1; but, when $x=1$, the sum is 0. In this example the remainder after the nth term is x^n, and therefore μ must be so chosen that $x^\mu<\epsilon$; that is, $\mu>\log 1/\epsilon \div \log 1/x$. The more closely x approximates to the value 1, the more nearly does the denominator approach 0 and the greater becomes the value of μ. The series is non-uniformly convergent in the interval ($x=0$ to $x=1$), and the convergence is infinitely slow when $1-x$ is infinitely small.

Example 3 (*Peano*). Let

$$f_n(x)=\frac{(nx)^2}{1+(nx)^2}\,\frac{e^{nx}-e^{-nx}}{e^{nx}+e^{-nx}} \quad (x \text{ real}).$$

When $x=0$,

$$\underset{n=\infty}{\mathrm{L}}\ f_n(x)=0,$$

and when x is greater or less than 0,

$$\underset{n=\infty}{\mathrm{L}}\ f_n(x)=1, \text{ or } -1.$$

The series $f_1(x)+\{f_2(x)-f_1(x)\}+\{f_3(x)-f_2(x)\}+\cdots$

is convergent for all values of x, but the convergence in an interval which includes $x=0$ is non-uniform.

§ 73. *Theorem.* If a series formed of one-valued and continuous functions $f_1(z), f_2(z), \cdots$ converge uniformly in a region which contains z_0, then

$$\lim_{z=z_0}\{f_1(z)+f_2(z)+\cdots\}=\lim_{z=z_0}f_1(z)+\lim_{z=z_0}f_2(z)+\cdots$$

The right-hand side may be written $\sum_1^\infty f_m(z_0)$, for a continuous function attains its limit. Let

$$\sum_1^\infty f_m(z)=\sum_1^n f_m(z)+\sum_{n+1}^\infty f_m(z)$$
$$=\sum_1^n f_m(z)+\rho_n(z).$$

Since the series $\sum_1^\infty f_m(z)$ is uniformly convergent, there exists a number μ such that, when $n>\mu$,

$$|\rho_n(z)|<\epsilon,$$

whatever be the position of z in the region under consideration. Let $z=z_0+h$, and let the absolute value of h be made sufficiently small to insure that

$$|f_\kappa(z)-f_\kappa(z_0)|<\epsilon/n\,(\kappa=1,2,\cdots n),$$

where n is finite. Hence, if $F(z)=\sum_1^\infty f_m(z)$, we have when $n>\mu$,

$$|F(z)-F(z_0)|=|\sum_1^n\{f_\kappa(z)-f_\kappa(z_0)\}+\rho_n(z)-\rho_n(z_0)|$$
$$<\sum_1^n|f_\kappa(z)-f_\kappa(z_0)|+|\rho_n(z)|+|\rho_n(z_0)|$$
$$<n\cdot\epsilon/n+\epsilon+\epsilon,<3\epsilon.$$

This proves that the limit of the sum of $f_1(z), f_2(z), f_3(z), \cdots$ is equal to the sum of the limits of $f_1(z)$, $f_2(z)$, $f_3(z), \cdots$, when z approaches z_0. This important theorem can be stated as follows: —

If a series of one-valued and continuous functions converge uniformly throughout a region Γ of the z-plane, the sum of the series is a continuous function of z throughout the region.

It is evident from this theorem that a point of the region of convergence at which the sum of the series is discontinuous is a point in the neighbourhood of which the series converges non-uniformly. Du Bois-Reymond has proved that a series may converge infinitely slowly in the neighbourhood of a point at which the sum of the series is continuous. Hence a non-uniformly convergent series may be continuous; but no uniformly convergent series can be discontinuous.

Dini (Fondamenti per la teorica delle funzioni di variabili reali, p. 103), and Darboux (Mémoire sur les fonctions discontinues, Ann. de l'Éc. Norm. Sup. Ser. 2, t. iv., p. 77) adopt a different definition. A series is said by them to be uniformly convergent when a number μ can be found such that

$$|f_{\mu+1}(z)+f_{\mu+2}(z)+\cdots|<\epsilon.$$

We refer the reader for further information on this matter to Tannery's Théorie des fonctions d'une variable, pp. 133–4.

Cauchy, in his Cours d'Analyse Algébrique, considered the sum of a series $\Sigma f_n(z)$, in which all the terms were continuous one-valued functions of the argument, but did not explicitly recognize the possibility that this sum might be discontinuous. In 1853 he published a note in the Comptes Rendus, t. xxxvi., pp. 456–8, in which he virtually enunciated the theory of uniform convergence. In connexion with Cauchy's earlier work, Abel gave an example of a non-uniformly convergent series, namely $\sin x - \sin 2x/2 + \sin 3x/3 \cdots$, whose sum is discontinuous when $x=(2n+1)\pi$ (Collected Works, t. i., p. 246). The writers who first cleared up the difficulties associated with this theory were Sir G. Stokes (On the Critical Values of the Sums of Periodic Series. Math. Papers, t. i., p. 280), and L. Seidel (Note ueber eine Eigenschaft von Reihen, welche discontinuirliche Functionen darstellen. Abh. d. Bayerischen Akademie, 1848). Thomé made use of the doctrine in a memoir published in Crelle's Journal, t. lxvi., p. 334, but a considerable amount of time elapsed before the importance of the distinction between uniform and non-uniform convergence was fully grasped by writers on the Theory of Functions.

To Heine is due the proof that a discontinuous function can never be represented as the sum of a uniformly convergent series (Crelle, t. lxxi.).

Co-existence of uniform and absolute convergence. There is nothing in the definition of uniform convergence which implies that the series must be absolutely convergent. The following is a simple test, by the aid of which it is often possible to determine when the convergence of a series is simultaneously uniform and absolute (Weierstrass, Abhandlungen aus der Functionenlehre, p. 70). Using the same notation, if, throughout a region of the z-plane, all the terms after a certain term f_κ be less, in absolute value, than the terms $a_{\kappa+1}$, $a_{\kappa+2}$, $a_{\kappa+3}$, $\cdots$ of an absolutely convergent series $\sum_1^\infty a_n$, the original series $\sum_1^\infty f_n$ is uniformly and absolutely convergent throughout the region.

Multiple Series.

§ 74. In the simple series $\sum_1^\infty u_n$, the general term depends on a single integer; in the multiple series $\Sigma u_{r,s,t,\ldots}$, the general term depends on several integers $r, s, t, \cdots$; these may take, independently of one another, all values from $-\infty$ to ∞, or they may be subjected to restrictive conditions. We shall, for the sake of simplicity, confine our attention at first to the double series

$$\Sigma u_{r,s} = u_{a,b} + u_{c,d} + u_{e,f} + \cdots,$$

where r, s take, independently of each other, all values from 1 to ∞. The extension of theorems on double series to multiple series of higher orders is, in most cases, immediate. The terms of the double series form the array

$$\begin{matrix} u_{1,1} & u_{1,2} & u_{1,3}\cdots \\ u_{2,1} & u_{2,2} & u_{2,3}\cdots \\ u_{3,1} & u_{3,2} & u_{3,3}\cdots \\ \cdot & \cdot & \cdot \quad \cdot \quad \cdot \end{matrix}$$

To give a meaning to the sum of a series formed from such an array, we must assume that the order of selection from the array proceeds according to some definite plan, so that to each term $u_{r,s}$ corresponds a single term $v_n = u_{r,s}$, and to each term v_n a single

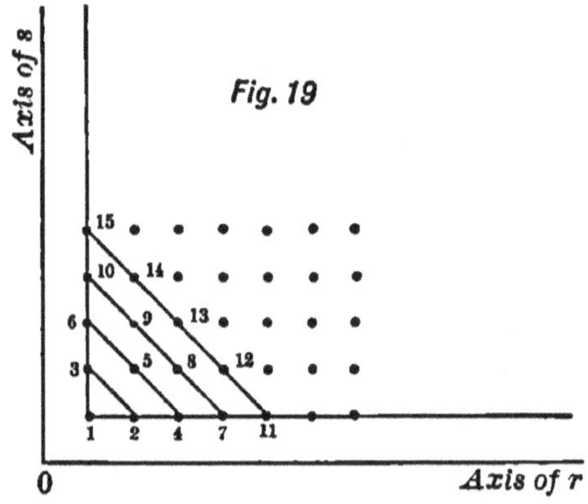

term $u_{r,s}$. Further, we shall suppose that r, s are finite when n is finite. The diagram shows one of the infinitely many ways of establishing a (1, 1) correspondence between the members of the system (r, s) and the system of positive integers.

For instance $\left.\begin{matrix} v_{14} = u_{2,4}, \\ v_{12} = u_{4,2}. \end{matrix}\right\}$

The double series $\Sigma u_{r,s}$, when converted into a simple series Σv_n by the method illustrated in the diagram,

$$= u_{1,1} + (u_{2,1} + u_{1,2}) + (u_{3,1} + u_{2,2} + u_{1,3}) + (u_{4,1} + u_{3,2} + u_{2,3} + u_{1,4}) + \cdots.$$

This mode of summation, which we shall call summation by diagonals, is only one of infinitely many; for with each unicursal correspondence is associated a mode of summation of the series $\Sigma u_{r,s}$.

Let the numbers of the system (1, 2, 3, ⋯) correspond unicursally to the members of the systems (r, s) and (r', s'). Then $\Sigma u_{r',s'}$ is formed from the series $\Sigma u_{r,s}$ by a commutation of the terms. Three cases arise: —

I. If $\Sigma |u_{r,s}|$ converge, $\Sigma u_{r,s} = \Sigma u_{r',s'}$; the series $\Sigma u_{r,s}$ is, in this case, absolutely and unconditionally convergent.

II. If $\Sigma |u_{r,s}|$ diverge, while $\Sigma u_{r',s'}$ converges, the latter series is conditionally convergent.

III. If the series $\Sigma u_{r,s}$, $\Sigma u_{r',s'}$, etc., be all divergent, no convergent series can be found from the array.

As in the case of simple series, unconditional convergence plays a more important part than conditional convergence.

From the terms of the array we construct the expression

$$\begin{array}{l} u_{1,1} + u_{1,2} + u_{1,3} + u_{1,4} + \cdots \\ + u_{2,1} + u_{2,2} + u_{2,3} + u_{2,4} + \cdots \\ + u_{3,1} + u_{3,2} + u_{3,3} + u_{3,4} + \cdots \\ + \cdot\ \cdot\ \cdot\ \cdot\ \cdot\ \cdot\ \cdot\ \cdot\ \cdot \\ + \cdot\ \cdot\ \cdot\ \cdot\ \cdot\ \cdot\ \cdot\ \cdot\ \cdot \end{array}$$

We have already explained what is meant by summation by diagonals; we shall however adopt a different definition for the sum of the expression just written down.

Let
$$\begin{aligned} S_{m,n} = {} & u_{1,1} + u_{1,2} + \cdots + u_{1,n} \\ & + u_{2,1} + u_{2,2} + \cdots + u_{2,n} \\ & + \cdot\ \cdot\ \cdot\ \cdot\ \cdot\ \cdot\ \cdot\ \cdot \\ & + \cdot\ \cdot\ \cdot\ \cdot\ \cdot\ \cdot\ \cdot\ \cdot \\ & + u_{m,1} + u_{m,2} + \cdots + u_{m,n}; \end{aligned}$$

also let S be a finite and determinate number. If, when m, n are made infinitely great, $LS_{m,n} = S$, independently of the order in which the two numbers tend to ∞, we shall speak of S as the sum of the double series. In this case the necessary and sufficient condition for convergence is that positive numbers μ, ν can be found such that, when $m > \mu$ and $n > \nu$,

$$|S_{m+p,n+q} - S_{m,p}| < \epsilon \quad (p, q = 0, 1, 2 \cdots).$$

The convergence of a double series is not inconsistent with the divergence of individual rows or columns. For instance, the first row of a semi-convergent series may be divergent, and the second row the same divergent series with signs changed. This case does not arise in absolutely convergent double series.

Double Series with Real and Positive Terms.

§ 75. In the convergent series

$$\begin{aligned} & a_{1,1} + a_{1,2} + a_{1,3} + \cdots \\ + & a_{2,1} + a_{2,2} + a_{2,3} + \cdots \\ + & a_{3,1} + a_{3,2} + a_{3,3} + \cdots \\ + & \cdot \quad \cdot \quad \cdot \quad \cdot \quad \cdot \quad \cdot \quad \cdot \end{aligned}$$

the horizontal rows evidently converge. Let their sums be $s_1, s_2, s_3, \cdots$; we shall prove that $\sum_1^\infty s_\kappa$ converges to S, where S is the sum of the double series. For if $\rho_1, \rho_2, \cdots, \rho_m$ be the remainders, after n terms, of the first m rows we have

$$s_1 + s_2 + \cdots + s_m = S_{m,n} + \rho_1 + \rho_2 + \cdots + \rho_m;$$

but $$\rho_1 + \rho_2 + \cdots + \rho_m < \epsilon \ (m > \mu,\ n > \nu),$$

by the condition of convergence; hence

$$s_1 + s_2 + \cdots + s_m = \underset{n=\infty}{\mathrm{L}}\, S_{m,n},$$

and $$\underset{m=\infty}{\mathrm{L}}\, \{s_1 + s_2 + \cdots + s_m\} = \underset{m=\infty}{\mathrm{L}}\ \underset{n=\infty}{\mathrm{L}}\, S_{m,n} = S.$$

Similarly, if $t_1, t_2, t_3, \cdots$ be the sums of the first, second, third, $\cdots$ columns,

$$\underset{n=\infty}{\mathrm{L}}\, (t_1 + t_2 + \cdots + t_n) = \underset{n=\infty}{\mathrm{L}}\ \underset{m=\infty}{\mathrm{L}}\, S_{m,n} = S.$$

Also the series summed diagonally, namely

$$a_{1,1} + (a_{1,2} + a_{2,1}) + (a_{1,3} + a_{2,2} + a_{3,1}) + \cdots,$$

has S for its sum. This is evident from the consideration that the sum of the latter series to κ terms is intermediate in value between $S_{h,h}$ and $S_{\kappa,\kappa}$, where h is the greatest integer in $\kappa/2$. Finally, since the series, summed diagonally, can be written as a single series $\Sigma a_{r,s} = \Sigma a_n$, where (r, s) take independently all values from 1 to ∞, and since $\Sigma a_{r,s} = \Sigma a_{r',s'}$, the double series is unconditionally convergent for derangements of the kind mentioned above.

Conversely, simple series can be converted into double series. If the simple series $\sum_1^\infty a_n$ be converted into the double series $\Sigma a_{r,s}$, and thence into

$$\begin{aligned} & a_{1,1} + a_{1,2} + a_{1,3} + \cdots \\ + & a_{2,1} + a_{2,2} + a_{2,3} + \cdots \\ + & a_{3,1} + a_{3,2} + a_{3,3} + \cdots \\ + & \cdot \quad \cdot \quad \cdot \quad \cdot \quad \cdot \quad \cdot \quad \cdot, \end{aligned}$$

the sums of the simple and double series are equal.

Absolutely Convergent Double Series with Positive and Negative Terms.

§ 76. Let $s_{m,n}$, $-t_{m,n}$ be the sums of the positive and negative terms of $S_{m,n}$. Let $S'_{m,n}$ be the sum when all the negative signs in $S_{m,n}$ are replaced by positive; then

$$\left.\begin{aligned} S_{m,n} &= s_{m,n} - t_{m,n} \\ S'_{m,n} &= s_{m,n} + t_{m,n} \end{aligned}\right\}$$

and, since $LS'_{m,n}$ is finite, the two positive quantities $s_{m,n}$ and $t_{m,n}$ must tend to finite limits, which we may call s and t. Hence

$$LS_{m,n} = s - t.$$

Absolutely Convergent Double Series with Complex Terms.

Let $\quad u_{r,s} = a_{r,s} + i\beta_{r,s}$, and let $\Sigma a_{r,s} = \rho$, $\Sigma\beta_{r,s} = \sigma$.

If $\Sigma | a_{r,s} + i\beta_{r,s} |$ converge, $\Sigma a_{r,s}$ and $\Sigma\beta_{r,s}$ must both converge;

for $\quad | a_{r,s} | < | a_{r,s} + i\beta_{r,s} |$ and $| \beta_{r,s} | < | a_{r,s} + i\beta_{r,s} |$.

Hence the series $\Sigma(a_{r,s} + i\beta_{r,s})$ converges and has for its sum $\rho + i\sigma$. As before, if $\rho_\kappa + i\sigma_\kappa$ be the sum of the κth row,

$$\sum_1^\infty \rho_\kappa = \rho \text{ and } \sum_1^\infty \sigma_\kappa = \sigma.$$

Hence $\quad \Sigma u_{r,s} = \sum_1^\infty \rho_\kappa + i\sum_1^\infty \sigma_\kappa = \rho + i\sigma.$

Similarly the sums by columns and diagonals $= \rho + i\sigma$.

As in the case of simple series, if the series $\Sigma u'_{m,n}$ be absolutely convergent, and if $L\,| u'_{m,n} | / | u_{m,n} | = 1$, when m, n tend in *any* manner to ∞, the series $\Sigma u_{m,n}$ is itself absolutely convergent.

To come to the more general case; we suppose the suffixes r, s to pass, independently of each other, through all values from $-\infty$ to $+\infty$; the sum $S_{m,n}$ may now be taken to represent the sum $\sum_{-m}^{m} r \sum_{-n}^{n} s\, u_{r,s}$. As before, if $S_{m,n}$ tend to a definite limit, irrespective of the order in which m, n tend to ∞, the double series is convergent and has for its sum S. Let each pair (r, s) be represented by a point whose Cartesian co-ordinates are r and s. $S_{m,n}$ is the sum of terms included within a rectangle whose centre is at the origin, and, in the limit, the rectangular contour extends to ∞ in the direction of both axes. It is evident that if the system of rectangles be replaced by another system of contours which extend to ∞ in all directions, the sum of an absolutely convergent series will not be affected. For a contour of the latter system can always be made to lie entirely within the region bounded by two rectangles of the first system which extend to ∞ in all directions simultaneously with the contour.

In absolutely convergent double series the form of the contour which extends to ∞ is immaterial, but this is not the case with conditionally convergent series.

To illustrate some of the points touched upon in the preceding paragraphs we shall select as an example the series

$$\Sigma\Sigma' 1/(m_1\omega_1 + m_2\omega_2)^\lambda,$$

where the summation extends over all integral values of m_1, m_2, from $-\infty$ to $+\infty$, exclusive of the combination $m_1 = m_2 = 0$. We assume that the ratio ω_2/ω_1 is complex.*

On the rim of the parallelogram whose vertices are $n(\pm\omega_1 \pm \omega_2)$, there are $8n$ points congruent to 0, with respect to ω_1 and ω_2; that is, there are $8n$ points of the set $m_1\omega_1 + m_2\omega_2$. Let ρ be the absolute value of that point of the rim which is nearest to 0, and let $\rho = n\rho'$. For all values of n, ρ' will be finite and different from 0; therefore the quantities ρ' must have a lower limit $\sigma > 0$. The series formed by the absolute values of the $8n$ terms on the rim

$$< \frac{1}{\sigma^\lambda}\left(\frac{8n}{n^\lambda}\right).$$

Hence the double series

$$\Sigma\Sigma' \left|\frac{1}{m_1\omega_1 + m_2\omega_2}\right|^\lambda < \frac{8}{\sigma^\lambda} \sum_1^\infty \frac{1}{n^{\lambda-1}}.$$

This series is convergent when $\lambda > 2$. Then, when $\lambda > 2$, the original doubly infinite series is absolutely convergent.

§ 77. *A test of convergence.* Cauchy's integral test for the convergence of simple series can be extended to double series. If the terms $a_{m,n}$ of $\Sigma a_{m,n}$ be all positive and capable of being represented by a function $\phi(x, y)$, which diminishes continuously as x, y increase numerically and tends ultimately to the limit 0, then the double series converges or diverges according as $\iint \phi(x, y)\,dx\,dy$, taken over the part of the plane exterior to a bounding curve C, has or has not a meaning. We shall reproduce Picard's elegant proof that the

* It will be shown in the chapter on elliptic functions that this assumption is necessary in order to avoid the occurrence of infinitely small values in the system $|m_1\omega_1 + m_2\omega_2|$.

necessary and sufficient condition for the absolute convergence of the double series $\Sigma\Sigma' \frac{1}{(m_1\omega_1 + m_2\omega_2)^\lambda}$ is $\lambda > 2$.

Let $\omega_1 = a + ib$, $\omega_2 = c + id$; the series of absolute values is

$$\Sigma\Sigma' \frac{1}{\{(m_1 a + m_2 c)^2 + (m_1 b + m_2 d)^2\}^{\lambda/2}}.$$

No denominator can vanish, for the combination $m_1 = m_2 = 0$ is excluded; moreover the ratio

$$\frac{(m_1^2 + m_2^2)^{\lambda/2}}{\{(m_1 a + m_2 c)^2 + (m_1 b + m_2 d)^2\}^{\lambda/2}}$$

is always finite. Hence the series in question will converge simultaneously with $\Sigma\Sigma' \frac{1}{(m_1^2 + m_2^2)^{\lambda/2}}$, but the latter series converges when $\int\int \frac{dx\,dy}{(x^2 + y^2)^{\lambda/2}}$ has a meaning; *i.e.* when $\int\int \frac{d\rho\,d\theta}{\rho^{\lambda-1}}$ is finite for infinite values of ρ, or, in other words, when $\lambda > 2$.*

Infinite Products.

§ 78. Let $u_1, u_2, u_3, \cdots$ be a series of quantities, real or complex, which are determined uniquely by a known law of construction,

and let $$P_n = (1 + u_1)(1 + u_2)\cdots(1 + u_n) = \prod_1^n (1 + u_\kappa).$$

We shall suppose that all the factors, with finite suffixes, are different from zero. If, when n increases indefinitely, P_n tend to a finite limit P which is different from zero, the value of the coresponding infinite product is said to be P. When the value towards which the infinite product tends is either 0 or ∞, the product is called divergent, and, when there is no definite limit, oscillating. It is often convenient to regard the term *divergent* as covering all cases of non-convergence.†

---

* Picard's Traité d'Analyse, t. i., p. 272. Simple proofs of Eisenstein's theorem that

$$\sum_{-\infty}^{+\infty}\sum \cdots' \frac{1}{(m_1^2 + \cdots + m_n^2)^\lambda}$$

converges when $2\lambda > n$, and of allied theorems, will be found in Jordan's Cours d'Analyse, t. i., p. 162.

† Some writers regard the product as convergent when $LP_n = 0$. The view adopted in the text is that of Pringsheim.

The necessary and sufficient condition for the convergence of the infinite product is that a number μ can be found such that when $n \geqq \mu$,

$$|(1+u_{n+1})(1+u_{n+2})(1+u_{n+3})\cdots(1+u_{n+p})-1|<\epsilon,$$

where ϵ is an arbitrarily small positive proper fraction, and p takes all integral values from 1 to ∞.

We shall first show that this condition is equivalent to the two conditions:—

(i.) However great n may be, P_n is a finite quantity which is different from 0,

(ii.) $\underset{n=\infty}{\mathrm{L}}\, P_{n+p} = \underset{n=\infty}{\mathrm{L}}\, P_n$, where p is any positive integer.

Let accents denote absolute values; we have

$$|P'_{\mu+p}/P'_{\mu}-1| \leqq |P_{\mu+p}/P_{\mu}-1| \ (p=1, 2, \cdots)$$

and $|P_{\mu+p}/P_\mu - 1|<\epsilon$; hence, if $\mu+p=n$, $P'_n/P'_\mu - 1$ lies between $-\epsilon$ and $+\epsilon$. As p increases, n increases; therefore LP'_n lies between $(1-\epsilon)P'_\mu$ and $(1+\epsilon)P'_\mu$. Thus LP_n is finite and different from 0.

We are therefore at liberty to multiply both sides of the inequality

$$|P_{n+p}/P_n-1|<\epsilon \quad (n \geqq \mu;\ p=1, 2, \cdots)$$

by P_n. The resulting inequality shows that

$$\underset{n=\infty}{\mathrm{L}}\, P_{n+p} = \underset{n=\infty}{\mathrm{L}}\, P_n.$$

The condition (ii.) shows that the product does not oscillate, and the condition (i.) shows that the value of the infinite product is neither 0 nor ∞.

Corollary. $|P_{n+1}/P_n - 1|<\epsilon$, therefore $L|u_n|=0$.

Two inequalities.

If $a_1, a_2, a_3, \cdots$ be all positive and less than 1, then

I. $(1+a_1)(1+a_2)\cdots(1+a_n)>1+(a_1+a_2+\cdots+a_n)$,

II. $(1-a_1)(1-a_2)\cdots(1-a_n)>1-(a_1+a_2+\cdots+a_n)$.

The proof of I. is immediate; for

$$(1+a_1)(1+a_2)=1+a_1+a_2+a_1a_2>1+a_1+a_2,$$

therefore

$$(1+a_1)(1+a_2)(1+a_3)>(1+a_1+a_2)(1+a_3)>1+a_1+a_2+a_3,$$

and so on. A similar proof can be given for II.; for

$$(1-a_1)(1-a_2)=1-(a_1+a_2)+a_1a_2>1-(a_1+a_2);$$

therefore

$$(1-a_1)(1-a_2)(1-a_3)>\{1-(a_1+a_2)\}(1-a_3)>1-(a_1+a_2+a_3),$$

and so on. These inequalities can be used to prove the following important theorem: —

The necessary and sufficient condition for the convergence of the products $\prod_1^\infty(1+a_n)$, $\prod_1^\infty(1-a_n)$, where the quantities a are real and positive, is the convergence of $\sum_1^\infty a_n$.

Let us suppose, in the first place, that the a's are all less than 1. When $\sum_1^\infty a_n$ is convergent, the second product is finite by the second inequality; and, when $\sum_1^\infty a_n$ is divergent, the first product is infinite by the first inequality. Now if

$$P_n=(1+a_1)(1+a_2)\cdots(1+a_n),$$

$$Q_n=(1-a_1)(1-a_2)\cdots(1-a_n),$$

the expression $Q_n<1/P_n$. For

$$1/(1+a_\kappa)>1-a_\kappa \quad (\kappa=1, 2, \cdots, n).$$

Hence when $\sum_1^\infty a_n$ is convergent, P_∞ is less than a finite quantity $1/Q_\infty$; and when $\sum_1^\infty a_n$ is divergent,

$$1/Q_\infty>P_\infty, \text{ or } Q_\infty=0.$$

It is evident from their composition that

$$P_1<P_2<P_3<P_4\cdots,$$

and

$$Q_1>Q_2>Q_3>Q_4\cdots,$$

and, therefore, that the sequences $(P_1, P_2, P_3, \cdots)$, $(Q_1, Q_2, Q_3\cdots)$ must define two numbers which are greater than P_1 and less than Q_1; but we have just proved that these numbers are distinct from 0 and ∞ when, and only when, $\sum_1^\infty a_n$ is convergent. Hence the theorem has been established for the case in which the a's are all less than 1. This restriction can be removed; for the presence, in the finite part of the infinite product, of quantities a which are greater than 1 cannot affect the state of convergence or divergence of the infinite products.

Theorem. The value of the convergent product $\prod_1^\infty(1+a_n)$, where the a's are all real and positive, is the same as the value of the sum

$$1+\sum_1^\infty a_n P_{n-1},$$

where $P_0=1,\ P_n=\prod_1^n(1+a_\kappa)$.

Because $P_\kappa=(1+a_\kappa)P_{\kappa-1}=P_{\kappa-1}+a_\kappa P_{\kappa-1}$,

we have
$$\begin{aligned}P_n&=P_{n-1}+a_nP_{n-1}=P_{n-2}+a_{n-1}P_{n-2}+a_nP_{n-1}\\&=P_{n-3}+a_{n-2}P_{n-3}+a_{n-1}P_{n-2}+a_nP_{n-1},\text{ and so on;}\end{aligned}$$

therefore $P_n=1+\sum_1^n a_\kappa P_{\kappa-1}$.

Therefore $$\prod_1^\infty(1+a_n)=1+\sum_1^\infty a_nP_{n-1}.$$

If the product converge, the series must converge; for the terms $P_1, P_2, \cdots$ are all finite and $\sum_1^\infty a_n$ is convergent. But further the series converges when the general term a_nP_{n-1} is separated into its component terms, by multiplying together the factors $a_n(1+a_1)\cdots(1+a_{n-1})$, inasmuch as each component term is positive. By re-arranging the terms of the resulting series, we can write it in the form $1+\sum_1^\infty a_{n_1}+\sum_1^\infty a_{n_1}a_{n_2}+\cdots$, where the signs of summation indicate that $n_1, n_2, \cdots$ are to take, independently of one another, all integral values from 1 to ∞. This proves that the value of the product is independent of the order of the factors; for a product which is formed from $\prod_1^\infty(1+a_n)$ by a re-arrangement of the factors is equal to a series whose value is precisely that of the double series

$$\begin{aligned}&1+a_1+a_2+\cdots\\&\quad+a_1a_2+a_1a_3+\cdots\\&\quad+a_1a_2a_3+\cdots\\&\quad+\quad.\quad.\quad.\quad.\\&=1+\sum a_{n_1}+\sum a_{n_1}a_{n_2}+\cdots.\end{aligned}$$

When a convergent product is subject to the commutative law, it is said to be unconditionally convergent.

§ 79. Let $u_1, u_2, u_3, \cdots$ be complex quantities; we shall prove that the infinite product $\prod_1^\infty(1+u_n)$ converges when $\sum_1^\infty u'_n$ converges; that is, when $\prod_1^\infty(1+u'_n)$ converges, where accents denote absolute values.

Let $\quad {}_pP_n = \prod_{n+1}^{n+p}(1+u_\kappa), \quad {}_pP'_n = \prod_{n+1}^{n+p}(1+u'_\kappa),$

and let $\prod_1^\infty(1+u'_n)$ be a convergent product. We have

$$\begin{aligned} |{}_pP_n - 1| &= |u_{n+1} + u_{n+2} + \cdots + u_{n+p} + u_{n+1}u_{n+2} + \cdots| \\ &\leqq u'_{n+1} + u'_{n+2} + \cdots + u'_{n+p} + u'_{n+1}u'_{n+2} + \cdots \\ &\leqq {}_pP'_n - 1, \end{aligned}$$

and therefore,

$$|{}_pP_n - 1| < \epsilon \ (n > \mu,\ p = 1, 2, 3, \cdots).$$

This inequality establishes the convergence of

$$\prod_1^\infty(1+u_n).$$

When $\prod_1^\infty(1+u'_n)$ converges, the infinite product $\prod_1^\infty(1+u_n)$ is said to be absolutely convergent. Thus (§ 78) the absolute convergence of $\sum_1^\infty u_n$ is a necessary and sufficient condition for the absolute convergence of $\prod_1^\infty(1+u_n)$.

Theorem. If $\sum_1^\infty u_n$ be absolutely convergent,

$$\prod_1^\infty(1+u_n) = 1 + \sum_1^\infty u_\kappa P_{\kappa-1}, \text{ where } P_0 = 1.$$

The proof is similar to that of § 78. We have

$$P_n = 1 + \sum_1^n u_\kappa P_{\kappa-1};$$

and the series on the right-hand side is absolutely convergent, when $n = \infty$. For $\prod_1^\infty(1+u'_n)$ is convergent, and therefore the series

$$1 + \sum_1^\infty u'_\kappa P'_{\kappa-1}$$

is also convergent. This means that the series

$$1 + \sum_1^\infty u_\kappa P_{\kappa-1}$$

converges absolutely. Hence n can be made infinite in the equation $\prod_1^n(1+u_\kappa) = 1 + \sum_1^n u_\kappa P_{\kappa-1}$. The series on the right-hand side remains absolutely convergent when the general term $u_\kappa P_{\kappa-1}$ is separated into its components

$$u_\kappa + u_1u_\kappa + \cdots + u_1u_2\cdots u_\kappa;$$

for the absolute values of the terms of the resulting series form an absolutely convergent double series. Hence the terms of the series may be re-arranged, and

$$\prod_1^\infty (1+u_n) = 1 + \sum_1^\infty u_{n_1} + \sum_1^\infty u_{n_1} u_{n_2} + \cdots.$$

This proves that $\prod_1^\infty (1+u_n)$ is unconditionally convergent, and therefore that the absolute convergence of an infinite product implies unconditional convergence.

To complete the proof that the terms *unconditional* and *absolute* are co-extensive as applied to infinite products, we must show that the unconditional convergence of an infinite product implies its absolute convergence. Pringsheim has established this fact by a purely algebraic method (Math. Ann., t. xxxiii., Ueber die Convergenz unendlicher Producte). When the theory of infinite products is developed from first principles, any proof which depends upon the use of transcendental functions must obviously be avoided. On the other hand, by the employment of logarithms, the discussion of infinite products can be reduced readily to that of infinite sums. We shall illustrate the method by using it to settle the outstanding question.

Let $1 + u_n, = \rho_n(\cos\theta_n + i \sin\theta_n)$, be the nth factor of an unconditionally convergent product. In order that $P_n = \prod_1^n (1+u_\kappa)$ may tend to a definite limit which is neither 0 nor ∞, it is necessary and sufficient that the two series

$$\sum_1^\infty \log \rho_n^2, \quad \sum_1^\infty \theta_n, \quad \text{where } L\rho_n = 1, \quad L\theta_n = 0,$$

be convergent; for

$$P_n = \rho_1\rho_2\cdots\rho_n\{\cos(\theta_1+\theta_2+\cdots+\theta_n) + i\sin(\theta_1+\theta_2+\cdots+\theta_n)\}.$$

For the unconditional convergence of the infinite product it is necessary that $\rho_1\rho_2\rho_3\cdots$ and $\sum_1^\infty \theta_n$ be unconditionally convergent; or, what amounts to the same thing, that $\sum_1^\infty \log\rho_n^2$, $\sum_1^\infty \theta_n$ be unconditionally convergent. The two series just written down cannot be unconditionally convergent without being also absolutely convergent. We have to show that the absolute convergence of the two series implies the absolute convergence of $\sum_1^\infty u_n$. Because $L(\rho_n^2 - 1)/\log\rho_n^2 = 1$, the absolute convergence of $\sum_1^\infty \log\rho_n^2$ is accompanied by the absolute convergence of $\sum_1^\infty (\rho_n^2 - 1)$. But, if $u_n = \alpha_n + i\beta_n$, we have

$$L\alpha_n = 0, \quad L\beta_n = 0; \quad \text{also } \rho_n^2 - 1 = 2\alpha_n + \alpha_n^2 + \beta_n^2.$$

Hence the series $\Sigma(2\alpha_n + \alpha_n^2 + \beta_n^2)$ is absolutely convergent. But since $|\sin\theta_n| < |\theta_n|$, and since $L\rho_n = 1$, the series $\Sigma\beta_n$, $= \Sigma\rho_n \sin\theta_n$, is absolutely convergent. Therefore $\Sigma\beta_n^2$ is absolutely convergent. It follows that $\Sigma(2+\alpha_n)\alpha_n$ must be absolutely convergent, or, since $L(2+\alpha_n) = 2$, $\Sigma\alpha_n$ must be absolutely convergent. Now, if both $\sum_1^\infty \alpha_n$ and $\sum_1^\infty \beta_n$ be absolutely convergent, we know that $\sum_1^\infty u_n$ is absolutely convergent; a result which implies that $\prod_1^\infty (1+u_n)$ is absolutely convergent. Thus the unconditional convergence of an infinite product implies its absolute convergence.

§ 80. *Uniform convergence of a product.*

If $f_1, f_2, f_3, \cdots$ be functions of z which are finite and one-valued within a region, throughout which the product $\prod_1^\infty (1+f_n)$ converges, the convergence is said to be uniform in that region when, to a positive number ϵ which may be arbitrarily small, there corresponds a finite number μ such that, when $n > \mu$,

$$|(1+f_{n+1})(1+f_{n+2})\cdots(1+f_{n+p}) - 1| < \epsilon \quad (p = 1, 2, 3, \cdots),$$

whatever be the position of z in the region.

In conclusion we state two theorems, which follow without difficulty from those already given.

Theorem I. If Σu_n be absolutely convergent,

$$\prod_1^\infty (1+u_n z) = 1 + (\sum_1^\infty u_{n_1})z + \{\sum_1^\infty u_{n_1}u_{n_2}\}z^2 + \cdots$$

Theorem II. If $\prod_1^\infty (1+u_n)$ be absolutely convergent, the only values of z for which $\prod_1^\infty (1+u_n z)$ can vanish are

$$z = -1/u_n \ (n = 1, 2, \cdots).$$

Coriolis seems to have been the first mathematician to state general rules for the convergence of infinite products. Cauchy gave the first detailed exposition of the chief properties of infinite products in his Cours d'Analyse Algébrique, Note 9; his proofs are simple and rigorous, but lie open to the objection that they depend upon logarithms. The proofs of § 78 are modifications of those used by Weierstrass in his memoir Ueber die Theorie der analytischen Facultäten, Abh. a. d. Functionenlehre. Dini gave the first proof of the theorem that unconditionally convergent products are, at the same time, absolutely convergent. We refer the student for further information to the valuable memoir by Pringsheim, referred to on the preceding page.

§ 81. *Doubly infinite products.* The discussion of double products is reducible to that of double series. The condition necessary and sufficient in order that $\Pi(1+u_{r,s})$ may be absolutely convergent is that $\Sigma u_{r,s}$ be absolutely convergent.

In conditionally convergent double products, the value depends upon the ultimate form of the contour which contains the terms, when r, s tend to ∞.

An example of this is the doubly infinite product

$$u\Pi\left[1+\frac{u}{a+m\omega+m'\omega'}\right],\quad m,\ m'=0,\ \pm 1,\ \pm 2,\cdots,$$

which depends upon the form of the bounding curve. See Cayley's Elliptic Functions, 1st ed., p. 303; also two papers by Cayley, Collected Works, t. i., 24, 25. The classical memoir on these products is that of Eisenstein, Crelle, t. xxxv., Genaue Untersuchung der unendlichen Doppelproducte, aus welchen die elliptischen Functionen als Quotienten zusammengesetzt sind. The student who wishes further information on double series is referred to a memoir by Stolz, Ueber unendliche Doppelreihen, Math. Ann., t. xxiv., and to Chrystal's Algebra, t. ii.

Integral Series.

§ 82. *The domain of a point.*

Let the series $\sum_1^\infty f_n(z)$, whose terms are one-valued and continuous functions of z throughout a region Γ, be uniformly convergent throughout Γ. If a positive number ρ can be found such that the series converges uniformly for all points within a circle whose centre is a and radius ρ, the series is said to converge uniformly in the *neighbourhood* of a, or *near* a. Let R be the greatest value of ρ for which this can be said; the region bounded by a circle whose centre is a and radius R is called the *domain* of a, and R is the radius of the domain. In the domain of a let a point b be selected. If the domain of b lie partially outside that of a, the outside portion belongs to the region of convergence. In this outside portion select a point c and form its domain. A repetition of this process will lead to a continuum within which the convergence is uniform. This may be only one of several separate continua.

Series in many variables. Suppose that $z_1, z_2, \cdots, z_n$ fill continuous regions $\Gamma_1, \Gamma_2, \cdots, \Gamma_n$ in their respective planes. These n regions may be called, for shortness, the region Γ, and each system of n values $z_1, z_2, \cdots, z_n$ is called a point z, or place z, of Γ. There is a theory of uniform convergence for series $\sum_{\kappa=1}^\infty f_\kappa(z_1, z_2, \cdots, z_n)$ as well as for series $\sum_1^\infty f_\kappa(z)$; the terms are supposed to be one-valued and continuous functions of $z_1, z_2, \cdots, z_n$. The series is said to be uniformly convergent within the region Γ when a number μ can be found such that for all points within Γ and for all values of $n > \mu$,

$$\left|\sum_{n+1}^{n+p} f_\kappa\right| < \epsilon, \ (p = 1, 2, 3, \cdots).$$

The definition of neighbourhood can be extended at once to functions of several variables. If $z_1, z_2, \cdots, z_n$ be n variables represented upon n planes, a neighbourhood of the system $(a_1, a_2, \cdots, a_n)$ is given by the n inequalities

$$|z_\kappa - a_\kappa| < \delta_\kappa \quad (\kappa = 1, 2, \cdots n).$$

That is, a neighbourhood is formed by the n circular regions $\Gamma_1, \Gamma_2, \cdots, \Gamma_n$ with centres a_κ and radii δ_κ $(\kappa = 1, 2, \cdots, n)$. These circular regions may be called, for shortness, the region Γ with centre a and radius δ. If a variable a_κ be situated at ∞, the corresponding quantity $z_\kappa - a_\kappa$ must be replaced by $1/z_\kappa$ (see § 27).

The theory of functions of many variables presents many difficulties, and has not yet been worked out in detail. The foundations have been laid by Weierstrass (Einige auf die Theorie der analytischen Functionen mehrerer Veränderlichen sich beziehende Sätze, Abh. a. d. Functionenlehre).

§ 83. *Integral series in one variable.* A special case of great importance is that in which the nth term is $u_n z^n$, where u_n is independent of z. The series

$$u_0 + u_1 z + u_2 z^2 + u_3 z^3 + \cdots + u_n z^n + \cdots$$

is known as an *integral series* or *power series.* The usual notation for such a series is $P(z)$. In the same notation the series

$$u_0 + u_1(z-c) + u_2(z-c)^2 + \cdots$$

is denoted by $P(z-c)$ or $P(z\,|\,c)$. When the series is known to begin with the term $u_\kappa(z-c)^\kappa$, we shall insert a suffix κ; thus,

$$P_\kappa(z-c) = u_\kappa(z-c)^\kappa + u_{\kappa+1}(z-c)^{\kappa+1} + \cdots,$$

where κ is any positive integer or zero. As $P(z)$ is merely a symbol for an integral series in general, it may denote several *different* series in the same investigation. When other symbols are needed for the purpose, we shall use $Q(z)$ or $P^{(\lambda)}(z)$ $(\lambda = 1, 2, 3, \cdots)$.

Integral series in n variables. Let $\lambda_1, \lambda_2, \cdots, \lambda_n$ take, independently of one another, all integral values from 0 to ∞; the series

$$\Sigma u_{\lambda_1, \lambda_2, \ldots, \lambda_n} z_1^{\lambda_1} z_2^{\lambda_2} \cdots z_n^{\lambda_n},$$

which may be denoted by $P(z_1, z_2 \cdots z_n)$, is an integral series in n variables. When it is absolutely convergent, the order of the terms is a matter of no consequence. To fix ideas we shall suppose that the terms of the κth order in $z_1, z_2, \cdots, z_n$ follow immediately after those of the $(\kappa - 1)$th order; that the terms of the κth order are arranged according to the descending powers of z_1; that the coefficients of $z_1^\kappa, z_1^{\kappa-1}, \cdots, z_1^1, z_1^0$ are arranged according to descending powers of z_2; and so on.

Theorem. If the absolute values at a of the terms of an integral series be less than a finite positive number μ, then the series converges for all points z, such that $|z_\kappa| < |a_\kappa|$, $\kappa = 1, 2 \cdots n$.

Let $\rho_1, \rho_2, \cdots, \rho_n$ be the absolute values of $z_1, z_2 \cdots z_n$; we have

$$|u_{\lambda_1, \lambda_2, \ldots, \lambda_n} a_1^{\lambda_1} a_2^{\lambda_2} \cdots a_n^{\lambda_n}| < \mu,$$

and therefore

$$|u_{\lambda_1, \lambda_2, \ldots, \lambda_n} z_1^{\lambda_1} z_2^{\lambda_2} \cdots z_n^{\lambda_n}| < \mu \left(\frac{\alpha_1}{\rho_1}\right)^{\lambda_1} \left(\frac{\alpha_2}{\rho_2}\right)^{\lambda_2} \cdots \left(\frac{\alpha_n}{\rho_n}\right)^{\lambda_n},$$

where $\quad |a_\kappa| = \alpha_\kappa (\kappa = 1, 2, \cdots n)$.

Hence $\quad |\Sigma u_{\lambda_1, \lambda_2, \ldots, \lambda_n} z_1^{\lambda_1} z_2^{\lambda_2} \cdots z_n^{\lambda_n}| < \mu \dfrac{1}{\left(1 - \dfrac{\alpha_1}{\rho_1}\right) \cdots \left(1 - \dfrac{\alpha_n}{\rho_n}\right)}.$

Thus the given series is convergent within a region bounded by circles. The theorem may be stated for one variable in the following way:—

Theorem. If the terms of the series $P(z)$, for the value $|z| = R$, be less in absolute value than a finite number μ, the series converges for every point z within the circle, centre 0, and radius R. This theorem was first stated by Abel (Remarques sur la série

$$1 + \frac{m}{1} x + \frac{m(m-1)}{2!} x^2 + \cdots,$$

Collected Works, ed. Sylow and Lie, t. i., p. 219; Crelle, t. i., p. 311).

Theorem. If $P(z)$ converge for a point z whose absolute value is R, it will converge for every point z whose absolute value is ρ, where $\rho < R$.

For if $U_\kappa = |u_\kappa|$ $(\kappa = 0, 1, 2, \cdots)$, the terms $U_\kappa R^\kappa$ have a finite upper limit M. The terms of the series $\sum_0^\infty U_\kappa \rho^\kappa$ are less than those of the convergent series $M\sum_0^\infty \rho^\kappa/R^\kappa$; therefore the integral series is absolutely convergent for all values of z which make $|z| < R$.

If $P(z_1, z_2, \cdots z_n)$ converge for a point a, it will converge for every point z such that $|z_\kappa| < |a_\kappa|$ $(\kappa = 1, 2, \cdots n)$.

Let $|z_\kappa/a_\kappa| = \alpha_\kappa$. The series

$$\left.\begin{aligned} 1 &+ \alpha_1 + \alpha_2 + \cdots + \alpha_n \\ &+ \alpha_1^2 + \alpha_1\alpha_2 + \cdots + \alpha_n^2 \\ &+ \alpha_1^3 + \alpha_1^2\alpha_2 + \alpha_1^2\alpha_3 + \cdots + \alpha_n^3 \\ &+ \cdots \end{aligned}\right\} \quad \cdots\cdots \quad (1)$$

converges if the sum of the first p lines, S_p suppose, be finite for all values of p.

Let $S_p^{(\kappa)}$ be the sum of the first p terms of

$$1 + \alpha_\kappa + \alpha_\kappa^2 + \alpha_\kappa^3 + \cdots,$$

then

$$S_p < S_p^{(1)} S_p^{(2)} \cdots S_p^{(n)}.$$

Since $\alpha_\kappa < 1$, $S_p^{(\kappa)}$ is finite for all values of p; therefore (1) is convergent. Multiplying the general term $\alpha_1^{\lambda_1}\alpha_2^{\lambda_2}\cdots\alpha_n^{\lambda_n}$ by the finite quantity $|u_{\lambda_1\lambda_2\ldots\lambda_n} a_1^{\lambda_1} a_2^{\lambda_2} \cdots a_n^{\lambda_n}|$, we see that the series whose general term is $|u_{\lambda_1\lambda_2\ldots\lambda_n} \cdot z_1^{\lambda_1} z_2^{\lambda_2} \cdots z_n^{\lambda_n}|$ is convergent; that is, $P(z_1, z_2, \cdots z_n)$ is absolutely convergent.

§ 84. *The circle of convergence.* The theorems just proved establish the fact that the region of convergence of $P(z)$ lies within a circle of radius R, where R is the upper limit of the values of $Z, = |z|$, which make $U_0, U_1Z, U_2Z^2, \cdots$ finite. For all points outside this circle some of the absolute values are greater than any finite number, however large; for such points the series is divergent. For points on the rim of the circle the question of convergence is left in doubt (see § 89).

We can determine R, which is called the *radius of convergence* of the given integral series, when we know (§ 58) the upper limit of the mass derived from the mass

$$|u_1|,\ |\sqrt{u_2}|, \cdots, |\sqrt[\kappa]{u_\kappa}| \cdots,$$

this mass being supposed to lie in a finite interval. For if G be the upper limit, there is an integer μ such that, when $n > \mu$,

$$|\sqrt[n]{u_n}| < G + \epsilon, \text{ or } |\sqrt[n]{u_n z^n}| < (G + \epsilon)Z.$$

Therefore (§ 68) the series is convergent when $(G + \epsilon)Z < 1 - \epsilon'$.

On the other hand, whatever term we take, there are terms beyond it such that $|\sqrt[n]{u_n}| > G-\epsilon$, or $|\sqrt[n]{u_n z^n}| > (G-\epsilon)Z$, and therefore the series is divergent when $(G-\epsilon)Z<1-\epsilon'$. Since ϵ and ϵ' are arbitrarily small, the series is convergent when $Z<1/G$, and divergent when $Z>1/G$; and therefore $R=1/G$ (see Hadamard, Liouville, Ser. 4, t. viii.; Picard, Traité d'Analyse).

A rule which suffices in many simple cases is the following: if $|u_\kappa/u_{\kappa+1}|$ have a limit when $\kappa=\infty$, this limit is R. For the series is convergent or divergent according as $L|u_{\kappa+1}z^{\kappa+1}/u_\kappa z^\kappa| \lesseqgtr 1$; that is, according as $Z \lesseqgtr L|u_\kappa/u_{\kappa+1}|$.

Examples. (i.) $1+z+z^2/2!+z^3/3!+\cdots$,

(ii.) $1+z+z^2+z^3+\cdots$,

(iii.) $z-z^2/2+z^3/3-\cdots$,

(iv.) $1+1!z+2!z^2+3!z^3+\cdots$,

are integral series with radii of convergence ∞, 1, 1, 0.

The series (i.) is not convergent at $z=\infty$. Such series as (iv.) do not define functions, and are therefore excluded from the subject.

In the case of n variables, let the absolute values of all terms of $P(z_1, z_2, \cdots z_n)$ be finite at a point a. The totality of circles $|z_\kappa|=|a_\kappa|$ is called a circle of convergence. It is clear that we cannot always assign an upper limit of the radius of convergence in any one plane, until we have assigned radii of convergence in the other planes.

§ 85. *Theorem.* The series $P(z)$ is not only absolutely but also uniformly convergent within its circle of convergence.

Let the radius of convergence be R, and let $|z|=Z<R_1<R$. The term $U_\kappa R_1^\kappa$, where $U_\kappa=|u_\kappa|$, must be less than some finite number C, for all values of κ. Hence

$$|u_{n+1}z^{n+1}+\cdots+u_{n+p}z^{n+p}|<U_{n+1}Z^{n+1}+\cdots+U_{n+p}Z^{n+p}$$

$$<C[Z^{n+1}/R_1^{n+1}+Z^{n+2}/R_1^{n+2}+\cdots \text{to } \infty]$$

$$<C\frac{Z^{n+1}}{R_1^{n+1}}\left\{\frac{1}{1-Z/R_1}\right\};$$

therefore $$|u_{n+1}z^{n+1}+\cdots+u_{n+p}z^{n+p}|<C\frac{\rho^{n+1}}{R_1^{n+1}}\frac{1}{1-\rho/R_1},$$

where $Z < \rho < R_1$. The expression on the right-hand side is now independent of Z; and, by the properties of the geometric series, a number μ can be found such that, when $n > \mu$, this expression can be made $< \epsilon$. As this number μ is finite and independent of z, the convergence is uniform within a circle of radius R_1, and the only limitation placed upon R_1 is $R_1 < R$.

Corollary. Since the integral series is uniformly convergent within the circle of convergence, it defines within that circle a continuous function of z.

In the series $P(z_1, z_2, \cdots, z_n)$ let z be within a circle of convergence A. Since the series is absolutely convergent, it may be re-arranged as an integral series in z_κ, which is convergent in A; the coefficients being integral series in the remaining z's, which are convergent in A. By the preceding theorem, the series in z_κ defines a continuous function. Therefore the series $P(z_1, z_2, \cdots, z_n)$ defines, within A, a function which is continuous with regard to each variable. We may therefore use the term 'neighbourhood of 0' as defined in § 82, for the region interior to a circle of convergence.

Theorem. The series

$$u_1 + 2\,u_2 z + 3\,u_3 z^2 + 4\,u_4 z^3 + \cdots$$

$$2\,u_2 + 3\cdot 2\,u_3 z + 4\cdot 3\,u_4 z^2 + 5\cdot 4\,u_5 z^3 + \cdots$$

. . . .

have the same circle of convergence as the original series.

For if M be the greatest term of the set $U_1, U_2R_1, U_3R_1^2, \cdots$, and if z lie within the original circle of convergence, then

$$U_1 + 2\,U_2 Z + 3\,U_3 Z^2 + \cdots = U_1 + 2\,U_2 R_1 \cdot \frac{Z}{R_1} + 3\,U_3 R_1^2 \frac{Z^2}{R_1^2} + \cdots$$

$$< M\left\{1 + 2\frac{Z}{R_1} + 3\frac{Z^2}{R_1^2} + \cdots\right\}$$

$$< \frac{M}{\{1 - Z/R_1\}^2};$$

this proves that the first series is absolutely convergent within the circle of radius R. In a similar manner it can be shown that the other series have the same circle of convergence. These series are all uniformly convergent within the circle, and represent, accordingly, continuous functions of z, which can be denoted be $f'(z)$, $f''(z), f'''(z), \cdots$, where $f(z) = \sum\limits_0^\infty u_n z^n$. To justify this notation we proceed to a proof of Cauchy's extension of Taylor's Theorem to integral series, from which the theorem

$$\lim_{h=0}\frac{f(z+h)-f(z)}{h}=u_1+2u_2z+3u_3z^2+\cdots$$

follows as an immediate corollary.

Cauchy's extension of Taylor's Theorem.

Let $f(z)=\sum_0^\infty u_nz^n$ have a radius of convergence R, and let $|z|+|h|<R$. Then Cauchy's theorem is

$$f(z+h)=f(z)+hf'(z)+\frac{h^2}{2!}f''(z)+\cdots.$$

Because $|z|+|h|<R$, the point $z+h$ lies inside the circle of convergence, and $f(z+h)$ is equal to the series

$$u_0+u_1(z+h)+u_2(z+h)^2+u_3(z+h)^3+\cdots;$$

but $|z|+|h|<R$, hence

$$u_0+u_1z+u_1h+u_2z^2+2u_2zh+u_2h^2+u_3z^3+\cdots$$

is absolutely convergent. Re-arranging it according to ascending powers of h, the value of the sum is unaltered; thus

$$\begin{aligned}f(z+h)&=(u_0+u_1z+u_2z^2+\cdots)+h(u_1+2u_2z+3u_3z^2+\cdots)\\&\quad+\frac{h^2}{2!}(2u_2+3\cdot 2u_3z+\cdots)+\cdots\\&=f(z)+hf'(z)+\frac{h^2}{2!}f''(z)+\cdots,\end{aligned}$$

for all points $z+h$, which lie within that circle, centre z, which touches internally the given circle of convergence.

A similar theorem holds good for integral series in n variables.

If $(z_1+h_1, z_2+h_2, \cdots, z_n+h_n)$ lie in the neighbourhood of $(z_1, z_2, \cdots, z_n)$,

$$f(z_1+h_1, z_2+h_2, \cdots, z_n+h_n)=\sum_{m_1\cdots m_n}\frac{f^{m_1, m_2, \cdots, m_n}(z_1, z_2, \cdots, z_n)}{m_1!\,m_2!\cdots m_n!}h_1^{m_1}h_2^{m_2}\cdots h_n^{m_n},$$

where the sign of summation indicates that $m_1, m_2, \cdots, m_n$ take, independently of one another, all integral values from 0 to ∞.

§ 86. *Theorem.* If $f(z)=u_0+u_1z+u_2z^2+\cdots$ be an integral series for which $u_0\neq 0$, there will always be a circle of finite radius round 0 as centre, within which the series never vanishes.

Let r be a positive number $<R$, the radius of convergence; let C be a finite number $>$ the absolute value of every term of $f(z)$ on the circle with centre 0 and radius r. We have $|u_m|r^m < C$, and, if $|z| = Z < r$,

$$|u_1z + u_2z^2 + \cdots| \leqq |u_1|Z + |u_2|Z^2 + |u_3|Z^3 + \cdots$$
$$< C\{Z/r + Z^2/r^2 + Z^3/r^3 + \cdots\}$$
$$< CZ/(r - Z).$$

However small the positive number $|u_0|$ may be, it will always be possible to find a value z_0 of z so small that

$$CZ_0/(r - Z_0) < |u_0|.$$

For all points such that $Z < Z_0$,

$$|u_1z + u_2z^2 + \cdots| < u_0|.$$

Thus, within a circle of radius Z_0, the integral series does not vanish.

Corollary. If the integral series be $P_\kappa(z) = u_\kappa z^\kappa + u_{\kappa+1}z^{\kappa+1} + \cdots$, there is a finite circle, with centre 0, within which the series does not vanish, except at the point 0. Hence if the integral vanish for infinitely many points of every circle however small, with 0 as a centre, it cannot begin with a term $u_\kappa z^\kappa$, where κ is finite. That is, the series vanishes identically.

The theorem of undetermined coefficients.

If the two integral series $P(z) = \sum_0^\infty u_n z^n$, $Q(z) = \sum_0^\infty v_n z^n$, be equal for infinitely many points $z_1, z_2, z_3, \cdots$ of a finite neighbourhood of 0, however small, then $u_n = v_n$ for all values of n. Suppose, if possible, that $u_0 = v_0$, $u_1 = v_1$, $\cdots$ $u_{\kappa-1} = v_{\kappa-1}$, $u_\kappa \neq v_\kappa$. We have, in consequence of this supposition,

$$P_\kappa(z) = (u_\kappa - v_\kappa)z^\kappa + (u_{\kappa+1} - v_{\kappa+1})z^{\kappa+1} + \cdots = 0,$$

for infinitely many values $z_1, z_2, z_3, \cdots$; some of these values must occur in every neighbourhood, however small, of 0. As this stands in contradiction to the preceding corollary, the supposition $u_\kappa \neq v_\kappa$ must be abandoned.

It results from this theorem that, whenever two integral series $P(z - c)$, $Q(z - c)$ are equal for infinitely many values within every circle of finite radius, however small, with c as centre, they are identically equal throughout their common region of convergence.

§ 87. *Theorem.* If $f(z)$ be an integral series which does not vanish identically, it is impossible that $f(z), f'(z), f''(z), \cdots$ shall all vanish at any point z_1 of the common domain.

For, if z_1 and $z_1 + h$ lie within the circle of convergence,

$$f(z_1 + h) = f(z_1) + hf'(z_1) + \frac{h^2}{2!}f''(z_1) + \cdots.$$

On the supposition that $f(z_1)$, $f'(z_1)$, $f''(z_1)$, $\cdots$ are all equal to 0, we deduce $f(z_1 + h) = 0$, where the only limitation on $z_1 + h$ is that the point is to lie within the circle of convergence. As this would imply that $f(z)$ vanishes identically, the supposition is untenable.

A further theorem follows from the one just proved: —

Theorem. If the integral series $P(z)$ vanish at infinitely many points z_1, z_2, $\cdots$ within a circle of finite radius contained within the circle of convergence, it must vanish identically.

For, since the area of the circle is finite and the number of zeros* is infinite, there must be a point of the area at which there is a nucleus of zeros, that is to say, there is a point within every neighbourhood of which, however small, infinitely many zeros are accumulated. Let z_0 be such a point, and let $f(z)$ be the integral series. Since z_0 is inside the circle of convergence, $f(z_0 + h)$ can be expanded according to ascending powers of h. The resulting integral series in h vanishes at infinitely many points of a circle round z_0, however small, and therefore identically for all points of the circle, centre z_0 and radius $R - |z_0|$. Let z_1 be any point of this circle; then for all points of a circle, centre z_1 and radius $R - |z_1|$, the series vanishes. By a continuation of this process it can be shown that the integral series vanishes identically for all points of its circle of convergence; and therefore all its coefficients vanish.

Corollary. If the integral series

$$u_0 + u_1 z + u_2 z^2 + \cdots + u_n z^n + \cdots$$
$$v_0 + v_1 z + v_2 z^2 + \cdots + v_n z^n + \cdots$$

converge within a circle R, and be equal for infinitely many distinct values z_1, z_2, z_3, $\cdots$, z_n, $\cdots$ whose absolute values are less than R_1, where R_1 is a finite radius $< R$, then the two series are identically equal. For the series

$$(u_0 - v_0) + (u_1 - v_1)z + (u_2 - v_2)z^2 + \cdots$$

is of the kind considered in the preceding theorem. An important consequence flows from this theorem: when a function can be represented by an integral series $P(z)$, the representation is unique.

* By a zero of $P(z)$ is meant a point at which $P(z)$ vanishes.

If two integral series $P(z_1, z_2, \cdots z_n)$, $Q(z_1, z_2, \cdots z_n)$, convergent in a neighbourhood Γ of the point a, be equal at all points within a region B whose centre is a, they are equal at all points of Γ. For let z be a point in B, and let z_κ, one of the variables, change while the others are constant. We have two integral series in z_κ, equal throughout the region B_κ, and therefore throughout Γ_κ. Similarly, any other variable may assume any position in Γ.

§ 88. Let there be a function $F(w, z_1, z_2 \cdots z_n)$, given by an integral series, which vanishes when w and all the z's are equal to zero. There will always be infinitely many points within a region Γ, formed by a circle $|w| = \rho$ in the w-plane and by equal circles $|z_\kappa| = \rho_1$ in the z-planes, which satisfy $F = 0$. Further, when $z_1, z_2, \cdots z_n = 0$, let $F(w, 0, 0 \cdots 0) = F_0(w) = P_\mu(w)$ (§ 83), so that when the z's $= 0$, μ values of $w = 0$; then for assigned values of the z's, lying within Γ, there are μ values of w in Γ such that $F = 0$.*

Write
$$\begin{aligned} F(w, z_1, z_2, \cdots z_n) &= F(w, 0, 0, \cdots 0) - F_1(w, z_1, z_2, \cdots z_n) \\ &= F_0 - F_1 \quad . \quad . \quad . \quad . \quad . \quad . \quad . \quad . \quad (1) \end{aligned}$$
so that $F_1 = 0$ for every value of w, when the z's are equal to 0. Since F_0 is not identically 0, there must exist a positive number ρ such that, within the circle ρ, the series $P_\mu(w)$ does not vanish, except when $w = 0$ (§ 86). Let ρ_0 be $< \rho$, and let the region formed by a ring of radii ρ_0, ρ in the w-plane and by circles of radius ρ_1 in the z-plane be called Γ'. As yet no limitation has been imposed upon ρ_1. It must be possible to choose the finite quantities ρ, ρ_1 sufficiently small to secure that $F(\rho, z_1, z_2, \cdots, z_n)$ converges for all values of $z_1, z_2, \cdots, z_n$ which lie within Γ'. Inasmuch as F_1 vanishes with the z's and F_0 is independent of the z's, it must further be possible to choose ρ_1 small enough to make $|F_1| < |F_0|$ for all points of Γ'.

In the region Γ' determined as above,
$$\frac{1}{F} = \frac{1}{F_0 - F_1} = \sum_{\lambda=0}^{\infty} \frac{F_1^\lambda}{F_0^{\lambda+1}},$$
and because
$$\frac{\delta F}{\delta w} = \frac{\delta F_0}{\delta w} - \frac{\delta F_1}{\delta w},$$
$$\begin{aligned} \frac{\delta F}{\delta w}\Big/F &= \frac{\delta F_0}{\delta w}\Big/F_0 + \sum_1^\infty F_1^\lambda \frac{\delta F_0}{\delta w}\Big/F_0^{\lambda+1} - \sum_1^\infty F_1^{\lambda-1} \frac{\delta F_1}{\delta w}\Big/F_0^\lambda \\ &= \frac{\delta F_0}{\delta w}\Big/F_0 - \sum_1^\infty \frac{1}{\lambda} \frac{\delta}{\delta w} (F_1/F_0)^\lambda. \end{aligned}$$

* Weierstrass, Abhandlungen aus der Functionenlehre, p. 107. See also a Thesis by M. Dautheville: Étude sur les séries entières à plusieurs variables. For the case in which F_0 vanishes identically, see Weierstrass, p. 113. Another proof of this important theorem will be found in Picard's Traité d'Analyse, t. ii., fascicule 1.

The series $\sum_1^\infty \frac{1}{\lambda}(F_1/F_0)^\lambda$ is absolutely convergent since, for all points in Γ', $|F_1| < |F_0|$. It may therefore be regarded as an integral series in w, which is absolutely and uniformly convergent. We may therefore write

$$\frac{\delta F}{\delta w} / F = \frac{\delta F_0}{\delta w} / F_0 - \frac{\delta}{\delta w} \sum_1^\infty \frac{1}{\lambda}(F_1/F_0)^\lambda \cdots.$$

By hypothesis $F_0 = P_\mu(w)$; therefore

$$(F_1/F_0)^\lambda = w^{-\mu\lambda}P_0(w),$$

the coefficients in $P_0(w)$ being absolutely convergent integral series in $z_1, z_2, \cdots, z_n$, which all vanish at the origin, since there $F_1 = 0$. Hence, uniting the various powers of w, we get

$$\sum_1^\infty \frac{1}{\lambda}(F_1/F_0)^\lambda = \sum_{-\infty}^\infty P^{(\kappa)}(z_1, z_2, \cdots z_n)w^\kappa,$$

where again each coefficient on the right vanishes at the origin.

Also $$\frac{\delta F_0}{\delta w}/F_0 = \mu/w + P(w).$$

Therefore, throughout Γ',

$$\frac{\delta F}{\delta w}/F = \mu/w + P(w) - \frac{\delta}{\delta w}\sum_{-\infty}^\infty{}_\kappa P^{(\kappa)}(z_1, z_2, \cdots z_n)w^\kappa \quad . \quad . \quad . \quad (2)$$

When the z's are all zero, the equation $F = 0$ has μ roots equal to zero, and these are the only solutions of the equation within the w-circle of radius ρ. When the z's take n assigned positions in Γ, let ν be the number of roots of $F = 0$, situated within Γ. We shall prove that $\nu = \mu$. In finding ν, the supposition is made that a multiple root is counted as often as is indicated by its order of multiplicity.

(1) ν must be different from zero. For if $F \neq 0$ for any value of w throughout Γ', when $z_1, z_2, \cdots, z_n$ take prescribed positions in Γ, distinct from $0, 0, \cdots 0$, the expression $\frac{\delta F}{\delta w}/F$ has no infinity in Γ for these values of the z's, and is, consequently, expansible for all values of w within the circle (ρ), as an integral series in the z's. This disagrees with (2), for the dexter of (2) contains at least one negative power μ/w, and Γ' is merely a part of Γ.

(2) Let $w_1, w_2, \cdots, w_\nu$ be the ν roots of w in Γ which satisfy the w-equation $F = 0$, for assigned values of the z's. The difference

$$\frac{\delta F}{\delta w}/F - 1/(w - w_1) - 1/(w - w_2) - \cdots - 1/(w - w_\nu)$$

is a function which is finite at $w_1, w_2, \cdots, w_\nu$, and therefore throughout Γ. Hence it can be expanded as an integral series in Γ. For all values of w which make $|w| < \rho$, but greater than $|w_1|, |w_2|, \cdots, |w_\nu|$, we have

$$\frac{\delta F}{\delta w}/F - 1/(w-w_1) - 1/(w-w_2) - \cdots - 1/(w-w_\nu) = P(w),$$

an integral series in w. But

$$1/(w-w_1) + \cdots + 1/(w-w_\nu) = \nu/w + s_1/w^2 + s_2/w^3 + \cdots,$$

where $s_\kappa =$ the sum of the κth powers of $w_1, w_2, \cdots, w_\nu$.

Hence $$\frac{\delta F}{\delta w}/F = \nu/w + P(w) + \sum_1^\infty s_\kappa/w^{\kappa+1} \quad . \quad . \quad . \quad . \quad (3).$$

Comparing (2) and (3), we see that $\mu = \nu$; and also that $s_1, s_2, s_3, \cdots$ are equal to integral series in $z_1, z_2, \cdots, z_n$ which vanish at the origin. Let the equation whose roots are $w_1, w_2, \cdots, w_\mu$ be

$$w^\mu + f_1 w^{\mu-1} + \cdots + f_\mu = 0 \quad . \quad . \quad . \quad . \quad . \quad . \quad . \quad . \quad (4),$$

then by Newton's theorem for the sums of the powers of the roots we have

$$f_1 = -s_1,$$
$$2f_2 = -s_2 - s_1 f_1,$$
$$3f_3 = -s_3 - s_2 f_1 - s_1 f_2,$$
$$\mu f_\mu = -s_\mu - s_{\mu-1} f_1 - s_{\mu-2} f_2 - \cdots - s_1 f_{\mu-1},$$

and therefore the coefficients f are integral series in $z_1, z_2, \cdots, z_n$, which certainly converge when these variables lie within the circles of radius ρ_1, and which vanish at the origin.

§ 89. *Behaviour of a series on the circle of convergence.*

We shall next examine the behaviour, when $|z| = 1$, of the integral series $\Sigma a_n z^n$, in which the coefficients are all real, and the ratio a_n/a_{n-1} can be expanded in a series

$$1 + c_1/n + c_2/n^2 + \cdots,$$

where the coefficients are independent of n.

We first suppose $c_1 = 0$. Let

$$c_1 = c_2 = \cdots = c_{\mu-1} = 0, \quad c_\mu \neq 0.$$

We shall show that $a_n = a + b_n/n^{\mu-1}$, where a is independent of n, and neither zero nor infinite, and b_n is finite for all values of n.

Let $\qquad a_n/a_{n-1} = 1 + k_n/n^\mu$, so that $Lk_n = c_\mu$.

If we begin with some fixed term, say the pth, such that no subsequent term vanishes, we have

$$a_{p+n} = a_{p-1}\left(1+\frac{k_p}{p^\mu}\right)\left(1+\frac{k_{p+1}}{(p+1)^\mu}\right)\cdots\left(1+\frac{k_{p+n}}{(p+n)^\mu}\right).$$

Since k_{p+n} is always finite, the series

$$\frac{k_p}{p^\mu} + \frac{k_{p+1}}{(p+1)^\mu} + \cdots \text{ to } \infty$$

be finite if

$$\frac{1}{p^\mu} + \frac{1}{(p+1)^\mu} + \cdots \text{ to } \infty$$

is finite, which is the case, since $\mu > 1$.

Therefore (§ 78) a_{p+n} has a finite, non-zero, limit a when $n = \infty$.

Now $a_n - a_{n-1} = k_n a_{n-1}/n^\mu$. If for n we write $n+1, \cdots n+r$ and add, we obtain

$$a_{n+r} - a_n = \mu_{n,r}\sigma_{n,r},$$

where

$$\sigma_{n,r} = \frac{1}{(n+1)^\mu} + \frac{1}{(n+2)^\mu} + \cdots + \frac{1}{(n+r)^\mu},$$

and $\mu_{n,r}$ is not less than the least, and not greater than the greatest, of the quantities $k_{n+1}a_n, \cdots k_{n+r}a_{n+r-1}$; $\sigma_{n,r}$ has a finite limit σ_n when $r = \infty$, and therefore $\mu_{n,r}$ has a limit μ_n when $r = \infty$, and this limit is finite because both k_{n+r} and a_{n+r-1} remain finite.

$$\text{Further, } \sigma_n = \frac{1}{(n+1)^\mu} + \frac{1}{(n+2)^\mu} + \cdots \text{ to } \infty$$

$$\overline{\overline{<}} \left\{ \frac{1}{(n+1)^2} + \frac{1}{(n+2)^2} + \cdots \right\} \frac{1}{(n+1)^{\mu-2}}$$

$$< \left\{ \frac{1}{n(n+1)} + \frac{1}{(n+1)(n+2)} + \cdots \right\} \frac{1}{n^{\mu-2}}$$

$$< \frac{1}{n^{\mu-1}}.$$

Therefore $a - a_n = \theta_n\mu_n/n^{\mu-1}$, where $0 < \theta_n < 1$, or, if $\theta_n\mu_n = -b_n$,

$$a_n = a + b_n/n^{\mu-1},$$

where b_n is finite however great n may be.

Next let $c_1 \neq 0$. The ratio of a_n/n^{c_1} to $a_{n-1}/(n-1)^{c_1}$ is

$$\frac{a_n}{a_{n-1}}(1-1/n)^{c_1}$$

$$= (1 + c_1/n + \cdots)(1 - c_1/n + \cdots)$$

$$= 1 - c_1^2/n^2 + \cdots;$$

therefore $$a_n/n^{c_1} = a + b_n/n,$$

or $$a_n = an^{c_1} + b_n n^{c_1 - 1}.$$

It follows at once that the series Σa_n is absolutely convergent when $c_1 < -1$, and divergent when $c_1 \geqq -1$. Also that the series $\Sigma a_n z^n$ is absolutely convergent at points of the circle of convergence other than $z = 1$, when $c_1 < -1$; and divergent when $c_1 \geqq 0$, since then $L \mid a_n z^n \mid > 0$. The case $0 > c_1 \geqq -1$ remains to be considered.

We have

$$(1 - z)\sum_0^n a_n z^n = a_0 + \sum_1^n (a_n - a_{n-1}) z^n - a_n z^{n+1}.$$

Now $L a_n z^{n+1} = 0$, and the series on the right hand is convergent when $n = \infty$, for

$$\frac{a_n - a_{n-1}}{a_{n-1} - a_{n-2}} = \frac{a_{n-1}}{a_{n-2}} \cdot \frac{a_n/a_{n-1} - 1}{a_{n-1}/a_{n-2} - 1}$$

$$= \left(1 + \frac{c_1}{n-1} + \frac{c_2}{(n-1)^2} + \cdots\right) \cdot \frac{\dfrac{c_1}{n} + \dfrac{c_2}{n^2} + \cdots}{\dfrac{c_1}{n-1} + \dfrac{c_2}{(n-1)^2} + \cdots}$$

$$= \left(1 + \frac{c_1}{n} + \cdots\right)\left(1 + \frac{c_2}{nc_1} + \cdots\right)\left(1 - \frac{c_1 + c_2}{nc_1} + \cdots\right)$$

$$= 1 + \frac{c_1 - 1}{n} + \cdots,$$

and $$c_1 - 1 < -1.$$

Therefore $(1 - z)\sum_0^\infty a_n z^n$ is finite, and therefore $\sum_0^\infty a_n z^n$ is convergent unless $z = 1$.

Summing up, the series in question, on the circle of convergence, is absolutely convergent when $c_1 < -1$, conditionally convergent when $0 > c_1 \geqq -1$ and $z \neq 1$, divergent when $0 > c_1 \geqq -1$ and $z = 1$, and divergent when $c_1 \geqq 0$.*

Example. Discuss the behaviour of the series $\sum_1^\infty z^n/n^m$, and of the expansion of $(1 + z)^m$, on the circle of convergence; m being real.

---

* This proof is based on that of Weierstrass for the integral series with complex coefficients (Abbandlungen a. d. Functionenlehre) and on Stolz's treatment of Weierstrass's theorem (Allgemeine Arithmetik). Pringsheim has shown that an integral series can be conditionally convergent at all points of the circle of convergence (Math. Ann., t. xxv.).

§ 90. The series $f(z)=z/1-z^2/2+z^3/3-\cdots$ has a radius of convergence $=1$. For the point $z=1$, the convergence is conditional. Let

$$\phi(z)=z/1+z^3/3-z^2/2+z^5/5+z^7/7-z^4/4\cdots.$$

when $|z|<1$, $f(z)=\phi(z)$; but, when $z=1$, the values of $f(1)$ and $\phi(1)$ are different, showing that one, at least, of the series $f(z)$, $\phi(z)$ must change its value abruptly as z passes along the real axis from a point inside the circle and very near 1, to the point 1 itself. The preceding considerations raise the question as to whether $f(1)$ is the limit of $f(x)$, when x tends to 1. (See § 62.) The following important theorem of Abel's settles the point:—

Let $f(x)=a_0+a_1x+a_2x^2+\cdots$ be a series with *real* coefficients, which converges within a circle of convergence of radius 1; further, let $a_0+a_1+a_2+a_3+\cdots$ be a convergent series whose sum is s. Then

$$\lim_{\xi=0} f(1-\xi)=a_0+a_1+a_2+\cdots,$$

where ξ is a quantity which is real and positive.

Let $$s_n=a_0+a_1+a_2+\cdots+a_n,$$

then $$f(x)=s_0+(s_1-s_0)x+(s_2-s_1)x^2+(s_3-s_2)x^3+\cdots.$$

If $x<1$, the series $=(1-x)(s_0+s_1x+s_2x^2+s_3x^3+\cdots)$.

Thus

$$f(1-\xi)=\xi[s_0+s_1(1-\xi)+s_2(1-\xi)^2+s_3(1-\xi)^3+\cdots+s_n(1-\xi)^n]$$
$$+\xi(1-\xi)^{n+1}[s_{n+1}+s_{n+2}(1-\xi)+s_{n+3}(1-\xi)^2+\cdots].$$

The series is divided into two parts. The number of the terms in the first part can be taken large enough to make $s_{n+1}, s_{n+2}, s_{n+3}, \cdots$ all $>s-\delta$ and $<s+\epsilon$, when δ, ϵ are arbitrarily small. ξ is still at our disposal. Choose it so as to make $\mathop{\mathrm{L}}_{n=\infty} n\xi=0$; *e.g.*, make $\xi=1/n^2$ or $1/n^3$. Then because

$$s_0+(1-\xi)s_1+\cdots+(1-\xi)^n s_n$$

is a finite expression for finite values of n,

$$\xi[s_0+s_1(1-\xi)+\cdots+s_n(1-\xi)^n]$$

must vanish, when ξ tends to 0. The remaining part lies between

$$\xi(1-\xi)^{n+1}(s-\delta)[1+(1-\xi)+(1-\xi)^2+\cdots \text{ to } \infty]$$

and $$\xi(1-\xi)^{n+1}(s+\epsilon)\text{ [the same series]};$$

i.e. between $(1-\xi)^{n+1}(s-\delta)$ and $(1-\xi)^{n+1}(s+\epsilon)$. Now $(1-\xi)^n$ lies between 1 and $1-n\xi$; hence $\lim_{\xi=0}(1-\xi)^{n+1}=1$. Thus, ultimately, $f(1-0)$ lies between $s-\delta$ and $s+\epsilon$, where δ and ϵ can be made arbitrarily small. This proves the theorem.

(i.) The theorem holds when the coefficients are complex, for then the series can be separated into two parts, one real and the other purely imaginary. As the theorem holds for each part separately, it must hold for the two parts in combination.

(ii.) Next the restriction that the variable is to be real can be removed.

$$f\{(1-\xi)(\cos\phi+i\sin\phi)\}=a_0+a_1(\cos\phi+i\sin\phi)(1-\xi)$$
$$+a_2(\cos 2\phi+i\sin 2\phi)(1-\xi)^2+\cdots$$

can be separated into two parts,

$$(a_0+a_1\cos\phi(1-\xi)+a_2\cos 2\phi(1-\xi)^2+\cdots),$$
$$i(a_1\sin\phi(1-\xi)+a_2\sin 2\phi(1-\xi)^2+\cdots).$$

By what has been already proved, these parts tend, separately, to the limits $a_0+a_1\cos\phi+a_2\cos 2\phi+\cdots$, $i(a_1\sin\phi+a_2\sin 2\phi+\cdots)$.

To return to the example with which we started: —

The series $z/1-z^2/2+z^3/3-z^4/4+\cdots$,

and $z/1+z^3/3-z^2/2+z^5/5+z^7/7-z^4/4+\cdots$,

are equal when $|z|<1$, but unequal when $z=1$, in spite of the fact that both series converge for that value. The following is the explanation of the paradox. Write $z=1-\xi$, and let ξ be real and positive; the two series are equal for all values of ξ, however small, but the convergence of the second series is infinitely slow in the neighbourhood of $\xi=0$, and the infinitely slow convergence is accompanied by discontinuity. The result can be expressed in this way: when ξ is real,

$$\lim_{\xi=0}\{(1-\xi)/1-(1-\xi)^2/2+(1-\xi)^3/3-\cdots\}=1-1/2+1/3-1/4+\cdots,$$
$$\lim_{\xi=0}\{(1-\xi)/1+(1-\xi)^3/3-(1-\xi)^2/2-\cdots\}=1-1/2+1/3-1/4+\cdots$$
$$\neq 1+1/3-1/2+1/5+1/7-1/4+\cdots.$$

§ 91. *The product of two integral series.*

Consider the two integral series $\sum_0^\infty u_m z^m$, $\sum_0^\infty v_m z^m$, with a radius of convergence $=1$. At a point within this circle the product is

$$\sum_0^\infty w_m z^m=u_0v_0+(u_0v_1+u_1v_0)z+\cdots+(u_0v_n+u_1v_{n-1}+\cdots+u_nv_0)z^n+\cdots.$$

If $\sum_0^\infty u_m$, $\sum_0^\infty v_m$ converge absolutely, we know that the expression just written down converges when $x=1$. In virtue of Abel's theorem we can extend this theorem; for

$$\lim_{x=1}\left(\sum_0^\infty u_m x^m\right)\lim_{x=1}\left(\sum_0^\infty v_m x^m\right)=\lim_{x=1}\left(\sum_0^\infty w_m x^m\right)$$

gives $\sum_0^\infty u_m\cdot\sum_0^\infty v_m=\sum_0^\infty w_m$, when $\sum_0^\infty u_m$, $\sum_0^\infty v_m$, $\sum_0^\infty w_m$ are convergent.

The theorem $\Sigma u_m\times\Sigma v_m=\Sigma w_m$, when Σu_m and Σv_m are *unconditionally convergent,* was stated by Cauchy in his Cours d'Analyse Algébrique. Mertens proved that the result still holds, when one of the two series is *conditionally convergent* (Crelle, t. lxxix., 1875). The theorem which we have just proved is due to Abel (Ueber die Binomialreihe, Crelle, t. i.). It will be noticed that Abel's theorem gives no information as to when Σw_m is convergent. For information on the product of *two conditionally convergent series* we refer the student to a memoir by Pringsheim (Math. Ann., t. xxi., pp. 327–378).

Weierstrass's Theory of the Analytic Function.

§ 92. *Series derived from an integral series.*

Let the integral series

$$u_0+u_1z+u_2z^2+u_3z^3+\cdots$$

have a radius of convergence R. Within the circle (R) it has properties analogous to those of integral polynomials, but without the circle it is divergent. If

$$f(z)=u_0+u_1z+u_2z^2+u_3z^3+\cdots=P(z),$$

$f(z)$ is also equal to an integral series in $z-z_0$ for all points inside the circle A whose centre is z_0 and radius $R-|z_0|$ (§ 87). This series in $z-z_0$ is denoted by $P(z\,|\,z_0)$, and is said to be *derived directly* from $P(z)$, with respect to the point z_0. For all points of A,

$$P(z\,|\,z_0)=P(z).$$

Let z_1 be a point interior to A. With centre z_1, describe a

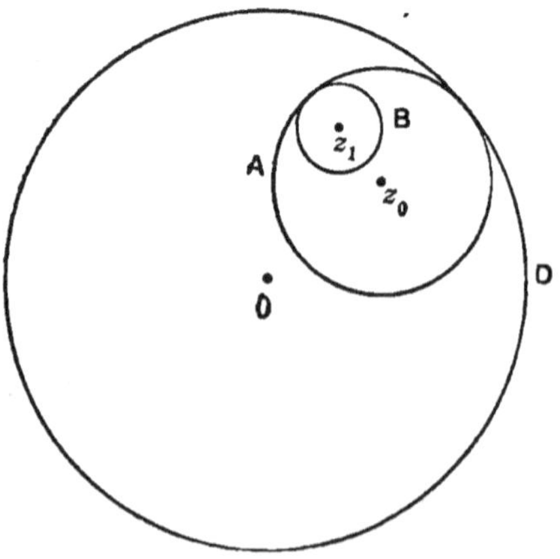

Fig. 20

circle B which touches A and lies entirely inside it. A series can be derived from $P(z)$ with respect to z_1, namely, $P(z|z_1)$, and its circle of convergence will include the whole of the circle B; but a series in $z-z_1$ can be derived from $P(z|z_0)$ with respect to z_1, namely, $P(z|z_0|z_1)$, and this latter series is convergent in B. The two series are identically equal in the circle B, for the former $=P(z)$, and the latter $=P(z|z_0)$, throughout B. There is an obvious extension of this theorem, namely

$$P(z|z_n)=P(z|z_0|z_1|\cdots|z_n),$$

when $z_0, z_1, z_2, \cdots$ lie inside D; $z_1, z_2, z_3, \cdots$ inside A; $z_2, z_3, \cdots$ inside B; and so on.

The importance of the result consists in the circumstance that a series *directly* derived from $P(z)$, with respect to a point inside the circle of convergence, has been shown to be the same integral series as that obtained by the *indirect* process described above.

The function defined by the integral series $P(z)$ has a meaning for all points within the circle of convergence. A method will now be described by which the integral series can be continued beyond this circle. It depends on the fact that the circles A, B, of Fig. 20, are not, in general, the full circles of convergence of the corresponding integral series $P(z|z_0)$, $P(z|z_1)$.

Let the circle of convergence of $P(z|z_0)$ be not A, but A'. Within A, $P(z)=P(z|z_0)$. The question is whether the two expressions are equal in the shaded region. Let a point z_1 be taken within A, and let the circle B' be drawn to touch the nearer of the two circles D, A'; and the circle B to touch A. Since B' lies within the circles of convergence of $P(z)$ and $P(z|z_0)$, the integral series $P(z|z_1)$ must be identically equal to $P(z)$, and $P(z|z_0|z_1)$ to $P(z|z_0)$, throughout B'. But we know that $P(z|z_0)$ is identically equal to $P(z)$ throughout B, a part of B'. Hence $P(z|z_1)$ and $P(z|z_0|z_1)$ are identically equal throughout B. These two integral series, which proceed according to powers of $z-z_1$, must be identically equal for all points for which either has a meaning (§ 86). But we know that

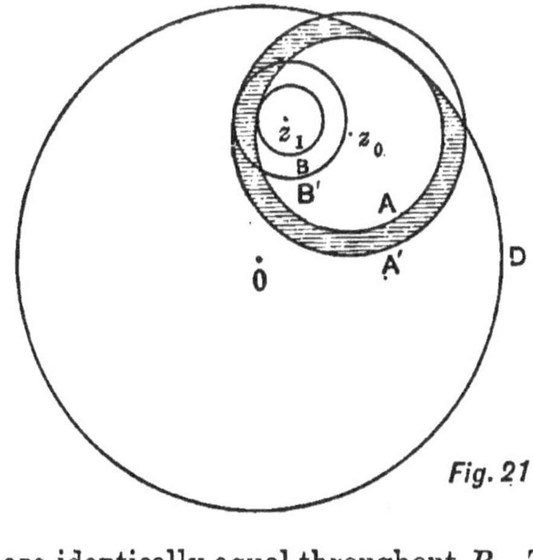

Fig. 21

the circle of convergence of $P(z \mid z_0 \mid z_1)$ extends at least as far as B'. Hence $P(z \mid z_1) = P(z \mid z_0 \mid z_1)$ *throughout* B'. Therefore $P(z)$ and $P(z \mid z_0)$ must be identically equal throughout B'. This proves the equality of the two series for that portion of the shaded region which is contained within B'. By adding on this portion to the circular region bounded by A, and by selecting suitably a new point z_1 within the region so extended, the equality can be affirmed for a further portion of the shaded region, and the process can be continued until the equality of the two integral series has been established for all points of the shaded region.

§ 93. *Theorems on derived series.*

Theorem I. Let $P(z \mid z_0)$, $P(z \mid z_0')$ be two integral series which are derived directly from the common element $P(z)$, where z_0, z_0' are points interior to D, the circle of convergence of $P(z)$; and let A, B be the circles of convergence of $P(z \mid z_0)$, $P(z \mid z_0')$. The two series are identically equal within that region, which is common to A, B, D. We have to prove that they are identically equal within the shaded region. Let us take a point c, near D (Fig. 22). Within the circle (not drawn in Fig. 22) which has centre c, and touches D,

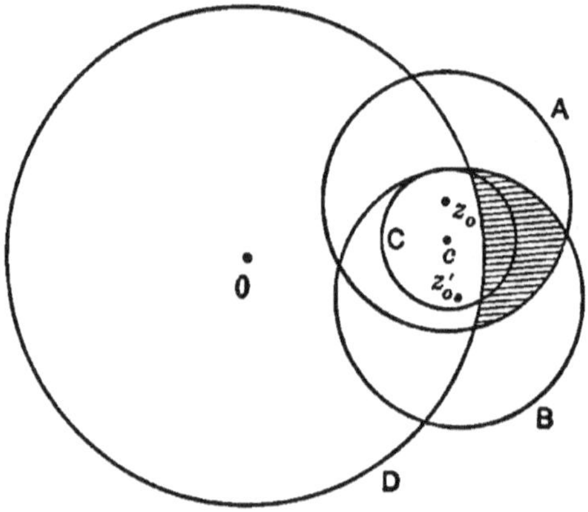

Fig. 22

$$P(z \mid z_0 \mid c) = P(z \mid z_0' \mid c) \text{ identically.}$$

As these two series proceed according to powers of $z - c$, and as $P(z \mid z_0 \mid c)$ converges for all points within the circle C of Fig. 22, the series

$$P(z \mid z_0 \mid c), \quad P(z \mid z_0' \mid c),$$

must be identically equal throughout C. Hence

$$P(z \mid z_0) = P(z \mid z_0')$$

for that portion of the shaded region which lies within C. Add this portion to the unshaded region common to A and B. By a suitable choice of a new point c, within the region so extended, a further extension can be effected; and the process can be continued until the equality of the series has been established for all the shaded region.

Theorem II. If the series $P(z \mid z_0 \mid z_1)$ be derived directly from $P(z \mid z_0)$, the series $P(z \mid z_0)$ can be derived from $P(z \mid z_0 \mid z_1)$.

If the series $P(z \mid z_0 \mid z_1)$ have the point z_0 within its circle of convergence, we have immediately,

$$P(z \mid z_0 \mid z_1 \mid z_0) = P(z \mid z_0),$$

and the theorem is proved by a direct derivation. This will always be possible if $|z_1 - z_0| < \rho/2$, where ρ is the radius of convergence of $P(z \mid z_0)$. If z_0 be without the circle of convergence of $P(z \mid z_0 \mid z_1)$, it is possible to interpolate, *between* z_0 *and* z_1, a series of points $a_1, a_2, \cdots, a_n$, where n is finite, such that a_1 lies inside the circle of convergence of $P(z \mid z_0 \mid z_1)$; a_2 inside the circle of convergence of $P(z \mid z_0 \mid z_1 \mid a_1)$; a_3 inside the circle of convergence of $P(z \mid z_0 \mid z_1 \mid a_1 \mid a_2) \cdots$; and finally, z_0 inside the circle of convergence of $P(z \mid z_0 \mid z_1 \mid a_1 \mid \cdots \mid a_n)$. From this last series can be deduced $P(z \mid z_0 \mid z_1 \mid a_1 \mid a_2 \mid \cdots \mid a_n \mid z_0)$, an integral series in $z - z_0$. As the latter integral series is identically equal to $P(z \mid z_0)$ for points within a neighbourhood of z_0 which is not infinitely small, the two series must be identically equal throughout the original circle of convergence. Therefore $P(z \mid z_0)$ can always be derived *indirectly* from $P(z \mid z_0 \mid z_1)$, when $P(z \mid z_0 \mid z_1)$ is *directly* derived from $P(z \mid z_0)$ (Pincherle, Giornale di Matematica, t. xviii., p. 348).

Theorem III. If c be a point of the region common to the two regions of convergence, A and B, of $P(z - z_0)$, $P(z - z_1)$, and if the two series derived directly from these, with regard to c, be identically equal:—

(i.) The given series can be derived, indirectly, the one from the other.

(ii.) The series derived from $P(z \mid z_0)$, $P(z \mid z_1)$, with regard to any other point common to A, B, are identically equal.

(i.) By supposition, $P(z \mid z_0 \mid c) = P(z \mid z_1 \mid c)$.

But $P(z \mid z_0)$ and $P(z \mid z_1)$ can be derived indirectly from $P(z \mid z_0 \mid c)$ and $P(z \mid z_1 \mid c)$; hence $P(z \mid z_0)$ can be derived, indirectly, from $P(z \mid z_1)$.

(ii.) Let c_1 be a point in the maximum circle which can be drawn from the centre c, without cutting A or B.

With regard to c_1, we can derive the two series

$$P(z \mid z_0 \mid c \mid c_1) \text{ and } P(z \mid z_1 \mid c \mid c_1);$$

and these series are identically equal to $P(z|z_0|c)$, $P(z|z_1|c)$ throughout the maximum circle. But $P(z|z_0|c)$, $P(z|z_1|c)$ are themselves identically equal throughout this circle; therefore $P(z|z_0|c|c_1)$ must be identically the same series as $P(z|z_1|c|c_1)$. The same must be the case for the series directly derived with respect to c_1, namely $P(z|z_0|c_1)$, $P(z|z_1|c_1)$. We can repeat the same reasoning for a point c_2, inside the maximum circle, described from c_1 as centre without cutting A, B, and prove that

$$P(z|z_0|c_2)=P(z|z_1|c_2).$$

By a continuation of the process, the theorem can be established for every point of the region common to A and B (Pincherle, loc. cit., p. 349).

§ 94. *The analytic function.* If $P(z)$ be convergent within a circle D, and if from this integral series be derived, directly, the integral series $P(z|z_0)$, where z_0 is a point of D, the circle of convergence A of $P(z|z_0)$ will, in general, cut the circle D. A function which is equal to $P(z)$ in the circle D; to $P(z|z_0)$ in the circle A, z_0 being any point in D; to $P(z|z_0|z_1)$ in its circle of convergence, z_1 being any point in A, and so on, is termed by Weierstrass an *analytic* function. The analytic function is defined by an integral series, whose radius of convergence is not zero, together with all possible continuations of that series. Each of the series is called an *element* of the function; thus $P(z)$ is an element of the function at 0, $P(z|z_0)$ an element at z_0, and so on. The value of any one of the totality of integral series at a point z of its region of convergence is *one* of the values of the analytic function at z.

In defining the analytic function the choice of the primitive element, from amongst the totality of integral series which serve to define the function, is perfectly arbitrary; for if $P(z|z_1)$ be derived from $P(z|z_0)$, the latter series can be derived from the former. Hence each series of the totality is an element of the function. In other words, any series selected from the totality will give rise to the other members of the totality, and will, in conjunction with them, completely define the analytic function.

§ 95. It must not be inferred from the preceding work that the analytic function is necessarily one-valued. Let $P(z|c)$ be an integral series. It may be continued beyond its circle of convergence until it passes into $P(z|z')$. Suppose this effected by the interpolation of n points. We have then a chain of $n+2$ points, and the element at each (the initial point c excepted) is to be directly derived from the preceding element. We may regard the n points as lying on a path L from c to z', the path being subjected to the restrictions that it is not to cut itself, and that each of the $n+1$ arcs into which it is divided by the n points is to lie within the circle of convergence of the element at the initial point of that arc. By what has been already proved (§ 92) the same final element is obtained when additional points are interpolated on L. And further the system of n points can be replaced by another system of m points on L, provided that each of the $m+1$ arcs so formed lies within the circle of convergence at its initial point. For the compound system of $n+m$ intermediate points leads to the final element $P(z \mid z')$, and we can regard the n points as interpolated among the new m points, and suppress them without affecting the final element.

When an analytic function is said to be continued along a path L, it is understood that L lies wholly within the chain of circles of convergence. What is proved is that the final element then depends uniquely on the initial element and the path L. It readily follows that the reverse path $-L$ restores the initial element.

On the other hand another path M from c to z' leads from the same initial element to a final element $Q(z \mid z')$, which may be different from $P(z \mid z')$. When the difference exists, the contour formed by $-L$ from z' to c, and M from c to z', leads from $P(z \mid z')$ to $Q(z \mid z')$. This is due to the presence, within the contour, of a *branch-point*, at which there is no element of the function. A point at which there is no element of an analytic function will be called a *singular point* of that function; and the branch-points of a function are included among its singular points. We shall return to the singular points, in connection with the present theory, in § 97. They are recommended to the reader's attention, inasmuch as the character of a function is betrayed by its behaviour at these exceptional points.

§ 96. *The coefficients of an integral series.* Consider the series $f(z) = \sum_0^\infty u_n z^n$, with a radius of convergence R, and let a be a primitive pth root of unity. Let $|z| = \rho < R$; then, on the circle (ρ), if $p > m$,

$$\frac{1}{p} \sum_1^p {}_\kappa \frac{f(a^\kappa \rho)}{a^{m\kappa} \rho^m} = u_m + u_{m+p}\rho^p + u_{m+2p}\rho^{2p} + \cdots.$$

When $p = \infty$, the right-hand side reduces to u_m. To prove this, we must show that

$$\lim_{p=\infty} (u_{m+p}\rho^p + u_{m+2p}\rho^{2p} + u_{m+3p}\rho^{3p} + \cdots) = 0.$$

The proof follows from the fact that $\sum_0^\infty u_n z^n$ is an absolutely convergent series for $|z|=\rho$, for we have

$$|(u_{m+p}\rho^p + u_{m+2p}\rho^{2p} + \cdots)\rho^m|$$

$$< |u_{m+p}|\rho^{m+p} + |u_{m+2p}|\rho^{m+2p} + \cdots$$

$$< |u_{m+p}|\rho^{m+p} + |u_{m+p+1}|\rho^{m+p+1} + \cdots \text{ to } \infty;$$

and this expression tends to the limit 0, when p increases indefinitely. Hence

$$u_m = \lim_{p=\infty} \frac{1}{p} \sum_1^p {}_\kappa f(\rho a^\kappa)/\rho^m a^{m\kappa}.$$

An important theorem of Cauchy's follows at once from the equation

$$u_m = \frac{1}{p} \sum_1^p {}_\kappa f(\rho a^\kappa)/\rho^m a^{m\kappa} - [u_{m+p}\rho^{m+p} + u_{m+2p}\rho^{m+2p} + \cdots].$$

For $|u_m| \leqq \frac{1}{p\rho^m} \sum_1^p |f(\rho a^\kappa)| + \epsilon$, where ϵ vanishes when $p=\infty$.

But $|f(\rho a^\kappa)| \leqq G$, where G is the upper limit of the absolute values of $f(z)$ along the circle (ρ).

Hence $$|u_m| \leqq G/\rho^m.$$

Integral series in $1/z$. If an integral series $P(z)$ converge within a circle of convergence of radius R, the series $P(1/z)$, obtained from $P(z)$ by writing $1/z$ for z, must converge for all points without a circle of radius $1/R$. Since $P(z-c)$ has a circle of convergence of centre c, it is natural to suppose that $P(z-\infty)$ has a circle of convergence, centre ∞; by what has been said $P(1/z)$ has such a circle of convergence, namely, the region *without* the circle of radius R. This agrees with what was said in § 82.

In such a series as $\sum_{-\infty}^{+\infty} u_n z^n$, the series of terms formed by the positive values of n has a radius of convergence R, while the series formed by the negative values of n, converges without a circle of radius R_1. Hence, if $R_1 < R$, the series $\sum_{-\infty}^{+\infty} u_n z^n$ converges absolutely within the ring formed by the two concentric circles of radii R, R_1. Let the sum be $f(z)$. If ρ be intermediate between R and R_1, it can be shown, by a proof strictly analogous to that already given, that

$$|u_m| \leqq G/\rho^m,$$

where G is the upper limit of the values of $|f(z)|$ on the circle ρ. Further, if two series $\sum_{-\infty}^{+\infty} u_n z^n$, $\sum_{-\infty}^{+\infty} v_n z^n$, have the same region of convergence, and if they be equal for all points of a neighbourhood, however small, of a point z_0 of this common region of convergence, then the theorem of undetermined coefficients holds good and $u_n = v_n$.

§ 97. *Discussion of the singular points.* Suppose that we know that a given integral series $P(z)$ converges for all points $|z| < R$, but that we do not know whether R is or is not the radius of convergence. If a be a point inside the circle (R), and if $f(z)$ be the analytic function defined by $P(z)$, then

$$f(a+h) = f(a) + hf'(a) + \frac{h^2}{2!}f''(a) + \cdots,$$

throughout the circle of convergence of the integral series in h; that is,

$$f(z) = f(a) + (z-a)f'(a) + \frac{(z-a)^2}{2!}f''(a) + \cdots = P(z|a),$$

when $|z-a| < R_a$, the radius of convergence of $P(z|a)$. Now let ρ be the lower limit of all the quantities R_a, when a is allowed to take every position within the circle (R); we shall prove the following theorem: —

The series $P(z)$ converges within the circle of radius $R+\rho$.

Let ρ_1 be a positive number $<\rho$, and let z' lie within the circular ring of radii R, $R+\rho_1$. A circle whose centre is z' and whose radius is ρ must intersect the circle (R). Let a be a point of the region Γ common to these two circles; since by hypothesis the radius of convergence at $a \not< \rho$, the point z' lies within the circle of convergence of $P(z|a)$, and the analytic function has at z' the value $P(z'|a)$. This value is not dependent on the choice of a. For if b be any other point of Γ, and if $c = (a+b)/2$, the two series $P(z|a|c)$, $P(z|b|c)$ are identical (Theorem I., § 93); and therefore the two series $P(z|a)$, $P(z|b)$ have the same values throughout their common region of convergence. In particular

$$P(z'|a) = P(z'|b).$$

The absolute values of the two series $P(a)$ and $P(z'|a)$ are both finite under the conditions $|a|, = A, < R$, and $|z'| < R + \rho_1$, and are less than an assignable finite number G.

The radius of convergence of $P(z' \mid a) > \rho_1$; therefore, by the theorem of the preceding paragraph,

$$\frac{1}{\kappa!}|f^\kappa(a)| < G\rho_1^{-\kappa},\ \kappa = 0, 1, 2, \cdots.$$

Let $\qquad P(z)$ be $\sum_{n=0}^{\infty} u_n z^n$;

then
$$f^\kappa(a) = \sum_{n=\kappa}^{\infty} n(n-1)\cdots(n-\kappa+1)u_n a^{n-\kappa},$$

a series whose radius of convergence $\not< R$ (§ 85); therefore, by a second application of the same theorem,

$$n(n-1)\cdots(n-\kappa+1)\,|u_n| < \kappa!\ G\rho_1^{-\kappa}A^{\kappa-n},$$

or
$$\frac{n(n-1)\cdots(n-\kappa+1)}{\kappa!}|u_n|A^n(\rho_1/R)^\kappa < G(A/R)^\kappa.$$

Write $\kappa = 0, 1, 2, \cdots, n$ and add the resulting inequalities; then

$$|u_n|A^n(1+\rho_1/R)^n < G\{1-(A/R)^{n+1}\}/(1-A/R),$$
$$< GR/(R-A).$$

Therefore the series $P(z)$ converges for all values of z such that

$$|z| < A(1+\rho_1/R),$$

and therefore, since $R - A$ and $\rho - \rho_1$ are arbitrarily small, the series converges within the circle whose radius is $R + \rho$. This circle is the true circle of convergence; for if the radius of convergence $> R + \rho$, the lower limit of the quantities R_a cannot be ρ. In the proof of this theorem of Weierstrass's we have followed Stolz, Allgemeine Arithmetik, t. ii., p. 180.

That the true circle of convergence must contain at least one point such that in any neighbourhood of the point, however small, ρ is infinitely small, is evident from the following considerations. The lower limit of the quantities R_a is zero, hence there must be at least one point, inside or on the rim, in every neighbourhood of which, however small, the lower limit is 0; and this point must lie on the rim, since the radius of convergence at an interior point is not less than the distance from the point to the rim.

The points on the rim can be divided into two classes according as they are or are not singular:—

A non-singular point z_1 lies inside the domain of some point a, interior to the circle of convergence of $P(z)$. The value of $P(z)$ at z_1 is $P(z_1 \mid a \mid z_1)$; for since $P(z)$ and $P(z \mid a \mid z_1)$ have a common region of convergence which includes the point a, the two series are equal to $P(z \mid a)$ throughout the region common to their circles of convergence, and in particular at the point z_1. Thus for non-singular points the limit to which $P(z)$ tends, when z tends to z_1, is the value $P(z_1 \mid a \mid z_1)$. The function defined by $P(z)$ can be continued over the rim at z_1. A singular point is one within every neighbourhood of which, however small, ρ is infinitely small. These points do not lie within the circle of convergence of any point a, interior to the circle of convergence of $P(z)$. As no elements of the function correspond to them, they are the singular points which we have already defined. By what we have already proved, no singular point of an analytic function can lie inside a circle of convergence, but there must be at least one on the rim.

As an example, consider the function $\dfrac{1}{1-z}$. Throughout the interior of the circle, centre 0, and radius 1,

$$\frac{1}{1-z} = 1 + z + z^2 + z^3 + \cdots,$$

but the point 1 is a singular point. If a be any point within the circle, the derived series with respect to a is

$$f(a) + (z-a)f'(a) + \frac{(z-a)^2}{2!}f''(a) + \cdots,$$

where $f(z)$ stands for the series. But

$$f(a) = 1 + a + a^2 + \cdots = \frac{1}{1-a},$$

$$f'(a) = 1 + 2a + 3a^2 + \cdots = \frac{1}{(1-a)^2},$$

$$f''(a) = 2 + 2\cdot 3a + 3\cdot 4a^2 + \cdots = \frac{2!}{(1-a)^3},$$

and so on.

The derived series $f(z|a)$ is, therefore,

$$\frac{1}{1-a}+\frac{z-a}{(1-a)^2}+\frac{(z-a)^2}{(1-a)^3}+\frac{(z-a)^3}{(1-a)^4}+\cdots,$$

and the radius of convergence is $|1-a|$.

But

$$\frac{1}{1-z}=\frac{1}{((1-a)-(z-a))}=\frac{1}{1-a}+\frac{(z-a)}{(1-a)^2}+\frac{(z-a)^2}{(1-a)^3}+\cdots,$$

throughout the circle, centre a, and radius $|1-a|$. Thus the integral series $1+z+z^2+z^3+\cdots$, together with its continuations, defines the analytic function $\frac{1}{1-z}$. In this example, if the successive series be derived with respect to points a_1, a_2, $a_3 \cdots a_\kappa$, 0, each point being within the circle of convergence of the point immediately preceding it, the final series is the same as the initial series. The figure is intended to show the way in which a function can be continued in the neighbourhood of an isolated singular point.

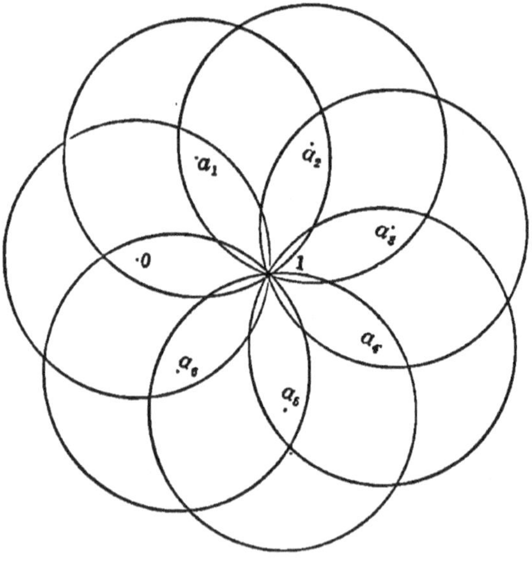

Fig. 23

§ 98. A one-valued analytic function $f(z)$ is said to be *regular* at a point c, if it can be developed, in the neighbourhood of c, in a convergent integral series $P(z-c)$.

If $f(z)$ be infinite at c, while $(z-c)^m f(z)$ is regular at c, m being a finite positive integer, c is said to be a *pole* of the function, and m is said to be the order of multiplicity, or simply the *order*, of the pole. If there be no such finite integer m, c is an *essential singularity.**

It is evidently a matter of some importance to determine, from the integral series which serves to define a function, the positions of the singular points on the circle of convergence. This problem is discussed by Hadamard in Liouville, Ser. 4, t. viii. In particular it appears that if the ratio $u_\kappa/u_{\kappa+1}$ have a limit a when $\kappa=\infty$, a is in general one of the singular points on the circle of convergence; that a is on the circle is evident from § 84. The idea of determining a singular point in this simple way is M. Lecornu's. Reference must also be made to Darboux, Sur l'approximation des fonctions de très-grands nombres, Liouville, Ser. 3, t. iv.

§ 99. When the circle of convergence of $P(z)$ covers the whole of the z-plane, with the single exception of the point $z=\infty$, the function defined by the series is nowhere infinite in the finite part of the plane, but has an essential singularity at $z=\infty$. The algebraic function $u_0+u_1z+u_2z^2+\cdots+u_\kappa z^\kappa$ is finite throughout the finite part of the plane, but has a polar discontinuity at ∞. The resemblance in properties caused Weierstrass to name the integral series $P(z)$, with an infinite radius of convergence, *a transcendental integral function.* As examples of transcendental integral functions we may instance those defined by the well-known integral series,

$$1+z+z^2/2!+z^3/3!+\cdots,$$

$$1-z^2/2!+z^4/4!-z^6/6!+\cdots,$$

$$z-z^3/3!+z^5/5!-z^7/7!+\cdots.$$

Such functions are necessarily one-valued, since there can be no branch-point at a finite distance from the origin. A *transcendental* function is a function which is not algebraic. The question of ascertaining whether a given integral series defines an algebraic or a transcendental function was considered by Eisenstein, who proved that when an integral series $P_1(z)$, with rational coefficients, defines an algebraic function, the coefficients can be rendered integers by the substitution of κz for z, where κ is a suitably chosen integer.

* The discussion of the singular points of a function resolves itself into that of branch-points, infinities, and essential singularities. These will be considered in the following chapters. The pole is a case of an infinity.

[Eisenstein, Monatsberichte der Akademie der Wiss. zu Berlin, 1852; Heine, Kugelfunctionen, Second Edition, t. i., p. 50. For a remarkably simple proof by Hermite, see Proc. London Math. Soc., t. vii., or Hermite's Cours d'Analyse, Third Edition, p. 174.]

§ 100. *The sum of an infinite number of integral series.*

Let $P^{(\kappa)}(z) = u_{0,\kappa} + u_{1,\kappa}z + u_{2,\kappa}z^2 + \cdots (\kappa = 1, 2, \cdots, \infty)$, be the κth term of the series $\sum_1^\infty {}^\kappa P^{(\kappa)}(z)$. We assume that the component terms, as well as the series itself, converge absolutely and uniformly within the same circle (R).

(i.) The infinite series

$$u_{n,1} + u_{n,2} + u_{n,3} + \cdots$$

converges to a definite value u_n, whatever be the positive integer n.

(ii.) For every value of z, within the circle (R), $\sum_0^\infty u_n z^n$ converges absolutely and $= \sum_1^\infty {}^\kappa P^{(\kappa)}(z)$.

Let ρ be a positive quantity $< R$. When $|z| = \rho$, there exists, by reason of the suppositions made above, a number μ such that

$$|P^{(n+1)}(z) + P^{(n+2)}(z) + \cdots + P^{(n+p)}(z)| < \epsilon \quad (n > \mu),$$

p being any positive integer. The expression

$$P^{(n+1)}(z) + \cdots + P^{(n+p)}(z)$$

can be written as an integral series, and in this series the absolute value of the coefficient of $z^\lambda \leqq \epsilon/\rho^\lambda$. Thus

$$|u_{\lambda,n+1} + u_{\lambda,n+2} + \cdots + u_{\lambda,n+p}| \leqq \epsilon/\rho^\lambda.$$

As this is true, however large p may be,

$$u_{\lambda,1} + u_{\lambda,2} + \cdots \text{ to } \infty$$

must be a convergent series. Suppose that its sum is u_λ. We have, if $|z| = Z < \rho$,

$$\sum_0^\infty {}^\lambda |u_{\lambda,n+1} + u_{\lambda,n+2} + \cdots| Z^\lambda \leqq \epsilon \sum_0^\infty {}^\lambda Z^\lambda/\rho^\lambda$$
$$\leqq \epsilon\rho/(\rho - Z).$$

The expression $\epsilon\rho/(\rho - Z)$ can be made as small as we please; hence $\sum_0^\infty {}^\lambda u_\lambda z^\lambda$ is absolutely convergent. Let the sum

$$\sum_0^\infty u_\lambda z^\lambda = f(z);$$

also let

$$P^{(1)}(z) + P^{(2)}(z) + \cdots = g(z).$$

We have to prove that $f(z) = g(z)$.

Now

$$P^{(1)}(z) + P^{(2)}(z) + \cdots + P^{(n)}(z) = (u_{0,1} + u_{0,2} + \cdots + u_{0,n})$$
$$+ (u_{1,1} + u_{1,2} + \cdots + u_{1,n})z$$
$$+ (u_{2,1} + u_{2,2} + \cdots + u_{2,n})z^2$$
$$+ \cdot \quad \cdot \quad \cdot \quad \cdot \quad \cdot \quad \cdot \quad \cdot \quad \cdot$$
$$\cdot \quad \cdot \quad \cdot \quad \cdot \quad \cdot \quad \cdot \quad \cdot \quad \cdot$$

Hence the equality of $f(z)$ and $g(z)$ will be proved if we can show that

$$\left| \sum_0^\infty{}^\lambda \rho_{\lambda, n} z^\lambda - \sum_{n+1}^\infty{}^\kappa P^{(\kappa)}(z) \right| < \epsilon_1,$$

where $\rho_{\lambda, n}$ is the value of $u_{\lambda, n+1} + u_{\lambda, n+2} + \cdots$ to ∞, and ϵ_1 is arbitrarily small. But we know that this inequality exists; for we have seen that

$$\left| \sum_0^\infty{}^\lambda \rho_{\lambda, n} z^\lambda \right| \leqq \epsilon \sum_0^\infty{}^\lambda \rho^{-\lambda} \cdot Z^\lambda$$
$$\leqq \epsilon\rho/(\rho - Z);$$

also $$| P^{(n+1)}(z) + P^{(n+2)}(z) + \cdots |$$

is, by the definition of uniform convergence, $< \epsilon$, when $n >$ a suitably chosen number μ. Thus $\sum\limits_0^\infty \rho_{\lambda, n} z^\lambda$ and $\sum\limits_{n+1}^\infty P^{(\kappa)}(z)$ are, simultaneously, arbitrarily small; and part ii. of the theorem has been established.

An application of the theorem. Let $P(z)$, $Q(z)$ be integral series, with radii of convergence R, R', and let R'' be a positive number $< R'$. For values of y such that $|y| \leqq R''$, the series $Q(y)$ converges uniformly. In

$$Q(y) = v_0 + v_1 y + v_2 y^2 + v_3 y^3 + \cdots,$$

let $P(z)$ be substituted for y; the resulting series

$$v_0 + v_1 P(z) + v_2 \{P(z)\}^2 + v_3 \{P(z)\}^3 + \cdots$$

converges, provided there is a number $\rho < R$, such that for all values of z which are less than ρ in absolute value

$$|P(z)| \leqq R''.$$

The expressions

$$\{P(z)\}^2, \ \{P(z)\}^3, \cdots$$

are integral series, with the *same* circle of convergence. Under these circumstances the series can be changed into an integral series. A necessary condition is $|v_0| < R'$.

§ 101. *The expression of a quotient as an integral series.*

Let the quotient be $\frac{Q(z)}{P_0(z)}$. We have, for sufficiently small values of $|z|$,

$$|u_0| > |u_1z + u_2z^2 + u_3z^3 + \cdots|,$$

and therefore, if $u_1z + u_2z^2 + u_3z^3 + \cdots = v$,

$$\frac{1}{P_0(z)} = \frac{1}{u_0 + v} = \frac{1}{u_0}\left\{1 - \frac{v}{u_0} + \frac{v^2}{u_0^2} - \cdots\right\}$$
$$= P(z).$$

Hence $\frac{Q(z)}{P_0(z)}$, being the product of two integral series, is itself an integral series. To determine this integral series, let

$$\frac{Q(z)}{P_0(z)} = \sum_0^\infty w_n z^n,$$

and, after multiplying both sides by $P_0(z)$, equate coefficients. The resulting equations determine uniquely the values of $w_0, w_1, w_2, \cdots$.

Corollary. The quotient of an integral series $P_0(z)$ by an integral series $P_\kappa(z)$ is $z^{-\kappa}\cdot Q_0(z)$.

Theorem. If $$w - w' = P_\kappa(z - z'),$$

then $$(w - w')^{1/\kappa} = (u_\kappa)^{1/\kappa} P_1(z - z').$$

For, if

$$w - w' = u_\kappa(z - z')^\kappa + u_{\kappa+1}(z - z')^{\kappa+1} + \cdots,$$

then

$$(w-w')^{1/\kappa} = (z-z')\{u_\kappa + u_{\kappa+1}(z-z') + u_{\kappa+2}(z-z')^2 + \cdots\}^{1/\kappa}$$
$$= (u_\kappa)^{1/\kappa}(z-z')\{1 + Q(z-z')\}^{1/\kappa}$$
$$= (u_\kappa)^{1/\kappa}(z-z')\left[1 + \frac{1}{\kappa}Q(z-z') + \frac{1-\kappa}{2!\,\kappa^2}\{Q(z-z')\}^2 + \cdots\right]$$
$$= (u_\kappa)^{1/\kappa}P_1(z-z').$$

The κ values are furnished by the κth roots of u_κ.

§ 102. *The reversion of an integral series.*

Let $$y = a_0 + a_1 z + a_2 z^2 + \dots \qquad (1),$$

be an integral series, whose radius of convergence is 1, and whose coefficient $a_1 \neq 0$. By a simple transformation any integral series can be changed into one with radius 1. When $(y - a_0)/a_1 = y_1$, the equation becomes

$$y_1 = z - A_2 z^2 - A_3 z^3 - A_4 z^4 - \dots \qquad (2).$$

The problem is to find an integral series for z, which has the form

$$z = y_1 + b_2 y_1^2 + b_3 y_1^3 + \dots \qquad (3).$$

If (3) be, in reality, a solution of (2), the relation

$$y_1 = (y_1 + b_2 y_1^2 + b_3 y_1^3 + \cdots) - A_2 (y_1 + b_2 y_1^2 + b_3 y_1^3 + \cdots)^2 - A_3 (y_1 + b_2 y_1^2 + b_3 y_1^3 + \cdots)^3 - \cdots$$

must be satisfied identically. Equating coefficients of the various powers of y_1, we find that

$$\left.\begin{aligned} &b_2 - A_2 = 0, \\ &b_3 - 2 A_2 b_2 - A_3 = 0, \\ &b_4 - A_2(b_2^2 + 2 b_3) - A_3 \cdot 3 b_2 - A_4 = 0, \\ &\dots\dots\dots\dots\dots \\ &\dots\dots\dots\dots\dots \end{aligned}\right\} \qquad (4);$$

equations which determine *uniquely* $b_2, b_3 \cdots$ as integral functions of the A's.

To determine whether these values of $b_2, b_3, \cdots$ make the right-hand side of (3) convergent within a domain whose radius is not infinitely small, let us replace, in the relations (4), the quantities A by their absolute values $|A|$. The $(\kappa - 1)$th of the relations (4) gives, in conjunction with the preceding $(\kappa - 2)$ equations, a value for b_κ in terms of the A's:

$$b_\kappa = \text{an integral expression in } A_2, A_3, \cdots, A_\kappa,$$

whose signs are all positive. Suppose this expression is

$$\phi(A_2, A_3, \cdots, A_\kappa).$$

The change of the A's into $|A|$'s cannot diminish the absolute value of the expression. *A fortiori* the change of all the A's into G, the maximum value of their absolute values, cannot diminish the value of $|\phi|$.

Let $b_2, b_3, \cdots, b_\kappa, \cdots$, after these changes, be $B_2, B_3, \cdots, B_\kappa, \cdots$. The series

$$z = y_1 + b_2 y_1^2 + b_3 y_1^3 + b_4 y_1^4 + \cdots,$$

will certainly converge in a domain of finite dimensions, if

$$z = y_1 + B_2 y_1^2 + B_3 y_1^3 + B_4 y_1^4 + \cdots$$

converge in the same domain. As a retracing of the steps by which the value

$$z = y_1 + b_2 y_1^2 + b_3 y_1^3 + \cdots$$

was obtained, leads back to

$$y_1 = z - A_2 z^2 - A_3 z^3 - \cdots,$$

it must follow that

$$z = y_1 + B_2 y_1^2 + B_3 y_1^3 + \cdots$$

arises from

$$\begin{aligned} y_1 &= z - Gz^2 - Gz^3 - Gz^4 - \cdots \\ &= z - Gz^2/(1-z), \end{aligned}$$

a quadratic equation in z. The two solutions are

$$z = \frac{(1+y_1) \mp \sqrt{(1+y_1)^2 - 4(G+1)y_1}}{2(1+G)},$$

each of which can be developed by the Binomial Theorem as an integral series. The plus sign gives an integral series in y_1 with a constant term $\neq 0$; but the minus sign gives a series of the form required. Since this series has a sum, it must be convergent within a circle of finite radius and, further, it is *unique*. It must accordingly be the same as

$$y_1 + B_2 y_1^2 + B_3 y_1^3 + \cdots.$$

Hence the series last written, and therefore also

$$y_1 + b_2 y_1^2 + b_3 y_1^3 + \cdots,$$

is convergent within a finite domain.

The case in which $a_1 = a_2 = \cdots = a_{m-1} = 0$, $a_m \neq 0$, will now be considered. Putting $(y - a_0)/a_m = y_1$, we get

$$(y_1)^{1/m} = z(1 - A_{m+1}z - A_{m+2}z^2 - \cdots)^{1/m},$$

where $(y_1)^{1/m}$ denotes *one* of the mth roots. But

$$(1 - A_{m+1}z - A_{m+2}z^2 - \cdots)^{1/m} = 1 - a_2 z - a_3 z^2 - a_4 z^3 - \cdots,$$

within a finite domain of $z = 0$. Hence

$$y_1^{1/m} = z - a_2 z^2 - a_3 z^3 - \cdots.$$

This series can be reversed in the way already indicated, the only difference being that $y_1^{1/m}$ takes the place of y_1. The final result is

$$z = (y_1)^{1/m} + \beta_2 (y_1)^{2/m} + \beta_3 (y_1)^{3/m} + \cdots;$$

and there are m expansions corresponding to the mth roots of y_1.

Expansions of this kind play an important part in the theory of algebraic functions. (See Chapter IV.)

§ 103. *Series whose regions of convergence consist of isolated pieces.*

There exist infinite series which represent, in different regions, different analytic functions of z. This fact seems to have been first noticed by Seidel (see a memoir by Pringsheim, Math. Ann., t. xxii.), but the full consequences which follow from it were pointed out by Weierstrass. The following simple example of such a series is due to Tannery: —

If
$$|z|<1,\ \underset{n=\infty}{\mathrm{L}}\frac{1}{1-z^n}=1,$$
$$|z|>1,\ \underset{n=\infty}{\mathrm{L}}\frac{1}{1-z^n}=0,$$
$$|z|=1,\ \underset{n=\infty}{\mathrm{L}}\frac{1}{1-z^n}\ \text{is non-existent.}$$

Let $m_0, m_1, m_2, \cdots$ be a series of integers which increase with m above all limits. The series

$$\begin{aligned}\psi(z) &= \frac{1}{1-z^{m_0}}+\left(\frac{1}{1-z^{m_1}}-\frac{1}{1-z^{m_0}}\right)+\left(\frac{1}{1-z^{m_2}}-\frac{1}{1-z^{m_1}}\right)+\cdots\text{to }\infty\\ &=\frac{1}{1-z^{m_\infty}}\\ &=1\text{ or }0,\end{aligned}$$

according as z lies within, or without the circle of convergence of radius 1. With the aid of this theorem it is easy to construct a function which shall represent κ arbitrary functions $\psi_r(z)$, within circles whose centres are $z_1, z_2, \cdots, z_\kappa$ and radii $\rho_1, \rho_2, \cdots, \rho_\kappa$, and shall represent another arbitrary function $\psi_{\kappa+1}(z)$ over the remaining portion of the plane. It is assumed that the circles do not intersect.

Such a function is

$$\psi\left(\frac{z-z_1}{\rho_1}\right)\cdot\psi_1(z)+\psi\left(\frac{z-z_2}{\rho_2}\right)\cdot\psi_2(z)+\cdots+\psi\left(\frac{z-z_\kappa}{\rho_\kappa}\right)\cdot\psi_\kappa(z)$$
$$+\left\{1-\sum_1^\kappa{}^r\,\psi\left(\frac{z-z_r}{\rho_r}\right)\right\}\psi_{\kappa+1}(z).$$

For points within the circle whose centre is z_r and radius ρ_r, all the terms except the rth vanish, and it equals $\psi_r(z)$. For points outside all the κ circles,

$$\psi\left(\frac{z-z_r}{\rho_r}\right)=0\ \text{without exception.}$$

* Weierstrass, Abh. a. d. Functionenlehre, p. 102.

A series $\sum_{1}^{\infty} f_\kappa(z)$, whose terms are rational functions of z, may have a region of convergence which consists of several isolated pieces A, B, C, $\cdots$. The concept of an analytic function is based upon a totality of integral series, each element of the totality being a continuation of some preceding element. The circles of convergence, which correspond to the elements of the totality, cover a continuum of points. The question is whether the analytic monogenic functions defined for the isolated continua A, B, C, $\cdots$ can be distinct monogenic functions, and not merely branches of one and the same monogenic function. The discovery made by Weierstrass was that the concept of a monogenic function of z is not co-extensive with the concept of functionality as expressed by an arithmetic expression, constructed with the help of the four arithmetic operations. Tannery's example proves conclusively that an arithmetic expression may represent different analytic functions in separate regions of the plane.

§ 104. *Lacunary Spaces.*

In Tannery's example an expression defines the monogenic function 1 when $|z| < 1$, and the monogenic function 0 when $|z| > 1$; while both functions exist for all values of z.

But there are functions which have no existence in regions of the z-plane. For example, consider the integral series $\sum_{0}^{\infty} b^n z^{a^n}$, where a is an integer > 1, and b is positive and < 1. The series $\sqrt[\kappa]{u_\kappa}$ or b^{n/a^n} has the upper limit 1, and therefore the radius of convergence is 1. On the circle of convergence there is certainly one singular point z_0; if we write $z = z'\ (\cos 2\mu\pi/a^\nu + i \sin 2\mu\pi/a^\nu)$, where μ and ν are integers, all but the first $\nu + 1$ terms are unaltered, and therefore the series in z' has the same circle of convergence and the same singular point z_0. Therefore the original series has the singular point $z_0(\cos 2\mu\pi/a^\nu + i \sin 2\mu\pi/a^\nu)$. By a proper choice of μ and ν this new singular point may be brought as near as we please to any assigned point on the circle. The circle is therefore *a singular line;* no circle of convergence of an integral series derived from the series in z can extend beyond this line, and the function defined by the series has no existence except for points within the circle. The rest of the plane is said to be a *lacunary space* of the function.*

* The above example is due to Weierstrass (Abh. a. d. Functionenlehre, p. 90); this proof is Hadamard's (Liouville, Fourth Series, t. viii., p. 115). Instances of such functions, defined by means of integrals, are given by Hermite (Acta Societatis Fennicæ, t. xii.); other important sources are Mittag-Leffler (Darboux's Bulletin, Second Series, t. v.) and Poincaré (American Journal, t. xiv.). Such functions occur naturally in the theory of modular functions. See Klein-Fricke, Modulfunctionen, t. i., p. 110.

§ 105. *The Exponential Function.*

The generalized exponential and circular functions $\exp z$, $\sin z$, $\cos z$ are defined by the integral series

$$1 + z + z^2/2! + z^3/3! + \cdots,$$

$$z - z^3/3! + z^5/5! - z^7/7! + \cdots,$$

$$1 - z^2/2! + z^4/4! - z^6/6! + \cdots,$$

whose radii of convergence are infinitely great. By multiplication of the two integral series and by the use of the Binomial Theorem, it can be proved without difficulty that

$$\exp z_1 \times \exp z_2 = \exp (z_1 + z_2).$$

When $z =$ a real quantity x, the series $\exp (z)$ becomes

$$1 + x + x^2/2! + x^3/3! + \cdots = e^x.$$

It is usual to write $\exp (z) = e^z$, when z is complex, but the reader must understand that e^z is then *defined* by a series.

From the addition theorem,

$$\exp z_1 \times \exp z_2 = \exp (z_1 + z_2),$$

the ordinary theorems of trigonometry can be readily deduced. For instance,

(i.)
$$\begin{aligned} \exp z &= \exp x \cdot \exp iy = e^x \cdot \exp (iy) \\ &= e^x (\cos y + i \sin y), \end{aligned}$$

since
$$\exp (iy) = \left(1 - \frac{y^2}{2!} + \frac{y^4}{4!} - \cdots\right) + i\left(y - \frac{y^3}{3!} + \frac{y^5}{5!} - \cdots\right).$$

(ii.)
$$\begin{aligned} \exp (z + 2\pi i) &= \exp z \cdot \exp (2\pi i) = \exp z (\cos 2\pi + i \sin 2\pi) \\ &= \exp z. \end{aligned}$$

(iii.)
$$\cos z + i \sin z = \exp (iz),$$

$$\cos z - i \sin z = \exp (-iz),$$

hence
$$\cos^2 z + \sin^2 z = 1.$$

(iv.) The addition theorems for sines and cosines can be proved by replacing the sines and cosines by their exponential values.

Enough has been said to show how the formulæ of Analytic Trigonometry can be established.*

---

* The reader will find a full treatment of the subject in Hobson's Trigonometry and Chrystal's Algebra, t. ii.

§ 106. *The Generalized Logarithm.*

The generalized logarithm $\log z$ is defined as the inverse function of $\exp z$. Let $z=\rho(\cos\theta+i\sin\theta)$; the equations

$$w=\log z,\quad z=\exp w,$$

express the same relation between w and z. Hence, if $w=u+iv$,

$$\rho(\cos\theta+i\sin\theta)=e^{u+iv}=e^{u}(\cos v+i\sin v),$$

and, therefore, $\rho=e^{u}$, $\theta+2n\pi=v$. If the real solution of $e^{u}=\rho$ be $u=\log\rho$, we have

$$\log z=\log\rho+i(\theta+2n\pi).$$

At a given point z_0, one value of $\log z$ is $\log\rho+i\theta$, but there are infinitely many others. If θ be regarded as capable of unlimited variation, these infinitely many values must not be regarded as distinct functions of z, but as branches of one and the same function, for after a description of the curve with initial point z_0, z returns to z_0, but $\log\rho+i\theta$ changes into $\log\rho+i(\theta+2\pi)$.

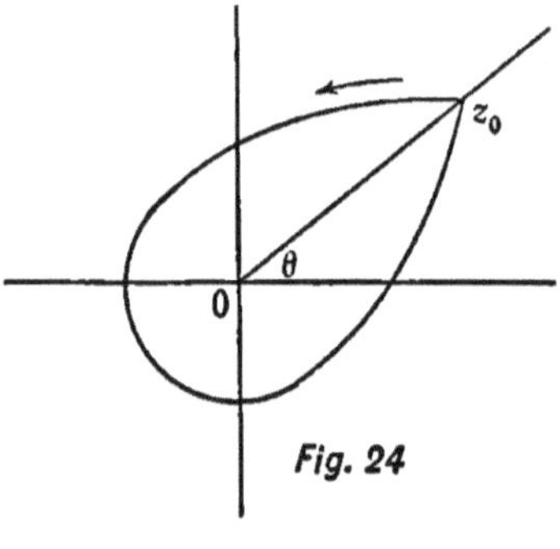

Fig. 24

If θ be limited in its range to an angle 2π, as for instance if

$$-\pi<\theta\leqq\pi,$$

it is possible to separate the infinitely many branches. The chief branch is $\log\rho+i\theta$, the nth branch $\log\rho+i(\theta+2n\pi)$. If $\log z$, ${}_n\log z$ denote respectively the chief branch and the nth branch, we have

$$\log z=\log\rho+i\theta\qquad(-\pi<\theta\leqq\pi),$$

$${}_n\log z=\log\rho+i(\theta+2n\pi)\qquad(-\pi<\theta\leqq\pi).$$

In particular, $\log i=\frac{\pi}{2}i$; $\log(-1)=\pi i$; $\log(-i)=-\frac{\pi}{2}i$.

Each of the components $\log\rho$, θ of a branch is continuous, except when $\rho=0$ or ∞, or when $z=$ a real negative quantity. In the latter case there is an abrupt change from $\theta=+\pi$ to $\theta=-\pi$, as a point crosses the negative part of the axis from above.

The series for the chief branch of log $(1+z)$.

Let us write

$$1+z=e^w=1+w+w^2/2!+\cdots,$$

where $|z|<1$, and let us assume the existence of an integral series $Q_1(z)=w$, in accordance with § 102.

Therefore $\qquad e^w=1+z$, and $e^w\cdot dw/dz=1$;

or $\qquad dw/dz=1/(1+z)=1-z+z^2-z^3+\cdots.$

This series converges when $|z|<1$, and is the derivative of

$$P_1(z)=z-z^2/2+z^3/3-z^4/4+\cdots.$$

Hence $\qquad w=z-z^2/2+z^3/3-z^4/4+\cdots.$

Writing $z-1$ for z we have for log z the series

$$P_1(z\mid 1)\equiv z-1-\tfrac{1}{2}(z-1)^2+\tfrac{1}{3}(z-1)^3-\cdots,$$

whose circle of convergence extends as far as the singular point 0.

Let us consider how, without the assumption of any knowledge of the exponential function, the complete idea of the logarithm can be deduced from the above element.

We first find the derived series $P(z\mid 1\mid z_1)$, where z_1 is any point within the first circle of convergence. From § 86,

$$P(z\mid 1\mid z_1)=P_1(z_1\mid 1)+P_1'(z_1\mid 1)(z-z_1)+P_1''(z_1\mid 1)(z-z_1)^2/2!+\cdots,$$

where accents denote differentiations;

and $\qquad P_1'(z\mid 1)=1-(z-1)+(z-1)^2-\cdots=1/z,$

$$P_1''(z\mid 1)=-1/z^2, \text{ etc.},$$

therefore when we give the name log z, first to the function so far as it is defined by the element $P_1(z\mid 1)$, and then to its continuation, we have

$$\log z=\log z_1+\frac{z-z_1}{z_1}-\frac{1}{2}\left(\frac{z-z_1}{z_1}\right)^2+\frac{1}{3}\left(\frac{z-z_1}{z_1}\right)^3-\cdots,$$

within a second circle whose centre is z_1 and which extends to the point 0.

In the same way when z_2 is any point within the second circle we have a series *of precisely the same form*, which defines log z within the circle whose centre is z_2 and radius $|z_2|$. In this way the function can be continued over the whole plane.

Example. Prove that $\exp P(z \mid 1 \mid z_1) = z$.

Let $C + C'$ be a contour which includes the point 0, and let it be symmetric with regard to the real axis (Fig. 25); further let the points $c_1, c_2, \cdots, c_n$ of C be such that c_κ lies within the circle whose centre is $c_{\kappa-1}$ and radius $|c_{\kappa-1}|$, where $\kappa = 1, 2, \cdots, n$, and $c_0 = 1$, and similarly let $c'_1, c'_2, \cdots, c'_n$ be points of C' such that c'_κ lies within the circle whose centre is $c'_{\kappa-1}$ and radius $|c'_{\kappa-1}|$, where $c'_0 = 1$.

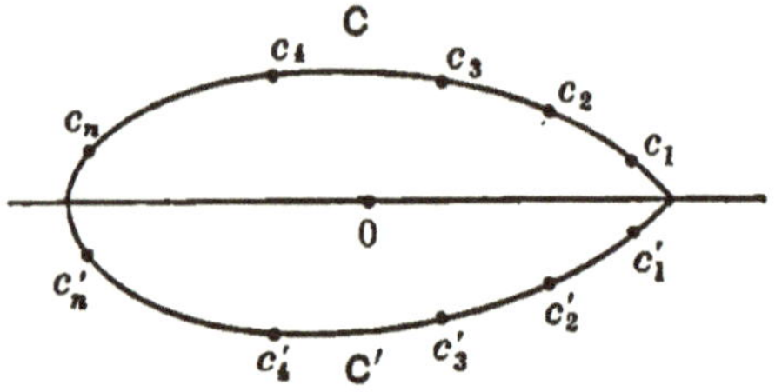

Fig. 25

We propose to show that when z_1 is a point on the real axis from $-\infty$ to 0, lying within the circles (c_n), (c_n'), the values of

$$P(x_1 \mid 1 \mid c_1 \mid c_2 \cdots \mid c_n) \text{ and } P(x_1 \mid 1 \mid c_1' \mid c_2' \cdots \mid c_n')$$

differ by a constant.

Let us suppose that c'_κ is conjugate to c_κ ($\kappa = 1, 2, \cdots, n$). The elements of the function at $c_1, c_2, \cdots, c_n$ are conjugate to the elements at $c_1', c_2', \cdots, c_n'$; hence the values at x_1 of the series $P(z \mid 1 \mid c_1 \; c_2 \cdots \mid c_n)$ and $P(z \mid 1 \mid c_1' \cdots \mid c_n')$ must be conjugate. Let us regard the line ($-\infty$ to 0) as a barrier which z is not allowed to cross, having a positive upper side and a negative lower side; then we may write the two elements at x_1 as $\log \overset{+}{x}_1$ and $\log \overset{-}{x}_1$, and what is proved is that these are conjugate. Let the paths C, C' be continued along the barrier by means of points $x_2, x_3, \cdots$, such that $\overset{+}{x}_{\kappa+1}$ lies within the domain of $\overset{+}{x}_\kappa$, and $\overset{-}{x}_{\kappa+1}$ within that of $\overset{-}{x}_\kappa$. Then, as before, $\log \overset{+}{x}_\kappa$ and $\log \overset{-}{x}_\kappa$ are conjugate. But

$$\log \overset{+}{x}_2 = \log \overset{+}{x}_1 + \frac{x_2 - x_1}{x_1} - \frac{1}{2}\left(\frac{x_2 - x_1}{x_1}\right)^2 + \frac{1}{3}\left(\frac{x_2 - x_1}{x_1}\right)^3 - \cdots,$$

and $\log \overset{-}{x}_2 = \log \overset{-}{x}_1 +$ the same series.

Therefore

$$\log \overset{+}{x}_2 - \log \overset{-}{x}_2 = \log \overset{+}{x}_1 - \log \overset{-}{x}_1,$$

and by the same argument

$$\log \overset{+}{x}_\kappa - \log \overset{-}{x}_\kappa = \log \overset{+}{x}_1 - \log \overset{-}{x}_1.$$

Therefore at opposite points of the barrier the values of the function differ by a constant, and further, since the values are conjugate, this constant is a pure imaginary.

To determine the constant, imagine $m-1$ points interpolated between the points 1 and -1, these points lying on a semicircle in the upper half of the plane, and forming with either of the end-points 1 and -1 the half of a regular $2m$-gon. If these points be $z_1, z_2, \cdots, z_{m-1}$, we have

$$z_1 - 1 = (z_2 - z_1)/z_1 = (z_3 - z_2)/z_2 = \cdots,$$

and when $m > 3$, each point lies within the circle of convergence of the preceding point. Therefore if $\log z_1 = P_1(z_1 - 1)$, we have

$$\begin{aligned} \log z_2 &= \log z_1 + P_1\left(\frac{z_2 - z_1}{z_1}\right) \\ &= \log z_1 + P_1(z_1 - 1) \\ &= 2P_1(z_1 - 1), \\ \log z_3 &= \log z_2 + P_1\left(\frac{z_3 - z_2}{z_2}\right) \\ &= 3P_1(z_1 - 1), \end{aligned}$$

and finally $\log(\overset{+}{-1}) = mP_1(z_1 - 1)$.

By taking the integer m large enough, the absolute value of $m(z_1 - 1)$, or the perimeter of the half polygon, is as nearly equal as we please to the length of the semicircle; while the amplitude of $m(z_1 - 1)$ is, as nearly as we please, a right angle. That is,

$$\lim_{m=\infty} m(z_1 - 1) = \pi i.$$

The limit of the ratio of any other term of the series $mP_1(z_1 - 1)$ to the first term is evidently zero; and we obtain

$$\log(\overset{+}{-1}) = \pi i.$$

Similarly $$\log(\overset{-}{-1}) = -\pi i,$$

and therefore the constant difference of $\log \overset{+}{x} - \log \overset{-}{x}$, where x is any negative real number, is $2\pi i$.

It follows that if we allow z to cross the barrier and return to the point 1 after a positive description of the contour $C + C'$, the value of the function at 1 is $2\pi i$, and the value near 1 is $2\pi i + P_1(z - 1)$.

To remove the restriction that $C+C'$ must be symmetric with regard to the real axis, we shall show that two curves L, M, which pass from 1 to z', without including the point 0, lead to the same series at z'. For let the dotted curve be at a finite but arbitrarily small distance from L. Let c_1, c_1' lie within the circle whose centre is 1; c_1', c_2', c_2 within the circle (c_2); and so on.

Also let the arcs $c_{\kappa-1}c'_{\kappa-1}$, $c'_{\kappa-1}c'_{\kappa}$, $c'_{\kappa}c_{\kappa}$, $c_{\kappa}c_{\kappa-1}$ lie within the circle (c_{κ}),

$$(\kappa=0, 1, 2, \cdots;\; c_0=1, c_0'=1).$$

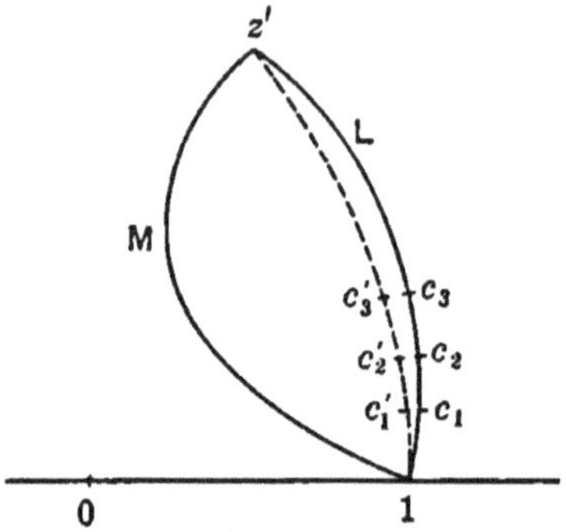

Fig. 26

Then
$$P(z\,|\,1\,|\,c_1) \quad = P(z\,|\,1\,|\,c_1'\,|\,c_2')$$
$$P(z\,|\,1\,|\,c_1\,|\,c_2) = P(z\,|\,1\,|\,c_1'\,|\,c_1\,|\,c_1'\,|\,c_2'\,|\,c_2)$$
$$= P(z\,|\,1\,|\,c_1'\,|\,c_2'\,|\,c_2),$$

and so on, the final equation being

$$P(z\,|\,1\,|\,c_1\,|\,c_2\,|\cdots|\,z') = P(z\,|\,1\,|\,c_1'\,|\,c_2'\,|\cdots|\,z').$$

The interpolation of additional points c on the curve L will not affect the final value; and the same remark applies to the dotted curve. By drawing, in this manner, a succession of dotted curves we can replace L by M, without affecting the value of the element at z_1. That is, the curve $L+M$ reproduces the initial element at 1. The results at which we have arrived may be summed up as follows:—

When the line from $-\infty$ to 0 is regarded as a barrier over which no curve can pass, the branches of $\log z$ are separate one-valued functions, and the values of any branch which may be selected differ at opposite points of the barrier by $2\pi i$. When curves are allowed to cross the barrier the branches pass continuously into one another; for instance, if a closed curve, which starts from 1, cross the barrier λ times from the positive half-plane to the negative half-plane and λ' times in the reverse direction, the tth branch, whose element at 1 is

$$2t\pi i + (z-1) - \tfrac{1}{2}(z-1)^2 + \tfrac{1}{3}(z-1)^3 - \cdots,$$

passes into the $(t+\lambda-\lambda')$th branch, whose element at 1 is

$$2(t+\lambda-\lambda')\pi i + (z-1) - \tfrac{1}{2}(z-1)^2 + \tfrac{1}{3}(z-1)^3 \cdots.$$

[See Weierstrass, Sur la theorie des fonctions elliptiques, Acta. Math. t. vi., p. 184; Thomae, Theorie der analytischen Functionen, p. 70.]

§ 107. *The Generalized Power.*

When both a and z are complex, the nth branch of a^z is defined as $\exp\,(z\cdot{}_n\log a)$, ${}_n\log a$ being the nth branch of $\log a$. If $z=x+iy$, $a=\rho\exp i\theta$, $-\pi<\theta\leqq\pi$, and if ${}_n\theta$ denote $\theta+2n\pi$,

$$\exp\,(z\cdot{}_n\log a)=\exp\,\{(x+iy)(\log\rho+i\cdot{}_n\theta)\}$$
$$=\exp\,(x\log\rho-y\cdot{}_n\theta)\{\cos\,(y\log\rho+x\cdot{}_n\theta)+i\sin\,(y\log\rho+x\cdot{}_n\theta)\}.$$

The totality of such branches, when n takes all integer values, is the generalized power a^z. The nth branch may be denoted by ${}_na^z$; when $n=0$ we have the chief branch.

It is evident that ${}_na^z\cdot{}_na^{z'}={}_na^{z+z'}$; but ${}_na^z\cdot{}_ma^{z'}$ when n and m are unequal integers is not equal to any branch of $a^{z+z'}$. Therefore the equation $a^z\cdot a^{z'}=a^{z+z'}$ is inexact.

Example. The equation $a^z\cdot b^z=(ab)^z$ holds; but the equation $(a^z)^{z'}=a^{zz'}$ is inexact (Stolz, Allgemeine Arithmetik).

References. On the general subject of this chapter the reader may consult: Biermann, Theorie der analytischen Functionen; Cauchy, Cours d'Analyse algébrique; Chrystal, Algebra, t. ii.; Hobson, Trigonometry; Méray, Nouveau Précis d'Analyse infinitésimale; Stolz, Allgemeine Arithmetik; Tannery, Théorie des Fonctions d'une Variable; Thomae, Theorie der analytischen Functionen; Weierstrass, Abhandlungen aus der Functionenlehre.

CHAPTER IV.

Algebraic Functions.

§ 108. In the present chapter we propose to discuss in some detail the expansions for the branches of an algebraic function, to sketch briefly Cauchy's theory of loops, and to explain Lüroth's system of grouping of the loops, as a preparation for the study of Riemann surfaces for algebraic functions.

Let $F(w^n, z^m)=0$ be an algebraic equation of orders n and m in w and z. We suppose the polynomial to be irreducible, for if $F(w, z)$ be expressible as a product of κ irreducible polynomials $F_1(w, z)$, $F_2(w, z), \cdots, F_\kappa(w, z)$, the reducible equation $F(w, z)=0$ can be replaced by κ irreducible equations $F_\iota=0$, $(\iota=1, 2, \cdots, \kappa)$.

Arranging $F=0$ according to descending powers of w, the equation takes the form

$$w^n f_0(z)+w^{n-1}f_1(z)+w^{n-2}f_2(z)+\cdots+f_n(z)=0,$$

in which the functions $f_0(z). f_1(z), \cdots, f_n(z)$ are integral polynomials in z of degrees not higher than m. To a given value of z correspond in all cases n branches of w, namely, $w_1, w_2, \cdots, w_n$, which are, in general, finite and distinct. Those points of the z-plane at which two or more branches are equal, or at which one or more branches are infinite, are named *critical* points. The finite points z, for which two or more values of w are equal and finite, are found by eliminating w from

$$F=0,\ \delta F/\delta w=0.$$

The points at which a branch is ∞ are given by $f_0(z)=0$. This equation is to be regarded as of the mth degree; if it appear to be of the degree m_1, it has $m-m_1$ infinite roots. A value of z which satisfies $f_0(z)=0$, but not $f_1(z)=0$, is a pole; a value which makes not only $f_0(z)$, but also $f_1(z), f_2(z), \cdots, f_\kappa(z)$ is an infinity at which $\kappa+1$ branches are equal.

The nature of the point $z=\infty$ can be determined by arranging the equation $F=0$ in powers of z and equating to zero the coefficient of z^m. We thus obtain an equation in w which is to be regarded as of degree n, having $n - n_1$ infinite roots, when the highest power of w therein is w^{n_1}. If this equation have equal roots, or infinite roots, $z = \infty$ is a critical point. Evidently $z = \infty$ is a critical point when the polynomial F does not contain the term $w^n z^m$.

It is worth remarking that the view taken of the equation $F = 0$ is different from that to which we are accustomed in Cartesian Geometry. Here we ask how many w's there are when z is assigned, and how many z's when w is assigned; in a word, it is the correspondence of w and z which primarily interests us. For example, take the equation $w^2z^2 + w^2 + z^2 - 3 = 0$. In Cartesian Geometry it would be said that this is a curve which is met by any line in four points, and therefore, when $z = 1$, the values of w are ± 1, ∞, ∞. But the present point of view is that we have a (2, 2) correspondence *always*, and that when $z = 1$ the values of w are ± 1.

§ 109. *Continuity of the branches.*

An immediate deduction from Weierstrass's theorem of § 88 can be made by writing $n = 1$; in this way we see that if μ roots of $F(w, z) = 0$ be equal to b when $z = a$, the equation has μ roots nearly equal to b when z is nearly equal to a. For write $z = a + z'$, $w = b + w'$, and suppress accents; when z is put equal to 0,

$$F(w, 0) = w^{\mu} \cdot R(w),$$

where $R(w)$ is an integral polynomial, which may reduce to its constant term. Hence

$$0 = w^{\mu} R(w) + (\text{a polynomial in } z,\, w, \text{ which vanishes when } z = 0).$$

By the reasoning of § 88 this equation is satisfied, in the immediate vicinity of $z = 0$, by μ values of w which are nearly 0.

Expansions of the branches at ordinary points. In the terminating or absolutely convergent infinite double series

$$F(w, z) = c_{10}z + c_{01}w + \frac{1}{2\,!}(c_{20}z^2 + 2\,c_{11}zw + c_{02}w^2) + \cdots,$$

assume $c_{01} \neq 0$. It is evident that $F(w, 0)$ is a series which starts with $c_{01}w$. Hence $\mu = 1$, and the equation

$$w^{\mu} + f_1 w^{\mu-1} + f_2 w^{\mu-2} + \cdots + f_{\mu} = 0$$

reduces to $w+f_1=0$, and, therefore, one and only one of the branches of w is equal to an integral series in z which has no term independent of z; for $f_1=-s_1$, and s_1 contains no term independent of z. This deduction is of great importance, for it establishes the existence, when $c_{01}\neq 0$, of a *single integral* series of the form

$$w=P_r(z) \quad (r>0).$$

Hence if a single branch w of the algebraic function given by $F(w^n, z^m)=0$ be equal to b when $z=a$, there is only one expansion of the form $w-b=P_r(z-a)$, $(r>0)$. If all the n roots $b_1, b_2, \cdots, b_n$ be distinct when $z=a$, there are n integral series

$$w-b_\kappa=P_r^{(\kappa)}(z-a) \quad (r>0,\ \kappa=1, 2, \cdots, n).$$

§ 110. *Discussion of the integral series $P_r(z)$.*

(i.) Let $r=1$, and let the series be written

$$w=a_1z+a_2z^2+a_3z^3+\cdots.$$

In the *immediate* neighbourhood of $z=0$, $w=a_1z$ approximately; and the z-plane near $z=0$ is conform with the w-plane near $w=0$.

(ii.) Let $r>1$. $\quad w=a_rz^r+a_{r+1}z^{r+1}+\cdots;$

when z describes a small circle round the z-origin, the approximate equation $w=a_rz^r$ shows that w makes r turns round the w-origin; in both cases (i. and ii.) the final and initial values of the series are the same.

§ 111. *The behaviour of a fractional series in the neighbourhood of a branch-point.*

Let w be a function, of which r values at a point z near $z=0$ are defined by the fractional series

$$w_{\kappa+1}=c_1(a^\kappa z^{1/r})^{q_1}+c_2(a^\kappa z^{1/r})^{q_2}+c_3(a^\kappa z^{1/r})^{q_3}+\cdots,$$

where $q_1, q_2, \cdots, r$ are integers,

$$q_1<q_2<q_3<\cdots,\ a=e^{2\pi i/r},\ \kappa=0,1,2, \cdots, r-1,$$

$z^{1/r}$ is the same for each term, and one at least of the exponents $q_1/r, q_2/r, \cdots$ is in its lowest terms. When z describes a small circle round the origin, $z^{1/r}$ changes into $az^{1/r}$, and the series for $w_1, w_2, \cdots, w_r$ change into the series for $w_2, w_3, \cdots, w_1$. Thus the r values $w_1, w_2, \cdots, w_r$

permute *cyclically* round $z=0$. When the description of a circle round a critical point permutes some of the branches, we shall call that critical point a *branch-point*. When a branch-point permutes only two branches it will be called a *simple* branch-point.

§ 112. The theory of algebraic expansions is purely analytic, depending merely on the nature of $F(w, z)=0$; but the understanding of what follows will be facilitated by the use of the language of Cartesian Geometry. By speaking of z and w as co-ordinates, it becomes possible to demonstrate the parallelism between the expansion-problem in the Theory of Functions and the problem of the resolution of higher singularities in the Theory of Higher Plane Curves. It is important, however, that the reader should bear in mind that ∞ is a *point* in the Theory of Functions, whereas in projective geometry all points at an infinite distance are regarded as situate upon a line.

The investigation resolves itself into an examination of the points of intersection of the curve $F(w^n, z^m)=0$ with a line $z=a$. When this line intersects the curve in n distinct points, which lie within the finite part of the plane, the n expansions for the n branches are all integral. When the line touches the curve or passes through a multiple point, there are still n expansions which may be integral or fractional according to circumstances. A finite pair of values which satisfies both $F=0$ and $\delta F/\delta w=0$ is in the Cartesian theory a point such that the line drawn through it parallel to the axis of w touches the curve or passes through a multiple point. In the parabola $w^2=z$, the critical point $z=0$ is a branch-point, for the tangent at $(w=0, z=0)$ has simple contact. The expansions $z^{1/2}$, $-z^{1/2}$ are terminating fractional series. In the curve $w^3=z$, the inflexional tangent at the origin yields a two-fold branch-point with a cyclic system $(w_1 w_2 w_3)$. This two-fold branch-point can be resolved into two coincident branch-points (see Fig. 27). More generally the tangent to $w^{\kappa+1}=z$ at the origin is vertical and yields a κ-fold branch-point, which can be resolved into κ simple branch-points. To examine the behaviour of the curve $F=0$ at the point $(z=a, w=b)$, it is customary to transform the origin to the

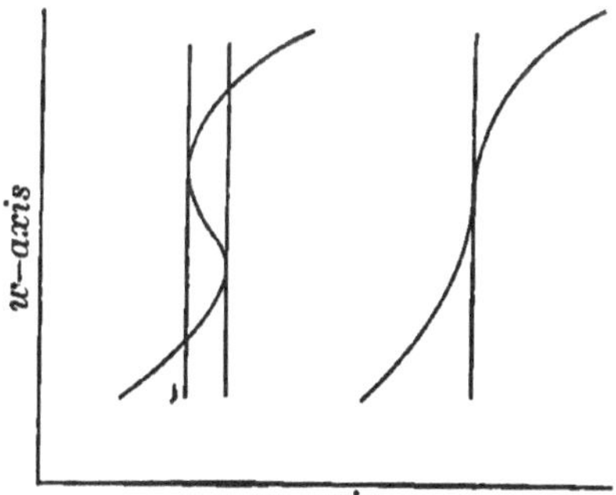

Fig. 27

point (a, b) by writing $z = z' + a$, $w = w' + b$, and thus change $F(w, z) = 0$ into $F'(w', z') = 0$, an equation free from a constant term. The problem of tracing the curve $F'(w', z') = 0$ at the new origin is equivalent to the examination of those values of w' which become infinitely small with z'. Omitting accents, it is evident that there is no loss of generality in considering the point $(0, 0)$ on the new curve instead of (a, b) on the original curve. The passage back to the original form is effected by writing $z - a$, $w - b$, for z, w.

§ 113. When the form of an algebraic curve near the origin is to be determined, the usual plan is to arrange the equation in the form

$$(w, z)_1 + (w, z)_2 + (w, z)_3 + \cdots = 0,$$

where the number of terms on the left-hand side is finite, and $(w, z)_r$ denotes the terms of the rth order in w, z combined. Let the equation be

$$c_{10}z + c_{01}w + \frac{1}{2!}(c_{20}z^2 + 2c_{11}zw + c_{02}w^2)$$
$$+ \frac{1}{3!}(c_{30}z^3 + 3c_{21}z^2w + 3c_{12}zw^2 + c_{03}w^3) + \cdots,$$

so that c_{fg} is the value of $\dfrac{\delta^{f+g}F(w, z)}{\delta z^f \delta w^g}$, when $z = w = 0$. When the lowest terms are of the order k ($k \geqq 1$), the origin is called a k-element. Preliminary to the discussion of the most general case, we shall examine the expansions when the origin is a 1-element, or a 2-element with no tangent parallel to the axes.

The 1-element.

(i.) Suppose $c_{10}, c_{01} \neq 0$. There is, by Weierstrass's theorem, one and only one expansion of the form $w = P_1(z)$.

(ii.) Next let $c_{10} = c_{20} = \cdots = c_{r-1,0} = 0$, $c_{r0}, c_{01} \neq 0$; then there is still a unique solution in the form of an integral series without a constant term; as the lowest power of z is z^r, this series is $w = P_r(z)$, which gives on reversion $z = P_1(w^{1/r})$.

(iii.) Finally, let $c_{10} \neq 0$, $c_{01} = c_{02} = \cdots = c_{0,k-1} = 0$, $c_{0k} \neq 0$; then, by reasoning similar to that just used,

$$z = P_k(w) \text{ and } w = P_1(z^{1/k}).$$

The geometric interpretations of these three cases are immediate:—

In (i.) the origin is an ordinary point with a tangent oblique to the axes; whereas in (ii.), (iii.), the tangent meets the curve in r or h consecutive points, and is parallel to the z-axis or w-axis. The origin yields in case (iii.) an $(h-1)$-fold branch-point, which is equivalent to $h-1$ simple branch-points. In case (ii.) the origin is an $(r-1)$-fold branch-point when z is regarded as a function of w.

The 2-element.

Here $c_{01}=c_{10}=0$. The equation of the curve is

$$0=\frac{1}{2\,!}(c_{20}z^2+2\,c_{11}zw+c_{02}w^2)+(w,\,z)_3+\cdots.$$

The nature of the singularity is dependent upon the factors of the terms

$$c_{20}z^2+2\,c_{11}zw+c_{02}w^2.$$

(i.) Let $\quad c_{20}z^2+2\,c_{11}zw+c_{02}w^2=c_{02}(w-\alpha z)\,(w-\beta z),$

where $c_{02},\ \alpha,\ \beta\neq 0$. The origin is a node with tangents oblique to the axes of z and w. When $w-\alpha z$ is put $=w_1z$ and the equation, after division by z^2, is re-arranged according to ascending powers of w_1 and z, the origin becomes a 1-element with a term w_1. Hence, if w' be the series connected with the factor $w-\alpha z$,

$$w_1=P_r(z)\;(r\geqq 1),$$

and $w'=\alpha z+P_{r+1}(z)$. Similarly the factor $w-\beta z$ contributes an integral series w'' with the first term βz. Analytically the two series w', w'' at the node are unaffected by the description of a small circle round the z-origin. Thus the node, with tangents oblique to the axes, is a critical point at which two roots become equal, but not a branch-point at which two roots permute. Geometrically it may be regarded as equivalent to a point at which two simple branch-points unite and destroy each other.*

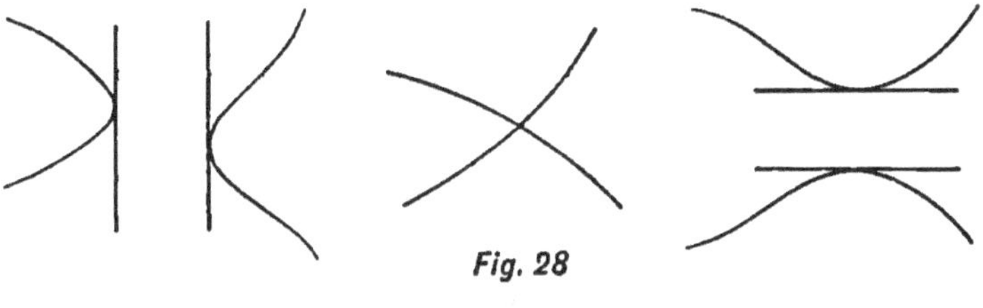

Fig. 28

(ii.) *The simple cusp, with a tangent inclined to the axes.*

Let $\beta=\alpha\neq 0$; and let the factors of $(w,\,z)_3$ be

$A(w-\lambda z)(w-\mu z)(w-\nu z)$ where $\lambda-\alpha,\ \mu-\alpha,\ \nu-\alpha\neq 0.$

* The node may equally be regarded as the union of two branch-points with regard to w.

The transformation $w - az = w_1 z$ leads, after division by z^2, to an equation in w_1, z, with a 1-element; the lowest power of z is the first, and the lowest power of w_1 the second; hence

$$w_1 = P_1(z^{1/2}) \text{ and } w - az = P_3(z^{1/2}).$$

Since $z^{1/2}$ is a two-valued function, two branches of w are represented by this series, and these branches permute, when z describes a small circle round the origin. Thus an ordinary cusp yields two fractional series in z, which are integral in $z^{1/2}$. Analytically the corresponding critical point behaves as a combination of a node and a branch-point. Geometrically the simple cusp, with a tangent oblique to the axes, has for its penultimate form a vanishing loop, and the branch-point corresponds to the vertical tangent to the loop.

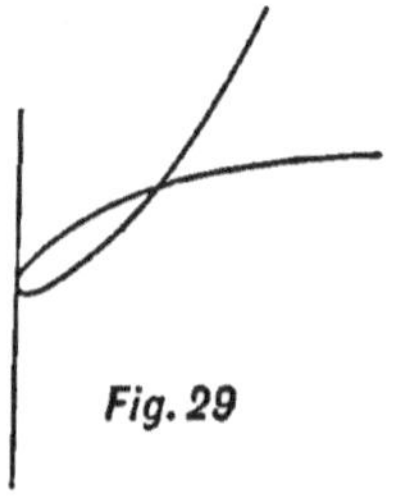

Fig. 29

Without stopping to examine the more complicated cases which may arise in the case of the 2-element, we pass on to some geometric considerations which will be of help in the treatment of the problem of expansions.

The Cartesian Problem.

§ 114. The resolution of a higher singularity of $F(w, z) = 0$ into simple singularities, namely, ν nodes, κ cusps, τ double tangents, ι inflexions, where ν, κ, τ, ι are integers which satisfy Plücker's equations, was first effected by Professor Cayley (Quarterly Journal of Mathematics, t. vii.), but the proof, which rested upon the nature of the expansions and the properties of the discriminant, was merely outlined. Afterwards Nöther, Stolz, H. J. S. Smith, Halphen, Brill, and others completed the theory. The determination of these numbers suggests the existence of a penultimate form in which these δ nodes, κ cusps, etc., are to be found separate from, but in infinitely close proximity to, one another, the actual form being derivable from the penultimate by a continuous deformation. Processes by which this can be effected have been described by Halphen and Brill (Math. Ann., t. xvi.); see the memoirs by Halphen, Mémoire sur les points singuliers des courbes algébriques planes, (t. xxvi. of the Mémoires présentés à l'Académie des Sciences); and Nöther (Math. Ann., t. xxiii.). These we shall not discuss, as all that we require is the expansions at the point, as determined from the given equation.

We shall have occasion to discuss two distinct methods for the determination of the expansions at a multiple point: one of these was used by Puiseux, the other by Hamburger and Nöther. The principle which underlies Nöther's process is the determination of the nature of a higher singularity on a given curve, by the employment of a series of Cremona transformations. His inference was that any higher singularity can be regarded as composed of one or more ordinary multiple points at which the tangents are all separate, combined usually with branch-points. The transformation used was the special quadric inversion with coincidence of two of the principal points, namely $w = w_1 z$. This arises from

$$\xi : \eta : \zeta = \xi'\zeta' : \xi'\eta' : \zeta'^2,$$

by writing

$$\xi = z, \quad \eta = w, \quad \zeta = 1,$$

$$\xi' = z, \quad \eta' = w_1, \quad \zeta' = 1.$$

No loss of generality results from the special form of the transformation, for every Cremona transformation is reducible to a succession of quadric inversions. It is not possible to replace, in this way, a curve with higher singularities by another with nodes only, but the penultimate form of a higher singularity with k separate tangents, not parallel to the axes, is easily conceivable. The singularity in question results from the union of $\frac{1}{2}k(k-1)$ nodes.

The principle that an equation $F(w, z) = 0$, with higher singularities at which the tangents are not separated, can be replaced, by means of Cremona transformations, by a new equation which has only ordinary multiple points with separated tangents, is useful for many purposes of the theory of algebraic functions. The theory of such transformations is given in Salmon, Higher Plane Curves, and in Clebsch, Vorlesungen über Geometrie.

Nöther's Method.

§ 115. We resume the discussion of the expansions in the neighbourhood of the non-critical and critical points of the algebraic function defined by the irreducible equation $F(w^n, z^m) = 0$. It has been explained already that no generality is lost by supposing the point (a, b), at which the expansions are required to be the origin; at the end w, z can be replaced by $w - b$, $z - a$, it being understood that $z - \infty$, $w - \infty$ are equivalent to $1/z$, $1/w$. It has been proved that, when $c_{01} \neq 0$, a root w, which vanishes when $z = 0$, can be expressed as an integral series

$$w = w'z + w''z^2/2! + w'''z^3/3! + \cdots,$$

the series having a finite region of convergence. The next question is to determine the coefficients w', w'', w''', …. These coefficients are the values of the derived functions dw/dz, d^2w/dz^2, … at (0, 0); and the problem is merely that of expressing the derived functions of w, when given by the implicit equation $F(w, z) = 0$, in terms of the partial derived functions. We may substitute for w the infinite series $w'z + w''z^2/2! + \cdots$, and equate the coefficients of the powers of z; or we may proceed as follows:—

Let w, z be functions of t, and let $F = F(w, z) = 0$. If accents denote differentiations with regard to t,

$$F' = w' \frac{\delta F}{dw} + z' \frac{\delta F}{\delta z} = D_1 F \text{ suppose.}$$

To find $\frac{d}{dt}(D_1 F)$ we observe that, while $\delta F/\delta w$ and $\delta F/\delta z$ are functions of w and z, w' and z' are functions of t, and that now

$$\frac{d}{dt} = w' \frac{\delta}{\delta w} + z' \frac{\delta}{\delta z} + \frac{\delta}{\delta t} = D_1 + \frac{\delta}{\delta t}.$$

Let $$\frac{\delta D_1}{\delta t} = w'' \frac{\delta}{\delta w} + z'' \frac{\delta}{\delta z} = D_2,$$

$$\frac{\delta D_2}{\delta t} = D_3, \text{ etc.}$$

Then $$F'' = \left(D_1 + \frac{\delta}{\delta t}\right) D_1 F = (D_1^2 + D_2) F,$$

$$F''' = \left(D_1 + \frac{\delta}{\delta t}\right) F'' = D_1 F'' + (2 D_1 D_2 + D_3) F,$$

$$= (D_1^3 + 3 D_1 D_2 + D_3) F.$$

$$F'''' = D_1 F''' + (3 D_1^2 D_2 + 3 D_2^2 + 3 D_1 D_3 + D_4) F,$$

$$= (D_1^4 + 6 D_1^2 D_2 + 4 D_1 D_3 + 3 D_2^2 + D_4) F.^*$$

Put $t = z$; then $z' = 1$, $z^{(\kappa)} = 0$ $(\kappa > 1)$,

and $$D_1 = \frac{\delta}{\delta z} + w' \frac{\delta}{\delta w}, \quad D_2 = w'' \frac{\delta}{\delta w}, \text{ etc.}$$

---

* The law of these expressions is that in $F^{(r)}$ we first find all positive integral solutions of

$$\sigma_1 s_1 + \sigma_2 s_2 + \cdots = r,$$

that is, all the unrestricted partitions for r; and then for each solution we form

$$\frac{r!}{\sigma_1! \, \sigma_2! \ldots (s_1!)^{\sigma_1} (s_2!)^{\sigma_2} \ldots} D_{s_1}^{\sigma_1} D_{s_2}^{\sigma_2} \ldots F,$$

and add the results.

In applying these formulæ of differentiation to our case, it is convenient to write $w = v_1 z$, and divide by z, so that $F = 0$ becomes

$$0 = c_{10} + c_{01}v_1 + \frac{1}{2\,!}z(c_{20} + 2\,c_{11}v_1 + c_{02}v_1^2) + \cdots,$$

or
$$0 = \phi_1 + z\phi_2 + \frac{z^2}{2\,!}\phi_3 + \cdots + \frac{z^g}{g!}\phi_{g+1},$$

g being an integer not greater than $m + n - 1$, and ϕ_r being a polynomial in v_1, of degree $\not> r$. The operator D_1 is now $v'_1\frac{\delta}{\delta v_1} + \frac{\delta}{\delta z}$; it may be replaced by $v'_1\frac{\delta}{\delta v_1} + E$, where E is an operator which changes ϕ_r into ϕ_{r+1} (where if $r > g$, ϕ_{r+1} must be zero), for

$$\frac{\delta F}{\delta z} = \phi_2 + z\phi_3 + \frac{z^2}{2\,!}\phi_4 + \cdots$$
$$= EF.$$

We have now only operations in which z does not occur, and we may put $z = 0$ before effecting the differentiations. Let $z = 0$, and let the values of v_1, v'_1, ⋯ then become v, v', ⋯. The equations $F^{(r)} = 0$ become

$$\phi_1 = 0,$$
$$D_1\,\phi_1 = 0,$$
$$(D_1^2 + D_2)\phi_1 = 0,$$
$$(D_1^3 + 3\,D_1D_2 + D_3)\phi_1 = 0,$$
$$(D_1^4 + 6\,D_1^2D_2 + 4\,D_1D_3 + 3\,D_2^2 + D_4)\phi_1 = 0,$$
$$\cdots\cdots\cdots\cdots$$

In these equations E must operate before $\frac{\delta}{\delta v}$. The two are not always commutative; for instance,

$$E\frac{\delta^2}{\delta v^2}\phi_1 = 0, \text{ while } \frac{\delta^2}{\delta v^2}E\,\phi_1 = \frac{\delta^2\phi_2}{\delta v^2}.$$

Further,

$$D_1^r\phi_1 = \phi_{r+1} + rv'\frac{\delta\phi_r}{\delta v} + \frac{r(r-1)}{2\,!}\,v'^2\,\frac{\delta^2\phi_{r+1}}{\delta v^2} + \cdots$$
$$+ \frac{r\,!}{s\,!\,\overline{r-s}\,!}\,v'^s\frac{\delta^s\phi_{r+1-s}}{\delta v^s},$$

where $s \not> r+1-s$, or $2s \not> r+1$. We have, accordingly,

$$
\text{I.}\left\{
\begin{array}{r}
\phi_1 = 0, \\
\phi_2 + v' \dfrac{\delta \phi_1}{\delta v} = 0, \\
\phi_3 + 2v' \dfrac{\delta \phi_2}{\delta v} + v'' \dfrac{\delta \phi_1}{\delta v} = 0, \\
\phi_4 + 3v' \dfrac{\delta \phi_3}{\delta v} + 3v'^2 \dfrac{\delta^2 \phi_2}{\delta v^2} + 3v'' \dfrac{\delta \phi_2}{\delta v} + v''' \dfrac{\delta \phi_1}{\delta v} = 0, \\
\phi_5 + 4v' \dfrac{\delta \phi_4}{\delta v} + 6v'^2 \dfrac{\delta^2 \phi_3}{\delta v^2} + 6v'' \left(\dfrac{\delta \phi_3}{\delta v} + 2v' \dfrac{\delta^2 \phi_2}{\delta v^2} \right) + 4v''' \dfrac{\delta \phi_2}{\delta v} + v'''' \dfrac{\delta \phi_1}{\delta v} = 0, \\
\cdots\cdots\cdots\cdots\cdots\cdots\cdots\cdots\cdots\cdots\cdots\cdots.
\end{array}
\right.
$$

It will be useful to write down the analogous results when the origin is a k-element; the equation $F=0$ becomes, after the transformation $w=vz$,

$$0 = \phi_k + z\phi_{k+1} + \frac{z^2}{2!}\phi_{k+2} + \cdots,$$

the suffix denoting in each case the degree in v. The equations are

$$
\text{II.}\left\{
\begin{array}{r}
\phi_k = 0, \\
\phi_{k+1} + v' \dfrac{\delta \phi_k}{\delta v} = 0, \\
\phi_{k+2} + 2v' \dfrac{\delta \phi_{k+1}}{\delta v} + v'^2 \dfrac{\delta^2 \phi_k}{\delta v^2} + v'' \dfrac{\delta \phi_k}{\delta v} = 0, \\
\phi_{k+3} + 3v' \dfrac{\delta \phi_{k+2}}{\delta v} + 3v'^2 \dfrac{\delta^2 \phi_{k+1}}{\delta v^2} + v'^3 \dfrac{\delta^3 \phi_k}{\delta v^3} \\
+ 3v'' \left(\dfrac{\delta \phi_{k+1}}{\delta v} + v' \dfrac{\delta^2 \phi_k}{\delta v^2} \right) + v''' \dfrac{\delta \phi_k}{\delta v} = 0, \\
\cdots\cdots\cdots\cdots\cdots\cdots\cdots\cdots\cdots.
\end{array}
\right.
$$

In the system of equations I., $\dfrac{\delta \phi_1}{\delta v}$ or $c_{01} \neq 0$. Hence the equations give in succession $v, v', v'', \cdots$ as rational fractional functions of the c's. The value of v_1, when z is not zero, is

$$v_1 = v + zv' + \frac{z^2}{2!}v'' + \cdots,$$

this series being legitimate near $z=0$, since we know that there is an integral series $P_\lambda(z)$ for w or $v_1 z$. Finally, since

$$w = zv_1 = zv + z^2 v' + \frac{z^3}{2!} v'' + \cdots,$$

the coefficients in the series for the branch of w are determined.*

§ 116. To return to the expansions at a k-element, we have now to discuss in detail the behaviour of the branches at such a point. If $k=1$, and if $c_{10} \neq 0$, $c_{0h} \neq 0$, $c_{01} = c_{02} = \cdots = c_{0,h-1} = 0$, we have proved that there is an expansion $z = P_h(w)$ and have shown how to find the coefficients $\frac{1}{h!}\left(\frac{d^h z}{dw^h}\right)$, $\frac{1}{h+1!}\frac{d^{h+1}z}{dw^{h+1}}$, etc. On reversion, this series gives a cyclic system of h series,

$$w = P_1(a^r z^{1/h}), \text{ where } a = e^{2\pi i/h} \text{ and } r = 0, 1, 2, \cdots, h-1.$$

For instance, if $w^3 = z + zw$, the roots in the neighbourhood of the origin are

$$w = az^{1/3} + \frac{a^2}{3} z^{2/3} - \frac{a}{81} z^{4/3} \cdots,$$

a being any cube root of unity. When $k > 1$, all the partial derived functions of the orders $1, 2, \cdots, k-1$ vanish, and the terms of the lowest order are

$$\frac{1}{k!}(c_{k0} z^k + c_{k-1 0} z^{k-1} w + \cdots + c_{0k} w^k), \text{ or } (w, z)_k.$$

With each factor of $(w, z)_k$ are associated one or more expansions for w in terms of z. Each factor must be examined in turn.

The simple factors of $(w, z)_k$.

(i.) Let $w - \beta_1 z$ be a simple factor such that $\beta_1 \neq 0$. Write $w - \beta_2 z = w_1 z$ in the equation to the curve and, after removal of the factor z^k, arrange the resulting equation according to ascending powers of w_1 and z. The origin of the new curve is a 1-element, in which w_1 occurs to the exponent 1. Hence there exists a *single* integral series for w_1 of the form $P_r(z)$, $(r > 0)$; *i.e.*

$$w = \beta_1 z + w_1 z = \beta_1 z + P_{r+1}(z) = P_1(z).$$

(ii.) Let w be itself a simple factor; *i.e.* let $c_{k0} = 0$, $c_{k-1,1} \neq 0$. The only difference from the preceding case arises from the circumstance that $\beta_1 = 0$. Thus $w = P_r(z)$, $(r > 1)$.

---

* See Plücker, Theorie der algebraischen Curven, p. 156. See also Stolz's memoir, Ueber die singulären Punkte der algebraischen Functionen und Curven, Math. Ann., t. viii., p. 415. The quadric transformation $w = v_1 z$ was first used by Cramer in connexion with the resolution of higher singularities (Analyse des Lignes Courbes).

(iii.) Let z be a simple factor of $(w, z)_k$; *i.e.* let $c_{0k}=0$, $c_{1,k-1}\neq 0$. The transformation to be used in this case is $z=z_1w$, called by Nöther a transformation of the second kind, the one previously used being a transformation of the first kind. After division by w^k, the resulting equation in z_1 and w has a 1-element at the origin, and $(z_1, w)_1$ reduces to z_1+aw, where a may be zero.

If $a\neq 0$, $z=z_1w=P_2(w)$, and reversion gives $w=P_1(z^{1/2})$. In this case z is not a factor of $(w, z)_{k+1}$. But if z be a factor in each of the groups $(w, z)_{k+1}$, $(w, z)_{k+2}$, $\cdots$, $(w, z)_{k+q-1}$, and not in $(w, z)_{k+q}$, then $z_1=P_q(w)$ and $z=z_1w=P_{q+1}(w)$. On reversion this gives $w=P_1(z^{1/(q+1)})$, a cyclic system of $q+1$ expansions. It will be noticed that the first power of w, which is not multiplied by z, is w^{k+q} and that the difference between the exponent $k+q$ and the order k of the element is precisely q; also that from the point of view of the Theory of Functions there is a q-fold branch-point associated with the cyclic system of $q+1$ expansions. Combining the preceding results, we see that to each simple factor $w-\beta_1 z$ of $(w, z)_k$ there corresponds one and only one integral series,

$$w-\beta_1 z=P_r(z),\quad (r>1);$$

and that to a simple factor z correspond $q+1$ series, where q is the difference between the exponent of the lowest separate power of w and the order of the element.

The Multiple Factors of $(w, z)_k$.

§ 117. We have now to examine the expansions connected with the multiple factors of $(w, z)_k$. Let $(w-\beta_1 z)^l$ be a multiple factor of these terms, and for simplicity assume $\beta_1\neq 0$. Apply the transformation of the first kind $w-\beta_1 z=w_1 z$, and let the transformed equation be divided by z^k and arranged according to ascending powers of w_1 and z. The lowest power of w_1 yielded by $(w, z)_k$ is w_1^l, and the new equation begins with a k_1-element, where $k_1\leqq l$. The order of the new element depends upon the manner in which the factor $w-\beta_1 z$ is involved in the groups $(w, z)_{k+1}$, $(w, z)_{k+2}$, $\cdots$, $(w, z)_{k+k_1}$.

It is necessary that it should occur in these groups raised to exponents not less than the numbers of the sequence k_1-1, k_1-2, $\cdots$, 1, and that, in one case at least, the exponent should be exactly equal to the corresponding number of the sequence.

By Nöther's reasoning the resolution has been *partially* effected: the k-element and the k_1-element contribute ordinary k-tuple and k_1-tuple points, while the absence of the terms w^{k_1}, w^{k_1+1}, $\cdots$, $w^{k_1+(l-k_1-1)}$, shows that there are $l-k_1$ branch-points. To account for the presence of the branch-points, all that we need do is to show that at an *ordinary* k_1-tuple point with separate tangents oblique to the axis of w_1, there are only k_1 values of w_1 which vanish when $z=0$, whereas in the present case there are l values of w_1.

Let the new equation $F_1(w_1, z)=0$ be written

$$(w_1, z)_{k_1}+(w_1, z)_{k_1+1}+(w_1, z)_{k_1+2}+\cdots=0.$$

(i.) If $(w_1, z)_{k_1}=0$ contain a simple factor $w_1-\beta_2 z$, the equation must be transformed by writing $w_1-\beta_2 z=w_2 z$. Divide by z^{k_1} and arrange according to ascending powers of w_2, z. The new origin is a 1-element, and the reasoning of § 113 shows that $w_2=P_r(z)$, $(r>0)$, or $w=\beta_1 z+\beta_2 z^2+P_r(z)$, $(r>2)$. The expansion associated with the pair of factors $w-\beta_1 z$, $w_1-\beta_2 z$ has been found.

(ii.) The necessary modifications when $\beta_1=0$ or $\beta_2=0$ can be easily worked out.

(iii.) If $(w_1, z)_{k_1}=0$ contain a multiple factor $(w_1-\beta_2 z)^{l_1}$, where β_2 is finite and different from 0, use the same transformation as before, namely $w_1-\beta_2 z=w_2 z$. After division by z^{k_1}, the equation in w_2, z begins with terms of order $k_2(\leqq l_1)$, say $(w_2, z)_{k_2}$. To each *simple* factor $w_2-\beta_3 z$ corresponds an integral expansion

$$w=\beta_1 z+\beta_2 z^2+\beta_3 z^3+P_r(z),\ (r>3)$$

and the process terminates. The existence of the k_2-element shows that the resolution of the higher singularity can be carried a stage further by combining the k-tuple point, the k_1-tuple point, and the $l-k_1$ simple branch-points, with an additional k_2-tuple point and l_1-k_2 simple branch-points. The further resolution of the k_2-element is accomplished by selecting a *multiple* factor $(w_2-\beta_3 z)^{l_2}$ and using the process already described. Assuming the truth of a theorem, — which will be proved below in order not to interfere with the continuity of the argument, — the process must terminate. That is to say, it is impossible to continue the expansion indefinitely by means of multiple factors $(w-\beta_1 z)^l$, $(w_1-\beta_2 z)^{l_1}$, $(w_2-\beta_3 z)^{l_2}$, $\cdots$. Sooner or later this process would lead to an equation $F_\nu(w_\nu, z)=0$, in which the origin is a k_ν-element, with simple factors $w_\nu-\beta_{\nu+1}z$, and possibly multiple factors in z alone.

All the expansions connected with the linear factors $w_\nu-\beta_{\nu+1}z$ are integral, and the process stops, so far as they are concerned, at this stage. It remains for us to examine the case in which one or more of the expressions $(w, z)_k$, $(w_1, z)_{k_1}$, $(w_2, z)_{k_2}$, $\cdots$ contain powers of z as factors. When, at any stage of the process, a power of z

makes its appearance as a factor, and when the expansions are to be continued by means of this factor, Nöther's second transformation $z=z_1w$ must be used. The effect of this transformation will be evident from what has been already said; z takes the place of w and w of z.

§ 118. The following is the general process: After m_1 transformations of the first kind suppose that a factor z is selected, and that the expansion associated with this factor is to be found. The part of this expansion due to the m_1 transformations is

$$\beta_1 z + \beta_2 z^2 + \beta_3 z^3 + \cdots + \beta_{m_1} z^{m_1} + w' z^{m_1}.$$

In the equation which connects w' with z, the terms of the lowest degree contain a factor z, and the expansion connected with this factor requires the transformation $z = z_1 w'$, to be followed by $z_1 - \gamma_2 w' = z_2 w'$, $z_2 - \gamma_3 w' = z_3 w'$, and so on. Thus

$$z = \gamma_2 w'^2 + \gamma_3 w'^3 + \gamma_4 w'^4 + \cdots + \gamma_{m_2} w'^{m_2} + w'' w'^{m_2},$$

the series stopping when the lowest terms in z_{m_2} and w' contain a factor w', and when the expansion is to be continued with regard to this factor. The next series of transformations is

$$w' = w_1' w'', \; w_1' = \delta_2 w'' + w_2' w'', \; w_2' = \delta_3 w'' + w_3' w'', \text{ and so on.}$$

The process terminates as soon as the factor to be considered is linear. For if $w^{(\iota+1)} - \lambda w^{(\iota)}$ be a linear factor of the lowest terms in $w^{(\iota+1)}, w^{(\iota)}$, we know that

$$w^{(\iota+1)} = P_r(w^{(\iota)}), \; (r > 0).$$

The system of expansions, when $w - \beta_1 z$ is the factor selected from $(w, z)_k$, is the following:

$$\text{I.} \left\{ \begin{array}{l} w = \beta_1 z + \beta_2 z^2 + \cdots + \beta_{m_1} z^{m_1} + w_1 z^{m_1}, \\ z = \gamma_2 w'^2 + \gamma_3 w'^3 + \cdots + \gamma_{m_2} w'^{m_2} + w'' w'^{m_2}, \\ w' = \delta_2 w''^2 + \delta_3 w''^3 + \cdots + \delta_{m_3} w''^{m_3} + w''' w''^{m_3}, \\ \cdot \quad \cdot \quad \cdot \quad \cdot \quad \cdot \quad \cdot \quad \cdot \quad \cdot \quad \cdot \quad \cdot \\ \cdot \quad \cdot \quad \cdot \quad \cdot \quad \cdot \quad \cdot \quad \cdot \quad \cdot \quad \cdot \quad \cdot \\ w^{(\iota-1)} = \kappa_2 (w^{(\iota)})^2 + \kappa_3 (w^{(\iota)})^3 + \cdots + \kappa_{m_{\iota+1}} (w^{(\iota)})^{m_{\iota+1}} + w^{(\iota+1)} (w^{(\iota)})^{m_{\iota+1}}. \end{array} \right.$$

In these expansions some of the constants may $= 0$.

Because $w^{(\iota+1)}$ is expansible as an integral series in $w_\iota (= \tau)$, it follows that $w^{(\iota-1)}, w^{(\iota-2)}, \cdots$ can be expressed as integral series in τ. Thus

$$\left. \begin{array}{l} z = a_0 \tau^\lambda + a_1 \tau^{\lambda+1} + a_2 \tau^{\lambda+2} + \cdots \\ w = b_0 \tau^\mu + b_1 \tau^{\mu+1} + b_2 \tau^{\mu+2} + \cdots \end{array} \right\}, *$$

* The two series are termed by Weierstrass a "functionenpaar." See Biermann, Analytische Functionen, ch. iv., sect. 1. It is understood that τ lies within the region of convergence of both series.

when λ, μ are positive integers and $\mu \geqq \lambda$. By reversion of series

$$\tau = A_0 z^{1/\lambda} + A_1 z^{2/\lambda} + \cdots,$$

and therefore $$w = B_0 z^{\mu/\lambda} + B_1 z^{(\mu+1)/\lambda} + \cdots.$$

If $(w, z)_\kappa$ contain a factor in z, and the corresponding expansion be required, the system I. must be replaced by

$$\text{II.}\ \begin{cases} z = \beta_2 w^2 + \beta_3 w^3 + \cdots + \beta_{m_1} w^{m_1} + w' w^{m_1}, \\ w = \gamma_2 w'^2 + \gamma_3 w'^3 + \cdots + \gamma_{m_2} w'^{m_2} + w' w'^{m_2}, \\ \cdot \quad \cdot \quad \cdot \quad \cdot \quad \cdot \quad \cdot \quad \cdot \quad \cdot \quad \cdot \quad \cdot \quad \cdot \\ w^{(t-1)} = \kappa_2 (w^{(t)})^2 + \kappa_3 (w^{(t)})^3 + \cdots + \kappa_{m_{t+1}} (w^{(t)})^{m_t+1} + w^{(t+1)} (w^{(t)})^{m_t+1}. \end{cases}$$

As before, $w^{(t+1)}$ is an integral series in $w^{(t)}$, and

$$\left. \begin{aligned} z &= a_0 \tau^\lambda + a_1 \tau^{\lambda+1} + \cdots \\ w &= b_0 \tau^\mu + b_1 \tau^{\mu+1} + \cdots \end{aligned} \right\},$$

but now the positive number $\mu \leqq$ the positive number λ. The series for w in terms of z is again

$$w = B_0 z^{\mu/\lambda} + B_1 z^{(\mu+1)/\lambda} + \cdots.$$

Returning to the point $(w = b,\ z = a)$ the corresponding expansion is

$$w - b = B_0 (z-a)^{\mu/\lambda} + B_1 (z-a)^{(\mu+1)/\lambda} + \cdots,$$

where μ is a positive integer.

If $b = \infty$ when $z = a$ we write for $w - \infty$, $1/w$; hence

$$1/w = B_0 (z-a)^{\mu/\lambda} + B_1 (z-a)^{(\mu+1)\lambda} + \cdots,$$

which gives $$B_0 w = (z-a)^{-\mu/\lambda} \{1 + C_1 (z-a)^{1/\lambda} + C_2 (z-a)^{2/\lambda} + \cdots\}.$$

When $a = \infty$, $z - a$ must be replaced by $1/z$.

§ 119. To complete the proof that at (a, b) $w - b$ can be expressed as an integral or fractional series, we have to show

(i.) That after a finite number of Nöther's transformations of the first kind there must result an equation $F_\nu(w_\nu, z) = 0$ in which the factors are either simple or powers of z.

(ii.) That t is in all cases finite.

If it be possible to continue the transformations of the first kind indefinitely, there must exist two or more branches of w which coincide, near $z = 0$, to any order of approximation. To see that this is impossible, form the equation $\phi(W, z) = 0$, whose roots are the squares of the differences of the n roots of $F(w^n, z^m) = 0$.

Let it be $$A_a W^{n(n-1)} + \cdots + L_\lambda W + M_\mu = 0,$$

where $A, B, \cdots, L, M$, are integral functions of z, of finite degrees $a, \cdots, \lambda, \mu$, and M is the discriminant, save as to a factor. If two series for w, near $z = a$, coincide as far as a term $(z-a)^q$ exclusive, one value of W is of the form $P_{2q}(z-a)$; and the equation shows that
$$M_\mu = P_\rho(z-a),$$

where ρ is certainly as great as $2q$. Hence $2q \leqq \mu$, and cannot be infinite. This shows that ν is finite.

It must next be proved that t is finite. We know that after a finite number g of transformations, the process either terminates or *must* be continued by a transformation of the second kind. Hence $m_1, \leqq g$, is finite. That the order of the element in the (w', z) equation $\not> k-1$, can be shown by the consideration that, after each transformation of the first kind, the order of the element is not increased, while the transformation of the second kind must diminish the order. For instance, if z be a factor of the k_1-element of § 117, k_1 must be less than l, and l is at most $= k$. Thus the order of the element in the (w', z) equation $\not> k-1$. Since, after each change in the character of the transformation, the order of the element is diminished by at least 1, the number t must be finite.

§ 120. *Résumé of the preceding results.*

The equation $F(w^n, z^m) = 0$ gives n expansions of w for each value of z. If the n branches for $z = a$ be finite, distinct, and equal to $b_1, b_2, \cdots, b_n$, there are n integral expansions
$$w - b_\lambda = P(z-a);$$

but if two or more of the values be equal, say $b_1 = b_2 = \cdots = b_\mu (= b)$, the corresponding expansions may be all fractional, some fractional and some integral, or all integral. To a given value $z = a$, there may belong several separate systems of equal roots, consisting of $\mu_1, \mu_2, \cdots$ members. Geometrically, these cases of equal roots occur when $z = a$ is a vertical line, which is a tangent to the curve at one or more points, or passes through one or more multiple points. In the most general case, the μ expansions of a system $(\mu \leqq n)$, which correspond to such a point are

$$w - b = (z-a)^{1/r_1} P((z-a)^{1/r_1}) \cdots r_1 s_1 \text{ expansions},$$
$$w - b = (z-a)^{1/r_2} P((z-a)^{1/r_2}) \cdots r_2 s_2 \text{ expansions},$$
$$\cdots\cdots\cdots\cdots\cdots\cdots,$$
$$\cdots\cdots\cdots\cdots\cdots\cdots,$$
$$w - b = (z-a)^{1/r_\lambda} P((z-a)^{1/r_\lambda}) \cdots r_\lambda s_\lambda \text{ expansions},$$

where $\mu = r_1 s_1 + r_2 s_2 + \cdots + r_\lambda s_\lambda$. The $r_\kappa s_\kappa$ expansions

$$(w-b) = (z-a)^{1/r_\kappa} P((z-a)^{1/r_\kappa})$$

form a set, which can be subdivided into s_κ sub-sets, each forming a cyclic system of r_κ expansions. The typical member of such a sub-set is

$$w - b = c_1\{a^\nu (z-a)^{1/r}\}^{q_1} + c_2\{a^\nu (z-a)^{1/r}\}^{q_2} + \cdots,$$

where $a = e^{2\pi i/r}$; $\nu = 0, 1, 2 \cdots r-1$,

and one at least of the fractions q_1/r, q_2/r, $\cdots$ is in its lowest terms.

In the special cases $a = \infty$, $b = \infty$, $z - a$ and $w - b$ are replaced by $1/z$, $1/w$.

The preceding treatment of higher singularities is based upon the memoirs of Hamburger (Zeitschrift für Mathematik und Physik, t. xvi.), Nöther (Math. Ann., t. ix.), and Stolz (Math. Ann., t. viii.).

§ 121. *Resolution of the singularity.*

When the branches of an algebraic function at a singular point are expanded in integral and fractional series in the way now explained, we are able to determine the number of nodes and of simple branch-points into which the singularity may be supposed to be resolved. It is convenient to define the relative order of two branches w_κ, w_λ as the least exponent of z in the difference of the two expansions of these branches. Geometrically speaking, if w_κ and w_λ be equal to integral series, the relative order = the order of $w_\kappa - w_\lambda$ = the number of intersections of the branches at the origin. The least exponent of z in the product of all the differences of the branches, or the sum of all the relative orders, may be called the total relative order.

Now a simple branch-point, at which there are two expansions

$$w_\kappa = P_1(z^{1/2}),\ w_\lambda = P_1(-z^{1/2}),$$

contributes 1/2 to the total relative order; and a node, at which $w_\kappa = a_\kappa z + P_2(z)$, $w_\lambda = a_\lambda z + Q_2(z)$, $a_\kappa \neq a_\lambda$, contributes 1. Therefore, if the origin can be resolved into δ nodes and β simple branch-points, we have

$$\delta + \beta/2 = l,$$

where l is the total relative order. But β is determinable at once from the expansions; for if r be the least common denominator of the exponents which enter into a cyclic system of expansions, that cyclic system contributes $r-1$ simple branch-points. Therefore $\beta = \Sigma(r-1)$, and then the preceding equation gives δ.

For example, suppose we have the single cyclic system

$$w = z^{4/3} + z^{5/2} + P_r(z^{1/6}),\ r > 15$$
$$= P_8(z^{1/6}).$$

Here $\beta = 5$, $l = 3 \times 5/2 + 12 \times 4/3 = 23\frac{1}{2}$, and therefore $\delta = 21$. The 5 branch-points indicate that there are in the penultimate form 5 small loops, and therefore the singularity is equivalent to 5 cusps and 16 nodes.

The number of branch-points at a multiple point ceases to be equal to the number of cusps if the w-axis be parallel to a tangent. To take the simplest case, while $w = z^{2/3}$ gives two branch-points and one node, $w = z^{3/2}$ gives one branch-point and one node. The explanation is given in Fig. 29. If the case occur, we may change the direction of the w-axis. It is sufficient to write, in the cyclic system in question, $z_1 + aw$ for z, where a is arbitrary.

§ 122. *Example* i. Let $(w - az)^3 + z^5 + (w - az)^4 z^2 = 0$. The origin is a 3-element; we wish to determine the expansions of w. Let $w - az = w_1 z$, and remove the factor z^3. The new equation is

$$w_1^3 + z^2 + w_1^4 z^3 = 0,$$

and the origin is a 2-element. Since z^2 is a factor of the terms of lowest degree, use a transformation of the second kind, $z = z_1 w_1$. Removing the factor w_1^2, $w_1 + z_1^2 + z_1^3 w_1^5 = 0$. The origin is now a 1-element, and $w_1 = P_2(z_1)$. Therefore, reversing the series,

$$z_1 = P_1(w_1^{1/2}),\quad \text{and}\quad z = z_1 w_1 = P_3(w_1^{1/2}).$$

Reversing again,

$$w_1^{1/2} = P_1(z^{1/3}),\quad w_1 = P_2(z^{1/3}),\quad w - az = w_1 z = P_5(z^{1/3}),$$

and finally, $$w = az + P_5(z^{1/3}).$$

The singularity at the origin is determined by $\beta = 2$, $\delta = 4$; therefore the penultimate form consists of four nodes and two vanishing loops, and the singularity is equivalent to two nodes and two cusps. Since the equation is of degree 4 in w, and since only three expansions have been found, there is one root which becomes infinite when $z = 0$. To obtain its expansion put $w = 1/u$. Then

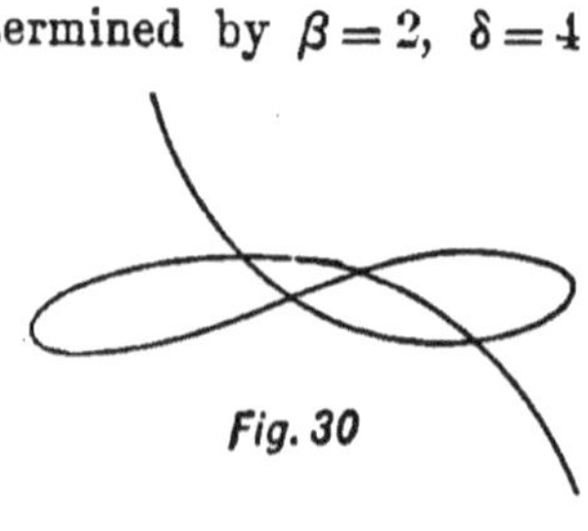

Fig. 30

$$(1 - azu)^3 u + z^2(1 - azu)^4 + z^5 u^4 = 0.$$

The origin is a 1-element, and $u = P_2(z)$.

Therefore $\quad 1/w = P_2(z)$, and $w = \frac{1}{z^2} P_0(z)$.

The following two examples have been supplied by Professor Charlotte A. Scott of Bryn Mawr College. For explanation of the method by which the penultimate forms are inferred, reference should be made to Professor Scott's paper in the American Journal, t. xiv.

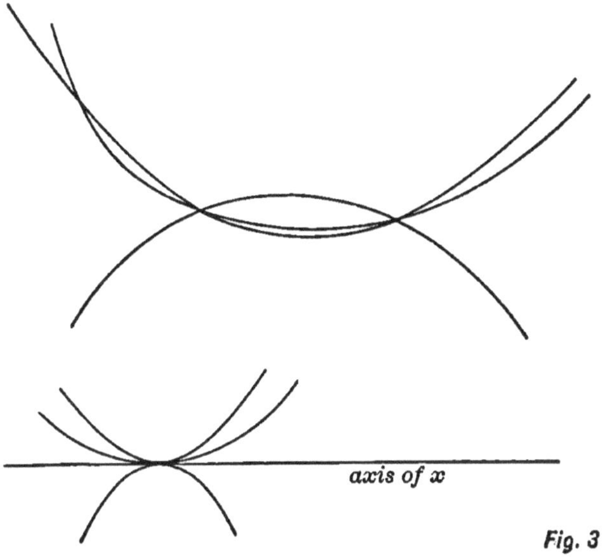

Fig. 31

Example ii. Instead of w, z use y, x, and let

$$y^3 - 3x^4y + 2x^6 - x^6y - 2x^8 + 9x^7y = 0;$$

we seek the expansions at the origin, which is a 3-element. Write $y = y_1x_1$, $x = x_1$. The new equation is

$$y_1^3 - 3x_1^2y_1 + 2x_1^3 - x_1^4y_1 - 2x_1^5 + 9x_1^5y_1 = 0.$$

The origin is a 3-element. First consider the simple factor $y_1 + 2x_1$. Write $y_1 + 2x_1 = x_2y_2$, $x_1 = x_2$. Then $y_1 = x_2(y_2 - 2)$, and the new equation is

$$y_2(3 - y_2)^2 - x_2^2y_2 - 9x_2^3(2 - y_2) = 0,$$

or $$9y_2 - 18x_2^3 - \cdots = 0.$$

Therefore $y = y_1x_1 = x_2^2(y_2 - 2) = x_2^2(-2 + 2x_2^3 +$ higher integral powers of x_2). Next consider the repeated factor $y_1 - x_1$. Let $y_1 - x_1 = x_2y_2$, $x_1 = x_2$; then $y_1 = x_2(1 + y_2)$. The resulting equation has for its lowest terms $3(y_2^2 - x_2^2)$, and the expansions associated with the two linear factors $y_2 - x_2$, $y_2 + x_2$ are integral in x_2. Thus

$$y = y_1 x_1 = x_2^2(1 + y_2)$$
$$= x_2^2(1 \pm x_2 + \text{higher integral powers}).$$

Therefore the expansions are

$$y = x^2 + x^3 + \cdots, \quad y = x^2 - x^3 + \cdots, \quad y = -2x^2 + 2x^5 + \cdots.$$

Here $\beta = 0$, $\delta = 7$, and the origin can be resolved into seven nodes. Fig. 31 shows the ultimate and penultimate forms.

Example iii. Show that the branches of y, given by

$$y^3 - 3x^4y + 2x^6 + 9x^7y = 0,$$

are

$$y = x^2 + 3ix^{7/2}\cdots,$$
$$y = x^2 - 3ix^{7/2}\cdots,$$
$$y = -2x^2 + 2x^5\cdots.$$

The ultimate and penultimate forms are given in Fig. 32.

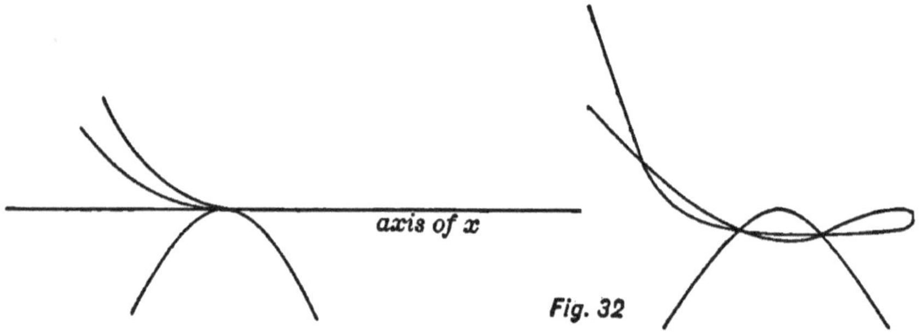

Fig. 32

Newton's Parallelogram.

§ 123. We shall next describe briefly a method of determining the expansions at a point which is historically interesting as emanating from Newton. In the hands of Puiseux it proved to be a potent instrument for the advancement of Cauchy's Theory of Functions. [For Puiseux's important memoirs, see Liouville, ser. 1, tt. xv., xvi., or Briot and Bouquet, Fonctions Elliptiques, 2d edition.]

Let us take, as a typical expansion,

$$w - b = c_1(z-a)^{q_1/r} + c_2(z-a)^{q_2/r} + c_3(z-a)^{q_3/r} + \cdots,$$

in which one, at least, of the integers $q_1, q_2, \cdots$ is prime to r, and $q_1 < q_2 < q_3 \cdots$. Let q_1/r, q_2/r, etc., when reduced to their lowest terms, be μ_1, $\mu_1+\mu_2$, $\mu_1+\mu_2+\mu_3$, $\cdots$. The expansion for $w-b$ is obtainable from the series of equations

$$\begin{aligned} w-b &= (z-a)^{\mu_1}(c_1+\zeta_1), \\ \zeta_1 &= (z-a)^{\mu_2}(c_2+\zeta_2), \\ \zeta_2 &= (z-a)^{\mu_3}(c_3+\zeta_3), \\ &\cdots\cdots\cdots, \end{aligned}$$

where $\zeta_1, \zeta_2, \cdots$ vanish when $z=a$. Newton's parallelogram offers a rapid and direct method for the calculation of the quantities $c_1, c_2, \cdots, \mu_1, \mu_2, \cdots$. For simplicity let $a=b=0$, and suppose that k branches of $w=0$ when $z=0$. Let the equation be written $\Sigma a_{fg} z^f w^g = 0$. Draw two rectangular axes *of*, *og*, and represent a term $a_{fg} z^f w^g$ by the point p, whose Cartesian co-ordinates are f, g. Let the points nearest to the origin on *og* and *of* be k and l. Assume $w = c_1 z^{\mu_1}$, and insert this value of w in the equation $\Sigma a_{fg} z^f w^g = 0$. The order of the term $a_{fg} z^f w^g$ in z is $\mu_1 g + f$. Now that line through p, which makes with *og* an angle $\tan^{-1}\mu_1$, intercepts on *of* a length $\mu_1 g + f$. Therefore all terms of the same order in z lie on a line. In any approximation we keep the terms of lowest order only, and of course there must be at least two such terms. Hence $w = c_1 z^{\mu_1}$ will not be a first approximation for w unless it makes at least two terms give the same minimum intercept on *of*. To find the admissible values of μ_1, suppose that pegs are placed at all the points p, and that a string *fog* is pulled in the directions *of*, *og*, until it is stopped by some of the pegs. The string has, between l and k, the form of a polygon, convex with regard to o, each side of which begins and ends at a peg, and may be in contact with intermediate pegs.

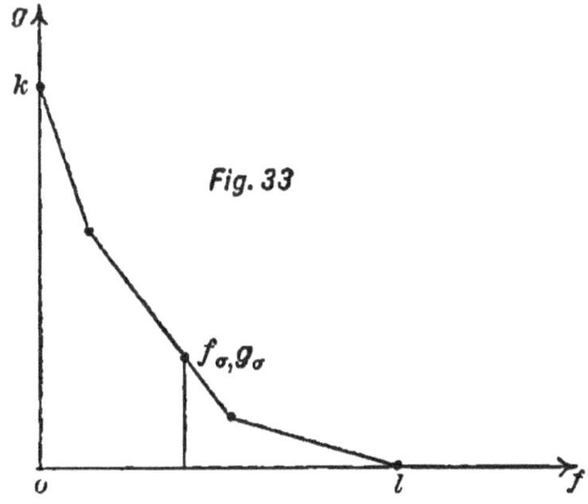

Fig. 33

(1) The number of these sides is the number of values which μ_1 can take.

(2) The value of μ_1 for each side is the tangent of the angle made by that side with *og*.

(3) The number of terms to be retained in the approximations which correspond to one of the sides is the number of pegs in contact with the side.

(4) The polygon begins at l and ends at k.

Let the pegs in contact with a side be p_σ, $\sigma = 1, 2, \cdots, \rho$.

Let $w = c_1 z^{\mu_1}$ for this side. We must have

$$\mu_1 g_1 + f_1 = \cdots = \mu_1 g_\sigma + f_\sigma = \cdots = \mu_1 g_\rho + f_\rho,$$

and therefore

$$\mu_1 = (f_1 - f_\sigma)/(g_\sigma - g_1), \quad (\sigma = 2, 3, \cdots, \rho)$$
$$= \kappa_1/\lambda_1, \text{ where } \kappa_1 \text{ is prime to } \lambda_1.$$

We suppose that g_1 is the least and g_ρ the greatest ordinate of the ρ pegs on the side in question. The equation, after division by $z^{\mu_1 g_\sigma + f_\sigma}$, is

$$\Sigma a_{f_\sigma g_\sigma} c_1^{g_\sigma} + \text{ terms involving } z = 0, \quad (\sigma = 1, 2, \cdots, \rho) \cdots (\text{i.}).$$

When z is zero, the values of c_1 are given by

$$\Sigma a_{f_\sigma g_\sigma} c_1^{g_\sigma} = 0.$$

Neglecting the g_1 zero values of c_1, there are $g_\rho - g_1$ values different from zero, and thus the number of expansions associated with the first side = the difference of the ordinates of the end pegs. The total number of expansions is therefore the difference of the ordinates of k and l, which shows that the figure accounts for all the k branches in question.

§ 124. The c_1-equation may be written

$$\Sigma a_{f_\sigma g_\sigma} c_1^{g_\sigma - g_1} = 0, \quad (\sigma = 1, 2, \cdots, \rho).$$

It is important to notice that $g_\sigma - g_1$ is a multiple of λ_1, and that therefore the equation may be written as an equation in $c_1^{\lambda_1}$. Owing to this, the branches of w which belong to the side break up into cyclic systems. Corresponding to each simple root of the equation in $c_1^{\lambda_1}$, there is a distinct cyclic system whose typical first term is $c_1 z^{\mu_1}$ or $(c_1^{\lambda_1} z^{\kappa_1})^{1/\lambda_1}$.

Corresponding to each k_1-fold root of $c_1^{\lambda_1}$ there are k_1 cyclic systems with the same first terms. To separate these systems, instead of writing $w = c_1 z^{\mu_1}$ write $w = (c_1 + \zeta_1) z^{\mu_1}$, where c_1 arises from the known k_1-fold root. When $z=0$, $\zeta_1=0$; and (i.) becomes an equation involving integral powers of ζ_1 and powers of z whose exponents have the denominator λ_1. Writing $z = z_1^{\lambda_1}$, we have an integral equation in ζ_1 and z_1. The terms independent of z_1 are obtained by writing $c_1 + \zeta_1$ instead of c_1 in the c_1-equation; and since k_1 values of c_1 are equal, the lowest power of ζ_1 in this equation is $\zeta_1^{k_1}$. Therefore, when $z_1 = 0$, k_1 values of ζ_1 are zero. We form a new polygon for the equation in ζ_1, z_1. The branches of ζ_1 break up into sets, which belong to the sides of the new polygon, and these sets break up into cyclic systems.

If $$\zeta_1 = c_2 z_1^{\kappa_2/\lambda_2}, \quad w = (c_1 + c_2 z^{\kappa_2/\lambda_1\lambda_2}) z^{\kappa_1/\lambda_1},$$

to the second order of approximation. The equation in $c_2^{\lambda_2}$ may have all its roots distinct, in which case the series arising from the multiple root of $c_1^{\lambda_1}$ are all separated, and form cyclic systems of $\lambda_1\lambda_2$ elements; or the equation may have multiple roots, in which case a further transformation must be effected. After a finite number of such transformations, all the expansions associated with the side of the original polygon become distinct.

§ 125. *Example* 1. $(w^2 - z)^2 - wz^2 - z^4 = 0.$

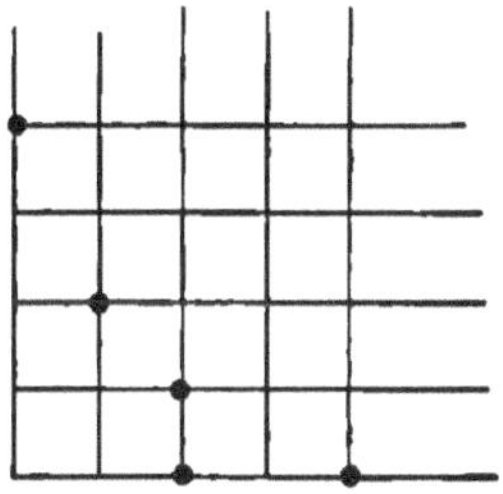

Fig. 34

The diagram shows that $\mu_1 = 1/2$, and, on substitution of $c_1 z^{1/2}$ for w, we have

$$(c_1^2 - 1)^2 = 0.$$

Thus the two roots in c_1^2 are equal. To discriminate between the two expansions for w, we must proceed to the second terms. Put $w = z^{1/2}(c_1 + c_2 z^{\mu_2})$. Then, remembering that $\mu_2 > 0$, the terms of lowest order that survive are

$$4c_1^2 c_2^2 z^{2+2\mu_2} - c_1 z^{5/2}.$$

Hence $$2 + 2\mu_2 = 5/2, \quad \mu_2 = 1/4, \quad 4c_1 c_2^2 = 1;$$

and to the second approximation

$$w = c_1 z^{1/2} + \frac{1}{2\sqrt{c_1}} z^{3/4}, \quad (c_1^2 = 1).$$

The four branches are approximately

$$w_1 = z^{2/4} + \tfrac{1}{2} z^{3/4}, \; w_2 = -z^{2/4} - \tfrac{1}{2} i z^{3/4}, \; w_3 = z^{2/4} - \tfrac{1}{2} z^{3/4}, \; w_4 = -z^{2/4} + \tfrac{1}{2} i z^{3/4};$$

the exact values are integral series in $z^{1/4}$.

Example 2. To see how equal roots of the c_1-equation introduce a new radical, take the general equation of the form just considered,

$$a_{04} w^4 + a_{12} z w^2 + a_{20} z^2 + a_{21} z^2 w + a_{40} z^4 = 0,$$

and first substitute $w = c_1 z^{1/2}$. This gives, if

$$\phi(c_1) = a_{04} c_1^4 + a_{12} c_1^2 + a_{20},$$
$$\phi(c_1) + a_{21} z^{1/2} + a_{40} z^2 = 0.$$

Now substitute, for c_1, $c_1 + c_2 z^{\mu_2}$. Then

$$\phi(c_1) + \phi'(c_1) c_2 z^{\mu_2} + \tfrac{1}{2}\phi''(c_1) c_2^2 z^{2\mu_2} + \cdots + a_{21} z^{1/2} + \cdots = 0.$$

When $\quad \phi(c_1) = 0, \;\; \phi'(c_1) \neq 0, \;$ we have $\; \mu_2 = 1/2$;

but when $\quad \phi(c_1) = 0, \;\; \phi'(c_1) = 0, \;\; \phi''(c_1) \neq 0,$

we have $\quad \mu_2 = 1/4, \;\; c_2^2 = -2a_{21}/\phi''(c_1).$

Enough has been said to show how Puiseux's method applies to any particular example.

The Theory of Loops.

§ 126. Let w be an algebraic function defined by the equation $F(w^n, z^m) = 0$. For a given value of z there will usually be n different finite values of w, which we denote by $w_1, w_2, \cdots, w_n$. Let two paths in the z-plane lead from $z^{(0)}$ to ζ. Will the two paths give rise to the same or different final values of w at ζ, when the same initial branch is chosen in both cases?

To make the question a definite one we must first show what is meant by the values of w_κ, $(\kappa = 1, 2, \cdots, n)$ along a path. For shortness we omit the suffix κ.

Let $z^{(0)}z^{(1)}$, $z^{(1)}z^{(2)}$, $\cdots$ be arbitrarily small elements of a path from $z^{(0)}$ to ζ. Assuming that $z^{(0)}$ is not a critical point, each branch is regular near $z^{(0)}$, and the values of w at $z^{(0)}$ are all distinct. If we choose an initial value $w^{(0)}$, the corresponding series $w - w^0 = P_r(z - z^{(0)})$, where $r > 0$, defines uniquely the value of w at $z^{(1)}$, so long as $z^{(1)}$ lies within the circle of convergence, and *a fortiori* so long as $|z^{(1)} - z^{(0)}| < \rho^{(0)}$, where $\rho^{(0)}$ is the distance from $z^{(0)}$ to the nearest critical point. This value of w is the value of the selected branch at $z^{(1)}$. In the same way the value at $z^{(2)}$ is determined uniquely so long as $|z^{(2)} - z^{(1)}| < \rho^{(1)}$, where $\rho^{(1)}$ is the distance from $z^{(1)}$ to the critical point nearest to it. No ambiguity arises so long as the path of z lies wholly within the chain of circles whose radii are $\rho^{(0)}, \rho^{(1)}, \cdots$. See § 95.

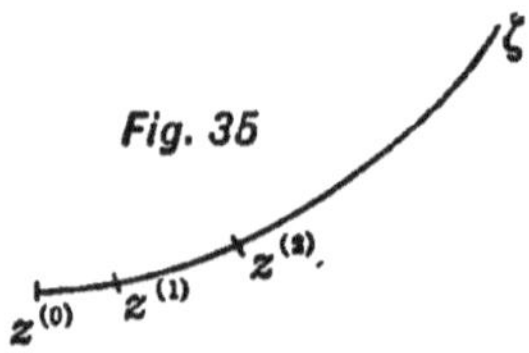

Fig. 35

The preceding argument does not apply to the case in which z passes through a pole or a branch-point, when the pole or branch-point affects the branch which we are considering. At a pole the indetermination can be readily removed by the use of the series in $1/w$. But at a branch-point a there is real indetermination. If μ

branches permute at a, when z passes through a, any one of these branches can, without breach of continuity, move in any one of μ directions.

To avoid this case, the z-path must be so chosen as not to pass through a branch-point. When the initial value of a branch is selected, the final value is now fully determined, for the particular path.

§ 127. *Theorem.* If a contour C include no critical point of the function, and if, starting from a point $z^{(0)}$ with the value $w_\kappa^{(0)}$ of w_κ, z describe the contour, the final value of w_κ, when z returns to $z^{(0)}$, is $w_\kappa^{(0)}$.

Since $z^{(0)}$ is not a critical point, there exists an integral series

$$w_\kappa - w_\kappa^{(0)} = P_r(z - z^{(0)}), \ (r > 0),$$

which within its circle of convergence defines w_κ as a one-valued continuous function of z. Let a circle smaller than but concentric with the circle of convergence meet C at the points $z^{(1)}$, $z^{(2)}$. The branch w_κ is one-valued along the contour $z^{(0)}z^{(1)}z^{(2)}z^{(0)}$; therefore the theorem holds for this contour. Let a circle, with centre $z^{(1)}$ and less than the circle of convergence whose centre is $z^{(1)}$, meet the contour $z^{(1)}\zeta z^{(2)}$ at $z^{(3)}$ and $z^{(4)}$; then the theorem is true for the contour $z^{(1)}z^{(3)}z^{(4)}z^{(1)}$. It is therefore true for $z^{(0)}z^{(1)}z^{(4)}z^{(1)}z^{(3)}z^{(4)}z^{(2)}z^{(0)}$, that is, for the contour $z^{(0)}z^{(3)}z^{(4)}z^{(2)}z^{(0)}$. Proceeding in this way with circles whose centres are $z^{(2)}$, $z^{(3)}$, $\cdots$, we prove the theorem for the contour C.

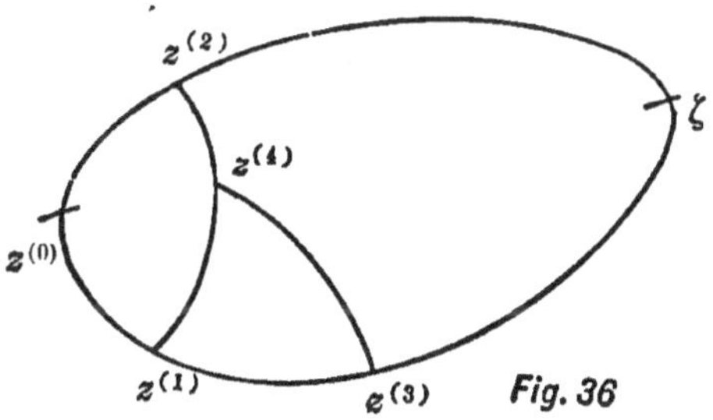

Fig. 36

Corollary. The paths $z^{(0)}z^{(1)}\zeta$ and $z^{(0)}z^{(2)}\zeta$ lead to the same final value of w_κ, the initial values being the same; that is, a path may be deformed in any manner provided it does not pass over a critical point.

It has appeared in § 46 and § 48 that a closed path which includes a branch-point may lead to a final value of w different from the selected initial value; therefore also two paths from $z^{(0)}$ to ζ may lead to different values of w. As infinitely many paths lead from $z^{(0)}$ to ζ, it is a matter of importance to reduce these to combinations of certain standard paths. This can be effected by means of the preceding theorem.

Let a small circle (c) be described with a critical point c as centre. Let L be a line from any point $z^{(0)}$, which is not a critical point, to the circle. The path from $z^{(0)}$ along L to the circle, round the circle, and back to $z^{(0)}$ along L, is called a *loop*. A loop is supposed neither to cross itself nor any other loop. It is positive or negative according as (c) is described positively or negatively; but it will be supposed to be positive unless the contrary is stated.

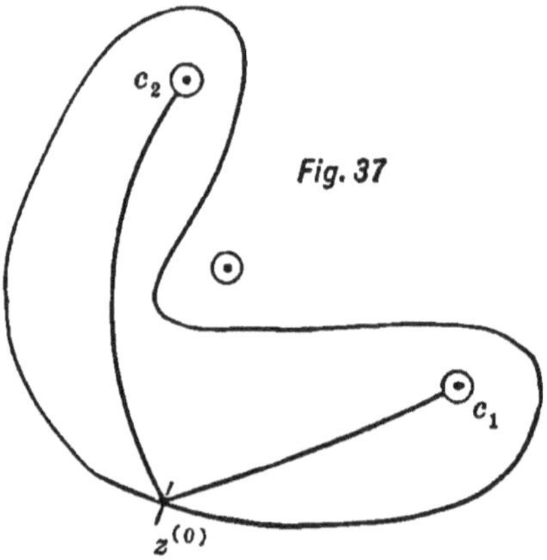

Fig. 37

If a contour C contain several critical points $c_1, c_2, \cdots, c_\lambda$, the path C which begins and ends at a point $z_{(0)}$ of the contour can be contracted, without passing over any critical point, into λ loops from $z^{(0)}$ to the points $c_1, c_2, \cdots, c_\lambda$; therefore by the preceding theorem the path C has the same effect on the branches of w as the successive λ loops.

A loop to an infinity which is not a branch-point will reproduce the initial value. For near such a point, say c, we have

$$\frac{1}{w_\kappa} = P_r(z-c), \quad r > 0,$$

or

$$w_\kappa = P_0(z-c)/z^r,$$

so that, when z describes the circle (c), w_κ does not change its value. Accordingly, in the present discussion, the loops to infinities which are not branch-points may be disregarded.

The nature of the changes produced by a loop to a branch-point a is to be decided from the expansions for w near that point. The selected value of w at $z^{(0)}$ leads by a given path to a definite value near a. If this value belong to one of the cyclic systems near a, it is changed by the description of the circle (a) into another, of the same system, which leads to a definite new value at $z^{(0)}$. If r be the least common denominator of the exponents which enter into the system of expansions, r descriptions of the loop will restore the initial value at $z^{(0)}$.

The sphere-representation (§ 27) shows that the effect of the description of a contour C may be determined as well from the external branch-points as from those inside; if the contour be described positively, the loops to the external branch-points must be negative. In particular, a contour which encloses all the branch-points in the finite part of the plane will reproduce the initial value of a branch, or not, according as the point $z=\infty$ is not, or is, a branch-point.

§ 128. *The prepared equation.*

It has been stated that the theory of quadratic transformations enables us to replace any given algebraic equation by another in which the multiple points have no coincident tangents. Further, if we project to infinity a line $t=0$ which is not a tangent, by means of the linear transformation $w'=w/t$, $z'=z/t$, we have an equation which has no contact with the line ∞. Lastly, if n be the order of this equation, a suitable change of axes yields an equation in which the term w^n occurs, and in which there are, parallel to the w-axis, only ordinary tangents with finite points of contact. The branch-points are now all simple and at a finite distance.

When an equation is brought to such a form it will be said to be *prepared*. The great advantage of such a form is that at a branch-point for which $z=a$, the expansions are all integral in $z-a$, except two which are integral in $(z-a)^{1/2}$.

Since there are no cusps, and since from the point at ∞ on the w-axis the tangents are all vertical, β, the number of branch-points, = the number of vertical tangents = the class of the curve. From Plücker's equations (Salmon, Higher Plane Curves, p. 65), if $\delta=$ the number of nodes and $p=$ the deficiency, we have

$$\beta=n(n-1)-2\delta,$$

$$p=\tfrac{1}{2}(n-1)(n-2)-\delta,$$

and therefore $$\beta=2p+2(n-1).$$

§ 129. *Deformation of loops.*

Let $F(w, z)=0$ be an irreducible prepared equation, and let β loops be drawn from $z^{(0)}$ to all the branch-points. Since all the branch-points are simple, if a loop lead from an initial value w_κ to a definite final value $w_{\kappa'}$, the initial value $w_{\kappa'}$ leads to the final value w_κ; and any other initial value is not affected by the loop. The negative description of a loop has the same effect as the positive description.

Beginning with an arbitrary loop, we assign to the branch-points the names $a_1, a_2, \cdots, a_\beta$ in the order in which their loops are met by a small circle described positively round $z^{(0)}$. We denote the corresponding loops by $A_1, A_2, \cdots, A_\beta$, and say that A_λ is in advance of A_μ when $\lambda<\mu$.

A loop A_λ to a branch-point a_λ can evidently be deformed so long as it does not pass over a branch-point. On the other hand, let two loops A_λ, A_λ' to the same branch-point a_λ enclose a single branch-point $a_{\lambda+1}$. In Fig. 38, the loop A_λ' can be deformed into the broken path without crossing a branch-point, and the broken path is equivalent to the loops $A_{\lambda+1}$, A_λ, $A_{\lambda+1}$ in succession.*

Let A_λ and $A_{\lambda+1}$ permute the branches w_ι, $w_{\iota'}$ and w_κ, $w_{\kappa'}$.

(1) If ι, ι' be different from κ, κ' the new loop A_λ' still permutes w_ι and $w_{\iota'}$. For, starting with w_ι, $A_{\lambda+1}$ leaves it unaltered, A_λ changes it to $w_{\iota'}$, $A_{\lambda+1}$ leaves $w_{\iota'}$ unaltered.

(2) If ι, ι' be the same as κ, κ' the loop A_λ' still permutes w_ι, $w_{\iota'}$.

(3) If $\kappa' = \iota$, $\kappa \neq \iota'$, A_λ' permutes $w_{\iota'}$ and w_κ. For $A_{\lambda+1}$ leaves $w_{\iota'}$ unaltered, A_λ changes it to w_ι, $A_{\lambda+1}$ changes w_ι to w_κ.

Hence, if we denote a loop which permutes w_ι, $w_{\iota'}$ by $(\iota\iota')$, we have the theorem: A loop $(\iota\iota')$ which retreats over a loop $(\iota\kappa)$ becomes $(\iota'\kappa)$.

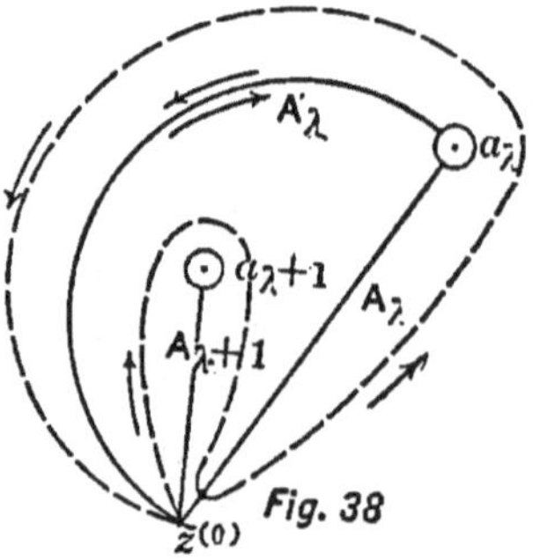

Fig. 38

That is, when adjacent loops affect a common branch, the deformed loop affects the other two branches.

Example. Show that the retreat of A_λ over $A_{\lambda+1}$, followed by the retreat of $A_{\lambda+1}$ over A_λ', does not restore the original situation. What has been said about the retreat of a loop applies equally to its advance.

§ 130. We are now in a position to prove some important theorems due to Lüroth and Clebsch. (Lüroth, Math. Ann., t. iv.; Clebsch, Math. Ann., t. vi.; Clifford, On the Canonical Form and Dissection of a Riemann's Surface, Collected Works, p. 241. See also Laurent, Traité d'Analyse, t. iv.)

We first choose $(n-1)$ fundamental loops in the following manner. However the loops are drawn, there must be at least one loop which permutes w_1 with one of the other branches, for otherwise w_1 would be one-valued. Call this the first fundamental loop, and let the roots affected by it be $(w_1 w_2)$. There must be a second loop which permutes either w_1 or w_2 with a third root w_3, for otherwise w_1, w_2 could only change into each other, and would form the roots of an irreducible equation of degree 2 in w. This would imply that

* If we were discussing an unprepared equation, we should have to attend to the point that the first description of $A_{\lambda+1}$ is negative.

$F(w, z) = 0$ is a reducible equation. Let this second loop be chosen as the second fundamental loop. In this way $(n-1)$ fundamental loops can be found: the first connects w_1 with w_2, the second w_1 or w_2 with w_3, the third w_1, w_2, or w_3 with w_4, and so on.

(i.) All the loops (1 2) which interchange w_1, w_2 can be placed in consecutive positions. The following is the way in which this is done. Let us call one of the loops (1 2) the first loop, and let the 2d, 3d, $\cdots$, $(r-1)$th loops be different from (1 2), while the rth is (1 2). A retreat of the $(r-1)$th, $(r-2)$th, $\cdots$, 2d loops over the rth will produce no change in the rth loop, for it has remained stationary. By a continuation of this process, all the loops (12) can be brought together into one group, which is conveniently represented by the notation [1 2].

(ii.) All the loops (1 3) can be gathered into a group [1 3] and placed next to [1 2]. For if m loops intervene between the first loop (1 3) and the last of the group [1 2], make these m loops retreat over (1 3), and thus bring the first loop (1 3) next to [1 2]; and continue the process until all the loops (1 3) are gathered into a single group [1 3] placed next to [1 2]. In this process a new loop (1 2) makes its appearance when a loop (2 3) retreats over a loop (13). It must be at once added to the group [12] by the process described above.

By following this plan, we can arrange all the β loops in successive groups

$$[1\ 2],\ [1\ 3],\ \cdots,\ [1\ n],\ [2\ 3],\ \cdots,\ [2\ n],\ \cdots,\ [n-1,\ n],$$

it being understood that members may be missing from the sequence.

(iii.) The groups have the following properties:

(A) Each group contains an even number of loops.

(B) The groups can be placed in any order without altering their effects.

(C) With the aid of the theory of fundamental loops, it is possible to make all the groups contain the index 1.

(D) It is possible to make all the groups, except the first, namely, [1 3], [1 4], $\cdots$, [1 n], consist of two loops each.

A. If the system $[1\ 2][1\ 3]\cdots[n-1,\ n]$ be described with initial value w_1, the final value must also be w_1 (§ 127). Now if [12] contain an odd number of loops, the effect of the group is to change w_1 into w_2, a value with respect to which the groups $[13]\cdots[1\ n]$ are inoperative. As the remaining groups are inoperative with respect to w_1, and as the initial value with which they are described is w_2, it is impossible that they should finally reproduce w_1. This proves that [12] contains an even number of loops, and that, therefore, it has no effect upon w_1. Hence the description of [13] begins with the value w_1. The same argument as that just used shows that [13] must contain an even number of loops. Assume the truth of the theorem for all the groups up to $[\iota\kappa]$, where $\iota < \kappa$. We can prove by induction that it holds for $[\iota\kappa]$, and hence for all the following groups. To prove that it holds for $[\iota\kappa]$ describe *all* the loops, beginning with the first loop of $[\iota\kappa]$ and with initial value w_ι. The final value must be w_ι. If possible, let the number of loops in $[\iota\kappa]$ be odd, then the value of w after the last loop of the group is w_κ. The remaining groups with index ι, namely $[\iota, \kappa+1]$, $[\iota, \kappa+2]$, $\cdots$, $[\iota n]$, are inoperative as regards w_κ, and, therefore, at the end of these the value is still w_κ. From thence onwards to $[n-1, n]$ the loops are inoperative as regards w_ι, and the value w_ι at the entrance of [12] is different from w_ι. Finally, the passage over the groups [12], [13], $\cdots$, up to $[\iota, \kappa]$, each with an *even* number of loops, produces no effect upon w_ι. That is to say, a complete circuit changes w_ι into w_ι. This is impossible, and accordingly $[\iota\kappa]$ contains an even number of loops.

B. The groups may be placed in any order without changing their effects.

For the whole effect of any group on a root is nil, since it contains an even number of loops; and therefore the retreat or advance of any number of single loops over a whole group produces no effect upon the single loops.

From this point onwards we assume that the process of collecting the β loops into groups has been completed. At all stages of the process it is possible to select $(n-1)$ fundamental loops, but a system which is fundamental at one stage need not be fundamental at succeeding stages.

C. Let $[\iota\kappa]$ be a group in which both ι and κ are greater than 1. It is a property of the fundamental loops that each of the roots w_ι, w_κ is connected by a fundamental loop with a w of lower index. Call this loop $(\theta\iota)$, $\theta < \iota$. Hence the group $[\theta\iota]$ certainly exists. Place it in front of $[\iota\kappa]$ and let the last of its loops retreat over $[\iota\kappa]$. This produces no change in the effect of either the loop $(\theta\iota)$ or the group $[\iota\kappa]$. Next let the group $[\iota\kappa]$ retreat over this single loop $(\theta\iota)$. Each loop of $[\iota\kappa]$ changes into $(\theta\kappa)$ and the new group is $[\theta\kappa]$. The original group $[\theta\iota]$ is restored by the double retreat, but instead of the sequence $[\theta\iota][\iota\kappa]$, there is now a sequence $[\theta\iota][\theta\kappa]$, that is, the index ι of one group has been lowered to θ without any accompanying change in the other groups. The process can be continued until each group has an index 1. At various stages of the process it may be necessary to change the fundamental loops.

D. If a group contain more than 2 loops, this number can be diminished by the addition of 2 loops to another group.

This can be proved by observing that the retreat of 2 loops $(\iota\kappa)$ over a loop $(\kappa\lambda)$ changes the order $(\iota\kappa), (\iota\kappa), (\kappa\lambda)$ into $(\kappa\lambda), (\iota\lambda), (\iota\lambda)$. The retreat of $(\kappa\lambda)$ over the pair $(\iota\lambda)$, $(\iota\lambda)$ produces no effect. Thus, by retreats of loops, the arrangement $(\iota\kappa)(\iota\kappa)(\iota\kappa)(\kappa\lambda)$ can be changed successively into

$$(\iota\kappa)(\kappa\lambda)(\iota\lambda)(\iota\lambda), \quad (\iota\kappa)(\iota\lambda)(\iota\lambda)(\kappa\lambda).$$

Now $(\iota\kappa)(\iota\lambda)(\iota\lambda)$ can, in a similar way (namely, by the retreat of $(\iota\kappa)$ over $(\iota\lambda)(\iota\lambda)$ followed by the retreat of $(\iota\lambda)(\iota\lambda)$ over $(\iota\kappa)$), be changed, first into $(\iota\lambda)(\iota\lambda)(\iota\kappa)$, and next into $(\iota\kappa)(\kappa\lambda)(\kappa\lambda)$; hence the arrangement $(\iota\kappa)(\iota\kappa)(\iota\kappa)(\kappa\lambda)$ can be replaced by

$$(\iota\kappa)(\kappa\lambda)(\kappa\lambda)(\kappa\lambda).$$

The table indicates, sufficiently, the successive changes:

$\iota\kappa$	$\iota\kappa$	$\iota\kappa$	$\kappa\lambda$
$\iota\kappa$	$\kappa\lambda$	$\iota\lambda$	$\iota\lambda$
$\iota\kappa$	$\iota\lambda$	$\iota\lambda$	$\kappa\lambda$
$\iota\lambda$	$\iota\lambda$	$\iota\kappa$	$\kappa\lambda$
$\iota\kappa$	$\kappa\lambda$	$\kappa\lambda$	$\kappa\lambda$

It is a consequence of the properties of the fundamental loops that it is always possible to unite any two loops $(\iota\kappa)$, $(\rho\sigma)$ by a chain of groups $[\iota\kappa][\kappa\lambda][\lambda\mu]\cdots[\rho\sigma]$. It has been proved that two loops can be removed from $[\iota\kappa]$ and added to $[\kappa\lambda]$. These, in turn, can be removed from $[\kappa\lambda]$ and added to $[\lambda\mu]$, and so on. Therefore two loops can be removed from a group which contains more than two, and added to another. When the process is continued sufficiently long, the groups $[13][14]\cdots[1n]$ can each be reduced to two loops, and then $[12]$ contains $\beta - 2\ (n-2)$ loops, or $2\ (p+1)$ loops (§ 128).

We have proved that the original groups $[12][13]\cdots[n-1, n]$ can be replaced by $(n-1)$ groups $[12][13]\cdots[1n]$. The same method shows that they can, equally well, be replaced by any system of $(n-1)$ groups which satisfy the condition that the $(n-1)$ different loops, entering into the system, are capable of forming a fundamental system.

Example. The following scheme shows the process for a system of 12 loops:

```
12 (23 · 14 · 24 · 34 · 13 · 24)(12) 23 · 24 · 34 · 34
12 · 12 · 13 (24)*(14) 34 · 23 (14) (23 · 24) 34 · 34
12 · 12 (13)(12) 24 · 34 · 23 · 23 · 24 · 12 · 34 · 34
12 · 12 · 12 (23 · 24 · 34 · 23 · 23 · 24) (12) 34 · 34
12 · 12 · 12 · 12 · 13 (14)*(34) 13 · 13 · 14 · 34 · 34
12 · 12 · 12 · 12 · 13 · 13 (14)(13 · 13) 14 · 34 · 34
12 · 12 · 12 · 12 · 13 · 13 · 13 · 13 (14 · 14)(34 · 34)
12 · 12 · 12 · 12 · 13 · 13 · 13 (13)*(34 · 34) 14 · 14
12 · 12 · 12 · 12 · 13 · 13 · 13 (14 · 14)*(13) 14 · 14
12 · 12 · 12 (12) (13 · 13 · 13)(13) (14 · 14 · 14) 14
12 · 12 · 12 · 12 · 12 (12) (13 · 13 · 13) 13 · 14 · 14
12 · 12 · 12 · 12 · 12 · 12 · 12 · 12 · 13 · 13 · 14 · 14
```

The first line gives the original order. The parentheses signify that the loops in the left-hand parenthesis retreat over those in the right-hand parenthesis. An asterisk denotes an advance. From the tenth stage on, use is made of theorem *D* alone.

This system of loops is used for another purpose in Clebsch and Gordan's Theorie der Abelschen Functionen, p. 100.

References. — For Weierstrass's treatment of the algebraic function, the reader is referred to Biermann, Analytische Functionen. Information on the subjects discussed in this chapter will also be found in Briot and Bouquet's Fonctions Elliptiques, Chrystal's Algebra, Clebsch's Geometry, Königsberger's Elliptische Functionen, and Salmon's Higher Plane Curves.

The Algebraic Function has been treated from the standpoint of the Theory of Numbers by Kronecker (Crelle, tt. xci., xcii.), Dedekind and Weber (ib. t. xcii.), and Hensel (Crelle, t. cix.).

CHAPTER V.

INTEGRATION.

§ 131. The theorems of this chapter will relate to functions which are one-valued within certain regions of the z-plane, bounded by simple or complex contours. A region with a complex contour, in the finite part of the plane, is one which is bounded by non-intersecting closed curves $C, C_1, C_2, \cdots, C_\kappa$, where $C_1, C_2, \cdots, C_\kappa$ lie inside C and outside one another; a region with a complex contour, which includes the point $z=\infty$, is one which lies outside non-intersecting curves $C_1, C_2, \cdots, C_\kappa$. Unless the contrary is stated, the region will usually be assumed to lie in the finite part of the plane, but sufficient directions will be given to enable the student to deal with the other case.

It will assist the reader if we anticipate some of the results of the chapter, and give here a brief statement of the behaviour of a function in the neighbourhood of its singular points, omitting all mention of the more unusual cases, which arise from clusters of singularities. The singular points are either places near which the function is one-valued or places near which it is many-valued. If c be a place of the former kind, in whose neighbourhood there is no other singular point, the function can be expressed in the form

$$P(z-c)+\frac{a_1}{z-c}+\frac{a_2}{(z-c)^2}+\cdots,$$

where, according as the singular point c is a pole or an essential singularity, the series of negative powers of $z-c$ does or does not terminate. If c be a place of the latter kind, and if there be no other singular point in the neighbourhood of c, branches of the function will be represented, near c, by one or more sets of series with *fractional* exponents; and c is said to be a branch-point with respect to these branches. If branches of the function be represented, near c, by a set of series

$$a_1(z-c)^{q_1/r}+a_2(z-c)^{q_2/r}+a_3(z-c)^{q_3/r}+\cdots,$$

where $q_1, q_2, \cdots, r$ are finite integers and r is positive, the branch-point is called *algebraic*. The point is an infinity of the r branches when q_1 is negative. In the expansions at ∞, $z-c$ must be replaced by $1/z$.

Definitions. If a function be one-valued and continuous, and possess a derivative, at all points within a region Γ, it is said to be *holomorphic* within Γ.

Such a function can have no singular points, of either kind, within Γ.

If a function be holomorphic within Γ, except at certain poles, it is said to be *meromorphic* within Γ.

For example, $\sin z$ is holomorphic throughout the finite part of the plane; $\tan z$ is meromorphic throughout the finite part of the plane, the poles being the zeros of $\cos z$.

The words "holomorphic" and "meromorphic," introduced by Briot and Bouquet, are intended to suggest that the functions behave respectively like the integral polynomial and the rational fraction. Halphen proposed that the terms be replaced by "integral" and "fractional." Cauchy's term "synectic" is synonymous with "holomorphic."

§ 132. *Curvilinear integrals.*

The properties of integrals of a real variable, taken over a curved path, are required as preliminary to the study of the integral of a function of z.

For our immediate purpose they may be stated very briefly: —

Let $U(x, y)$ be a continuous function of two real variables, and let L be a path defined by the equations $x = \phi_1(t)$, $y = \phi_2(t)$. We suppose that $\phi_1(t)$, $\phi_2(t)$ are continuous functions of the real variable t, such that t increases constantly from t_0 to T, as the point (x, y) describes the path L from z_0 to Z. For instance, t may be the length of the arc of the curve. While it is supposed that L is an unbroken path from z_0 to Z, it is not assumed that the various portions of the path are represented by arcs of one and the same curve. That is to say, while $\phi_1(t)$, $\phi_2(t)$ are to be continuous along L, they need not be represented throughout the path by one and the same pair of arithmetic expressions. We may assume, accordingly, that $\phi_1'(t)$, $\phi_2'(t)$ are subject to a finite number of discontinuities of finite amount.

Suppose (x_0, y_0) and (X, Y) to be the terminal points z_0, Z, and let (x_1, y_1), (x_2, y_2), $\cdots$, (x_{n-1}, y_{n-1}) be intermediate points on the path L, while $t_1, t_2, \cdots, t_{n-1}$ are the corresponding values of t. Let us construct the sum

$$U(\xi_1, \eta_1)(x_1 - x_0) + U(\xi_2, \eta_2)(x_2 - x_1) + \cdots + U(\xi_n, \eta_n)(X - x_{n-1}),$$

where $(\xi_1, \eta_1), \cdots, (\xi_n, \eta_n)$ are points on the n arcs which connect the points $(x_0, y_0), (x_1, y_1), \cdots, (X, Y)$, and $\tau_1, \tau_2, \cdots, \tau_n$ are the corresponding values of t. When the number n is increased, and each arc diminished indefinitely, this sum tends to a definite limit, whatever be the mode of interpolation of the points (x_κ, y_κ) and $(\xi_\kappa, \eta_\kappa)$. This limit is the same as that of

$$U\{\phi_1(\tau_1), \phi_2(\tau_1)\}\{\phi_1(t_1) - \phi_1(t_0)\} + \cdots,$$

or that of
$$U\{\phi_1(\tau_1), \phi_2(\tau_1)\}\{\phi_1'(t_0) + \epsilon\}(t_1 - t_0) + \cdots,$$

where $|\epsilon|$ vanishes with $|t_1 - t_0|$. This limit, by the properties of ordinary definite integrals of a single real variable t,

$$= \int_{t_0}^{T} U(\phi_1, \phi_2)\phi_1' dt.$$

This is the meaning of the curvilinear integral $\int_L U dx$. Similarly, if $V(x, y)$ be a continuous function of x, y, the meaning to be attached to $\int_L V dy$ is $\int_{t_0}^{T} V(\phi_1, \phi_2)\phi_2' dt$; and generally by $\int_L U dV$ is understood $\int_{t_0}^{T} U \frac{dV}{dt} dt$, where U, V are expressed as functions of t by means of the equations $x = \phi_1(t), y = \phi_2(t)$.

The conception of the definite integral of a one-valued function of a real variable, as the limit of the sum of infinitely many elements, is capable of extension to functions of z. Let $f(z)$ be a function which is one-valued and continuous along a path L, of finite length l, which leads from z_0 to Z; and let $z_1, z_2, \cdots, z_{n-1}$ be intermediate points on L.

Further, let $\zeta_1, \zeta_2, \cdots, \zeta_n$ be points situated on the arcs

$$(z_0, z_1), (z_1, z_2), (z_2, z_3), \cdots, (z_{n-1}, Z).$$

We shall prove that the sum

$$\Sigma f(\zeta_\kappa)(z_{\kappa+1} - z_\kappa) = f(\zeta_1)(z_1 - z_0) + f(\zeta_2)(z_2 - z_1) + \cdots + f(\zeta_n)(Z - z_{n-1})$$

tends towards a perfectly determinate limit, when the number n increases indefinitely and each interval $z_{\kappa+1} - z_\kappa$ tends to zero. This limit will be called the integral of $f(z)$ with regard to the path L from z_0 to Z, and will be written $\int_{z_0}^{Z} f(z) dz$. That the absolute value of the integral is finite is evident from the consideration that for all values of n, and all positions of $z_1, z_2, \cdots, z_{n-1}, \zeta_1, \zeta_2, \cdots, \zeta_n$,

$$|\Sigma f(\zeta_\kappa)(z_{\kappa+1}-z_\kappa)| \leqq M\{|z_1-z_0|+|z_2-z_1|+\cdots+|Z-z_{n-1}|\},$$

where M is the greatest absolute value of $f(z)$ upon the path L. When n tends to ∞, the coefficient of M becomes the length l of the path L. This proves that the absolute value of the definite integral is finite. To show that the sum tends to a *limit*, let us write $u+iv$ for $f(z)$ and $x+iy$ for z. Thus $\lim \Sigma f(\zeta_\kappa)(z_{\kappa+1}-z_\kappa)$

$$= \lim \Sigma [u(\xi_\kappa, \eta_\kappa)+iv(\xi_\kappa, \eta_\kappa)](\overline{x_{\kappa+1}-x_\kappa}+i\overline{y_{\kappa+1}-y_\kappa})$$

$$= \int_L (udx - vdy) + i\int_L (udy + vdx),$$

an expression whose value does not depend upon the mode of interpolation of $z_1, z_2, \cdots, z_{n-1}, \zeta_1, \cdots, \zeta_n$.

It is very important to notice that the limit has been proved to exist, for a given path between z_0 and Z, on the two conditions that $f(z)$ is one-valued and continuous along that path. If $f(z)$ be a many-valued function, one of its values must be selected at z_0, and then the succession of values along the path from z_0 to Z is completely determined by the principle of continuity, provided the path does not traverse a singular point at which the selected value becomes discontinuous or equal to another value.

It follows from the definition that, for one and the same path,

$$\int_{z_0}^{Z} f(z)dz = -\int_{Z}^{z_0} f(z)dz,$$

since the element, $(z_{\kappa+1}-z_\kappa)f(\zeta_\kappa)$, of the first integral is replaced by $(z_\kappa - z_{\kappa+1})f(\zeta_\kappa)$ in the second. As the integral $\int_{z_0}^{Z} f(z)dz$ is reducible to real curvilinear integrals, the ordinary processes such as integration by parts, change of the independent variable, and so on, are still applicable. When z is changed to a new variable z', the path L in the z-plane must, of course, be replaced by the corresponding path L' in the z'-plane; z' must be so chosen as to be one-valued and continuous along the path L'.

§ 133. The problem which arises is to find the effect of a change of path from z_0 to Z upon the value of $\int_{z_0}^{Z} f(z)dz$. Cauchy discovered, in this connexion, a theorem which is of vital importance in the Theory of Functions.

Cauchy's Theorem. If $f(z)$ and its derivative $f'(z)$ be one-valued and continuous on the boundary of and within a region Γ, the integral $\int_C f(z)\,dz$ taken in the positive direction over the complete boundary C is equal to 0.

For simplicity we shall postpone, for the present, the examination of the case in which the bounding contour is complex.

Many proofs have been given of this fundamental theorem in the Theory of Functions. The one which we shall give is due to Goursat (Démonstration du Théorème de Cauchy. Acta Math., t. iv., p. 197). This proof is instructive inasmuch as it makes direct use of the property of monogenic functions that the value of the derivative of $f(z)$ at z_1 is independent of the path along which z approaches z_1.*

Let the region Γ, with a simple contour, be divided by two systems of equidistant lines, parallel to two rectangular axes, into a network of small regions. The interior regions are squares of area h^2, while exterior regions are *portions* of squares. The integral

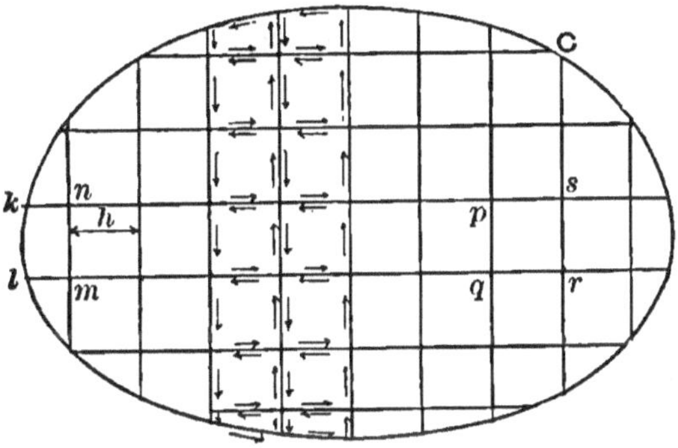

Fig. 39

$\int_C f(z)\,dz$ is equal to the sum of the integrals taken positively over the rim of each region, for the figure shows that each line other than C is traversed twice, in *opposite* directions, with the same value of $f(z)$. We shall first examine the value of an integral taken round a complete square, such as *pqrs*, whose contour may be denoted for shortness by C_ι. If z_ι be any point inside C_ι, we have, for each point of C_ι,

$$\frac{f(z)-f(z_\iota)}{z-z_\iota}=f'(z)+\epsilon,$$

where $|\epsilon|$ varies along C_ι, acquiring a maximum value ϵ_ι.

* We reserve for Chapter VI. a proof by Riemann of the generalized theorem which relates to multiply connected surfaces. Riemann's proof has justly become classical, but it is open to the objection that it employs double integrals in the proof of an elementary theorem.

Hence $$\int_{C_\iota} f(z)\,dz = \int_{C_\iota} \epsilon(z - z_\iota)\,dz,$$

because $\int_{C_\iota} dz$ is zero from the definition of an integral, and if

$$z^2 = t,\ \int_{C_\iota} z\,dz = \tfrac{1}{2}\int dt = 0.$$

Thus the absolute value of $\int_{C_\iota} f(z)\,dz$

$$\leqq \int_{C_\iota} |\epsilon(z - z_\iota)\,dz| < \epsilon_\iota \cdot h\sqrt{2} \cdot 4h$$
$$< 4\sqrt{2}\,\epsilon_\iota(\text{area of square}) \quad . \quad . \quad . \quad . \quad . \quad (A).$$

We next examine the value of $\int_{D_\iota} f(z)\,dz$, where D_ι is the contour of the incomplete square $klmn$. As before, let z_ι be a point inside D_ι,

then $$\int_{D_\iota} f(z)\,dz = \int_{D_\iota} \epsilon(z - z_\iota)\,dz,$$

but $$\left|\int_{D_\iota} \epsilon(z - z_\iota)\,dz\right| < \epsilon_\iota h\sqrt{2}\int_{D_\iota} |dz|$$
$$< \epsilon_\iota h\sqrt{2}\{\text{perimeter of complete square} + \text{arc } kl\}$$
$$< \epsilon_\iota\sqrt{2}\{4 \cdot \text{area of complete square}\}$$
$$+ \epsilon_\iota h\sqrt{2} \cdot (\text{length of arc } kl) \quad . \quad . \quad . \quad . \quad . \quad (B).$$

The result of adding up all the right-hand sides of the set of inequalities A and B is an arbitrarily small quantity, for the sum is

$$< \mu\sqrt{2}\{4 \text{ sum of all the } \textit{complete} \text{ squares } C_\iota,\ D_\iota$$
$$+ (\text{the length of } C)h\},$$

when μ is the greatest of the quantities ϵ_ι. But, if μ become arbitrarily small with h, this expression vanishes in the limit. Hence

$$\int_C f(z)\,dz = 0.$$

To complete the proof, it must be shown that a quantity h can always be found such that $\mu <$ an arbitrarily assigned small quantity η. By supposition $f(z)$ and $f'(z)$ are one-valued and continuous within Γ and upon C; hence a positive number λ can be found such that

$$\left|\frac{f(z+\zeta) - f(z)}{\zeta} - f'(z)\right| < \eta,$$

provided $|\zeta| \leqq \lambda$. Now if z be within a small square, and $z + \zeta$ be on the perimeter, the maximum value of $|\zeta|$ is $h\sqrt{2}$. Therefore if h be taken $< \dfrac{\lambda}{\sqrt{2}}$, the absolute values of the expressions ϵ will all be less than η, which completes the proof.

§ 134. Cauchy's theorem can be stated in the following form: If a path L_1 from z_0 to z_1 can be deformed continuously into a path L_2 from z_0 to z_1, without passing over any singular point of $f(z)$ in the process, the integral $\int_{z_0}^{z_1} f(z)dz$ is the same for both paths.

For let L_1, L_2 be represented by z_0az_1 and z_0bz_1, and let $I(z_0, a, z_1)$, $I(z_0, b, z_1)$ stand for the integral $\int_{z_0}^{z_1} f(z)dz$ with regard to these two paths. We have

$$I(z_0, a, z_1) + I(z_1, b, z_0) = 0,$$

and
$$I(z_1, b, z_0) = -I(z_0, b, z_1);$$

therefore
$$I(z_0, a, z_1) = I(z_0, b, z_1).$$

This result, which is of fundamental importance, permits the reduction of any path from z_0 to z_1 to a path composed of loops issuing from z_0, and of some standard path from z_0 to z_1.

Theorem. Let $g(z)$ be holomorphic within a region Γ bounded by a simple contour, and let the path of integration lie entirely within Γ; then, if $\gamma(z_1) = \int_{z_0}^{z_1} g(z)dz$, $\gamma(z_1)$ is a holomorphic function which has $g(z_1)$ for its derivative throughout the region Γ.

In the first place $\gamma(z_1)$ is one-valued, for its value is independent of the path from z_0 to z_1, so long as the path lies in Γ. Let $z_1 + h$ be a point near z, and let the path $(z_0, z_1, z_1 + h)$ lie entirely within Γ. Then

$$\gamma(z_1 + h) - \gamma(z_1) = \int_{z_0}^{z_1+h} g(z)dz - \int_{z_0}^{z_1} g(z)dz = \int_{z_1}^{z_1+h} g(z)dz.$$

To determine the value of $\int_{z_1}^{z_1+h} g(z)dz$, take for the path ($z_1$ to $z_1 + h$) the join of the two points. This can be done because the value of the integral is the same for all paths in Γ which connect z_1 and $z_1 + h$. If the point $z_1 + h$ approach z_1, the element of the integral is $(g(z_1) + \epsilon)dz$, where $|\epsilon|$ tends to zero with $|h|$, and the integral itself takes the form

$$\int_{z_1}^{z_1+h} g(z_1)dz + \int_{z_1}^{z_1+h} \epsilon dz = hg(z_1) + \int_{z_1}^{z_1+h} \epsilon dz.$$

But the absolute value of an integral is less than the sum of the absolute values of the elements of the integral; therefore the absolute value of $\int_{z_1}^{z_1+h} \epsilon dz$ is less than $|\eta||h|$, where $|\eta|$ is the greatest value of $|\epsilon|$ on the straight line (z_1 to $z_1 + h$). Hence,

$$\frac{\gamma(z_1 + h) - \gamma(z_1)}{h} = g(z_1) + \eta_1,$$

where $|\eta_1|$ tends to zero with $|h|$. This proves that $\int_{z_0}^{z} g(z)dz$ is a holomorphic function of z within Γ, for it is one-valued, continuous, and possesses a first derivative $g(z)$.

§ 135. Cauchy's theorem can be extended to functions which are holomorphic within a region bounded by more than one closed contour.* For simplicity, we shall consider a region bounded by only two rims; but the argument is general. In the figure, the shaded region is bounded by two contours abc, $a'b'c'$. Let $g(z)$ be a function which is holomorphic within the shaded region, and let the contours be joined by a cut aa'. Because the simple contour $abcaa'b'c'a'a$ bounds a region, we have, by Cauchy's theorem,

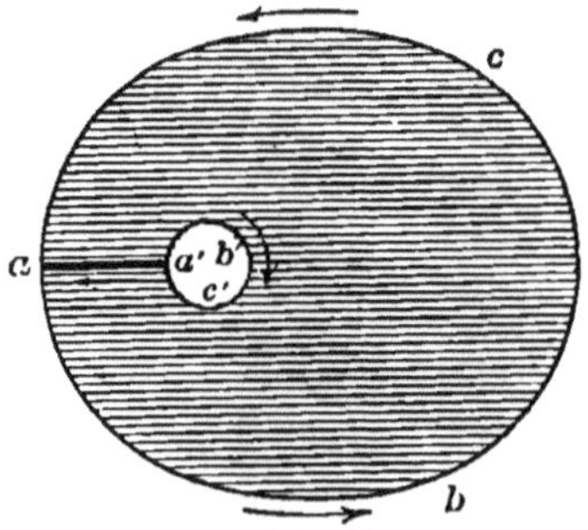

Fig. 40

$$I(a, b, c, a) + I(a, a') + I(a', b', c', a') + I(a', a) = 0;$$

and because the function $g(z)$ is holomorphic within the shaded region

$$I(a, a') = -I(a', a).$$

Hence $$I(a, b, c, a) + I(a', b', c', a') = 0.$$

The contours $abca$, $a'b'c'a'$ are both described *positively* in the direction of the arrows, for the moving points have the shaded region to the left in both cases. As examples of functions holomorphic within a region with two rims, we may instance $\dfrac{1}{(z-a_1)(z-a_2)}$ and either branch of $\sqrt{(z-a_1)(z-a_2)}$, where a_1, a_2 lie inside $a'b'c'$.

In a similar manner it is easy to establish the general theorem:

Theorem. If $g(z)$ be holomorphic within a region Γ, bounded externally by C, and internally by $C_1, C_2, \cdots, C_n$, the sum of the integrals $\int g(z)dz$, taken over all these contours, in the positive direction with regard to the region bounded by them, is equal to zero,

or $$\int_C g(z)dz + \int_{C_1} g(z)dz + \cdots + \int_{C_n} g(z)dz = 0,$$

where the integrals are taken in the directions of the arrows.

* Throughout this chapter, when we speak of taking the integral of a function along the boundary of a region Γ, within which the function is holomorphic, it is to be understood that Γ is itself part of a larger region within which the function is holomorphic, so that the boundary of Γ lies wholly in this larger region.

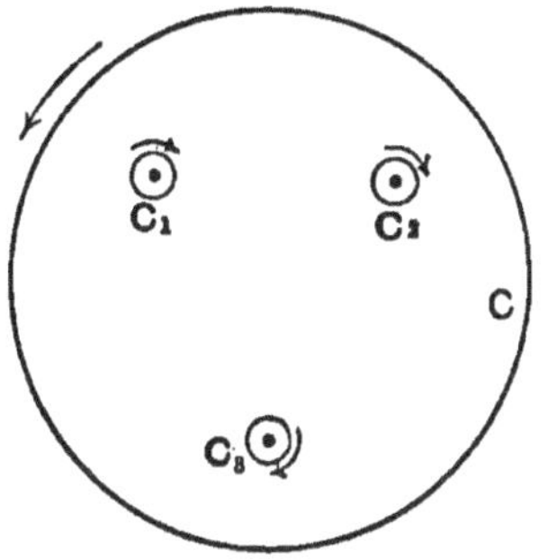

Fig. 41

Corollary. $\int_C g(z)\,dz = \sum_1^n{}_\kappa \int_{C_\kappa} g(z)\,dz,$

when the arrows are all drawn in the same direction.

§ 136. When C_κ encloses no singular point, $\int_{C_\kappa} g(z)\,dz = 0$. Suppose that C_κ encloses one and only one singular point, c_κ. If the function $g(z)$ be a branch of a many-valued function, and if z be confined to the region Γ, there are four conceivable cases:—

(i.) c_κ is a branch-point of the many-valued function, which is operative with regard to $g(z)$.

(ii.) c_κ is a branch-point which is inoperative with regard to $g(z)$.

(iii.) c_κ is a pole.

(iv.) c_κ is an essential singularity.

The hypotheses (i.) and (ii.) must be rejected. In the former case $g(z)$ would change its value after a description of C_κ, and would not be holomorphic in Γ; while, in the latter case, c_κ is not a singular point of the branch $g(z)$. In cases (iii.), (iv.), $\int_{C_\kappa} g(z)\,dz$ may or may not be zero. The following theorem supplies a test:—

Theorem. The integral of any function $f(z)$, taken round a circle whose radius is ρ and centre c, vanishes with ρ if

$$\lim_{z=c} (z-c)f(z) = 0.$$

Let us in future denote by (c) a circle with centre c, so small as to include no singular point other than c. Let μ be the maximum absolute value of $(z-c)f(z)$ on such a circle, whose radius is ρ.

Then $$z - c = \rho e^{i\theta},\quad dz = i\rho e^{i\theta}d\theta,$$

$$\int_{(c)} f(z)\,dz = \int_{(c)} (z-c)f(z)\cdot i\,d\theta.$$

Therefore $$\left|\int_{(c)} f(z)\,dz\right| \leqq \int_{(c)} \mu\,d\theta \leqq 2\pi\mu,$$

and $$\lim_{\rho=0} \int_{(c)} f(z)\,dz \leqq \lim_{\rho=0} 2\pi\mu = 0.$$

Similarly, the limit when $\rho=\infty$ of $\int_{(\infty)} f(z)dz$, where the path of integration is a circle C, with centre 0, so large as to include all singular points in the finite part of the plane, is zero when

$$\lim_{z=\infty} zf(z)=0.$$

On the sphere the circle C is a small circle enclosing the point ∞.

These theorems remain true when the complete perimeters of the circles are replaced by arcs.

§ 137. *The integral* $\int_C \frac{dz}{z-c}$.

Let the contour C contain c. The function $\frac{1}{z-c}$ is not holomorphic within C, and consequently Cauchy's theorem is not immediately applicable. Draw a circle (c) of radius ρ. Within the region bounded by C and (c), $\frac{1}{z-c}$ is holomorphic, and therefore

$$\int_C dz/(z-c)=\int_{(c)} dz/(z-c),$$

where the integrals are taken in the same directions.

Write $z-c\doteq\rho e^{i\theta}$; then

$$\int_{(c)} dz/(z-c)=\int_0^{2\pi} id\theta=2\pi i.$$

Therefore $$\int dz/(z-c)=2\pi i.$$

The contour C divides the sphere into two parts Γ and Γ'. Let c be in Γ. If C be described positively with regard to Γ, it is described negatively with regard to Γ'. The integral with regard to Γ' is not zero although $\frac{1}{z-c}$ is holomorphic in Γ'. The reason is that Γ' contains the point ∞ at which $\lim \frac{z}{z-c}=1\neq 0$.* The value of the integral taken positively round Γ' is of course equal to the integral taken negatively round Γ, or $=-2\pi i$.

The remark is made lest the student should suppose that there is a peculiar virtue in those singular points of an integral which lie inside a contour. The contour C merely divides the critical points

* Generally in considering $\int U dV$ where U and V are functions of z, we must attend to the singular points of V as much as to those of U, in accordance with the equation $UdV+VdU=d(UV)$. The two sets form the *singular points of the integral.*

into two sets, lying in Γ and Γ'; the value of the integral along C depends on the values of the integrals round *either* set of singular points, and in the problem of determining its value we choose the more convenient set; and the advantage of the region which does not contain $z=\infty$ is that the singular points of the integral are the same as those of the function.

Example. Determine the value of $\int z dz/(z-a)(z-b)$ along any given simple contour, from the consideration of the singular points of the integral which lie inside, or outside, the contour.

When n is an integer >1, $\int_C dz/(z-c)^n = 0$, though

$$\lim_{z=c}(z-c)f(z) \neq 0.$$

This instance shows that the condition of the preceding paragraph, though *sufficient* for the vanishing of $\int_C f(z)dz$, is not *necessary*. More generally $\int_C g'(z)dz=0$, where C is any contour lying in a region within which $g(z)$ is holomorphic. If $\int \frac{dz}{z-c}$ taken from any initial point a along a given path to a point b, be equal to I, we can, by making z turn n times round c, give to the integral at b the value $I+2n\pi i$; that is, the integral has at b infinitely many values all congruent to the modulus $2\pi i$. If, however, we draw any line from c to ∞ and regard this as a barrier which z must not cross, the value of the integral is fully determined when the value at any point is given. For instance let $c=1$, and let the barrier lie along the real axis. Let the value of the integral at 2_+ be 0; then the value at 2_- is $2\pi i$, and $\int_{2_+}^{z} \frac{dz}{z-1} = \log|z-1| + i\theta_0$, log denoting the real logarithm, and θ_0 the acute angle $21z$. Compare § 106.

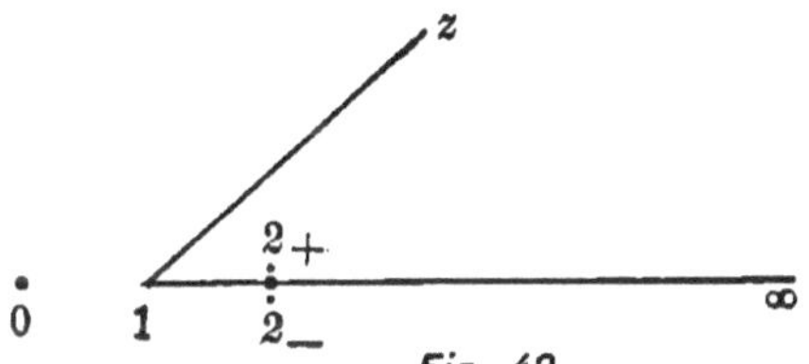

Fig. 42

§ 138. *An example of Cauchy's treatment of the integral of a many-valued function.*

Though the subject of the integral of a many-valued function will be treated in detail later by the help of Riemann surfaces, it seems advisable to give an illustration of the way in which, when by means of loops such a function has been rendered one-valued, Cauchy's theorem may be used.

Let us consider a branch of the two-valued function defined by $w^2 = 1/(1-z^2)$. The points ± 1 are both branch-points and infinities. Let w_1 be that branch which $= 1$, when $z = 0$. As z moves along $0p$ in the direction of the arrow, w_1 remains positive. After the description of the small circle pqr, its value at r is $-1/|\sqrt{1-z^2}|$, and along $r0$ the value is different from that along $0p$. Within the region bounded by C and pqr the branch w_1 is not holomorphic, and we are not justified in saying that

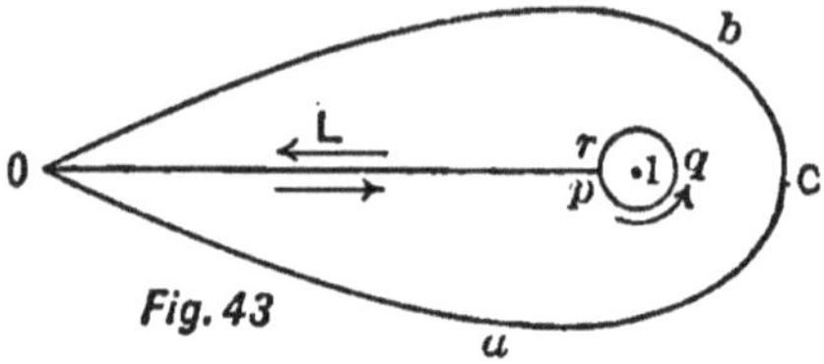

Fig. 43

$$\int_C w_1 dz = \int_{pqr} w_1 dz.$$

Within the region bounded by the simple rim $0ab0rqp0$, w_1 is holomorphic; for $0r$ must now be regarded as distinct from $0p$, and a complete circuit round 1 is prevented by the barrier L.

Hence $\quad I(0, a, b, 0) = I(0, p, q, r, 0)$

$$= \int_0^p \frac{dz}{\sqrt{1-z^2}} + I(p, q, r) + \int_r^0 \frac{dz}{-\sqrt{1-z^2}}$$

$$= 2\int_0^1 \frac{dz}{\sqrt{1-z^2}} + I(p, q, r)$$

$$= \pi + 0 = \pi,$$

since

$$\lim_{\delta=0} \int_0^{1-\delta} \frac{dx}{\sqrt{1-x^2}} = \frac{\pi}{2},$$

and

$$\int \frac{dz}{\sqrt{1-z^2}} \text{ along } pqr = 0$$

in the limit, owing to the fact that

$$\lim_{z=1} (z-1) \cdot \frac{1}{\sqrt{1-z^2}} = 0.$$

More generally, if $c_1, c_2, \cdots, c_n$ be those branch-points of $f(z)$ which are situated within a closed curve C, and if there be no critical point in the region Γ contained between C and the small circles round $c_1, c_2, \cdots, c_n$, a branch of the function $f(z)$ need not be one-valued in the region Γ, but must be one-valued within the region bounded by the single rim, composed of C and the n loops from a point of C to the branch-points. But, if m poles $d_1, d_2, \cdots, d_m$ be contained within Γ, $\int_C f(z)\,dz$ – [the sum of the integrals taken positively over the n loops to the branch-points] $= \sum_1^m \int_{(d_\kappa)} f(z)\,dz.$

It is interesting to observe the way in which Cauchy's theorem accounts for all the values of the integral $\int_0^z \frac{dz}{\sqrt{1-z^2}}$. Any path from 0 to z can be contracted, without passing over any singular point of $1/\sqrt{1-z^2}$, into a path formed by single or repeated descriptions of loops from 0 to $1, -1$ and of a definitely chosen line from 0 to z, as for instance the join of the two points. Let $u, -u$ be the values of the integral along this latter line when the initial values are $w_1, -w_1$. A single positive description of the loop to 1 changes w_1 into $-w_1$, and the corresponding value of the integral $\int_0^z w_1 dz$ is $\pi - u$. A description of the loop to 1, followed by a description of the loop to -1, is equivalent to a description of the first loop with the value w_1 and of the second with the value $-w_1$, the final value being w_1, and therefore the corresponding path from 0 to z with initial value w_1 leads to the value $2\pi + u$ of the integral. By a consideration of all possible combinations of the loops, it is easy to see that $\int_0^z dz/\sqrt{1-z^2}$ has infinitely many values, which are all included in the formulæ

$$u + 2n\pi, \quad \pi - u + 2n\pi,$$

where n is any integer, positive or negative. The resulting theorems in Trigonometry are

$$\begin{cases} \sin(u + 2n\pi) = \sin u, \\ \sin(\pi - u + 2n\pi) = \sin u. \end{cases}$$

A similar process of reasoning would establish the double periodicity of the function $z = \operatorname{sn} u$, defined by the equation

$$u = \int_0^z \frac{dz}{\sqrt{1-z^2 \cdot 1-\kappa^2 z^2}};$$

but we shall consider the elliptic functions (chapter vii.) when the general theory of periodicity has been given in the chapter on Riemann surfaces. We refer the reader who wishes for further information on Cauchy's Theory of Periodicity to the great treatise of Briot and Bouquet, Théorie des Functions Elliptiques; to Clebsch and Gordan, Abelsche Functionen, ch. iv.; and to Jordan's Cours d'Analyse, t. ii.

§ 139. *A second theorem of Cauchy's.*

Let $g(z)$ be holomorphic within and on the boundary of a region Γ, bounded by a simple contour C, and let t be any point within the region. The function $g(z)/(z-t)$ becomes infinite when $z=t$. The function $g(z)/(z-t)$ is holomorphic in the region bounded by C and (t). Hence

$$\int_C \frac{g(z)dz}{z-t} = \int_{(t)} \frac{g(z)dz}{z-t}.$$

Now, since $g(z)$ has a derivative within Γ, $\{g(z)-g(t)\}/(z-t)$ has a finite and determinate limit when $z=t$, and therefore is a holomorphic function throughout Γ. Cauchy's theorem gives

$$\int_{(t)} \frac{g(z)-g(t)}{z-t}dz = 0,$$

or

$$\int_{(t)} \frac{g(z)dz}{z-t} = g(t)\int_{(t)} \frac{dz}{z-t} = 2\pi i g(t).$$

Cauchy's integral for $g(t)$ is $\dfrac{1}{2\pi i}\displaystyle\int_C \frac{g(z)dz}{z-t}$.

From this result can be deduced expressions for the successive derivatives. By subtraction, and after division by Δt, there results from the two formulæ

$$g(t) = \frac{1}{2\pi i}\int_C \frac{g(z)dz}{z-t};\quad g(t+\Delta t) = \frac{1}{2\pi i}\int_C \frac{g(z)dz}{z-t-\Delta t},$$

the third formula
$$g'(t) = \frac{1}{2\pi i}\int_C \frac{g(z)dz}{(z-t)^2}.$$

But, everywhere on C, $z-t$ differs from 0; therefore $g(t)$ has a single first derivative, and this derivative is holomorphic throughout Γ. $g'(t)$ has, in turn, a holomorphic first derivative $g''(t)$, and so on. Hence when a function is holomorphic in the región Γ, all its derivatives are holomorphic in the same region.

§ 140. *Taylor's theorem.* Let $g(z)$ be holomorphic in and on a circle C; and let c be the centre of the circle, $c+t$ any inside point. We have

$$\begin{aligned} g(c+t) &= \frac{1}{2\pi i}\int_C \frac{g(z)}{z-c-t}dz \\ &= \frac{1}{2\pi i}\int_C g(z)dz\left\{\frac{1}{z-c} + \frac{t}{(z-c)^2} + \cdots + \frac{t^n}{(z-c)^{n+1}}\right. \\ &\qquad\left. + \frac{t^{n+1}}{(z-c)^{n+1}(z-c-t)}\right\}, \\ &= g(c) + tg'(c) + t^2\frac{g''(c)}{2!} + \cdots + t^n\frac{g^n(c)}{n!} + R_n, \end{aligned}$$

where $$R_n = \frac{1}{2\pi i}\int_C g(z)\,dz\,\frac{t^{n+1}}{(z-c)^{n+1}(z-c-t)}.$$

Now $$|R_n| \leqq \frac{1}{2\pi}|t|^{n+1}\int_C\left|\frac{g(z)}{(z-c)^{n+1}(z-c-t)}\right|d\theta.$$

Let $\left|\frac{g(z)}{(z-c)^{n+1}(z-c-t)}\right|$ take its maximum value at the point ζ on C; hence

$$|R_n| \leqq \left|\frac{t}{\zeta-c}\right|^{n+1}\left|\frac{g(\zeta)}{\zeta-c-t}\right|.$$

Since $\left|\frac{t}{\zeta-c}\right| < 1$, $|R_n|$ diminishes as n increases, and, when n is chosen sufficiently great, can be made less than any assignable quantity. Thus Taylor's theorem

$$g(c+t) = g(c) + tg'(c) + \frac{t^2}{2!}g''(c) + \cdots$$

is established.

When the region Γ includes 0, we have, putting $c = 0$, Maclaurin's theorem, namely, $g(t) = g(0) + tg'(0) + \frac{t^2}{2!}g''(0) + \cdots$. As this result expresses that every function $g(z)$, which is holomorphic within a region which includes 0, is expansible as an integral series, it forms a bridge between the theories of Cauchy and Weierstrass.

These important extensions of Taylor's and Maclaurin's theorems to the field of the complex variable were due to Cauchy (Mémoire sur le Calcul des résidus et le Calcul des limites, Turin, 1831. See also the Comptes Rendus, 1846).

When we speak, in what follows, of the whole z-plane, we shall imply that $z = \infty$ is included, whereas, when we speak of the finite part of the plane, we shall exclude $z = \infty$, by drawing on the sphere an infinitely small circle round the point.

Theorem. It is impossible to find a function $g(z)$, which is holomorphic over the whole plane, and not a constant.

For if it were possible, we should have

$$|g(z)| = |g(0) + zg'(0) + \frac{z^2}{2!}g''(0) + \cdots|$$

finite for every point z on the plane. Let G be the greatest value which $|g(z)|$ can take. By the theorem of § 96, the absolute value of the coefficient of $z^n < G/R^n$, where $R(>|z|)$, can be made as great as we please. That is, the absolute value of a constant coefficient is less than a quantity which can be made arbitrarily small. Such a constant coefficient must be zero. In this way it can be proved that if G be finite, all the terms, except the first, in Maclaurin's expansion for $g(z)$ vanish. This important theorem (due to Liouville), shows that if a function $g(z)$ be holomorphic throughout the finite part of the plane, and also at ∞, it must be constant. A function $g(z)$, which is holomorphic over the finite part of the plane, is of course no exception. Its property is that it has no infinity *at a finite distance* from 0.

§ 141. *Laurent's Theorem.* Let $g(z)$ be holomorphic in the ring bounded by two concentric circles C', C'', with centre c; and let $c+t$ be any point in the ring. Laurent's theorem states that $g(c+t)$ can be expanded in a convergent series of the form $\sum\limits_{-\infty}^{+\infty} a_m t^m$. Describe a small circle C''' with centre $c+t$. Then

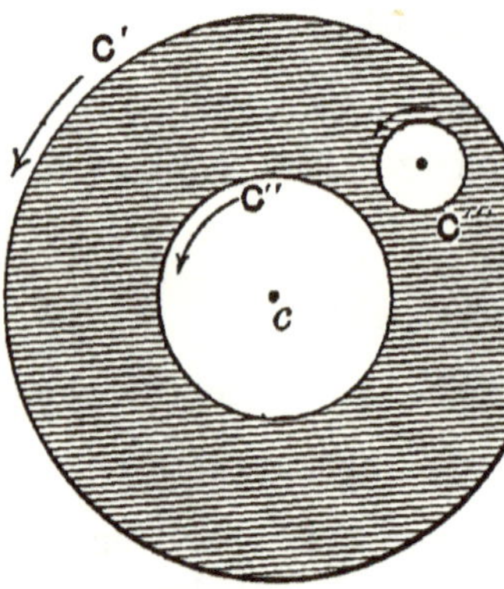

Fig. 44

$$g(c+t)=\frac{1}{2\pi i}\int_{C'''}\frac{g(z)dz}{z-c-t},$$

or, since $g(z)$ is holomorphic in the shaded region,

$$g(c+t)=\frac{1}{2\pi i}\int_{C'}\frac{g(z)dz}{z-c-t}-\frac{1}{2\pi i}\int_{C''}\frac{g(z)dz}{z-c-t},$$

where the integration is performed in the sense of the arrows. On the circle C', $|z-c|>t$,

$$\frac{1}{z-c-t}=\frac{1}{z-c}+\frac{t}{(z-c)^2}+\frac{t^2}{(z-c)^3}+\cdots+\frac{t^n}{(z-c)^{n+1}}$$
$$+\frac{t^{n+1}}{(z-c)^{n+1}(z-c-t)},$$

and $$\frac{1}{2\pi i}\int_{C'}\frac{g(z)dz}{z-c-t}=a_0+a_1t+a_2t^2+\cdots+a_nt^n+R_n,$$

where

$$a_\kappa=\frac{1}{2\pi i}\int_{C'}\frac{g(z)dz}{(z-c)^{\kappa+1}},\quad R_n=\frac{1}{2\pi i}\int_{C'}\frac{t^{n+1}g(z)dz}{(z-c)^{n+1}(z-c-t)}.$$

Now it can be proved precisely as in § 140, that the limit of R_n, when $n=\infty$, is zero.

Again, on the circle C'', $|t|>|z-c|$,

$$\frac{1}{c+t-z}=\frac{1}{t}+\frac{z-c}{t^2}+\frac{(z-c)^2}{t^3}+\cdots+\frac{(z-c)^n}{t^{n+1}}+\frac{(z-c)^{n+1}}{t^{n+1}(c+t-z)},$$

and
$$\frac{1}{2\pi i}\int_{C''}\frac{g(z)\,dz}{c+t-z}=\frac{a_{-1}}{t}+\frac{a_{-2}}{t^2}+\cdots+\frac{a_{-n-1}}{t^{n+1}}+R_n,$$

where

$$a_{-\kappa}=\frac{1}{2\pi i}\int_{C''}(z-c)^{\kappa-1}g(z)\,dz,\ R_n=\frac{1}{2\pi i}\int_{C''}\frac{g(z)\,dz(z-c)^{n+1}}{t^{n+1}(c+t-z)},$$

but
$$|R_n|\leqq\frac{1}{2\pi}\cdot\frac{1}{|t|^{n+1}}\int_{C''}\frac{|g(z)||z-c|^{n+1}}{|c+t-z|}d\theta$$

$$\leqq\left|\frac{\zeta-c}{t}\right|^{n+1}\left|\frac{g(\zeta)}{c+t-\zeta}\right|,\text{ where } \zeta \text{ is some point on } C'';$$

therefore, since $\left|\frac{\zeta-c}{t}\right|<1$, $\lim_{n=\infty}R_n=0.$

Hence
$$g(c+t)=\sum_{-\infty}^{\infty}a_m t^m.$$

It is clear that this formula is not proved for the case when $c+t$ is on the *boundary* of a ring-shaped domain within which $g(z)$ is holomorphic. The coefficients of the various powers of t are integrals which extend over C' and C'', but they might equally well be taken over concentric circles D', D'' which lie within the former ring and still include the point $c+t$.

§ 142. *Integration and differentiation of infinite series.*

If $f(z)=\sum_1^{\infty}f_m(z)$ be a uniformly convergent series within a region Γ, and if the path of integration L lie within Γ, then

$$\int_a^b f(z)\,dz=\sum_1^{\infty}\int_a^b f_m(z)\,dz.$$

For let

$$f(z)=f_1(z)+f_2(z)+\cdots+f_n(z)+\rho_n(z);$$

then, however small ϵ may be, it will always be possible to choose n so that throughout the path L $|\rho_n(z)|<\epsilon$. If n be so chosen, .

$$\int_a^b f(z)\,dz=\sum_1^{n}\int_a^b f_m(z)\,dz+\int_a^b\rho_n(z)\,dz,$$

but
$$\left|\int_a^b\rho_n(z)\,dz\right|\leqq\int_a^b|\rho_n(z)||dz|\leqq\int_a^b\epsilon|dz|\leqq\epsilon(\text{length of } L).$$

Hence, when n tends to ∞, $\int_a^b \rho_n(z)dz$ tends to zero, and

$$\int_a^b f(z)dz = \sum_1^\infty \int_a^b f_m(z)dz.$$

An immediate deduction can be made:—

If $$\phi(z) = \phi_1(z) + \phi_2(z) + \phi_3(z) + \cdots$$

be convergent within a certain region, and if

$$\psi'(z) = \phi_1'(z) + \phi_2'(z) + \phi_3'(z) + \cdots$$

be uniformly convergent within the same region, then

$$\psi(z) = \phi(z).$$

That Laurent's series is unique can be shown readily by the use of the above theorem on integration. For if

$$g(c+t) = \sum_{-\infty}^{+\infty} a_m t^m = \sum_{-\infty}^{+\infty} a_m' t^m,$$

we must have

$$\sum_{-\infty}^{+\infty} (a_m' - a_m) t^m = 0.$$

Divide by t^{n+1}, and integrate along any circle, centre c, which lies within the ring. The integral $\int t^{m-n-1} dt$ is zero, except when $m = n$; in this exceptional case $\int \frac{dt}{t} = 2\pi i$. Hence, $2\pi i(a_n' - a_n) = 0$, and $a_n' = a_n$, whatever integer n may be.

Laurent's expansion consists of two parts,

$$a_0 + a_1 t + a_2 t^2 + a_3 t^3 + \cdots,$$

and $$a_{-1} t^{-1} + a_{-2} t^{-2} + a_{-3} t^{-3} + \cdots.$$

The former has a circle of convergence C', and is convergent for every point within that circle. In the latter put $t^{-1} = \tau$; the series

$$a_{-1} t^{-1} + a_{-2} t^{-2} + a_{-3} t^{-3} + \cdots$$

is convergent for all points within the ring C', C''; that is,

$$a_{-1}\tau + a_{-2}\tau^2 + a_{-3}\tau^3 + \cdots$$

is convergent for values τ which arise from points near C''. But if the series be convergent for these values of τ, it is also convergent for all values of τ with smaller absolute values, that is, for all points

t outside C''. Thus the two parts of Laurent's expansion have different domains of convergence, namely, the region inside C' and the region outside C''. The region common to these two domains is the ring.

Laurent's theorem shows that in the neighbourhood of an isolated singular point c, a one-valued function $f(z)$ consists of two parts; an integral series in $z-c$ and an integral series in $\frac{1}{z-c}$. For z may be taken nearer to c than the nearest of the other singular points, and circles C', C'', with centre c, can be so drawn as to include z and no singular points; in other words there is a ring round c which contains z, and within which $f(z)$ is holomorphic. Writing $z=c+t$, we may apply Laurent's expansion. The larger circle C' must stop short of the nearest of the other singular points.

Functions of Several Variables.

§ 143. Let $\phi(z_1, z_2)$ be a function of two independent variables, which is holomorphic when z_1, z_2 are confined to regions Γ_1, Γ_2 bounded by contours C_1, C_2. If c_1, c_2 lie within these regions, we have

$$\frac{1}{2\pi i}\int_{C_1}\frac{\phi(z_1, z_2)dz_1}{z_1-c_1}=\phi(c_1, z_2),$$

$$\frac{1}{2\pi i}\int_{C_2}\frac{\phi(c_1, z_2)dz_2}{z_2-c_2}=\phi(c_1, c_2),$$

and therefore
$$\left(\frac{1}{2\pi i}\right)^2\int_{C_1}\int_{C_2}\frac{\phi(z_1, z_2)dz_1dz_2}{(z_1-c_1)(z_2-c_2)}=\phi(c_1, c_2).$$

More generally, if

$$f(z_1, z_2)=\frac{\delta^{n_1}\phi}{\delta z_1^{n_1}},\quad f(c_1, z_2)=\frac{n_1!}{2\pi i}\int_{C_1}\frac{\phi(z_1, z_2)dz_1}{(z_1-c_1)^{n_1+1}},$$

$$\left(\frac{\delta^{n_2}f}{\delta z_2^{n_2}}\right)_{c_2}=\frac{n_2!}{2\pi i}\int_{C_2}\frac{f(z_1, z_2)dz_2}{(z_2-c_2)^{n_2}},$$

and
$$\frac{\delta^{n_1+n_2}\phi}{\delta c_1^{n_1}\delta c_2^{n_2}}=\frac{n_1!\, n_2!}{(2\pi i)^2}\int_{C_1}\int_{C_2}\frac{\phi(z_1, z_2)dz_1dz_2}{(z_1-c_1)^{n_1+1}(z_2-c_2)^{n_2+1}}.$$

This expression for $\frac{\delta^{n_1+n_2}\phi}{\delta c_1^{n_1}\delta c_2^{n_2}}$ shows that the order of differentiation is indifferent, and also that each partial derivative of $\phi(z_1, z_2)$ is holomorphic throughout the regions Γ_1, Γ_2. By an easy extension of the method of proof used in finding Taylor's expansion, we have

$$\phi(c_1+t_1, c_2+t_2)=\frac{1}{(2\pi i)^2}\int_{c_1}\int_{c_2}\frac{\phi(z_1, z_2)dz_1dz_2}{(z_1-c_1-t_1)(z_2-c_2-t_2)}$$

$$=\phi(c_1, c_2)+\frac{\delta\phi}{\delta c_1}t_1+\frac{\delta\phi}{\delta c_2}t_2+\frac{1}{2\,!}\left(\frac{\delta^2\phi}{\delta c_1{}^2}t_1{}^2+2\frac{\delta^2\phi}{\delta c_1\delta c_2}t_1t_2\right.$$

$$\left.+\frac{\delta^2\phi}{\delta c_2{}^2}t_2{}^2\right)+\cdots.$$

§ 144. *Some general theorems on holomorphic and meromorphic functions.*

A function $g(z)$ which is holomorphic in a region Γ and constant along a curve of finite length, lying in the region, is constant throughout Γ, however small the length may be.

This follows at once from § 86 and § 140.

Hence it results that two functions which are holomorphic in Γ and are equal for infinitely many points of Γ, are equal throughout Γ. with

This theorem marks off Cauchy's monogenic function (Weierstrass's analytic function) from the function used by Dirichlet. In the historical account of the growth of the present views on the nature of functions, we showed that, adopting Dirichlet's definition, no connexion need exist between the values of a function for different values of the variable. The monogenic function $u+iv$ of z is not so general, for it is fettered by the differential equations

$$\frac{\delta u}{\delta x}=+\frac{\delta v}{\delta y},\quad \frac{\delta u}{\delta y}=-\frac{\delta v}{\delta x};$$

but the reader must not suppose that these restrictions interfere with the usefulness of the Theory of Functions. Without them $u+iv$ would cease to possess a definite differential quotient, and fundamental results, such as the expansion by Taylor's theorem, Cauchy's theorem, etc., would fall to the ground.

By § 87, a holomorphic function $g(z)$, which vanishes when $z=c$ and is not identically zero, is, near c, of the form $P_m(z-c)$; that is, it is equal to $(z-c)^m\times$ (a holomorphic function which does not vanish at c). The integer m is called the order of multiplicity, or simply the *order*, of the zero c.

Differentiation shows that

$$g'(z)=P_{m-1}(z-c),\quad g''(z)=P_{m-2}(z-c),$$

and so on. Since a region containing c can be assigned whose area is not infinitely small, within which there is no zero of $g(z)/(z-c)^m$, it follows that the number of zeros of $g(z)$, which lie in Γ, is finite if Γ lie in the finite part of the plane.

Next let us consider a function $h(z)$, meromorphic in Γ. Let c be a pole; then by definition there is a positive integer m, such that $(z-c)^m h(z)$ is holomorphic near c; therefore

$$h(z)=(z-c)^{-m}P_0(z-c),$$

and $$1/h(z)=(z-c)^m P_0'(z-c),$$

and c is a zero of $1/h(z)$, of the order m.

Since there is a region containing c, of area not infinitely small, within which $P_0(z-c)$ is not zero and is not ∞, it follows that the number of poles of $h(z)$ which lie in Γ is finite, if Γ lie in the finite part of the plane, and that a zero and a pole cannot be infinitely near.

At a pole C of order m, the function $h(z)$

$$=(z-c)^{-m}P_0(z-c)=\frac{A_1}{(z-c)^m}+\frac{A_2}{(z-c)^{m-1}}+\cdots+\frac{A_m}{z-c}+P(z-c).$$

The importance of this expression lies in the division of $h(z)$ into a purely algebraic part

$$\frac{A_1}{(z-c)^m}+\cdots+\frac{A_m}{z-c},$$

which accounts for the polar discontinuity, and into a function holomorphic at c.

§ 145. *Theorem.* A function $h(z)$ which has, within a region Γ in the finite part of the plane, a finite number of poles, can be expressed as the sum of a rational fraction and a function holomorphic throughout Γ.

For let $h(z)$ have, within Γ, ν poles $c_1, c_2, \cdots, c_\nu$ of orders $m_1, m_2, \cdots, m_\nu$. In the neighbourhood of c_1,

$$h(z)=\frac{A_1}{(z-c_1)^{m_1}}+\frac{A_2}{(z-c_1)^{m_1-1}}+\cdots+\frac{A_{m_1}}{z-c_1}+h_1(z),$$

where $h_1(z)$ is not infinite at c_1, but *is* infinite at $c_2, c_3, \cdots, c_\nu$ with orders $m_2, m_3, \cdots$. As before, it can be proved that

$$h_1(z)=\frac{B_1}{(z-c_2)^{m_2}}+\frac{B_2}{(z-c_2)^{m_2-1}}+\cdots+\frac{B_{m_2}}{z-c_2}+h_2(z),$$

where $h_2(z)$ is not infinite at c_1, c_2, but *is* infinite at $c_3, c_4, \cdots, c_\nu$. Proceeding in this way, we arrive finally at a function $h_\nu(z)$ which is holomorphic throughout Γ. Hence

$$h(z) = \frac{A_1}{(z-c_1)^{m_1}} + \frac{A_2}{(z-c_1)^{m_1-1}} + \cdots + \frac{A_{m_1}}{z-c_1}$$
$$+ \frac{B_1}{(z-c_2)^{m_2}} + \frac{B_2}{(z-c_2)^{m_2-1}} + \cdots + \frac{B_{m_2}}{z-c_2}$$
$$+ \cdot \cdot \cdot \cdot \cdot \cdot \cdot \cdot \cdot \cdot \cdot \cdot \cdot \cdot$$
$$+ \text{ a function which is holomorphic throughout } \Gamma.$$

Theorem. If $g(z)$ be holomorphic throughout the finite part of the plane, with a pole at ∞ of order m, it must be an integral polynomial of degree m in z.

For the one-valued function $g\left(\frac{1}{z}\right)$ is holomorphic everywhere except at the origin, where it has a pole of order m. It must be of the form
$$\frac{a_m}{z^m} + \frac{a_{m-1}}{z^{m-1}} + \cdots + \frac{a_1}{z} + \phi(z);$$
but $\phi(z)$ has no singularity in the plane, and reduces to a constant a_0 (§ 140); hence
$$g(z) = a_0 + a_1 z + a_2 z^2 + \cdots + a_m z^m.$$

Theorem. If a function $h(z)$ have no infinities in the plane other than a finite number of poles, it must be a rational fraction.

For $h(z) =$ a rational fraction $+$ a function which is holomorphic throughout the plane $=$ a rational fraction $+$ a constant.

Theorem. Two meromorphic functions $h(z)$ and $h_1(z)$, which have the same zeros and the same infinities, with the same orders of multiplicity, are in a constant ratio to each other. For $h_1(z)/h(z)$ can be infinite at no places other than the infinities of the numerator and the zeros of the denominator; but, by hypothesis, each infinity of the numerator is neutralized by an infinity of the denominator, and each zero of the denominator by a zero of the numerator. Hence $h_1(z)/h(z)$ is holomorphic throughout the plane, and must be a constant.

Residues.

§ 146. Let c be an isolated singular point of a one-valued function $f(z)$; the function can, as has been said, be expanded in positive and negative powers of $z-c$, when z is near c, by means of Laurent's theorem. If we integrate round c, the integral of each term is zero, excepting the term in $\frac{1}{z-c}$. Let the coefficient of this term be a;

therefore $$\int_{(c)} f(z)\,dz = 2\pi i a.$$

The coefficient a is called (after Cauchy) the *residue* of $f(z)$ at the point c.

The integral being taken in the positive direction with regard to c, we have in the case when c is ∞

$$\int_{(\infty)} f(z)\,dz = -2\pi i a',$$

where a' is the coefficient of $\frac{1}{z}$ in the expansion at ∞. There is accordingly a residue at ∞ when it is a regular point, while at other regular points of the function the residue is zero; the reason being that the residue belongs to the integral rather than to the function. If $\lim_{z=c} (z-c)f(z)$ be finite, this limit is the residue at c, and if $\lim_{z=\infty} zf(z)$ be finite, this limit, with sign changed, is the residue at ∞.

If we cannot apply Laurent's theorem at a point c, as may happen when c belongs to a cluster of essential singularities, we may draw a contour C including the cluster, and define the residue as

$$\frac{1}{2\pi i}\int_C f(z)\,dz.$$

It is evident, by differentiation, that the only infinities of $h'(z)$, in Γ, are poles of orders $m_1+1, \cdots, m_\nu+1$, at $c_1, \cdots, c_\nu$.

Cauchy's theorem (§ 135) may now be stated as follows: If a function $f(z)$ be holomorphic in a region Γ, bounded by a simple contour C, except at isolated singular points $c_1, c_2, \cdots, c_n$, and if the residues at these points be $a_1, a_2, \cdots, a_n$, then

$$\int_C f(z)\,dz = 2\pi i \sum_1^n a_\kappa.$$

For $$\int_C f(z)\,dz = \sum_1^n \int_{(c_\kappa)} f(z)\,dz,$$

and $$\int_{(c_\kappa)} f(z)\,dz = 2\pi i a_\kappa.$$

The theorem is true for both the regions into which C divides the plane, if the conditions apply to both. Therefore, for a function holomorphic over the whole plane, except at isolated singular points, the sum of the residues is zero; the residue at infinity being included.

Example. Verify that the sum of the residues is zero in the cases: —

$$(1)\ \frac{z^n}{(z-a)(z-b)(z-c)}, \quad (2)\ e^{1/z}.$$

When $h(z)$ has a pole of order m at c, $g(z)D\log h(z)$ has a residue $-mg(c)$; $g(z)$ being any function holomorphic in Γ, which does not vanish at c, and D standing for $\frac{d}{dz}$.

For $\qquad h(z)=(z-c)^{-m}P_0(z-c),\ g(z)=g(c)+P_1(z-c),$

and therefore $\qquad g(z)D\log h(z)=-\dfrac{mg(c)}{z-c}+P(z-c).$

Similarly at a zero c' of order m' the residue of $g(z)D\log h(z)$ is $m'g(c')$. Therefore, if C be any simple contour lying in Γ, we have, by the preceding theorem,

$$\int_C g(z)D\log h(z)=2\pi i\Sigma m'g(c')-2\pi i\Sigma mg(c),$$

the summations being for all zeros and poles of $h(z)$ which lie in C.

The theorem is equally true if $g(z)$ vanish at any of the points c or c'.

Corollary. If μ' be the number of zeros, and μ the number of poles, within a region Γ bounded by a contour C,

$$\frac{1}{2\pi i}\int_C D\log h(z)=\mu'-\mu,$$

provided a zero of order m' is counted as m' zeros and a pole of order m as m poles.

Example. Show from this formula that the number of zeros of the integral polynomial $z^n+a_1z^{n-1}+a_2z^{n-2}+\cdots+a_n$ is n.

*Essential Singularities.**

§ 147. If $a_0+a_1z+a_2z^2+\cdots$ be a series whose circle of convergence is infinitely great, the series

$$a_0+a_1/z+a_2/z^2+a_3/z^3+\cdots$$

---

* See § 99.

will represent a function which is holomorphic throughout the plane, except at the point 0. The singularity at this exceptional point is of an entirely different nature from that at a pole, or, to state the matter in another way, the behaviour of

$$a_0 + a_1/z + a_2/z^2 + a_3/z^3 + \cdots \text{ to } \infty,$$

at $z = 0$, is entirely different from that of

$$a_0 + a_1/z + \cdots + a_n/z^n,$$

at the same point. Consider the function $e^{1/z}$. This function is holomorphic throughout the plane, the point $z = 0$ excepted, and

$$= 1 + 1/z + 1/2!\,z^2 + 1/3!\,z^3 + \cdots.$$

If z be near the origin, we may put $z = \rho e^{\theta i}$, where ρ is small. The function $e^{1/z}$ becomes $e^{\cos\theta/\rho} \times e^{-i\sin\theta/\rho}$, and the absolute value of $e^{1/z} = e^{\cos\theta/\rho}$. The value of $e^{\cos\theta/\rho}$ depends upon the direction θ in which z approaches the origin. If $\cos\theta$ be negative, $e^{\cos\theta/\rho} = 0$ when $\rho = 0$; if $\cos\theta$ be positive, $e^{\cos\theta/\rho} = \infty$, when $\rho = 0$; if $\cos\theta$ be zero, $e^{\cos\theta/\rho} = 1$. There is also indetermination as regards the amplitude $-\sin\theta/\rho$. It is easy to show that $e^{1/z}$ takes an arbitrarily prescribed value $\alpha + i\beta$ at infinitely many points within a circle, whose centre is at the origin, no matter how small the radius may be; for the equation

$$e^{1/z} = \alpha + i\beta = ce^{\nu i}$$

gives $\quad 1/z = \log c + i(\nu + 2m\pi)$, where m is any integer;

or $\quad x + iy = 1/(A + iB)$, where $A = \log c$, $B = \nu + 2m\pi$.

Thus $\quad x = A/(A^2 + B^2)$, $y = -B/(A^2 + B^2)$;

that is, the absolute values of x and y can, by varying m, be made to take infinitely many values less than assigned positive quantities, however small. The argument fails when $A = \infty$; that is, when $c = 0$ or $c = \infty$.

Next let us consider the function $1/\sin(1/z)$. At a point $z = 1/m\pi$, where m is an integer, the function has a polar infinity, for the reciprocal function $\sin 1/z$ vanishes. This is true however large the integer m may be; therefore, however small be the radius of a circle, centre 0, it is always possible to find an integer μ such that for $m = \mu$, and for all greater integers, $1/m\pi$ is less than the radius of the circle. Thus $1/\sin 1/z$ has infinitely many poles accumulated in the neighbourhood of $z = 0$, and the function cannot be expanded in the neighbourhood of $z = 0$.

The singularity of the many-valued function $\log z$ at $z=0$ is of a different nature. Within the circle $|z|=\rho$, the absolute value of $\log z \overline{\gtrless} \log \rho$, and therefore $\log z$ cannot take all assigned values within the circle. The origin is here a branch-point at which infinitely many branches unite.

§ 148. A point c is a pole or ordinary singularity of the one-valued function, when we can transform the function by multiplying by a power of $z-c$, into one which presents no singularity at $z=c$; otherwise $z=c$ is an essential singularity. In the reciprocal of a one-valued function $f(z)$, poles become zeros, but essential singularities remain essential singularities. When a function has infinitely many polar and essential singularities scattered over the whole plane, a complication may arise by the presence of infinite accumulations of these singularities at special points of the plane. This case has received treatment in some important memoirs (notably that of Mittag-Leffler, Acta Math., t. iv.), but the simpler case is that in which the places of discontinuity in the *finite* part of the plane are isolated. It is this latter case which we shall consider.

Transcendental functions can be classified according to the number of their essential singularities. Those with a finite number are more nearly allied to rational functions than those with an infinite number. After Weierstrass, we use the notation $G(z)$ to denote an *integral* transcendental function (see Ch. iii., § 99). We have used $g(z)$ for functions holomorphic over the finite plane, or a region of it; while $G(z)$ is reserved for functions holomorphic throughout the finite part of the plane. We shall understand by $G\left(\frac{1}{z-c}\right)$ the function defined by the series $\frac{a_1}{z-c}+\frac{a_2}{(z-c)^2}+\cdots$ to ∞, where $a_1z+a_2z^2+\cdots$ defines an integral transcendental function.

A theorem of Weierstrass's. A one-valued function $f(z)$ which has only polar discontinuities in the finite part of the plane, and which has a single essential singularity at ∞, can be represented as the quotient of two integral functions.

The number of polar infinities may be infinitely great, but only a finite number of these are supposed to be scattered over the finite part of the plane. Let the poles be of orders $m_1, m_2, \cdots$ at points $c_1, c_2, \cdots$, such that

$$|c_1| \leqq |c_2| \leqq |c_3| \cdots; \quad L\,|c_n| = \infty.$$

If a function $G_1(z)$ could be constructed with zeros of orders m_1, $m_2, \cdots$ at $c_1, c_2, \cdots$, the functions $G_1(z)\cdot f(z)$ would be finite for all finite values of z, and would therefore be an integral function $G_2(z)$, say. Hence $f(z)=\frac{G_2(z)}{G_1(z)}$. That the function $G_1(z)$ can always be constructed will be proved presently. In the special case when $\Sigma\frac{1}{c_\lambda}$ is a convergent series, Weierstrass has given the following very simple proof:—

Write, tentatively,

$$G_1(z)=(z-c_1)^{m_1}(z-c_2)^{m_2}(z-c_3)^{m_3}\cdots;$$

then
$$\frac{G_1(z)}{G_1(z_0)}=\prod_{\lambda=1}^{\infty}\left(\frac{z-c_\lambda}{z_0-c_\lambda}\right)^{m_\lambda}.$$

This product converges when $\Pi\left(\frac{z-c_\lambda}{z_0-c_\lambda}\right)^M$ converges, where M is the greatest of the quantities m. This second product converges with $\Pi\left(\frac{z-c_\lambda}{z_0-c_\lambda}\right)$, that is, with $\Pi\left(1+\frac{z-z_0}{z_0-c_\lambda}\right)$ or with $\Sigma\frac{z-z_0}{z_0-c_\lambda}$. But $\Sigma\frac{z-z_0}{z_0-c_\lambda}$ converges when $\Sigma\frac{1}{c_\lambda}\cdot\frac{1}{1-z_0/c_\lambda}$ converges, and therefore also when $\Sigma\frac{1}{c_\lambda}$ converges; for, after a sufficient number of terms,

$$\left|\frac{1}{1-\frac{z_0}{c_\lambda}}\right|$$

will always be less than some fixed finite number.

§ 149. *Statement of the theorems of Weierstrass and Mittag-Leffler.*

I. *Factors.* This example suggests the problem: to represent as an infinite product any function which is *holomorphic* throughout the finite part of the plane, and has infinitely many zeros, of which only a finite number lie within any finite distance from the origin. This problem has been completely solved by Weierstrass, who has proved that such a holomorphic function can be expressed in the form

$$\frac{G(z)}{G(0)}=e^{G_0(z)}\Pi\left\{\left(1-\frac{z}{c_\kappa}\right)e^{G_s(z/c_\kappa)}\right\},$$

where $c_1, c_2, c_3, \cdots$ are the zeros of $G(z)$ arranged in ascending order of absolute value, and

$$G_s(z)=z+\frac{z^2}{2}+\cdots+\frac{z^s}{s}.$$

The simplest special case of Weierstrass's theorem is offered by the integral polynomial of degree n, which can be represented as the product of n simple factors. Another example, which will be proved later, is afforded by the formula

$$\sin \pi z = \pi z \prod_{-\infty}^{+\infty}{}' \left(1 - \frac{z}{n}\right) e^{\frac{z}{n}}. *$$

II. *Partial fractions.* We know that any algebraic fraction of the form $\frac{a_0 z^n + a_1 z^{n-1} + \cdots + a_n}{b_0 z^m + b_1 z^{m-1} + \cdots + b_m}$ can be converted into partial fractions, and also that any meromorphic function $h(z)$, with a finite number of poles at $c_1, c_2, \cdots, c_n$ in a region $\Gamma, = \sum_1^n{}_\kappa R\left(\frac{1}{z - c_\kappa}\right) +$ a function $g(z)$ which is holomorphic throughout Γ, where $R\left(\frac{1}{z-c}\right)$ represents an integral polynomial in $1/(z-c)$. This theorem is capable of great extension. Mittag-Leffler has shown that a one-valued function can be constructed with infinitely many polar or essential singularities of assigned characters at places $c_1, c_2, c_3, \cdots$,

such that $\qquad |c_1| \leqq |c_2| \leqq c_3, \cdots, L|c_n| = \infty,$

and that the analytic expression for the function is

$$G(z) + \sum_1^\infty \left\{ G_\kappa\left(\frac{1}{z - c_\kappa}\right) - f_\kappa(z) \right\}.$$

In this theorem the number of discontinuities is infinite in the whole plane, but finite in the *finite* part of the plane. Weierstrass in a brilliant memoir (Zur Theorie der eindeutigen analytischen Functionen) proved the corresponding theorem for the special case of a finite number of singularities.

It is to be noticed in Weierstrass's and Mittag-Leffler's Theorems that, when c_κ is a pole, the series $G_\kappa\left(\frac{1}{z - c_\kappa}\right)$ ends after a finite number of terms, whereas, when c_κ is an essential singularity, the series is infinite but convergent throughout every region of the plane which excludes the c_κ's, and $G_\kappa\left(\frac{1}{z - c_\kappa}\right)$ expresses a transcendental function. Further it must be understood that the points $c_1, c_2, \cdots$, in the *finite* part of the plane are at finite distances apart, so that ∞ is the only limiting point of the infinite system $c_1, c_2, \cdots$.

---

* The conditions for convergence may or may not require that s shall be constant. Laguerre suggested a classification of integral functions based on this theorem. When s is constant he called $G(z)$ a function of 'genre' s. According to this definition $\sin z$ is of 'genre' 1. Hermite, Cours d'Analyse.

Proof of Mittag-Leffler's Theorem.

§ 150. We have to construct a one-valued function $f(z)$ which has for its singular points $c_1, c_2, c_3, \cdots$, and is discontinuous at these points like $G_\kappa\left(\frac{1}{z-c_\kappa}\right)$, $(\kappa=1, 2, 3, \cdots)$. Expand $G_\kappa\left(\frac{1}{z-c_\kappa}\right)$, by Maclaurin's theorem, in the form

$$a_{0,\kappa}+a_{1,\kappa}z+a_{2,\kappa}z^2+\cdots+a_{p,\kappa}z^p+\lambda_p=f_\kappa(z)+\lambda_p, \text{ say,}$$

the circle of convergence extending as far as c_κ. For a given value of z we can, by taking p large enough, make the remainder λ_p as small as we please. Let $\epsilon_1, \epsilon_2, \cdots, \epsilon_n, \cdots$ be positive quantities forming a convergent series, and let p be taken so great that

$$\left|G_\kappa\left(\frac{1}{z-c_\kappa}\right)-f_\kappa(z)\right|=|\lambda_p|<\epsilon_\kappa, \text{ where } |z|<|c_\kappa|.$$

Consider now the series

$$\sum_1^\infty\left|G_\kappa\left(\frac{1}{z-c_\kappa}\right)-f_\kappa(z)\right|.$$

Whatever finite value $|z|$ may have, there is a finite number of singular points c_κ which are less than z in absolute value, and an infinite number which are greater than z in absolute value. Suppose c_n to be the first singular point, such that $|z|\leqq|c_n|$.

The first $n-1$ terms of the series

$$\sum_1^\infty\left|G_\kappa\left(\frac{1}{z-c_\kappa}\right)-f_\kappa(z)\right|$$

form a finite series with a finite sum. Removing these terms, we have a series whose terms are less than

$$\epsilon_n+\epsilon_{n+1}+\cdots, \text{ for all values of } |z|\leqq|c_n|,$$

since, if $|z|\leqq|c_n|$, it must, *a fortiori*, $\leqq|c_{n+1}|$, etc.

The series $\epsilon_n+\epsilon_{n+1}+\epsilon_{n+2}+\cdots$ being, by hypothesis, convergent, so is the series

$$\sum_n^\infty\left|G_\kappa\left(\frac{1}{z-c_\kappa}\right)-f_\kappa(z)\right|,$$

proving that $\sum_1^\infty\left[G_\kappa\left(\frac{1}{z-c_\kappa}\right)-f_\kappa(z)\right]$ is an unconditionally and uniformly convergent series, when z is not situated at one of the points c.

Now in a part of the plane which contains only the singular point c_κ we have assumed that $f(z)=G_\kappa\left(\frac{1}{z-c_\kappa}\right)+$ a function which is holomorphic in the domain of c_κ. Hence, by subtracting the function $G_\kappa\left(\frac{1}{z-c_\kappa}\right)$ from $f(z)$, we render the point c_κ an ordinary point. Instead of subtracting $G_\kappa\left(\frac{1}{z-c_\kappa}\right)$, subtract

$$G_\kappa\left(\frac{1}{z-c_\kappa}\right)-f_\kappa(z);$$

the point c_κ is equally an ordinary point. Hence

$$f(z)-\sum_1^\infty\left[G_\kappa\left(\frac{1}{z-c_\kappa}\right)-f_\kappa(z)\right]$$

is a function $G(z)$, free from singular points in the finite part of the plane. We have, then, the general expression for every one-valued function with isolated singularities; namely,

$$f(z)=G(z)+\sum_1^\infty\left[G_\kappa\left(\frac{1}{z-c_\kappa}\right)-f_\kappa(z)\right].$$

This theorem of Mittag-Leffler's has been stated in the following form by Casorati (Aggiunte a recenti lavori dei Sig[i] Weierstrass e Mittag-Leffler. Annali di Matematica, serie ii., t. x.):—

Given a series of functions of z, which are finitely or infinitely many-valued, let one value be selected for each when z takes a given initial value z_0, and when z describes a path let the function-values at a neighbouring point to z_0 on the z-path be those which succeed, by the law of continuity, those already selected. In this way we single out an infinite series of branches $\phi_1(z)$, $\phi_2(z)$, $\cdots$. Suppose that these branches can be represented by integral series in z, with radii of convergence ρ_1, ρ_2, ρ_3, $\cdots$ such that

$$\rho_1 \leqq \rho_2 \leqq \rho_3 \cdots, \; L\rho_n=\infty;$$

then, by a suitable choice of the integral polynomials,

$$f_1(z), f_2(z), \cdots,$$

the series

$$\sum_1^\infty(\phi_\kappa(z)-f_\kappa(z))$$

can be made unconditionally and uniformly convergent throughout every part of the plane from which the singular points of $\phi_1(z)$, $\phi_2(z)$, $\cdots$, are excluded.

Proof of Weierstrass's Factor-Theorem.

§ 151. Let c_1, c_2, c_3, $\cdots$, differ from one another by finite amounts, and let

$$|c_1| \leqq |c_2| \leqq |c_3| \cdots, \; L|c_n|=\infty;$$

also let $$G_\kappa(z) = \log\left(1 - \frac{z}{c_\kappa}\right),$$

where that branch is selected which

$$= -\frac{z}{c_\kappa} - \frac{z^2}{2c_\kappa^2} - \frac{z^3}{3c_\kappa^3} - \cdots,$$

a series whose radius of convergence is $|c_\kappa|$. Then, by Mittag-Leffler's Theorem,

$$\chi(z) = \sum_1^\infty \left\{ \log\left(1 - \frac{z}{c_\kappa}\right) - f_\nu(z) \right\}$$

can be made to converge unconditionally and uniformly throughout every part of the plane which contains none of the points c_1, c_2, c_3, $\cdots$. This series will represent a branch of a many-valued function of z, for when z describes positively a closed contour round several of the points c_κ, each of the corresponding terms increases by $2\pi i$. As the various branches of $\chi(z)$ differ merely by multiples of $2\pi i$, $e^{\chi(z)}$ must be one-valued.

Hence $$G(z) = e^{\chi(z)} = \prod_1^\infty \left(1 - \frac{z}{c_\kappa}\right) e^{-f_\kappa(z)}$$

is a one-valued integral transcendental function, which vanishes exclusively at the points c_1, c_2, $\cdots$. In this way we have formed a product which is unconditionally and uniformly convergent, and whose factors $\left(1 - \frac{z}{c_\kappa}\right) e^{-f_\kappa(z)}$, called by Weierstrass 'primary factors,' vanish at the assigned points c_1, c_2, $\cdots$ [Casorati, loc. cit.].

To complete the discussion let us consider a mode of construction of the functions $f_\kappa(z)$, $(\kappa = 1, 2, \cdots)$. The problem is to determine integral polynomials $f_\kappa(z)$ such that

$$D \log G(z) = \Sigma \left\{ \frac{1}{z - c_\kappa} - f_\kappa(z) \right\}$$

shall be unconditionally and uniformly convergent. We have

$$\frac{1}{z - c_\kappa} = -f_{\kappa, s}(z) + \frac{z^s}{c_\kappa^s (z - c_\kappa)},$$

where $$f_{\kappa, s}(z) = \frac{1}{c_\kappa} + \frac{z}{c_\kappa^2} + \frac{z^2}{c_\kappa^3} + \cdots + \frac{z^{s-1}}{c_\kappa^s}.$$

Thus $$\Sigma \left\{ \frac{1}{z - c_\kappa} + f_{\kappa, s}(z) \right\} = \Sigma \frac{z^s}{c_\kappa^s (z - c_\kappa)},$$

and the problem will be solved if this latter series be convergent. This will be the case for a *fixed* value of s, if $\Sigma \frac{1}{c_\kappa^{s+1}}$ be absolutely convergent. For many distributions of the points c_κ it is possible to find a fixed value for s which will make the series absolutely convergent. That there are distributions for which s must depend upon κ, is shown by the series

$$\frac{1}{(\log 3)^{s+1}} + \frac{1}{(\log 4)^{s+1}} + \cdots,$$

which is divergent, however great s may be. Let us, then, suppose s a function of κ; as, for instance, $s = \kappa - 1$. The series of absolute values is now

$$\sum_1^\infty \left| \frac{z^{\kappa-1}}{c_\kappa^{\kappa-1}(z - c_\kappa)} \right|,$$

and the ratio of the $(\kappa + 1)$th to the κth term is

$$\left| \frac{z c_\kappa^{\kappa-1}}{c_{\kappa+1\kappa}} \right| \left| \frac{z - c_\kappa}{z - c_{\kappa+1}} \right|,$$

which is evidently zero when $\kappa = \infty$. Thus the series is convergent. That the convergence of $\sum_1^\infty \frac{z^{\kappa-1}}{c_\kappa^{\kappa-1}(z - c_\kappa)}$ is uniform throughout every part of the plane, which excludes the points c, is evident from the consideration that, for each finite value of z, the series can be decomposed into two parts,

$$\sum_1^n \frac{z^{\kappa-1}}{c_\kappa^{\kappa-1}(z - c_\kappa)}, \quad \sum_{n+1}^\infty \frac{z^{\kappa+1}}{c_\kappa^{\kappa-1}(z - c_\kappa)},$$

such that n is finite and $z < |c_{n+1}|$. For all values of z within a region of the z-plane, which contains none of the points c_κ, the latter series is convergent and expressible as an integral series, and the former series consists merely of a finite number of terms.

Now that a satisfactory form has been found for $f_{\kappa,s}(z)$, namely,

$$\frac{1}{c_\kappa} + \frac{z}{c_\kappa^2} + \frac{z^2}{c_\kappa^3} + \cdots + \frac{z^{\kappa-2}}{c_\kappa^{\kappa-1}},$$

we write
$$D \log G(z) - \Sigma \left\{ \frac{1}{z - c_\kappa} + f_{\kappa,s}(z) \right\} = G_0'(z),$$

where $G_0'(z)$ is holomorphic throughout the finite part of the plane. Multiply by dz and integrate; the resulting equation is

$$\frac{G(z)}{G(0)} = \exp G_0(z)\, \Pi \left\{ \left(1 - \frac{z}{c_\kappa}\right) \exp\left(\frac{z}{c_\kappa} + \frac{z^2}{2 c_\kappa^2} + \cdots + \frac{z^{\kappa-1}}{(\kappa - 1) c_\kappa^{\kappa-1}} \right) \right\},$$

where $G_0(0) = 0$.

The reader should notice that the expression for a holomorphic function as an infinite product is fully determined, save as to an exponential factor which nowhere vanishes. Also that when r of the numbers c are equal, the corresponding primary factor must be raised to the exponent r, and that when r of the numbers $c=0$, the preceding reasoning applies to the function $\dfrac{G(z)}{z^r}$, so that $G(z)$ now contains a factor z^r.

§ 152. *An example of the factor-theorem.*

Let $G(z)=\sin \pi z/\pi z$. The zeros of $G(z)$, namely $z=\pm 1, \pm 2, \cdots$, are distributed so that the series

$$\frac{1}{|c_1|}+\frac{1}{|c_2|}+\frac{1}{|c_3|}+\cdots \text{ is divergent,}$$

but

$$\frac{1}{|c_1|^2}+\frac{1}{|c_2|^2}+\frac{1}{|c_3|^2}+\cdots \text{ convergent.}$$

Hence $s=1$, and

$$G(z) = \Pi(1-z/c_\kappa)e^{z/c_\kappa}=\prod_{-\infty}^{+\infty}{}'(1-z/n)e^{z/n},$$

the accent signifying that $n=0$ is omitted. By combining each term $(1-z/n)e^{z/n}$ with the corresponding term $(1+z/n)e^{-z/n}$, we get

$$G(z) = \prod_1^\infty (1-z^2/n^2),$$

or $$\sin \pi z = \pi z(1-z^2/1^2)(1-z^2/2^2)(1-z^2/3^2)\cdots,$$

or $$\sin z = z(1-z^2/\pi^2)(1-z^2/2^2\pi^2)(1-z^2/3^2\pi^2)\cdots.$$

The presence of the exponential in the primary factor ensures the unconditional convergence of the product. The value of $\prod_{-m}^{n}(1-z/\kappa)$ depends upon the ratio n/m, when n, m tend to ∞.*

An Example of Mittag-Leffler's Theorem.†

The series $\Sigma\dfrac{1}{z-n\pi}\,(n=0, \pm 1, \pm 2, \cdots)$ is divergent, but if by Σ' we denote that $n=0$ is not to be included in the summation,

$$\Sigma'\left\{\frac{1}{z-n\pi}+\frac{1}{n\pi}\right\}=\Sigma'\frac{z}{n\pi(z-n\pi)}$$

---

* Cayley, Collected Works, t. i., p. 156.

† See M. De Presle, Bulletin de la Société Mathématique, t. xvi.; also Hermite's Cours d'Analyse.

is convergent. Hence $\cot z - 1/z - \Sigma' z/n\pi(z - n\pi) =$ a holomorphic function $G(z)$. If it can be proved that $G(z)$ does not become infinite as $|z|$ tends to the limit ∞, $G(z)$ must reduce to a constant (§ 140).

(i.) Let $z = x + iy$, where $y \neq 0$. In the expression

$$\frac{\cot x \cot iy - 1}{\cot x + \cot iy} - \frac{1}{x + iy} - \Sigma' \frac{z}{n\pi(z - n\pi)}$$

the first two terms do not tend to ∞ with z, and the third term is finite.

(ii.) Let $y = 0$. The first term is infinite only when $x = m\pi$ (m an integer), the second term is infinite only when $x = 0$, while the third term tends to ∞ only for values of x which tend to $m\pi$. To examine the behaviour in the neighbourhood of $m\pi$, write $x = x' + m\pi$. The limit of the function

$$\cot(x' + m\pi) - \frac{x' + m\pi}{m\pi x'} \text{ is } -\frac{1}{m\pi}, \text{ when } x' = 0.$$

This shows that $\cot x - \dfrac{1}{x} - \Sigma' \dfrac{x}{n\pi(x - n\pi)}$ is finite for every value of x.

(iii.) By combining (i.) and (ii.) we see that $G(z)$ is finite throughout the plane and reduces to a constant. That this constant $= 0$ is evident from the consideration that $\cot z - 1/z - \Sigma' \dfrac{z}{n\pi(z - n\pi)}$ changes sign with z.

Hence $$\cot z = 1/z + \sum_{-\infty}^{\infty}{}' \frac{z}{n\pi(z - n\pi)}.$$

Examples. Prove that

$$\csc z = \frac{1}{z} + (-1)^n \sum_{-\infty}^{\infty}{}' \frac{z}{n\pi(z - n\pi)}.$$

$$\tan z = \Sigma \frac{z}{\frac{r\pi}{2}\left(\frac{r\pi}{2} - z\right)}, \quad r = \pm 1, \pm 3, \pm 5, \cdots.$$

$$\sec z = 1 + \Sigma\, i^{r+1} \frac{z}{\frac{r\pi}{2}\left(z - \frac{r\pi}{2}\right)}, \quad r = \pm 1, \pm 3, \pm 5, \cdots.$$

§ 153. *Theorem.* In the neighbourhood of an essential singularity c, a one-valued function approaches, as nearly as we please, every arbitrarily given magnitude (Weierstrass).

Let A be any arbitrarily assigned magnitude. We have to prove that within any circle, however small, round c as centre, there must be infinitely many points such that the value of the function at each of these differs from A by as small a quantity as we please. If infinitely many roots of the equation $f(z) = A$ lie within the circle, the theorem is clearly true. If, on the contrary, the number of roots be finite, describe round c, as a centre, a circle C in which $f(z)$ is nowhere $= A$. The function $\frac{1}{f(z) - A}$ has an essential singularity at $z = c$, and no pole in the neighbourhood of c. Hence we may use Laurent's theorem:

$$\frac{1}{f(z) - A} = P(z - c) + P\{1/(z - c)\},$$

throughout the interior of C, the point c exclusive. The second series converges throughout the plane. Writing $1/(z - c) = x$, the second series becomes an integral series $P(x)$ convergent for every point within a circle of radius R concentric with C. But, however great R may be, there are points outside C, at which the absolute value is greater than an arbitrarily assigned value. Let x be such a point; to it corresponds a point $z = c + 1/x$, arbitrarily close to c, at which the series is greater, in absolute value, than any assigned quantity. Consequently $f(z)$ has values which differ arbitrarily little from A. See Picard's Cours d'Analyse.

Picard has further shown that, as in the case of $e^{\frac{1}{z}}$ (§ 147), near an essential singularity of a one-valued function there are infinitely many points at which the function is exactly equal to A, except possibly for two values of A. Comptes Rendus, t. lxxxix., p. 745; Bulletin des Sciences Math., 1880.

The first theorem of § 145 can be extended to any one-valued functions. If $f_1(z)$, $f_2(z)$, two one-valued functions with arbitrarily many poles and essential singularities, coincide in value along a line L of length not infinitely small, they must be identically equal.

For $f_1(z) - f_2(z)$, $= \phi(z)$, is zero along L. Let C be a closed curve which does not include a discontinuity of $f_1(z)$, $f_2(z)$, but does include a portion L_1 of L. Within C, $\phi(z)$ is holomorphic; but it $= 0$ along L_1, therefore it must $= 0$ throughout the interior of C. Now C may be drawn as close as we please to a pole c_1 of $f_1(z)$ or $f_2(z)$. The place c_1 cannot be a pole of $\phi(z)$, for this would imply an abrupt change in value from 0 to ∞. Assume c_1 to be an essential singularity of each of the functions $f_1(z)$, $f_2(z)$. All the points inside C and close to c_1 make $\phi(z) = 0$, whereas, near c_1, $\phi(z)$ should be completely indeterminate. It follows that c_1 is not an essential singularity of $\phi(z)$. Hence $\phi(z)$, having neither poles nor essential singularities, reduces to zero. (This proof is due to Picard. It is to be found in Hermite's Cours d'Analyse.)

§ 154. An n-valued function of z, which has in the finite part of the plane only algebraic singularities, is the root of an algebraic equation whose coefficients are meromorphic functions of z.

Let $w_1, w_2, \cdots, w_n$ be the n branches. In the preceding chapter it was shown that near any finite point $z = a$ there exists for each branch an expansion of the form $P(z-a)^{1/r}/(z-a)^{\lambda/r}$ where r and and λ are positive integers. From the behaviour of a cyclic system at an algebraic singularity a it follows that if μ be an integer the sum of the μth powers of all the branches, S_μ, is unaltered when z describes the circle (a). Therefore S_μ is developable in the form $P(z-a)/(z-a)^\nu$, where ν is a positive integer; that is, S_μ is a one-valued function, with no finite singularities except poles; that is, S_μ is meromorphic.

In the equation

$$(w - w_1)(w - w_2)\cdots(w - w_n) = 0 \quad . \quad . \quad . \quad . \quad \text{(i.)},$$

the coefficients are expressible as integral functions of

$$S_\mu, \ (\mu = 1, 2, \cdots, n)$$

and are therefore also meromorphic functions of z.

If the equation in w be irreducible, a closed path can be found which leads from a given initial branch w_1 to any other branch.

For if it be possible to pass from w_1 to $w_2, w_3, \cdots, w_\kappa$, $\kappa < n$, and to no other branch, what has been said about all the branches applies to the system $w_1, w_2, \cdots, w_\kappa$, and we have an equation

$$(w - w_1)\cdots(w - w_\kappa) = 0,$$

with meromorphic coefficients. That is, the equation of degree n admits a factor of the same form but of degree $< n$, and is, therefore, reducible.

If the number of singularities be finite, and if $z = \infty$ be also an algebraic singularity, each coefficient in the equation (i.) has a finite number of poles, and is (§ 145) a rational fraction. In this case w is an algebraic function of z.

§ 155. *Applications of Cauchy's theorems to definite integrals.*

(1) Consider $\int_C e^{iaz}dz/(b^2+z^2)$, where a and b are real and positive, and let the contour C be the semicircle and diameter of Fig. 45.

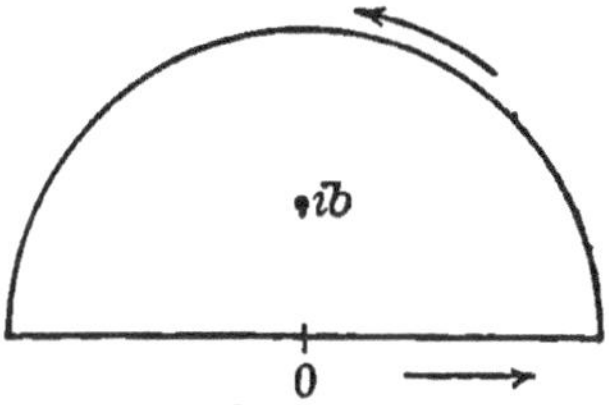

Fig. 45

In the first place,

$$\lim_{z=ib}(z-ib)e^{iaz}/(b^2+z^2)=e^{-ab}/2ib;$$

hence $$\int_{(ib)}=\pi e^{-ab}/b.$$

Secondly,

$$\lim_{z=\infty} z\cdot e^{iaz}/(b^2+z^2)=0,$$

since $|e^{iaz}|=e^{-ay}$, a quantity finite when $y \geqq 0$.

Therefore, when the radius of the semicircle is ∞, the integral along it is zero. There remains the integral along the real axis, which must be equal to the integral round ib.

Hence $$\int_{-\infty}^{\infty} e^{iax}dx/(b^2+x^2)=\pi e^{-ab}/b,$$

and, on separation of the real and imaginary parts,

$$\int_{-\infty}^{\infty}\cos ax dx/(b^2+x^2)=\pi e^{-ab}/b,$$

$$\int_{-\infty}^{\infty}\sin ax dx/(b^2+x^2)=0.$$

(2) Let $$I_1=\int_{-\infty}^{\infty}e^{ax}dx/(1+e^x),$$

$$I_2=\int_{-\infty}^{\infty}e^{ax}dx/(1-e^x),^*$$

where a is real and $0<a<1$.

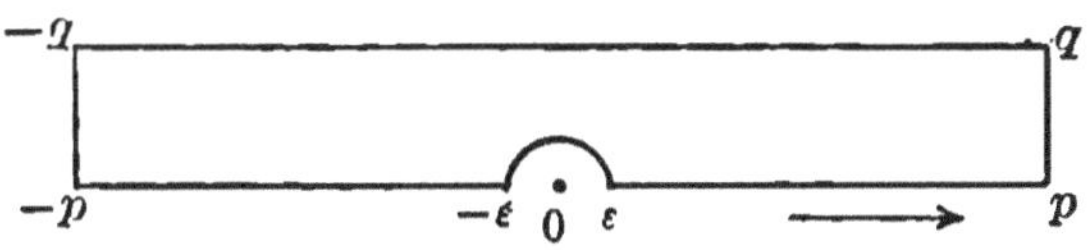

Fig. 46

* In I_2 the path of integration passes through the pole $x=0$ of the integrand. In such a case we define the *principal value* of the integral $\int_a^b f(z)dz$ as the limit of

$$\int_a^{c'} f(z)dz+\int_{c''}^b f(z)dz,$$

where c', c'' are the points in which the path of integration is cut by a small circle with centre at the pole, when the radius of the circle tends to zero. In the example, I_2 is the principal value of the integral.

Let the contour C be as in the figure, the breadth of the rectangle being π, and the length ∞. The integral

$$\int_C e^{az}dz/(1-e^z)=0,$$

since the pole $z=0$ is excluded by the semicircle. Over the paths from p to q and from $-q$ to $-p$ which are of finite length, the integrand vanishes, and therefore the integral is zero. The residue at 0 being $\lim_{z=0} ze^{az}/(1-e^z)$ or -1, the integral along the semicircle $=\pi i$. From q to $-q$, $z=x+\pi i$, and therefore the integral is

$$\int_\infty^{-\infty} e^{ax+\pi i a}dx/(1+e^x) \text{ or } -e^{\pi i a}I_1.$$

Therefore $I_2+\pi i-e^{\pi i a}I_1=0,$

or $\qquad I_2-\cos\pi a\cdot I_1=0,\quad \pi-\sin\pi a\cdot I_1=0,$

or $\qquad I_1=\pi\csc\pi a,\quad I_2=\pi\cot\pi a.$

It may be remarked that if we write the last result in the form

$$\int_{-\infty}^{\infty}\frac{e^{ax}-e^{bx}}{1-e^x}dx=\pi(\cot\pi a-\cot\pi b),\ 0<b<1,$$

there is no longer a pole at the origin, and the statement is not restricted to the principal value.

If we write $e^x=t$, we have

$$\int_0^\infty t^{a-1}dt/(1+t)=\pi\csc\pi a,$$

$$\int_0^\infty (t^{a-1}-t^{b-1})dt/(1-t)=\pi(\cot\pi a-\cot\pi b).$$

(3) In order to make the function $\log\dfrac{\beta-z}{z-a}$ one-valued, we draw a line from a to β, and regard this line as a barrier which z must not cross. Let a and β be real, and let $a<\beta$; the barrier may be drawn along the real axis. Assign a real value of the logarithm to a point x_+ on the positive side of the barrier; then at any point z the value of the logarithm is

$$\log\left|\frac{\beta-z}{z-a}\right|-i(\text{supplement of angle } az\beta).$$

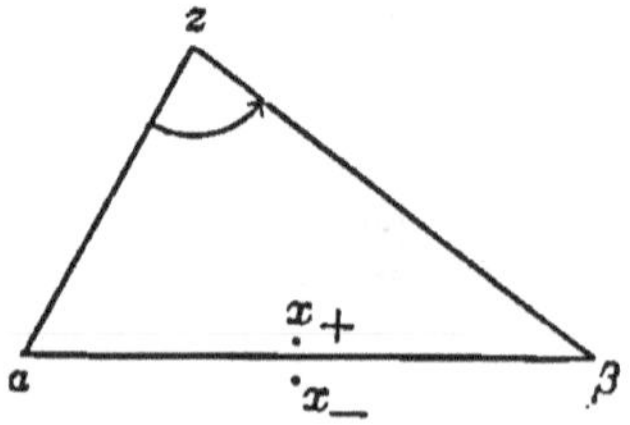

Fig. 47

Consider the integral

$$\int_C\frac{\phi(z)}{\psi(z)}\log\frac{\beta-z}{z-a}dz,$$

where $\dfrac{\phi(z)}{\psi(z)}$ is a rational fraction; for simplicity we suppose that if n be the degree of $\psi(z)$, that of $\phi(z) < n-1$, and that $\psi(z)$ has n simple zeros $z_1, z_2, \cdots, z_n$, none of which lie on the barrier.

Let C consist of the two sides of the barrier and two small circles round a and β. Since the residue at z_r is $\frac{\phi(z)}{\psi'(z_r)}$, we have

$$\int_C = 2\pi i \Sigma \frac{\phi(z_r)}{\psi'(z_r)} \log \frac{\beta - z_r}{z_r - a}.$$

Again, the residues at a and β are zero, therefore $\int_{(a)}$ and $\int_{(\beta)}$ are zero, and $\int_C$ may be written

$$\int_a^\beta f(x_+)dx_+ + \int_\beta^a f(x_-)dx_-,$$

where
$$f(x) = \frac{\phi(x)}{\psi(x)} \log \frac{\beta - x}{x - a},$$

or
$$\int_a^\beta (f(x_+) - f(x_-))dx,$$

or, since the logarithm loses $2\pi i$ in passing from x_+ to x_-,

$$2\pi i \int_a^\beta \frac{\phi(x)}{\psi(x)} dx.$$

Accordingly
$$\int_a^\beta \frac{\phi(x)}{\psi(x)} dx = \Sigma \frac{\phi(z_r)}{\psi'(z_r)} \log \frac{\beta - z_r}{z_r - a},$$

the formula for the integration of the rational fraction $\frac{\phi(x)}{\psi(x)}$.

If by $\log \frac{z-\beta}{z-a}$, we mean that branch which is real when z is real and either $>\beta$ or $<a$, $\log \frac{z-\beta}{z-a} = \log \frac{\beta - z}{z-a} + \pi i$, and the formula may be written

$$\int_a^\beta \frac{\phi(x)}{\psi(x)} dx = \Sigma \frac{\phi(z_r)}{\psi'(z_r)} \log \frac{z_r - \beta}{z_r - a},$$

for
$$\Sigma \frac{\phi(z_r)}{\psi'(z_r)} = 0.$$

(4) In the same way, if $f(z) = \frac{\phi(z)}{\psi(z)} \left(\log \frac{\beta - z}{z - a}\right)^2$,

$$2\pi i \Sigma \frac{\phi(z_r)}{\psi'(z_r)} \left(\log \frac{\beta - z_r}{z_r - a}\right)^2 = \int_a^\beta (f(x_+) - f(x_-))dx$$

$$=\int_a^\beta \frac{\phi(x)}{\psi(x)}\left\{\left(\log\frac{\beta-x}{x-a}\right)^2-\left(\log\frac{\beta-x}{x-a}-2\pi i\right)^2\right\}dx$$

$$=4\pi i\int_a^\beta \frac{\phi(x)}{\psi(x)}\log\frac{\beta-x}{x-a}dx+4\pi^2\int_a^\beta\frac{\phi(x)}{\psi(x)}dx,$$

and therefore

$$\int_a^\beta \frac{\phi(x)}{\psi(x)}\log\frac{\beta-x}{x-a}dx=\tfrac{1}{2}\Sigma\frac{\phi(z_r)}{\psi'(z_r)}\left\{\left(\log\frac{\beta-z_r}{z_r-a}\right)^2+2\pi i\log\frac{\beta-z_r}{z_r-a}\right\}$$

$$=\tfrac{1}{2}\Sigma\frac{\phi(z_r)}{\psi'(z_r)}\left(\log\frac{z_r-\beta}{z_r-a}\right)^2.$$

Examples. (1) Prove that if a be real and $0 < a < 1$,

$$\int_{-\infty}^{\infty}\sin axdx/x=\pi/2.$$

(2) By means of $\int\log\frac{z^2+1}{2z}\cdot\frac{dz}{z}$, taken over the circle $|z|=1$, prove $\int_0^{\pi/2}\log\cos\theta d\theta=-\pi\log 2/2.$

(3) $$\int_a^\beta\log\frac{\beta-x}{x-a}\frac{dx}{x}=\tfrac{1}{2}\left(\log\frac{\beta}{a}\right)^2,$$

where a, β are real and positive, and $a<\beta$.

§ 156. *Differential equations.* The following paragraphs give a succinct account of a method, due to Cauchy, of proving the existence of integrals of differential equations. The mode of presentation is taken from Briot and Bouquet's Traité des Fonctions Elliptiques, second edition, 1875 (Livre V., Fonctions définies par des Équations Différentielles).

It was proved in § 143 that, when $\phi(z_1, z_2)$ is holomorphic in the regions Γ_1, Γ_2 bounded by c_1, c_2, each partial differential quotient can be expressed as a definite integral. Let $\phi(z, w)$ be a function of the two variables z, w, which is holomorphic in two circular regions Γ, Γ_1, of radii ρ, ρ_1; here z, w are regarded as independent variables. The value of $\frac{\delta^{r+s}\phi(z,w)}{\delta z^r\delta w^s}$ at z_0, w_0 (the centres of the two circles)

is given by $$\frac{\delta^{r+s}\phi}{\delta z_0{}^r\delta w_0{}^s}=\frac{r!\,s!}{(2\pi i)^2}\int_{c_1}\int_{c_2}\frac{\phi(z,w)dzdw}{(z-z_0)^{r+1}(w-w_0)^{s+1}}.$$

Write $z-z_0=\rho e^{i\theta}$, $w-w_0=\rho_1 e^{i\theta_1}$, then

$$\frac{\delta^{r+s}\phi}{\delta z_0{}^r\delta w_0{}^s}=\frac{r!\,s!}{4\pi^2\rho^r\rho_1{}^s}\int_0^{2\pi}\int_0^{2\pi}\phi(z_0+\rho e^{i\theta},\ w_0+\rho_1 e^{i\theta_1})e^{-i(r\theta+s\theta_1)}d\theta d\theta_1;$$

and $\left|\frac{\delta^{r+s}\phi}{\delta z_0{}^r \delta w_0{}^s}\right| < r!\ s!\frac{\mu}{\rho^r \rho_1{}^s}$, where μ is the greatest absolute value of $\phi(z, w)$ for the regions Γ, Γ_1. For purposes of comparison we shall need another function,

$$\Phi(w, z) = \frac{\mu}{\left(1 - \frac{z - z_0}{\rho}\right)\left(1 - \frac{w - w_0}{\rho_1}\right)};$$

the partial differential quotient at z_0, w_0 is

$$\frac{\delta^{r+s}\Phi}{\delta z_0{}^r \delta w_0{}^s} = r!\ s!\frac{\mu}{\rho^r \rho_1{}^s},$$

a quantity which is real and positive; and therefore

$$\left|\frac{\delta^{r+s}\phi}{\delta z_0{}^r \delta w_0{}^s}\right| < \frac{\delta^{r+s}\Phi}{\delta z_0{}^r \delta w_0{}^s} \quad \ldots\ldots\ldots\ldots (1).$$

This inequality plays an important part in what follows. Let us now consider the differential equation

$$dw/dz = \phi(w, z),$$

where ϕ is the function discussed above. Suppose that, when $z = z_0$, a value of w is w_0; there will be no loss of generality if we write $z_0 = 0$, $w_0 = 0$, for this merely amounts to a change of both origins. We have to prove the existence of a solution $w = P_1(z)$.

Write tentatively $w = P_1(z)$, so that $w = 0$ when $z = 0$. The coefficients in $P_1(z)$ must be the values of dw/dz, $\frac{1}{2!}d^2w/dz^2$, $\frac{1}{3!}d^3w/dz^3, \cdots$, when $z = 0$ and $w = 0$. We have for their determination a system of equations which are *free from negative terms*,

$$\left.\begin{aligned} w' &= \phi, \\ w'' &= \phi_1 + w'\phi_2, \\ w''' &= \phi_{11} + 2\,w'\phi_{12} + w'^2\phi_{22} + w''\phi_2, \\ &\ldots\ldots\ldots\ldots\ldots, \end{aligned}\right\} \quad \ldots\ldots (2),$$

where $\quad \phi_1 = \delta\phi/\delta z,\ \ \phi_2 = \delta\phi/\delta w,\ \ \phi_{rs} = \delta^{r+s}\phi/\delta z^r \delta w^s.$

The series which is obtained in this way, namely,

$$w_0'z + w_0''z^2/2! + w_0'''z^3/3! + \cdots,$$

has, so far, merely formal significance as an integral. For it has not been proved that it converges in the neighbourhood of (0, 0), or that it actually satisfies the differential equation. It will be sufficient if we show that the series is convergent for a finite domain of (0, 0); for it results, from the method of formation of the successive differential quotients of w, that ϕ and w', together with all their differential quotients with regard to z, are equal at (0, 0). The convergence for a finite domain is found by comparing the series

$$w = P_1(z), \text{ derived from } w' = \phi(w, z),$$

with the series

$$W = Q_1(z), \text{ derived from } W' = \Phi(W, z),$$

where $$\Phi(W, z) = \mu/(1 - W/\rho_1)(1 - z/\rho).$$

The coefficients in $Q_1(z)$ are obtained by equations

$$\left.\begin{aligned} W' &= \Phi, \\ W'' &= \Phi_1 + W'\Phi_2, \\ &\ldots\ldots, \end{aligned}\right\} \quad \ldots\ldots \quad (3),$$

which are precisely similar to the equations (2); and since Φ and all its partial differential quotients are real and positive at (0, 0), all these coefficients are real and positive. Further the inequalities

$$\left.\begin{aligned} |w'| &= |\phi_1|, \\ |w''| &< |\phi_1| + |w'\phi_2|, \\ &\ldots\ldots, \end{aligned}\right\}$$

which are derived from (2) show, with the help of the inequalities (1), that, when $z = 0$, $w = 0$, $W = 0$,

$$\left.\begin{aligned} |w'| &< \Phi_1, \\ |w''| &< \Phi_1 + |w'|\Phi_2, \\ &\ldots\ldots, \end{aligned}\right\}$$

and therefore $|w_0^{(\kappa)}| < W_0^{(\kappa)}$, $\kappa = 1, 2, 3, \cdots$.

It follows that when the series $Q_1(z)$ is convergent, the series $P_1(z)$ is also convergent. It will therefore suffice if we prove the convergence of $Q_1(z)$, or

$$W_0'z + W_0''z^2/2! + W_0'''z^3/3! + \cdots,$$

for a neighbourhood of $z = 0$ which is not infinitely small. This is easily done, for the equation in W can be solved by the separation of the variables, one of the ordinary methods of differential equations. Let $W = 0$ when $z = 0$; then the solution is

$$W - \frac{W^2}{2\rho_1} = -\mu\rho \log\left(1 - \frac{z}{\rho}\right),$$

and the two values of W are found by solving a quadratic.

That branch of W which vanishes when $z=0$ is

$$W=\rho_1-\rho_1\sqrt{1+\frac{2\mu\rho}{\rho_1}\log\left(1-\frac{z}{\rho}\right)},$$

the radical being supposed to reduce to $+1$ when $z=0$. By expanding $\log(1-z/\rho)$, and afterwards the radical, as an integral series, we find that W is equal to an integral series $Q_1(z)$ whose circle of convergence extends to the nearest singular point. This singular point is the branch-point given by

$$1+\frac{2\mu\rho}{\rho_1}\log\left(1-\frac{z}{\rho}\right)=0,$$

that is, $$z=\rho(1-e^{-\rho_1/2\mu\rho})=\rho'(<\rho).$$

For all points within the circle (ρ'), the branch of W in question is given by the series $Q_1(z)$, whose coefficients are all real and positive; therefore

$$|W|\leqq Q_1(|z|).$$

This last expression increases with $|z|$ and is equal to ρ_1 when $|z|=\rho'$. Therefore, for all points within the circle (ρ'), $|W|<\rho_1$.

It follows that the series $P_1(z)$ is convergent within the same circle, and that $|w|<\rho_1$ within this circle. Hence the solution is a true one.

Briot and Bouquet prove, by the following considerations, that the differential equation admits no other solution which vanishes when $z=0$. Suppose that w is the holomorphic integral just found, and that $w+\omega$ is a second integral which vanishes when $z=0$.

Hence $$d\omega/dz=\phi(w+\omega, z)-\phi(w, z).$$

Since the right-hand side vanishes for $\omega=0$ independently of the value of z, it must contain an integral power of w. Thus

$$d\omega/dz=\omega^{\kappa}\psi(z),$$

where $\psi(z)$ is made a function of z only, by writing for w, ω, their values in terms of z. On separating the variables, this equation can be integrated along a curve (0 to ζ) and gives

$$\frac{1}{\kappa-1}(\omega_0^{1-\kappa}-\omega^{1-\kappa})=\int_0^{\zeta}\psi(z)dz.$$

Since the integral is finite and $\omega_0=0$, the equation is untrue. There remains the case $\kappa=1$; the value of ω is then

$$\omega=\omega_0\, exp^{\int_0^{\zeta}\psi(z)dz}.$$

Since $\omega_0 = 0$, ω must vanish identically. Thus the solution of the kind proposed is not only existent, but also unique.

The analytic function associated with the solution $w = P_1(z)$ can be found by the method of continuation. To find a value at a place z', draw a path from 0 to z' and interpolate points between z and z' in the manner explained in Chapter III., using each point as the centre for a circle of convergence. Progress towards z' may be barred by the presence of singular points in the region to be traversed.

§ 157. With the aid of the preceding theorem it is possible to prove the existence and one-valuedness of w as a function of z when

$$dw/dz = \sqrt{(w - a_1)(w - a_2)(w - a_3)(w - a_4)},$$

a_1, a_2, a_3, a_4 being four constants, supposed unequal. Suppose that $w = w_0$ when $z = 0$, w_0 being different from the constants a_1, a_2, a_3, a_4. Also let the sign of the radical be assigned when $w = w_0$. The equation has an integral which is holomorphic in the neighbourhood of $z = 0$. By the Weierstrassian method of continuation, this element defines an analytic function. When a place z_0 is reached, for which w is equal to a_1, a_2, a_3, a_4, or ∞, the expression on the right-hand side of the equation for dw/dz ceases to be holomorphic. That the integral of the equation is holomorphic in the neighbourhood of a point z_0, for which $w = a_1$, can be shown by the transformation $w = a_1 + W^2$. The right-hand side of the resulting equation,

$$2\,dW/dz = \sqrt{(a_1 - a_2 + W^2)(a_1 - a_3 + W^2)(a_1 - a_4 + W^2)},$$

is holomorphic in the neighbourhood of $W = 0$, and therefore W and w are holomorphic in the neighbourhood of this point z_0. Similarly the transformations $w = 1/W$, $z = z_0 + z'$ enable us to determine what happens at a point z_0 for which $w = \infty$. For the equation

$$dW/dz' = -\sqrt{(1 - a_1 W)(1 - a_2 W)(1 - a_3 W)(1 - a_4 W)},$$

when coupled with the initial conditions $z' = 0$, $W = 0$, has an integral W which is holomorphic in the neighbourhood of $z' = 0$, and $z = z_0$ is a pole of w. Thus, whatever be the values of w at a point $z = z_0$, the integral of the differential equation is one-valued in the neighbourhood of $z = z_0$.

Picard has pointed out (Bulletin des Sciences mathématiques, 1890; Traité d'Analyse, t. ii., fasc. 1) that the above considerations do not prove that the function w is a one-valued function of z

throughout the finite part of the z-plane. It has been assumed that the value of the integral is *determinate* at each point of the z-plane. He makes the method of continuation apply rigorously to the whole plane by proving that the continuation can be effected by a circle of fixed radius ρ. Round the points $a_1, a_2, a_3, a_4, \infty$ describe circles $(a_1), (a_2), (a_3), (a_4), (\infty)$. Within the region bounded by these five circles there is a finite radius of convergence at each point, and therefore a lower limit, which does not vanish, for the radii of convergence. When w lies within (a), the transformation $w = a_1 + W^2$ transforms (a_1) into a new circle, and so long as W lies within this new circle, there is a lower limit of the radii of convergence for W. Similarly for the other points a_2, a_3, a_4, ∞. The least of these six lower limits of the radii of convergence of the series which define w is a number ρ, different from 0; and when we assign to z, w any arbitrary values z_0, w_0, z_0 being finite, w is one-valued and determinable within the circle whose centre is z_0 and whose radius is the fixed number ρ. From this the desired conclusion follows.

If we attempt to apply the same argument to the equation

$$dw/dz = \sqrt{(w-a_1)(w-a_2)\cdots(w-a_{2p+2})},$$

we have, when $w = 1/W$,

$$dW/dz = -\sqrt{(1-a_1W)(1-a_2W)\cdots(1-a_{2p+2}W)}/W^{p-1},$$

and the function on the right is not holomorphic in the neighbourhood of $W=0$, when $p>1$. The inference is that w is not a one-valued function of z throughout the finite part of the z-plane; in point of fact, w is an infinitely many-valued function of z, ~~as will~~ appear later.

References. On the general subject of this chapter, the reader should consult: Briot and Bouquet, Fonctions Elliptiques; Cauchy's Works; Hermite's Cours d'Analyse; Jordan's Cours d'Analyse; Laurent's Traité d'Analyse.

Cauchy's fundamental theorem has been extended to double integrals of functions of two independent complex variables by Poincaré (Acta Math., t. ix.). Algebraic functions of two variables are discussed in Picard's prize memoir (Liouville, ser. iv., t. v.).

CHAPTER VI.

Riemann Surfaces.

§ 158. Hitherto we have considered the z-plane as a single plane sheet upon which the variable z is represented. In the Argand diagram, to each point z are attached all the corresponding values of w. As long as w is a one-valued function of z, no difficulty arises; for every path from z_0 to z in the Argand diagram leads from the initial value w_0 to the same final value w. But if w be a many-valued function, a definitely selected initial value at z_0 does not determine which of the values is to be chosen at z, since different paths from z_0 to z may lead to different values of w at z. A familiar instance is afforded by the algebraic function given by the irreducible equation $F(w^n, z^m) = 0$. Here each point z has attached to it n values w. When $n > 1$, two paths from z_0 to z can always be found which will lead from the same initial value w_0 at z_0 to different values at z.

In Riemann's method of representation one value of w, and only one, corresponds to each point of a surface. Before considering the general problem we shall show how to form a Riemann surface in some simple special cases.

Let $w = \sqrt{z}$. To one z in the Argand diagram correspond two values of w, whereas we wish to make two points correspond to two values of w. To these two points the same complex variable z should be attached.

Instead of a single z-plane we take two indefinitely thin sheets, one of which lies immediately below the other. For convenience of description we suppose them to be horizontal.* To the one value of z correspond two places in these sheets, one vertically below the other. Every place in the upper or lower sheet has one, and only one, of the two values $\sqrt{z}$, $-\sqrt{z}$, permanently attached to it. If, for a given z, $\sqrt{z}$ be attached to the upper sheet, then $-\sqrt{z}$ must be attached to the lower sheet at the point z; but we cannot infer, from

* To avoid confusion we shall speak of a point of a Riemann surface as a *place*. A pair of numbers (z, w) is attached to each place, and serves to name the place; but it is frequently convenient to use a single letter for the pair (z, w). When n sheets are spread over the Argand diagram for z, the vertical line through the *point* z meets the surface in n *places* (z, w).

this alone, the value for the upper sheet at a second point z'. When $z=0$ or ∞, the values of w are equal, and only one place is needed to represent them; hence we regard the sheets as hanging together at 0 and ∞. But we require a further connexion between the two sheets. In the Argand diagram a closed path, which starts from z and passes once around the origin, leads from $\sqrt{z}$ to $-\sqrt{z}$. Therefore, in the Riemann representation, if the path start from a place in the upper sheet, to which a value $\sqrt{z}$ is attached, it must lead to that place, in the lower sheet, which lies vertically below the initial place. It follows that either we must give up the idea of having only one w-value attached to each place of the two sheets, or else we must make a connexion or bridge between the sheets, over which every path, which goes once round the origin, must necessarily pass. If the bridge stretch from 0 to ∞, without intersecting itself, this condition will be satisfied for any finite path. In this example it should be noticed that 0 and ∞ are branch-points in the z-plane, and that the bridge extends from branch-point to branch-point.

Secondly we know that the same path, described twice in the same direction, leads from $\sqrt{z}$ to $\sqrt{z}$, *i.e.* from a place of the upper sheet to the same place of the upper sheet, assuming that the initial place lies in the upper sheet. The first circuit takes the place into the lower sheet; in order that the second circuit may restore the place to the upper sheet, it is necessary that it should pass along a bridge from the lower sheet to the upper. This condition and the preceding one are satisfied by a double bridge, or *branch-cut*, which extends from 0 to ∞. Figure 48, a, is a section of the surface made by a plane perpendicular to the bridge.

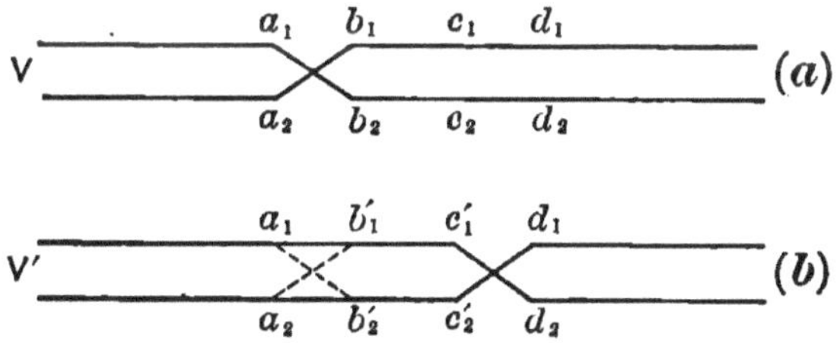

Fig. 48

Finally a closed path in the Argand diagram, which includes no branch-point, restores the initial value of w. It remains then for us to show that such a path gives a closed path in the Riemann surface. There is often an advantage in distinguishing by suffixes the places in which a vertical line cuts the sheets. In this notation z_1 and z_2 mean corresponding places z in the first and second sheets.

A closed path which starts at z_1 must end at z_1 and not at z_2. With this notation, Fig. 49 shows that a closed path in the Argand diagram which does not enclose 0, is a closed path in the Riemann surface. For instance, a point which starts at a place a_1 and describes the path in Fig. 49, in the direction of the arrow, passes at b into the lower sheet, at c into the upper sheet, at d into the

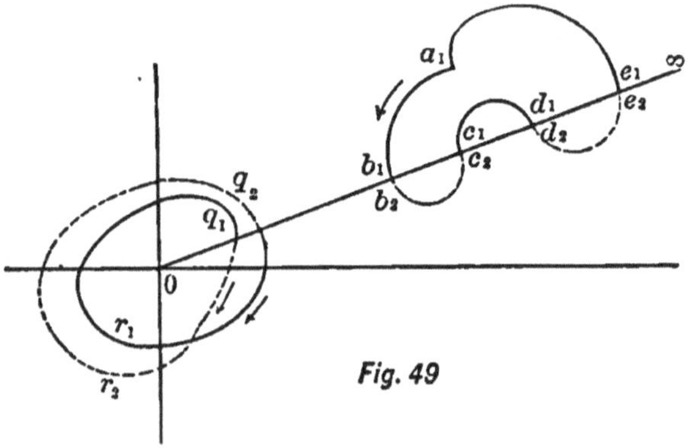

Fig. 49

lower sheet, and finally at e into the upper sheet, returning to the initial place a_1. The figure also shows that the path $q_1r_2q_2$ leads from q_1 to q_2, whereas $q_1r_2q_2r_1q_1$ is closed.

The Riemann surface for $w=\sqrt{z}$ is of the form given in Fig. 50. Looking from 0 to a, the sheet 1 on the left continues along $0a$ into the sheet 2 on the right, and the sheet 1 on the right into the sheet 2 on the left. The branch-cut $0a$ extends from 0 to ∞, but is not necessarily straight; the sheets 1 and 2 are nearly parallel, infinitely close to one another, and infinitely extended in two dimensions. The figure illustrates the way in which the surface winds round the point 0.

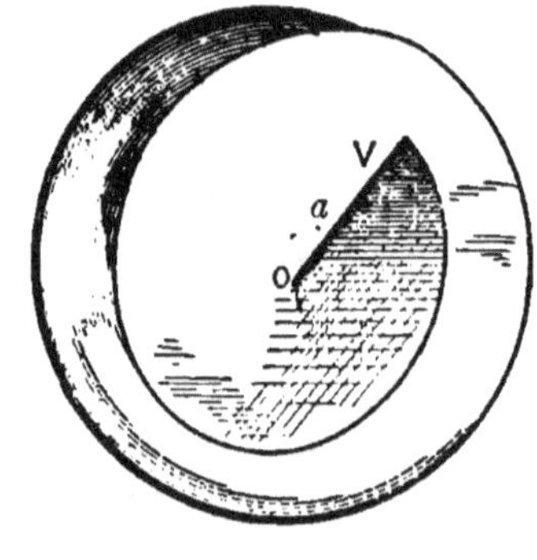

Fig. 50

The displacement of a branch-cut.

The places a_1, b_2, are consecutive in Fig. 48, a; but a_1, b_1, or a_2, b_2, are not consecutive. Thus if the values of the function at a_1, a_2, be $\sqrt{a}$, $-\sqrt{a}$, continuity requires that the values at b_1, b_2, shall be $-\sqrt{b}$, $\sqrt{b}$; then the values at c_1, d_1, ... are $-\sqrt{c}$, $-\sqrt{d}$, ..., and so on. The position of the branch-cut or bridge has been defined only to this extent, that it must pass from 0 to ∞, and never cut itself. Naturally any other branch-cut, subject to these conditions, will equally serve our purpose. Let the cut be deformed from V to V', and consider the same vertical section of the surface as in Fig. 48. Fig. 48, b, represents the new state of things. A comparison of this figure with Fig. 48, a, shows that the points which have changed sheets (*e.g.* b_1, b_2, which have become b_2', b_1') carry the values of w with them. For the value at a_1, in Figs. 48, a, and 48, b, being $\sqrt{a}$, the value at b_2 in Fig. 48, a, is $\sqrt{b}$, and the value at b_1' in Fig. 48, b, is also $\sqrt{b}$. Viewing the whole surface, Fig. 50, when the branch-cut is shifted to V', the part of the lower sheet between V and V' becomes part of the upper sheet, and *vice versâ*. These considerations show what is involved in the shifting of a branch-cut.

Let us next examine the function $w = \sqrt{(z-a)(z-b)}$. This function is two-valued; we wish to make one value of w correspond to one place of the Riemann surface, and must therefore have a two-sheeted surface. The branch-points are a and b. In the Argand diagram, we know that a circuit round either branch-point, with an initial value w, leads to a final value $-w$; while a circuit which includes both branch-points restores the initial value. This suggests that we draw a bridge from a to b. Any path, on meeting the bridge, has to change from one sheet to another, but the sheets are unconnected except along the bridge.

§ 159. *The Riemann sphere.* We have now had two simple illustrations of a Riemann surface. Before giving other examples and the general theory, we point out that, as in the case of the Argand diagram, there is often a great advantage in substituting for a surface formed of infinite plane sheets a closed surface. Any surface may be selected which has a (1, 1) correspondence with the plane surface, the simplest closed surface being the sphere.

Let us suppose, then, the Riemann surface inverted with regard to an external point 0′, which lies vertically below 0 at unit distance. A two-sheeted Riemann surface becomes two infinitely near spheres, joined together at the branch-points. These two spheres touch each other at 0′, but by a slight displacement, this connexion can be destroyed, except when ∞ is a branch-point in the plane surface; in fact, in the last example, the only connexion is along a line extending from the branch-point a to the branch-point b, whereas the Riemann sphere for the function $w = \sqrt{z}$ has branch-points at 0 and 0′, which are connected by a bridge lying along any line, as for instance the half of a great circle, from 0 to 0′.

It is now clear that the function $\sqrt{z}$ is only a case of the more general function $\sqrt{(z-a)(z-b)}$. We may regard the sphere, in the second case, as stretched, without tearing, until the branch-points are diametrically opposite, and then developed, by inversion, upon the tangent plane at either branch-point. It should be noticed that, in the case of $\sqrt{(z-a)(z-b)}$, the method suggested by the first example, namely, the drawing of branch-cuts from a, b, to ∞, does not differ essentially from that adopted here; for ∞ is an ordinary point on the sphere, and the two cuts, from a, b, to ∞, amount merely to a single cut from a to b.

§ 160. *The function* $w = z^{1/n}$. To a given z correspond n values of w. We wish to make one value of w correspond to one place of the Riemann surface; we must therefore have an n-sheeted surface. Let the n values of w, for a given z, be $w_1 = z^{1/n}$, $w_2 = aw_1$, $w_3 = a^2 w_1$, $\cdots$, $w_n = a^{n-1}w_1$, where $a = \cos\frac{2\pi}{n} + i\sin\frac{2\pi}{n}$. The way in which we number the sheets is immaterial. Let, then, $w_1, w_2, \cdots, w_n$ belong, initially, to the first, second, $\cdots$, nth sheets respectively. In the Argand diagram, one positive circuit round the origin changes $z^{1/n}$ into $az^{1/n}$, *i.e.* w_1 into w_2. A second circuit changes w_2 into w_3, $\cdots$, an nth changes w_n into w_1.

Hence if, with the previous notation, $z_1, z_2, \cdots, z_n$ be n places in the same vertical line, to all of which z is attached, the first circuit must lead from z_1 to z_2, the second from z_2 to z_3, $\cdots$, the nth from z_n

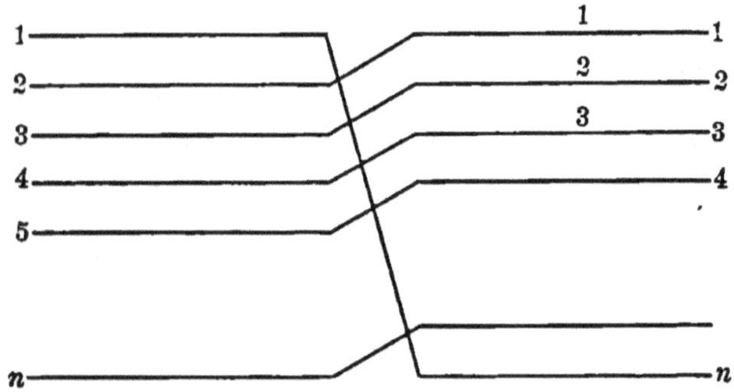

Fig. 51

to z_1. A bridge must be drawn, as in Fig. 49. A closed path which passes once round the origin is an open curve in the Riemann surface. Suppose that this path starts at the point z_1 and proceeds in the positive direction. The first description leads from z_1 to z_2, the second from z_2 to z_3, $\cdots$, the nth from z_n to z_1. A negative description of the path leads from z_1 to z_n, from z_n to z_{n-1}, from z_{n-1} to z_{n-2}, and so on. By drawing the bridge, it is easy to see that any closed path in the Argand diagram, which does not include the branch-point 0, gives a closed path on the Riemann surface, for such a path must cross the bridge an even number of times (see Fig. 49 for the special case $n=2$); let a, b be two consecutive points where the path crosses the bridge. If, at a, the point pass from the κth to the λth sheet, it remains in the λth sheet till it reaches b, and then passes back to the κth sheet. Thus, on the whole, it returns to the original sheet, and the path is closed. Fig. 51 represents a section of the surface made by a plane perpendicular to the direction of the bridge, and viewed from the origin.

§ 161. The function $w=\sqrt{z(1-z)(a-z)(b-z)(c-z)}$. The function is two-valued, with branch-points at $0, 1, a, b, c, \infty$. The Riemann surface must be two-sheeted. It is required to find the connexions between the sheets. From the theory of loops we know that a closed curve which surrounds an even number of the six branch-points (infinity included), or which passes an even number of times round a single branch-point, will reproduce the initial value of w; whereas, if the curve enclose branch-points an odd number of times, the initial value of w is reproduced with a change of sign. The branch-cuts in the Riemann surface must be so constructed that a path, which is closed in the Argand diagram, shall lead from z_1 to z_1 or z_2, according as it includes an even or an odd number of branch-points, where, in estimating the number, we must count each branch-point as often as the path encircles it.

A figure would show, immediately, that cuts along straight lines from $0, 1, a, b, c$, to ∞ satisfy these requirements, but a simpler arrangement is to join 0 to 1, a to b, c to ∞, by branch-cuts. These branch-cuts must not cross themselves or one another. In Fig. 52 they are represented by straight lines. This figure shows that paths

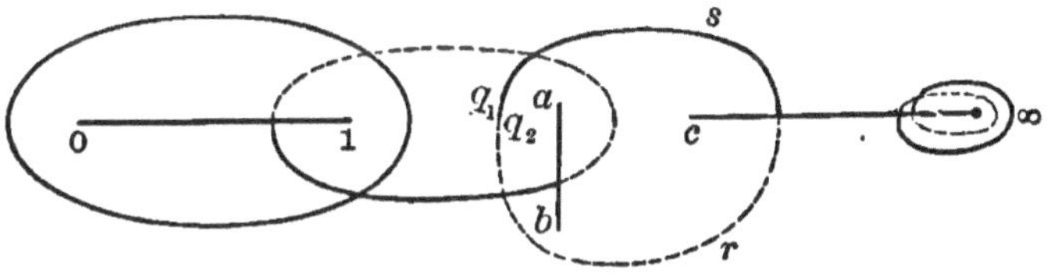

Fig. 52

which enclose an even number of branch-points in the z-plane are closed on the Riemann surface, whereas q_1srq_2, which encloses three branch-points, begins in the upper sheet and ends in the lower.

In the same way the Riemann surface can be found for the square root of a rational integral function of degree $2n-1$ or $2n$. It should be noticed that the former function is a special case of the latter, created by one branch-point moving off to infinity.

§ 162. *Riemann surfaces for irreducible algebraic functions.*

We are now in a position to construct a Riemann surface for the general algebraic function. Let the function be n-valued. Take n infinitely thin sheets, lying infinitely near one another. We first assign to the n sheets the branches which correspond to a given z, say $z = z^{(0)}$, which is not a critical point. When z describes a path from $z^{(0)}$ to a branch-point a, we determine by the principle of continuity those values into which the several initial values of w pass; by observing those which belong to the cyclic systems near a we decide the connexion of the sheets at a. Through the sheets to which one and the same cyclic system of values is attached, a branch-cut is drawn from a to ∞, and this is done for each cyclic system at a; the branch-cuts being distinct, though they may interlace (as in Fig. 53).

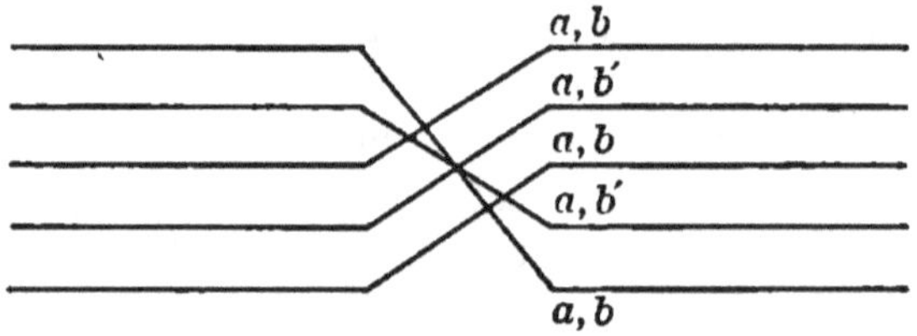

Fig. 53

Starting afresh from $z^{(0)}$, we repeat the process for every other branch-point.

It will be convenient to suppose that neither the paths from $z^{(0)}$ to the branch-points, nor the branch-cuts from the branch-points to ∞, intersect one another; also that none of the paths encounters a branch-cut. It is clear, as in § 159, that a new system of paths may lead to quite different connexions of the sheets. When the connexions of the sheets have been ascertained, the branch-cuts cannot be drawn at random, but must satisfy the requirements of the cyclic system or systems at ∞. That is, a very large circuit which begins at any place of the surface must produce the same effect as in the theory of loops; this effect being nil when ∞ is not a branch-point.

The system of branch-cuts adopted is merely one of infinitely many which are available when the connexion of the sheets has been established. For example, the cuts may be drawn to any finite point instead of to ∞, regard being paid to the requirements of the cyclic systems at that point and to the branch-point at ∞. Suppose that a non-critical point z_0 is selected in the Argand diagram and that cuts are made in the z-plane from z_0 to the branch-points of the function. Let these cuts, $V_1, V_2, \cdots, V_r$, be regarded as having positive and negative banks, which are met in the order $V_1^+ V_1^- V_2^+ \cdots V_r^+ V_r^-$ on description of a small circle round z_0. By superposing $n-1$ similarly cut sheets, called sheets $2, 3, \cdots, n$, on the z-sheet, and by joining the n positive banks of the cuts V_κ to the n negative banks of the same cuts, the n sheets become connected, and the sheets $1, 2, \cdots, n$ pass into the sheets $\alpha_1, \alpha_2, \cdots, \alpha_n$, where $(\alpha_1, \alpha_2, \cdots, \alpha_n)$ is a permutation of $(1, 2, \cdots, n)$. Let the resulting substitutions along the r cuts V_κ be $S_1, S_2, \cdots, S_r$; then if the new surface is to be a Riemann surface in which it is possible to pass from any one place to any other by a continuous path, it is necessary that $S_1 S_2 \cdots S_r$ be equal to 1. When an n-sheeted surface is constructed by the process just described, the requisite data, in order that it may be a Riemann surface, are the positions of z_0, of the branch-points, and of the lines V_κ, together with the substitutions $S_1, S_2, \cdots, S_r$, where $S_1 S_2 \cdots S_r = 1$. See Hurwitz, Riemann'sche Flächen mit gegebenen Verzweigungspunkten, Math. Ann., t. xxxix.

A branch-point a in Cauchy's theory of loops is on the Riemann surface one or more branch-places (a, b) with or without ordinary places (a, c); all these places lying in the same vertical. A loop to a from $z^{(0)}$, in Cauchy's theory, is a vertical section of the surface, consisting of n *loops on the surface* which are closed or unclosed according as they pass near (a, c) or (a, b). When a branch-cut has been drawn from every branch-place to ∞, any path C on the surface which begins and ends at places in the same vertical can be contracted into loops on the surface, precisely as in § 127. For there has arisen nothing to invalidate the theorem of that article. A closed loop on the surface is of no effect, and may be omitted.

To each place of the surface there corresponds a pair of numbers (z, w) satisfying the equation $F=0$. How many places of the surface correspond to each pair of numbers? Evidently only those places which lie in the same vertical, and which also bear the same value of w. First, when z is near a, let r values of w become nearly equal to b, and form a cyclic system. As in § 160 the r sheets in which these nearly equal values lie are regarded as hanging together at one place (a, b), whether or not these sheets are consecutive. Next, let there be a node at a; then not only are two values of w equal to b when z is equal to a, but also two values of z are equal to a when w is equal to b. The series near a for these branches are integral (§ 113), and a circuit round a, which starts from a place near (a, b), must lie wholly in one sheet. There is therefore no connexion between the two sheets near (a, b). And, in general, when the pair (a, b) constitute a higher singularity, the sheets to which are attached those branches which are given, near (a, b), by integral series, have no connexion at (a, b) with other sheets.*

We have already had instances in which branch-cuts, instead of proceeding to the common point ∞, are drawn directly from one branch-place (a, b) to another (a', b'). We have now to examine when this is possible. Since branch-cuts are not allowed to cross one another, the same sheets must hang together at the two branch-places, and a path C, which starts in one of these sheets and encloses these branch-places and no others, must run entirely in the one sheet. If, then, on crossing the original branch-cut $a\infty$, C pass from the κth to the λth sheet, it must on crossing the original branch-cut $a'\infty$, pass back from the λth to the κth sheet.

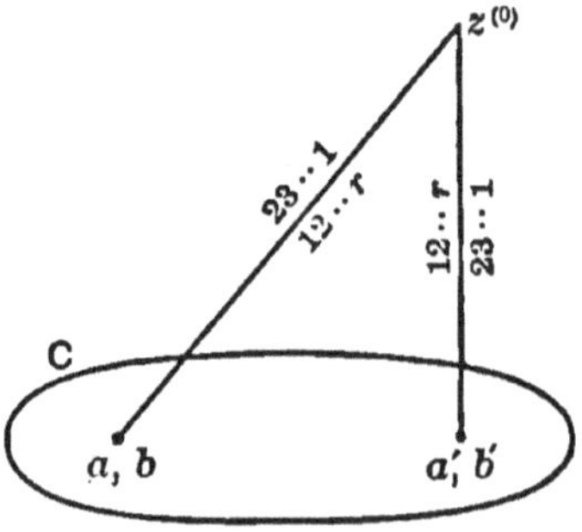

Fig. 54

This shows that if the positive loop from $z^{(0)}$ to a permute the branches in the cyclic order $(w_1 w_2 \cdots w_r)$, the positive loop from $z^{(0)}$ to a' must permute the same branches in the cyclic order $(w_1 w_r w_{r-1} \cdots w_2)$; that is, the substitutions due to the branch-cuts at $z^{(0)}$ must be inverse. If not, aa' is not a permissible branch-cut. When the surface is two-sheeted, the condition is satisfied by any pair of branch-points.

§ 163. *Examples of Riemann surfaces for algebraic functions.*

The construction of a Riemann surface, in the study of the relation of two complex variables, is an exercise similar to the tracing of a curve in the field of real variables. The present problem is to map the w-plane on the z-plane, when a relation $F(w^n, z^m) = 0$ is assigned. It should be noticed that there is not only an n-sheeted z-surface, but also an m-sheeted w-surface, the places (z, w) and (w, z) on the two surfaces being in (1, 1) correspondence, except possibly at special points. It is often advisable to construct both surfaces. The continuity of each branch, and the angular relations near special points, as summed up in § 111, are the most important elements in the solution of the problem. In determining the path of w in its

* Instead of a penultimate form in the Cartesian sense, we can imagine a horizontal displacement of some of the sheets, which necessitates possibly a stretching of these sheets, until there lie, in any one vertical, only two equal values of w.

plane, for a given path of z in its plane, we must attend specially not only to the branch-points in the z-plane, but also to those points in the z-plane which correspond to branch-points in the w-plane. Corresponding to a simple branch-point b in the w-plane, there is a point a in the z-plane near which $w-b=P_2(z-a)$; so that when z moves towards a in a given direction, makes a half-turn round a, and proceeds in the same direction, a branch of w moves towards b in a certain direction, makes a complete turn round b, and proceeds in the opposite direction. We shall speak of such a point as a turn-point. If (a, b) be any pair of finite values, and if z describe a small circle round a, w turns round $w=b$ with the same direction of rotation. We shall suppose that the branches $w_1, w_2, \cdots, w_n$, are initially assigned to the upper, second, $\cdots$, lowest sheets.

Ex. (1). $$w=\left(\frac{z+1}{z-1}\right)^{1/4}.$$

Initially let $z=0$, $w_1=\exp(\pi i/4)$, $w_2=iw_1$, $w_3=iw_2$, $w_4=iw_3$. The series for w near $z=-1$ are $w=P_1(z+1)^{1/4}$. Therefore when z, starting from 0, describes positively the circle whose centre is -1, which by § 127 is equivalent to a loop from 0 to -1, the branches permute in the cyclic order $(w_1w_2w_3w_4)$. Similarly, near $z=1$, the series for w are $1/w=P_1(z-1)^{1/4}$, and when z describes positively a circle whose centre is 1, the branches permute in the order $(w_1w_4w_3w_2)$. A four-sheeted surface, with a branch-cut from -1 to 1, extending through all the sheets, satisfies the requirements. Looking from -1 to 1, the permutations, either at 1 or -1, require that the right-hand portions of sheets 1, 2, 3, 4 shall pass into the left-hand portions of sheets 2, 3, 4, 1. This completes the construction of the surface. A contour on the surface, which encloses both 1 and -1, restores the initial value (§ 127).

Ex. (2). $$w+1/w=2z.$$

For a given z there are two values of w, say w_1 and w_2, connected by the relation $w_1w_2=1$. Hence the surface has two sheets. The branch-places are $(1, 1)$ and $(-1, -1)$.

If w_1 describe an arc of a circle from -1 to $+1$, the other root w_2 describes the remaining arc, and $z, =(w_1+w_2)/2$, describes an arc from -1 to $+1$ in its own plane (Fig. 55). If w_1 describe the circle C, whose centre is 0 and radius 1, so does w_2; and z moves along the real axis from -1 to $+1$. Let this line be chosen as the branch-cut. Since the two sheets are joined solely along the straight line from -1 to $+1$, and since this line maps into the unit circle round the w-origin, it follows that to all points of one of the z-sheets correspond all points *inside* the circle C in the w-plane, and to all points of the other z-sheet, all points *outside* this circle. In the case of a circle in the w-plane through -1 and 1, to the arc inside C corresponds an arc in the one sheet; to the remaining arc outside C corresponds the same arc in the other sheet.

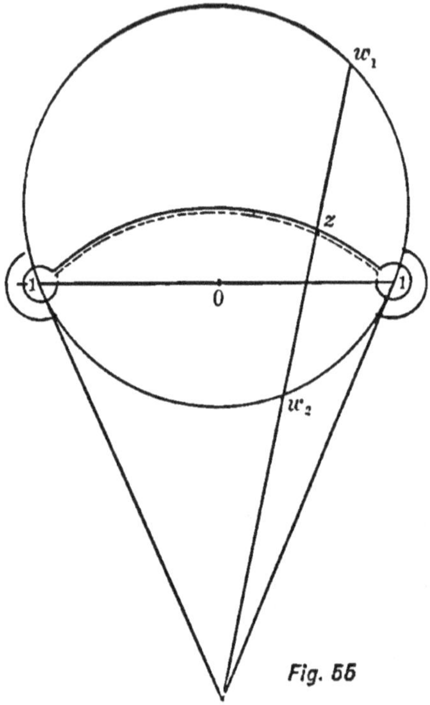

Fig. 55

To the set of circles in the w-plane with limiting points ± 1, correspond circles with limiting points ± 1, described in both sheets of the z-plane. Again, if $w_1 = \rho e^{i\theta}$, we have

$$w_2 = e^{-i\theta}/\rho, 2x = (\rho + 1/\rho)\cos\theta,$$
$$2y = (\rho - 1/\rho)\sin\theta;$$

therefore to circles round the origin in the w-plane, correspond ellipses with foci ± 1; also to lines through the origin in the w-plane, correspond the confocal hyperbolas. The w-plane is a conform representation of the two-sheeted Riemann surface.

The reader will find interesting discussions of this and allied examples in Holzmüller's Theorie der Isogonalen Verwandtschaften.

Ex. (3). The relation considered here is a form of the relation

$$aw^2 + 2bw + c + z(a_1w^2 + 2b_1w + c_1) = 0,$$

which defines the general (2, 1) correspondence. We shall show that this general relation can be brought, by (1, 1) transformations of both w and z, to the form $w'^2 = z'$.

The branch-points z_1, z_2 are given by the equation

$$(a + a_1z)(c + c_1z) = (b + b_1z)^2.$$

Let $(z - z_1)/(z - z_2) = z'$; in the new equation between w and z', the equal values of w are given by $z' = 0$ and $z' = \infty$; and the equation is of the form

$$(a'w + b')^2 = z'(a_1'w + b_1')^2.$$

If then we write $(a'w+b')/(a_1'w+b_1')=w'$, the proposed reduction is effected.

Without going into details, we can infer at once that in the general (2, 1) relation, when we choose as branch-cut a circular arc in the z-plane, the two sheets of the z-surface map respectively into the interior and exterior of a circle in the w-plane. By the last example, any circle in the w-plane maps into the inverse of a conic in the z-plane (compare § 23); and by § 47, any circle in the z-plane maps into a special bicircular quartic in the w-plane, namely the inverse of a Cassinian. The z-circle is a double circle, lying in both sheets; when it includes one branch-point it is a unipartite curve and maps into a unipartite quartic; but when it includes neither or both of the branch-points, it is bipartite, and maps into two distinct ovals.

Ex. (4). $w^3 - 3w = 2z$.

Here $\delta F/\delta w = 0$ gives $3w^2 - 3 = 0$, *i.e.* $w = \pm 1$.

The equal values of w arise from the z-values $1, -1, \infty$; and the values of w which correspond to these special z-points are $-1, -1, 2$; $1, 1, -2$; ∞, ∞, ∞. We have to determine the manner in which the sheets of the Riemann surface hang together.

At the origin $w=0, \sqrt{3}, -\sqrt{3}$. Let these values be attached in this order to the three sheets. Calling the branches w_1, w_2, w_3, it is clear, from the Theory of Equations, that the following lemmas hold:—

(1) $w_1+w_2+w_3=0$.

(2) When w is small, an approximate value is $w=-2z/3$.

(3) All the roots are real, when z is real and $|z|<1$.

Using a w-plane, let z, starting from 0, move along the real axis to -1; by lemma 3, w_1 starting from 0 moves along the real axis, and, by lemma 2, it moves to the right. In order to maintain the centroid at 0 (lemma 1), w_3, which starts from $-\sqrt{3}$, must remain to the left of 0. Hence w_1 and w_2 unite when $z=-1$. In the same way it can be proved that w_1 and w_3 unite when $z=1$. Accordingly, in the Riemann surface, sheets 1 and 2 join at $z=-1$, sheets 1 and 3 at $z=1$. At $z=\infty$ all the sheets join.

§ 164. Ex. (5). *The* (3, 1) *correspondence.*

Let $zU+V=0$, where U and V are cubics in w. Any two cubics when rendered homogeneous by writing w/t for w, may be written in the forms $\delta Q/\delta w$, $\delta Q/\delta t$, where Q is a homogeneous quartic in w

and t (Salmon, Higher Algebra, Fourth Edition, § 217), and the relation may be written

$$z\delta Q/\delta w + \delta Q/\delta t = 0,$$

the 7 effective constants of the original relation being reduced to 4 by means of the 3 disposable constants in the (1, 1) transformation effected on w. In this form z is the third polar of w with regard to a fixed quartic, and accordingly the correspondence of a point and its third polar with regard to a quartic is the general (3, 1) correspondence. Let Q be brought to the canonical form $w^4 + 6cw^2t^2 + t^4$ (§ 44); then the correspondence is defined by

$$z(w^3 + 3cw) + 3cw^2 + 1 = 0.$$

The turn-points in the w-plane are given by the Hessian of Q (§ 36), that is by

$$H \equiv c(w^4 + 1) + (1 - 3c^2)w^2 = 0.$$

The roots of H being a, $-a$, $1/a$, $-1/a$, we shall take the case when one root is real; it follows that all the roots are real. Let a be that root which is positive and < 1.

We have when $\quad z = 0,\ \ w = \pm\sqrt{-1/3c},\ \infty,$

and when z is near 0, $\quad 18c^2(w \mp \sqrt{-1/3c}) = (1 - 9c^2)z,$

or $\quad w = -3c/z.$

To real values of z correspond real values of w, and also the points of the curve whose equation is

$$\frac{w^3 + 3cw}{3cw^2 + 1} = \frac{\bar{w}^3 + 3c\bar{w}}{3c\bar{w}^2 + 1},$$

where w and $\bar{w}$ are conjugate,

or $$3c(w^2\bar{w}^2 + 1) + w^2 + \bar{w}^2 + w\bar{w}(1 - 9c^2) = 0,$$

or, in polar co-ordinates,

$$3c(\rho^4 + 1) + 2\rho^2 \cos 2\theta + \rho^2(1 - 9c^2) = 0.$$

Regarding a as given, c is determined by

$$3c^2 - 1 = c(a^2 + 1/a^2),$$

and the two values of c have opposite signs. Selecting the negative value of c, let w_1, w_2, w_3 be the 3 branches of w, and let w_1 be that branch which is positive when $z = 0$, w_2 that which is ∞ when $z = 0$, and w_3 that which is negative. Also let these values be assigned to the first, second, and third sheets respectively at $z = 0$.

When z (Fig. 56), starting from the origin, moves to the right along the real axis, the approximate values show that w_1 and w_3 move to the right, inasmuch as $1 - 9c^2 > 0$; while w_2 moves to the left. Also the point at which w_1 starts lies between α and $1/\alpha$. Therefore, at the first branch-point β, sheets 1 and 2 join. When z describes

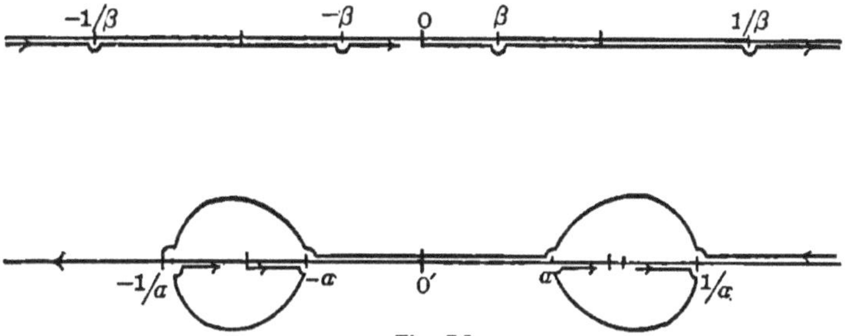

Fig. 56

a half-turn round β and proceeds along the real axis, w_1 and w_2 make quarter-turns round the point $1/\alpha$, and then describe the right-hand oval, meeting again at $w = \alpha$. Therefore again, at the second branch-point $1/\beta$, sheets 1 and 2 join. When z, after making a half-turn round $1/\beta$, proceeds to the right, w_2 proceeds to the left to meet w_3, while w_1 moves to the right. Therefore, at the branch-point $-1/\beta$, sheets 2 and 3 join. When z, after a half-turn round $-1/\beta$, moves to $-\beta$, w_2 and w_3 describe the left-hand oval, and meet again at $w = -1/\alpha$. Therefore sheets 2 and 3 join at $-\beta$. The branch-cuts may be drawn from β to $1/\beta$ in sheets 1 and 2, and from $-\beta$ to $-1/\beta$ in sheets 2 and 3. When they are drawn along the real axis, the whole of the first sheet corresponds to the interior of the right-hand oval, the whole of the third sheet to the interior of the left-hand oval, and the whole of the second sheet to the remainder of the w-plane.

§ 165. Ex. (6). $\qquad a^2(w^2z^2 + 1) = (w + z)^2. \quad \ldots \ldots \quad (1).$

The branch-points in the z-plane are given by

$$z^2 = (a^2z^2 - 1)(a^2 - z^2),$$

and are $\alpha, -\alpha, 1/\alpha, -1/\alpha$, where

$$\alpha^2 + 1/\alpha^2 = a^2.$$

We consider the case in which one, and therefore all branch-points, are on the real axis, and take $\alpha < 1$.

We have, if w_1 and w_2 be the branches,

$$\left.\begin{aligned}(a^2z^2-1)(w_1+w_2)&=2z,\\(a^2z^2-1)w_1w_2&=a^2-z^2.\end{aligned}\right\} \quad \ldots\ldots \quad (2).$$

When $w_1=w_2$, there follows, on elimination of a^2, either

$$w+z=0, \quad \text{or} \quad wz^3-1=0.$$

The case $w+z=0$ is extraneous, as it requires $a=0$; and therefore, the symmetry of (1) being observed, the turn-points in either plane, corresponding to the branch-points a, $-a$, $1/a$, $-1/a$ in the other plane, are $1/a^3$, $-1/a^3$, a^3, $-a^3$.

When $z=0$, $\quad w=\pm a$, and near $z=0$, $w=\pm a-z$.

When $z=\infty$, $\qquad w=\pm 1/a$.

Since a is real, to a real value of z correspond either two real or two conjugate values of w. Therefore to the real z-axis corresponds part or all of the real w-axis, together with a curve whose equation is obtained by regarding w_1 and w_2 as conjugate, and eliminating z from (2).

The equation is

$$4(a^2+w_1w_2)(a^2w_1w_2+1)=(w_1+w_2)^2(1-a^4)^2,$$

and the curve is a central bicircular quartic.

Starting from the point 0 to the right, let z describe the whole of the real axis, making half-turns round all critical points (Fig. 57).

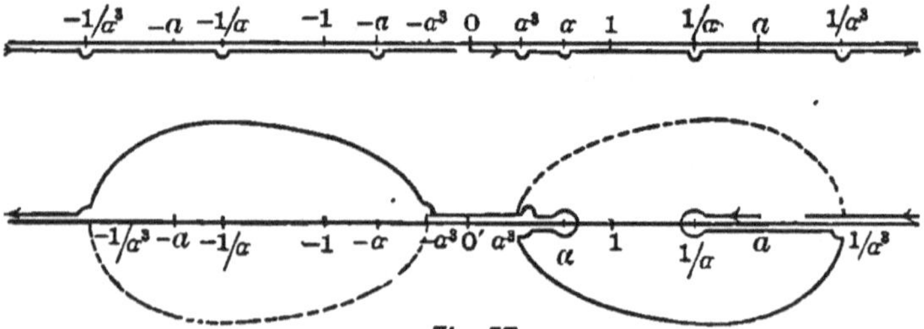

Fig. 57

The figure shows the path described by the branch whose initial value is a.*

Let the branch-cuts in the z-plane be drawn along the real axis from $-a$ to a, and from $-1/a$ viâ ∞ to $1/a$. A careful consideration of the figure shows that to these parts of the real z-axis correspond

* When z is near $-1/a^3$, the branch of w in question is not near $-a$; hence to the half-turn in the z-plane round $-1/a^3$ corresponds a half-turn in the w-plane.

the parts from $-1/\alpha$ viâ ∞ to $1/\alpha$, and from $-\alpha$ to α, of the real w-axis. Let then the two-sheeted w-surface, which represents the dependence of z on w, have the same branch-cuts as those chosen for the z-surface; whenever (z, w) changes sheets, so does (w, z). In this case, *one* surface serves to represent both the dependence of w on z, and that of z on w; and this surface is mapped on itself by the relation (1) without alteration of the branch-cuts.

The most general quadri-quadric correspondence between w and z involves 8 effective constants. The general symmetric correspondence involves 5 effective constants, and can therefore be obtained from the unsymmetric relation by a $(1, 1)$ transformation of z alone, inasmuch as the $(1, 1)$ transformation places 3 effective constants at our disposal. The symmetric relation may be written

$$aw^2z^2 + 2\,bwz(w+z) + c(w^2+z^2+4\,wz) + 2\,d(w+z) + e + \mu(w-z)^2 = 0,$$

where the first five terms are the second polar of w with regard to the quartic

$$az^4 + 4\,bz^3 + 6\,cz^2 + 4\,dz + e\,;$$

and if this quartic be brought to its canonical form

$$Q \equiv z^4 + 6\,c'z^2 + 1,$$

then also, by the same transformation applied to both w and z, the $(2, 2)$ relation becomes

$$w^2z^2 + c'(w^2 + z^2 + 4\,wz) + 1 + \mu'(w-z)^2 = 0.$$

The branch-points are given by

$$(z^2 + c' + \mu')\,[(c' + \mu')z^2 + 1] = (2\,c' - \mu')^2z^2\,;$$

that is, by . $H + \mu'Q = 0,$

where H is the Hessian of Q.

[Halphen, Fonctions Elliptiques, t. ii., ch. ix.; Cayley, Elliptic Functions, ch. xiv.]

§ 166. Ex. (7). $w^5 - 5w = z$.

The branch-points, together with the values of w which become equal at them, are

$$z = 4,\ w = -1, -1\,;\ z = i4,\ w = -i, -i\,;\ z = -4,\ w = 1, 1\,;$$
$$z = -i4,\ w = i, i\,;\ z = \infty,\ w = \infty, \infty, \infty, \infty, \infty.$$

When $z = 0$, the values of w are 0 and the four fourth roots of 5, and the approximate values are

$$w = -z/5,\ \text{and}\ w - 5^{1/4} = z/20.$$

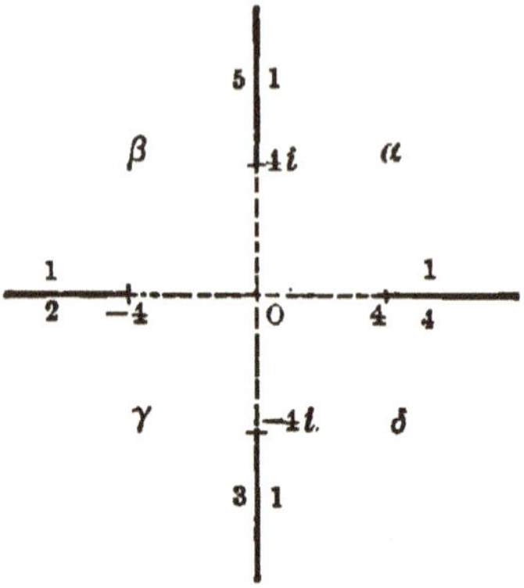

5 sheeted z–plane

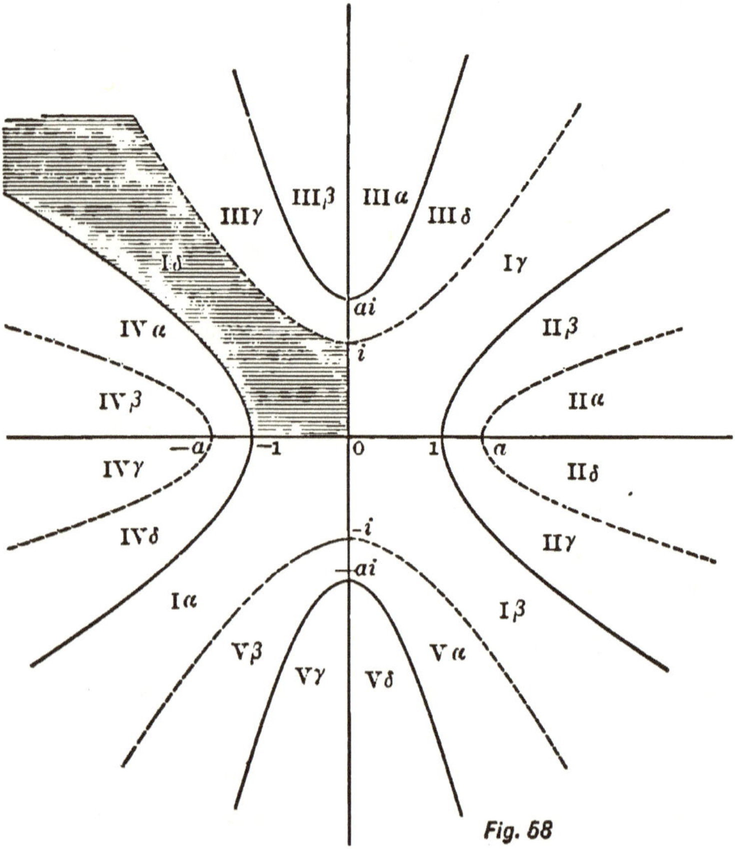

Fig. 58

w–plane

First, we trace the w-curves which correspond to the real z-axis. These are the real axis and the curve

$$w^4 + \bar{w}^4 + w\bar{w}(w^2 + w\bar{w} + \bar{w}^2) = 5,$$

where w, $\bar{w}$ are conjugate. In polar co-ordinates this curve assumes the form

$$\rho^4 = 5/(1 + 4\cos\theta\cos 3\theta).$$

The curve which corresponds to the imaginary axis is obtained by turning the curve already considered through a right angle. This may be seen directly, since, for each place (z, w) there is another place (iz, iw). In Fig. 58 the second curve is dotted. Let us write a for the positive fourth root of 5. Assign the values $w=0$, a, ia, $-a$, $-ia$ to the first, second, . . ., fifth sheets. When z moves from 0 to the right, the approximate values show that the branch w_1 moves to the left, while the other branches move to the right; also, when z moves from 0 into any quadrant, the branch w_1 moves, in its own plane, into the opposite quadrant. As z continues to move from 0 towards 4, the branch w_1 continues to move to the left along the real axis, and the branch w_4, whose initial value is $-a$, to the right along the same axis; and, when z is near 4, the two branches in question come near to each other, so that a description of a circle round $z=4$ must lead from the first into the fourth sheet. We therefore make a branch-cut from 4 to $+\infty$ along the real axis. If z move from 0 to the left, the branches w_1 and w_2 interchange round $z=-4$, and a second branch-cut must be drawn to connect the first and second sheets. Let this extend to $-\infty$. Similarly, branch-cuts from $4i$ to ∞i, and from $-4i$ to $-\infty i$, connect respectively the sheets 1, 5 and the sheets 1, 3. Since the complete w-plane corresponds to all five sheets of the Riemann surface in the z-plane, there will correspond to the twenty separate quadrants twenty separate regions of the w-plane, and to the boundary between two consecutive quadrants, the corresponding lines in the w-plane. When two straight lines in the z-plane meet at a branch-point at which two branches permute, so as to enclose an angle of 180°, the paths in the w-plane of the two branches in question intersect at right angles. Hence, when z describes that side of the positive part of the real axis which lies in the fourth quadrant and the first sheet, the corresponding branch w_1 describes that part of the line (0 to -1) which lies in the second quadrant; and, when z describes a small semicircle round 4 in the negative direction, afterwards continuing the description of the upper side of the positive real axis, w_1 describes a small quadrant round -1 in the negative direction, afterwards describing the lower half of the curve through -1. Hence the straight line from 0 to -1, and the lower half of the curve through -1, form part of the boundary of that portion of the w-plane which corresponds to the first quadrant of the first sheet of the five-sheeted Riemann surface in the z-plane. Similarly, the remaining boundaries can be determined. Let a point starting in the quadrant marked δ in the first sheet of the z-plane describe a large positive circuit five times. Since, when z is large, the approximate value of w is $z^{1/5}$, w describes a large positive circuit once in the w-plane; and since, whenever z moves from one quadrant to another, w must cross one of the curves of the figure, the parts of the w-surface which correspond to the quadrants of the z-surface are readily determined. For instance, the region Iδ, in the w-plane, lies in the second quadrant, and has for its boundary two portions of the axes and two portions of curves.

It is necessary to test the system of branch-cuts by seeing that they produce the right results for circuits about the point ∞. In the present instance, we have seen that when z is at a great distance along the positive direction of its real axis, and just above this axis, w_1 has assumed the position $|\sqrt[5]{z}|\exp(6\pi i/5)$, and it can be similarly seen that, simultaneously, w_2 is $|\sqrt[5]{z}|$, w_3 is $w_2\cdot\exp(2\pi i/5)$, w_4 is $w_2\cdot\exp(4\pi i/5)$, and w_5 is $w_2\cdot\exp(8\pi i/5)$; that is, the branches are in the cyclic order $(w_2w_3w_4w_1w_5)$. A large positive circuit in the z-plane should therefore permute the branches in this order, and the figure shows that this is what takes place.

Ex. (8). $$w^3-3wz+z^3=0.$$

Let $w=tz$; then $z=3t/(1+t^3)$. The surface for t is also the surface for w, for to each pair (t, z) corresponds one w, and when two values of t become equal, so do two values of w. When $z=a$, (where $a=\sqrt[3]{4}$), $t=a/2,\ a/2,\ -a$; when $z=va$, (where $v=\exp(2\pi i/3)$), $t=va/2,\ va/2,\ -va$; when $z=v^2a$, $t=v^2a/2,\ v^2a/2,\ -v^2a$. When $z=0$, $t=\infty,\ \infty,\ 0$; and near $z=0$, $t=z/3$ or $\pm\sqrt{3/z}$.

Finally at $z=\infty$, $t=-1,\ -v,\ -v^2$. The real axis gives

$$\frac{3t}{1+t^3}=\frac{3\bar{t}}{1+\bar{t}^3},\ \text{or}\ t\bar{t}(t+\bar{t})=1,$$

i.e. $$\rho^3=\tfrac{1}{2}\sec\theta.$$

Let c be a point on the positive side of the real axis near $z=0$; assign $t_1=c/3$ to sheet 1, $t_2=\sqrt{3/c}$ to sheet 2, and $t_3=-\sqrt{3/c}$ to sheet 3. Now let z move from c to a; at a sheets 1 and 2 must hang

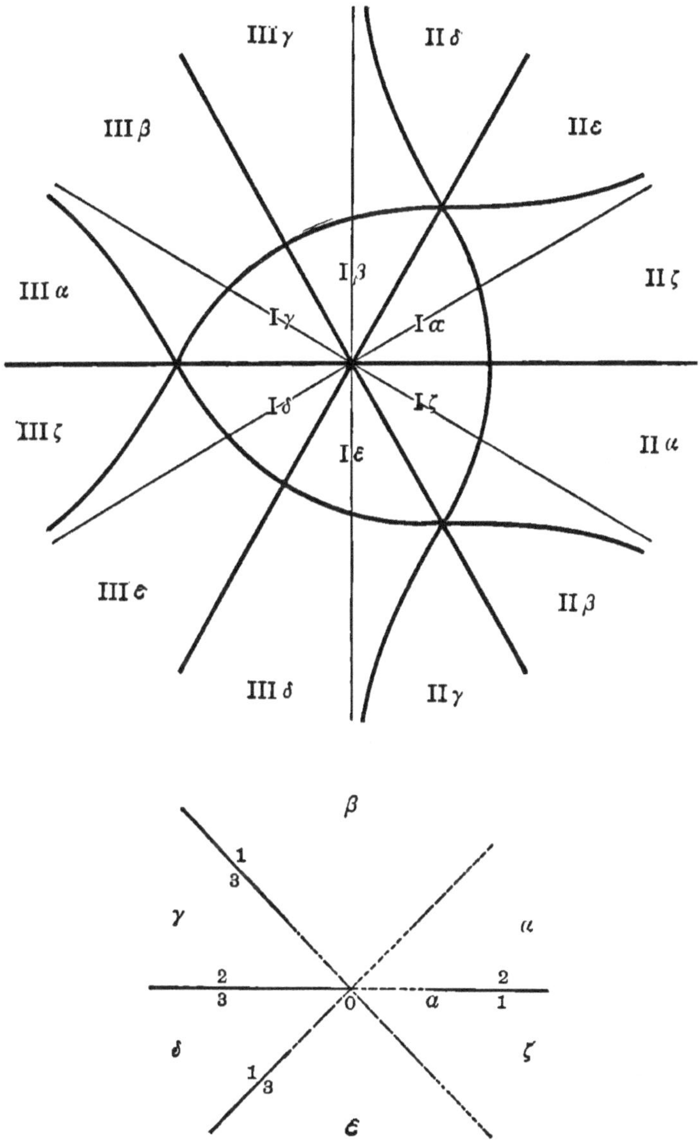
III γ
II δ
III β
II ε
I β
III α
II ζ
I γ
I α
I δ
I ζ
III ζ
II α
I ε
III ε
II β
III δ
II γ
β
1
3
γ
α
2
3
2
0
a
1
δ
ζ
1
3
ε

Fig. 59

together. Let z, starting from c, make a turn of $2\pi/3$ round 0; then t_2, t_3 turn through $-\pi/3$. Next, let z move to νa. We see that t_1, t_3 unite at $\nu a/2$. Therefore at νa sheets 1 and 3 hang together. Similarly at $\nu^2 a$, sheets 1 and 3 hang together. The branch-cuts may be drawn from the branch-points to ∞, as in Fig. 59. It is necessary that the cut $0 \cdots \infty$ pass between νa and $\nu^2 a$, in order that a large circuit may be closed, as it should be since ∞ is not a branch-point. The parts of the t-plane which map into the various parts of the surface are determined by observing that to a small circle round $z=0$, starting from c, correspond a small circle round $t=0$ starting from t, and two large negative semicircles.

§ 167. Ex. (9). $\quad 32\,wz = (w+z)^5.$

Let $\quad w + z = t,$

then $\quad F = 32\,z(t-z) - t^5,$

and the Riemann surface, which represents t as a function of z, will equally serve for w. The equations $F=0$, $\delta F/\delta t = 0$, give

$$t = 0,\ \alpha a;\ z = 0,\ 4\,\alpha a/5;$$

where $a = \sqrt[3]{2^7/5^2} = 1.7 \cdots$, and α is any cube root of 1.

The approximate values of t are

(i.) When z is small, $t = z$ or $2(2\,z)^{1/4}$;

(ii.) When $z - 4\,\alpha a/5$ is small, $t - \alpha a = \{-3\,\alpha a(z - 4\alpha a/5)/8\}^{1/2}$.

The equations $F = 0$, $\delta F/\delta z = 0$, give

$$z = 0,\ \alpha;\ t = 0,\ 2\,\alpha;$$

and the approximate values of z are

(i.) When t is small, $z = t$ or $t^4/32$;

(ii.) When $t - 2\,\alpha$ is small, $z - \alpha = \{-3\,\alpha(t - 2\,\alpha)/2\}^{1/2}$.

When $z = \infty$, t is also ∞, and when $t = \infty$, $z = \infty$; the approximate relation is $\quad 32\,z^2 = -t^5.$

When z is a small positive quantity, let $t_1 = z$, $t_2 = 2\sqrt[4]{2\,z}$, $t_3 = it_2$, $t_4 = -t_2$, $t_5 = -it_2$; and let these values be assigned to the first, second, ⋯, fifth z-sheets. Then at the origin the sheets 2, 3, 4, 5 must hang together, and when z turns round the origin it must change from sheet to sheet in the order 2, 3, 4, 5, 2. Accordingly we draw a branch-cut through the sheets from 0 to ∞; let this cut lie along the negative half of the real axis.

As z moves along the real axis to the right, t_1, t_2 move along the real axis to the right, the point t_2 leading. When z is near 1, it is t_2 which is near 2; and when z makes a positive half-turn round 1, t_2 makes a positive turn round 2. Thus, when z proceeds from 1 to the right, t_2 moves to the left, while t_1, having made only a half-turn, continues its motion to the right. When z is near $4a/5$, t_1 and t_2 are near a; therefore the sheets 1 and 2 must be connected at $4a/5$.

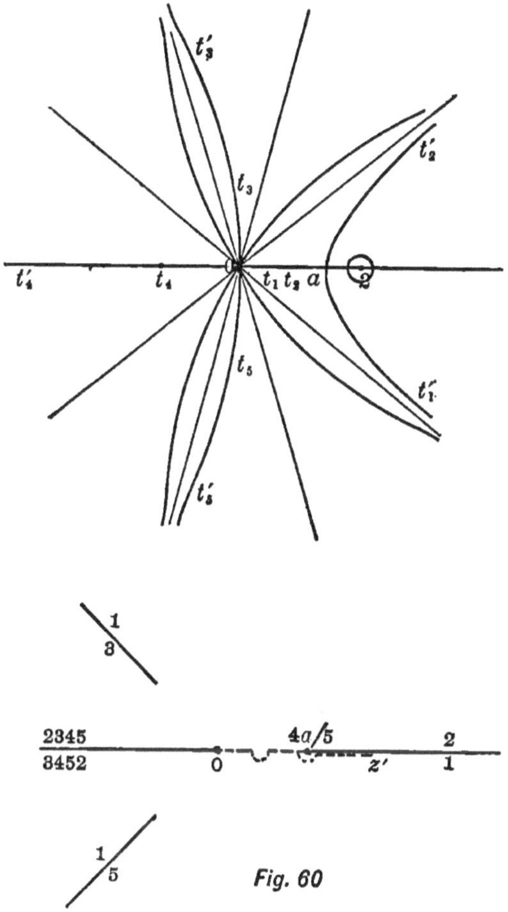

Fig. 60

Starting afresh, let z make a turn $2\pi/3$ round 0, and then move directly towards the point v; t_1 makes an equal turn, but the other roots in the t-plane make turns $\pi/6$, and therefore t_3 moves after t_1 towards the point $t = va$. As before, sheets 1 and 3 join at $z = v$. Similarly, sheets 1 and 5 join at $z = v^2$. Let the branch-cuts be drawn as in the figure. We have to see that these satisfy the requirements at the point $z = \infty$.

At 0, let the dotted path in the figure be inclined at an infinitely small positive angle to the real axis. When z describes this dotted path to the position z', the five roots move to the positions t_1', t_2', $\cdots$, t_5'. For instance, t_2 turns round 2, returns on its former path towards 0, turns through nearly 90° when near a, and proceeds along the curve to t_2'.

It follows that a large circuit in the z-plane ought to lead from sheet to sheet in the order 135241, since $32z^2 = -t^5$ at ∞. This requirement is met by the branch-cuts as drawn in the figure. It is to be noticed that the cut from 0 to ∞ does not pass between the points $4va/5$ and $4v^2a/5$. Other possible systems of branch-cuts for the same initial assignment of values are shown in figure 61.

Hitherto the equation $F(w^n, z^m) = 0$ has been considered irreducible, and, in this case, it is possible to pass from any one place of the z-surface (or w-surface) to any other place of that surface by a continuous path which lies wholly in that surface (§ 154).

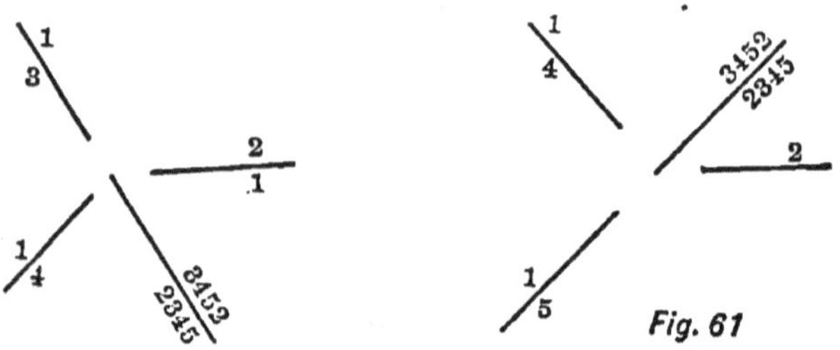

Fig. 61

If $F(w^n, z^m) = 0$ be reducible, it can be replaced by two or more irreducible equations; for instance, by two irreducible equations of degrees κ, $n - \kappa$ in w. In this case, the n sheets of the Riemann surface T will separate into sets of κ and $n - \kappa$ sheets which are unconnected; in other words, we have two z-surfaces, one for each of the irreducible equations.

The important property of w, namely, its one-valuedness upon the surface T, is shared with it by all rational algebraic functions $w', = R(w, z)$, of w and z. For example, $w', = w^3 + z^2$, is one-valued at a place (z, w) of T. This function w' will be, in general, n-valued in z; but it may be only n_1-valued where n_1 is a factor of n (1 included). For example, let $w^6 = z$; the function $w' = w^5 + z$ is a six-valued function of z, which is one-valued upon the surface T. The deduced relation $(w' - z)^6 = z^5$ requires the same surface T. But $w' = w^2 + z$, which is also one-valued upon T, leads to a relation $(w' - z)^3 = z$, which requires merely a three-sheeted surface T′. In this case, T arises by the superposition of T′ on T′.

§ 168. *The transformation of an n-ply connected Riemann surface into a simply connected surface.*

The following definitions and propositions are illustrated by figures in one plane, but are applicable to surfaces in space.

We consider a continuous surface bounded by one or more closed rims, the totality of rims constituting the boundary. The surface is supposed to be two-sided. We thus exclude the unilateral surfaces such as that of Möbius, which is constructed as follows: Take a rectangular strip of paper *abcd*, and, keeping one end *ab* fixed, turn the other round the axis which bisects *ab* and *cd*, through the angle π. Now *ab* and *dc* are parallel; bring them into coincidence, and glue them together. The two-sided rectangle has become a one-sided or unilateral surface, *ad* and *bc* now forming the rim.*

Let us understand by a *cross-cut*, a cut which, starting from a point on the boundary, and ending at a point on the same, runs entirely within the region. When such a cut is made, its two banks are to be regarded as forming part of the boundary of the new surface. A surface is called *simply connected* when every cross-cut severs the surface. The surfaces contained within a rectangle, a circle, etc., are simply connected. A surface is called *doubly connected* when a cross-cut can be made which will change it into a simply connected surface; for instance, the surface contained between two concentric circles is doubly connected. A surface is called *triply connected* when a cross-cut can be made which renders the surface doubly connected, and so on for surfaces of higher connexion. The surface contained between two non-intersecting circles and a circle which encloses them is triply connected; when two cuts are made from the rims of the inner circles to the outer rim, the surface becomes simply connected. Another example of a triply connected surface is the anchor-ring after a puncture has been made in it; for, after a cut has been made along a circle of latitude, starting and ending at the puncture, the ring can be deformed into the surface of a cylinder, while a second cut from rim to rim reduces the cylinder to a rectangle.

Different forms of cross-cuts, on a surface with three separate rims, are shown in Fig. 62.

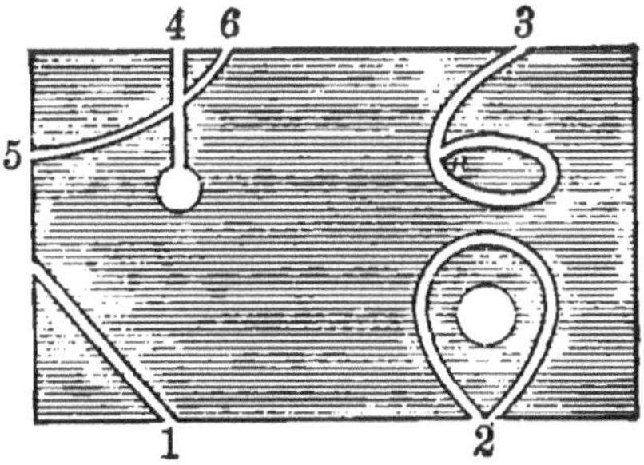

Fig. 62

(1) begins and ends at the same rim.

(2) begins and ends at the same point of the same rim.

(3) begins at the boundary and ends at *a*. As the cut proceeds, its two edges become boundary, so that *a* is on the boundary. The cut terminates when it reaches the point *a* on the boundary.

(4) begins and ends at different rims.

The cut (4) reduces the connexion of the surface without dividing

* There is good reason to regard the ordinary plane of projective geometry as unilateral. See Klein, Math. Ann., t. vii., p. 549, where it is shown that Riemann's theory of the connexion of surfaces can be effectively employed in the projective field.

it into 2 separate parts; we shall speak of it as a cross-cut of the second kind. The cuts (5) and (6), if made *after* the cut (4), must be treated as distinct cross-cuts. A unilateral surface is not necessarily severed by a cross-cut of the first kind.

The assumption has been made that the surfaces under consideration have rims; but there are surfaces, such as the sphere, which are completely closed. Cut out an infinitely small region of circular contour; from this puncture the first cross-cuts must start. In the case of a sphere one such cross-cut will sever the surface, and accordingly the connexion of the punctured sphere must be simple. Another example of a closed surface is a Riemann sphere with n sheets. By puncturing the surface along a small closed curve round any point, a boundary can be created. If the point be a branch-point at which r roots permute, the new rim goes r times round the point. The order of connexion of the punctured surface will be determined later.

Lemma. A simply connected surface is separated by any κ cross-cuts into $\kappa+1$ separate simply connected surfaces. This lemma requires no proof.

Theorem 1. If a surface, or a system of surfaces, be divided by κ cross-cuts into σ simply connected pieces, and by κ' cross-cuts into σ' simply connected pieces, then $\kappa-\sigma=\kappa'-\sigma'$.

Two points are to be clearly noticed in the proof of this theorem. The first is the obvious fact that a cut possesses two, and only two, ends; the second is that, by definition, as a cut proceeds its sides are turned into boundary. A cross-cut is represented geometrically by a single line, but this line must be regarded as having two sides; hence, if a cross-cut be intersected by a line, at the place of intersection, there are two points of intersection a, b which may be geometrically undistinguishable, but a belongs to one side, b to the other, of the cross-cut.

First draw the κ cross-cuts, and secondly the κ' lines which coincide with the second system of κ' cuts. Let the κ' lines meet the κ lines in λ points. Each of these λ points counts as two boundary points; and also the κ' lines meet the original boundary in $2\kappa'$ points. Thus, on the whole, $2\kappa'+2\lambda$ boundary points have been furnished by the κ' lines. Now making the second system of cross-cuts along the κ' lines, we have $\kappa'+\lambda$ new cross-cuts. These, being applied to σ separate simply connected pieces, give $\sigma+\kappa'+\lambda$ separate pieces. But the final result due to the superposed systems of cross-cuts will clearly be independent of the order in which we make the κ and κ'

cuts. If we make the κ' cross-cuts first and then the κ cross-cuts, the number of separate simply connected pieces will be $\sigma' + \kappa + \lambda$. Hence $\sigma + \kappa' + \lambda = \sigma' + \kappa + \lambda$, and $\kappa - \sigma = \kappa' - \sigma'$. The number $\kappa - \sigma$ is therefore independent of the number of cross-cuts.

The number $\kappa - \sigma + 2$ may be called the *index* of the surface or system of surfaces.*

A simple surface is divided by one cross-cut into two pieces. Therefore the index is $1 - 2 + 2$ or 1; a doubly connected surface has the index 2, and generally when only one surface is considered, we may use the term "order of connexion" in place of "index."

The index of a system of σ simple surfaces is $2 - \sigma$; for here $\kappa = 0$.

Theorem 2. Let there be μ separate surfaces, whose indices are $\nu_1, \nu_2, \cdots, \nu_\mu$. The number of cross-cuts necessary to render the rth surface simply connected is $\nu_r - 1$. Hence the number of cross-cuts necessary to render all the surfaces simple is $\sum\limits_{r=1}^{\mu}(\nu_r - 1)$, or $\sum\limits_{1}^{\mu}\nu_r - \mu$. Hence the index of the system is $\Sigma\nu_r - 2\mu + 2$.

Theorem 3. The index ν of a system Σ becomes $\nu - \alpha$, when α cross-cuts have been made. Let a single cross-cut change Σ into Σ', and let κ more cross-cuts change Σ' into σ simple surfaces. Then the index of Σ' is $\nu' = \kappa - \sigma + 2$, and since $\kappa + 1$ cross-cuts have changed Σ into the σ simple surfaces, $\nu = (\kappa + 1) - \sigma + 2$. Therefore $\nu' = \nu - 1$, a result which shows that a simple cross-cut reduces the index by unity. Each further cross-cut does the same; hence the theorem.

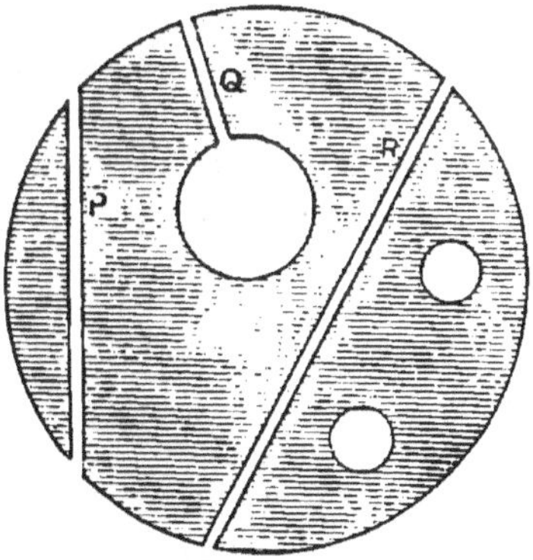

Fig. 63

For example, let T be a quadruply connected surface (Fig. 63). The surface can be made simple by three cross-cuts, and the index of a simple surface is 1; hence, by the converse of Theorem 3, the original index was 4. The cross-cut P gives two surfaces of indices 1 and 4; by Theorem 2 the index of the system is $5 - 4 + 2$, or 3. Again, if in the original surface T the cut R be made, T separates into two surfaces of indices 2 and 3; and again the index of the system is 3.

* In this definition we follow Neumann. Klein (Math. Ann., t. vii.) and Schläfli (Crelle, t. lxxvi.) define the *grund-zahl* as $\kappa - \sigma + 1$.

§ 169. *The retrosection.* This is a cut which starts from a point within a surface and returns to that point, without having crossed itself or met any rim.

Theorem 4. A retrosection does not alter the index of a system of surfaces. Let T be the surface in which the retrosection is made, T′ the surface or surfaces arising from T when the retrosection is made; also let ν, ν' be the indices of T, T′. From a point a on the retrosection draw a cut C to any point of the original boundary. This cut, together with the retrosection, forms a cross-cut of the surface T, and therefore reduces ν to $\nu-1$; but C is a cross-cut of T, and therefore reduces ν' to $\nu'-1$. Hence $\nu-1=\nu'-1$, and $\nu=\nu'$. This proves the theorem. A cut formed by a retrosection and a cross-cut, as for instance (3) in Fig. 62, is a cross-cut of the original surface; cuts of this kind are commonly named sigma-shaped cross-cuts.

Consider a limiting case of the retrosection, namely, when, at an ordinary point of a surface T, an infinitely small region is separated. The index of this region is of course 1; the rest of the surface is what we termed the punctured surface.

Let ν be the index of T, ν_1 that of the punctured surface. We have (by writing $\mu=2$, $\nu_2=1$ in the formula of Theorem 2) the equation $\nu'=\nu_1+1-4+2$, where ν' is the index of T′; and by the present theorem $\nu=\nu'$. Therefore $\nu=\nu_1-1$; and the index of a punctured surface is greater by unity than that of the original surface.

Theorem 5. Every cross-cut either increases or diminishes the number of rims by unity.

For a cross-cut of the second kind, such as (4) in Fig. 62, makes two rims become one, whereas each of the cross-cuts (1), (2), (3), gives rise to a new rim.

Theorem 6. The number of rims of an n-ply connected surface is either n, or $n-2\kappa$, where κ is a positive integer.

For let m be the number of rims. Since the surface is n-ply connected, it can be made simply connected by $n-1$ cross-cuts. Let ϵ_μ denote the ± 1 added to the number of rims by the cross-cut μ.

Therefore $$m+\sum_1^{n-1}\epsilon_\mu=1.$$

Let κ be the number of positive ϵ's; the equation just written down gives

$$m+\kappa-(n-1-\kappa)=1, \quad \text{or} \quad m=n-2\kappa.$$

Since the punctured Riemann sphere has one rim, its order of connexion must be an odd number. We shall now determine this number.

§ 170. *The index of a Riemann surface.* Let the surface consist of n sheets which hang together at s branch-places a_r $(r = 1$ to $s)$; and let $m_1, m_2, \cdots, m_r, \cdots, m_s$ be the numbers of sheets which hang together at these points. Assume, at first, that no two branch-places lie in the same vertical. The Riemann surfaces under consideration, namely those associated with algebraic functions, are without rims.* We suppose a puncture P made at an ordinary point, the puncture extending through only one sheet. Through P draw a cylinder cutting the remaining $(n-1)$ sheets along $(n-1)$ retrosections. These retrosections do not alter the index ν. Let Q be an infinitely small closed curve at some ordinary point of the surface, and let the n sheets be cut along n retrosections by a cylinder through Q. Project P, Q, and the branch-places on a plane parallel to each of the sheets. The projection will consist of P, Q, and the points $a_1, a_2, \cdots, a_s$. Draw lines from P to Q in such a way that the plane is divided into s compartments, each of which contains one of the points a_r. With the walls of these compartments as bases, construct cylinders, cutting through the n sheets. These cylinders cut the surface along ns cross-cuts. The surface has now been divided into simply connected portions. The cylinder P detaches $n-1$ portions of surface; the cylinder Q, n portions; but the cylinder (a_r) not n but $n - m_r + 1$ portions, since m_r sheets are connected at a_r. Hence

$$\begin{aligned} \nu &= ns - \{n + (n-1) + \sum_{r=1}^{s}(n - m_r + 1)\} + 2 \\ &= \sum_{r=1}^{s} m_r - 2n - s + 3. \end{aligned}$$

We have assumed that $a_1, a_2, \cdots, a_s$ have different projections. If, however, two branch-places lie in the same vertical, the difficulty can be evaded by distorting the Riemann surface slightly, and thus making the two branch-places have distinct projections. Such a distortion will not alter the index of the surface, and therefore the formula holds in this case. This proof, which involves merely considerations of Analysis Situs, is given by Neumann, Abel'sche Integrale, p. 168.

* In connexion with fundamental polygons Klein introduced the idea of rimmed Riemann surfaces, Math. Ann., t. xxi.; in this memoir the student will find important extensions of the usual modes of Riemann representation.

In § 128 it was shown that in the case of the prepared equation

$$\beta = 2p + 2(n-1),$$

where β is the number of branch-points. We have just seen that, *whatever be the nature of the branchings supplied by an algebraic equation*, the number of simple branch-points, viz. $\sum_{r=1}^{\bullet}(m_r - 1)$, is equal to $2(n-1)+(\nu-1)$. Hence, in the case of the prepared equation,

$$2p + 2(n-1) = (\nu - 1) + 2(n-1),$$

and $$\nu = 2p + 1.$$

Thus the number of cross-cuts required to reduce to simple connexion the Riemann surface which arises from a prepared equation is precisely $2p$, where p is the deficiency of the Cartesian curve associated with the equation. *Whatever be the nature of the basis-equation*, let p be defined as half the number of cross-cuts required to reduce the corresponding Riemann surface to simple connexion.

The number p was introduced by Riemann. Its importance in the theory of curves was pointed out by Clebsch. The above definition applies to basis-curves with higher singularities. Suppose that the higher singularities of a curve have been resolved into nodes, cusps, double tangents, and inflexions, the numbers of nodes and cusps being determined as in ch. iv., and the numbers of double tangents and inflexions by considerations of duality. A problem which suggests itself is to prove Plücker's equations,

$$\alpha = \beta(\beta-1) - 2\Sigma\nu - 3\Sigma\kappa, \quad \beta = \alpha(\alpha-1) - 2\Sigma\tau - 3\Sigma\iota,$$

$$p = \tfrac{1}{2}(\alpha-1)(\alpha-2) - \Sigma\nu - \Sigma\kappa, \quad p = \tfrac{1}{2}(\beta-1)(\beta-2) - \Sigma\tau - \Sigma\iota,$$

given that α, β, are the order and class of a curve and that $\Sigma\nu$, $\Sigma\kappa$, $\Sigma\tau$, $\Sigma\iota$ are the numbers supplied by the singularities of the curve. This question belongs to Higher Plane Curves rather than to the Theory of Functions, but it is interesting to notice that H. J. S. Smith, in his demonstration of Plücker's formulæ, makes use of the Riemann definition of p, and is thus able to avail himself of a theorem (to be proved later on) that p is invariant with regard to birational transformations, and of Neumann's perfectly general result that the number of branchings is equal to $2p + 2(n-1)$. [Smith, On the Higher Singularities of Plane Curves, Lond. Math. Soc., t. vi., p. 161.]

Let us consider Riemann's form of the equation $F(w, z) = 0$; namely, $f_0 w^n + f_1 w^{n-1} + \cdots + f_n = 0$, where the coefficients of the various powers of w contain z to the exponent m; and let us assume that in the finite part of the plane there are ν nodes and κ cusps, but no higher singularities. The line at infinity meets the Cartesian curve $F = 0$ in an m-point on the w-axis, and in an n-point on the z-axis. The formula in Salmon's Higher Plane Curves (3rd ed., § 82) for the class of a curve applies, since the nodes and cusps are distinct, and since at the m-point and n-point the tangents are separated. Hence the class of the curve

$$=(m+n)(m+n-1)-n(n-1)-m(m-1)-2\nu-3\kappa.$$

The number of tangents which can be drawn from the m-point is equal to the class of the curve $-2m$. Since each of these tangents gives a simple branch-point, and since each of the κ cusps likewise gives a simple branch-point, the number of branchings is

$$\{(m+n)(m+n-1)-n(n-1)-m(m-1)-2\nu-3\kappa\}-2m+\kappa,$$

i.e. $\quad 2m(n-1)-2\nu-2\kappa.$

But $\quad 2p+2(n-1)=$ the number of branchings,

therefore $\quad p=(m-1)(n-1)-\nu-\kappa.$

[This expression for p was given by Riemann in his memoir Abel'sche Functionen, Ges. Werke, p. 106.]

§ 171. *The canonical dissection of a Riemann surface.*

In § 130 we proved two theorems due to Lüroth and Clebsch. By means of these theorems we found that the n sheets of a Riemann surface can be connected in such a way that the branch-cuts between successive sheets are single links, except between the first and second sheets. In this exceptional case the number of branch-cuts is $p+1$, when p is the deficiency of the algebraic curve $F(w, z)=0$, obtained by treating w and z as Cartesian coördinates; it is assumed that the equation is prepared, and therefore contributes merely simple branch-points. There are thus p superfluous connexions between the first two sheets. We have now to see that this n-fold sphere can be transformed, without tearing, into the surface of a body with p perforations.

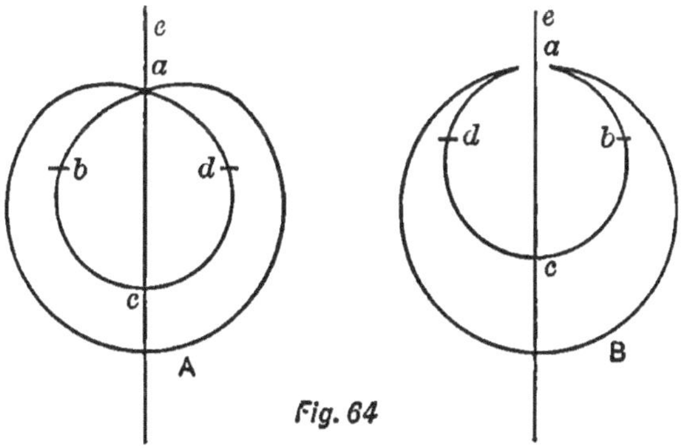

Fig. 64

First take the case of a twofold sphere, whose two sheets are connected by a single bridge. In Fig. 64, the bridge is perpendicular to the plane of the paper, and a is its middle point. Let every point of the inner sheet be reflected in the diametral plane which contains the bridge. Then the two halves of the inner sphere are interchanged, and the figure changes from A to B. This can be effected by a process of continuous deformation, the one hemispherical surface passing through the other, without any severing of the Riemann sphere, and the order of connexion is unchanged. To an observer at e the appearance is that of a sphere with a hole in it. The inner sphere can now be pulled through the hole, and stretched, without tearing, into the form of a one-sheeted sphere, or of a doubly-sheeted flat plate without a hole.

Next take the case of a two-sheeted sphere with $p+1$ bridges; the surface can be stretched, without tearing, until the bridges lie along one great circle. Apply to the inner sheet the same process of reflexion as before; then the two spheres are connected along $p+1$ holes. One of the holes may be stretched, as before, so as to form the outer rim of a two-sheeted flat plate; and in the plate there remain p holes. The surface has become the surface of a solid box perforated by p holes.

In the general case the inner sphere, and the one next to it, are connected by one bridge. The two can be replaced by a single spherical sheet. This sheet is connected with the third sheet by a single bridge. Repeat the process until the first $n-1$ sheets are replaced by a single sheet. This single sheet is connected with the outside sheet by $p+1$ bridges; and the two can be replaced by a box with p holes through it. It is thus proved that the general Riemann surface arising from an algebraic equation of deficiency p can be deformed, without tearing, into the surface of a box with p holes.

The principle of this method is due to Lüroth; the form of proof to Clifford (Collected Papers, on the canonical form and dissection of a Riemann's surface, p. 241). See also Hofmann's tract, Methodik der stetigen Deformation von zweiblättrigen Riemann'schen Flächen (Halle, 1888). The flat two-sided sheet with $p+1$ rims is used by Schottky in his important memoir on conform representation, Crelle, t. lxxxiii.

The following diagrams will illustrate these processes:

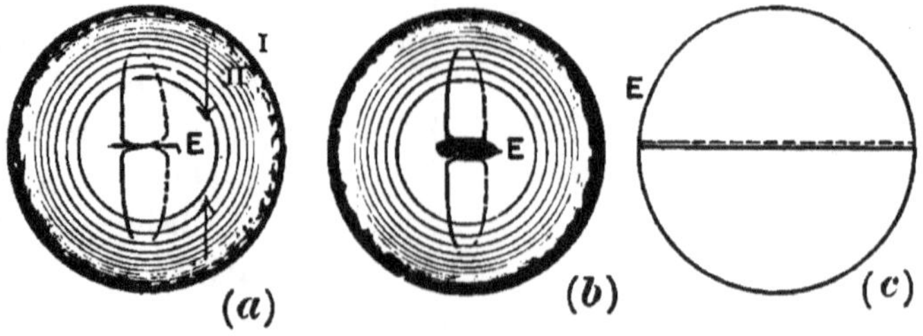

Fig. 65

Figs. (*a*), (*b*), (*c*) show stages of the passage from the double Riemann sphere, with one branch-cut (Fig. *a*), into the double flat plate (Fig. *c*). In Fig. *b*, the reflexion has changed the bridge into a hole which connects surface II. with surface I. In Fig. *c*, the rim E of the hole has been stretched until it takes the position in the figure, the hole has disappeared, and the *interior* sphere II. has become the *upper* flat sheet, the *exterior* sphere I. the *lower* flat sheet.

The process of conversion of a thin flat surface with two sheets, connected by p holes, into a double Riemann sphere, can be illustrated in a similar manner. We leave the drawing of the figures to the reader. Keeping the outer rim E fixed, we can make the two sheets pass from the flat form into one in which the two sheets form two infinitely close hemispheres connected along E and along the p holes. These holes may be regarded as places at which the outer hemisphere passes into the inner, and E is a place of like nature. By a contraction of E into a small circle, these two hemispheres can be made to pass into two infinitely close spheres connected by $p+1$ holes, namely, the original holes and the new one supplied by E. Arrange these $p+1$ holes along a great circle, and employ the reflexion process as described above. Each hole changes into a bridge, and the final form is a two-sheeted Riemann sphere with $p+1$ bridges.

§ 172. *Klein's normal surface.*

The box with p holes can be deformed, without tearing, into a sphere with p handles, a convenient normal form used by Klein.

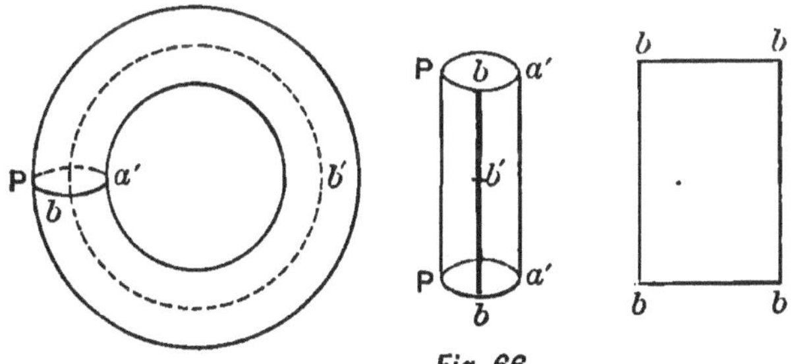

Fig. 66

When $p=0$, we have the ordinary sphere; when $p=1$, the anchor-ring can be selected as the normal surface, for it can be deformed into a sphere with one handle.

Fig. 67*a* is the normal surface for the case $p=2$; the curves A, B are latitude and meridian curves respectively. The rim of

Clifford's box has become the curve E in the plane of the paper, which can be called the equator.

Dissection of the normal surface. — It is to be understood that throughout what follows the surfaces employed are punctured at a point; otherwise there is no boundary, and the definition of multiple connexion has no meaning.

A surface is said to be dissected when it has been made simply connected by a system of cross-cuts. Let us consider the mode of

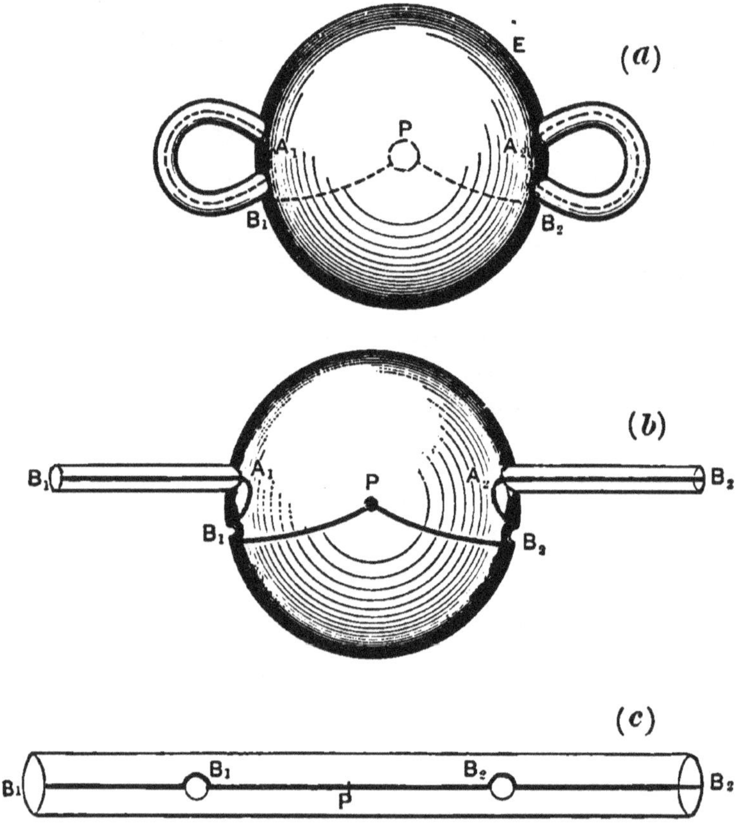

Fig. 67

dissection of the anchor-ring (Fig. 66). Let P be the puncture. Make the cross-cut Pa' along a meridian curve. The cut surface can now be deformed into a cylinder (as in § 168). As a second cross-cut, take $bb'b$; the cut cylinder can now be deformed into a rectangular sheet. Thus two cross-cuts, along a meridian curve and a latitude curve, have made the surface simple.

In exactly the same way, the surface of a sphere with p handles can be rendered simple by drawing a meridian cut through each handle, and a latitude cut round each handle. We have p pairs of cuts A_κ, $B_\kappa (\kappa = 1, 2, \cdots, p)$. To dissect the surface, take a point s_κ on either A_κ or B_κ, say on B_κ; make cuts C_κ to the p points s_κ from the puncture. The cuts have now become cross-cuts; p of these are sigma-shaped, being formed by the combination of C_κ and B_κ; and there are also p cross-cuts A_κ. In all there are $2p$ cross-cuts.

Fig. 67 shows the process in the case $p = 2$.

After two meridian cuts B_1, B_2, the section along the equator of the sphere with two handles takes the form represented in Fig. 67 (b). The latitude-cuts, and the joins to P, pass into the heavily marked lines of the figure. The surface is now dissected; for Fig. 67 (c) can be derived from Fig. 67 (b) by pulling out the tubes in the directions of the arrows, and the cylinder, if cut along the heavily marked line, is evidently dissected.

In the dissection of the box with p holes, cuts were made along curves A, B, C, the curves A, B being respectively latitude and meridian curves. In dealing with n-valued algebraic functions, it is often necessary to retain the n-sheeted sphere, and to dissect it by $2p$ cross-cuts.* The easiest way of finding out how this can be done is to retrace our steps and change the multiply connected surface with p holes into the Riemann sphere with n sheets, connected by branch-cuts as in § 171.

Now the p cuts A must change into curves round branch-cuts. Also the branch-cuts round which these curves pass can only be those connecting the first and second sheets; for a cut round the single branch-cut which connects the first and third sheets would sever the third, since the third is linked to the remaining sheets merely by this bridge. Thus the cuts A must pass into curves such as A (Fig. 68). The curves B passed originally through the holes for which the A's were latitude curves. Hence the B's, after the change, must traverse the p bridges round which the A's have been drawn. This shows how it comes to pass that the $(n-2)$ bridges between the first and third, fourth, $\cdots$, nth sheets, are unrepresented in the diagrams of the cross-cuts. The equator E becomes a branch-cut over which all the B's pass.

Clifford's box is deformed by the same process; the p holes pass into p bridges, the rim into a $(p+1)$th bridge. The meridian curves can be drawn in many ways from hole to hole; the most

* As before, the equation between w and z is in the prepared form.

natural construction is to draw each round one hole and the rim. Returning to the Riemann sphere, this gives cross-cuts $B_1, B_2, \cdots, B_p$, which all intersect one and the same bridge, namely, the $(p+1)$th; thus the cut B_r passes over the rth and $(p+1)$th bridges

$$(r=1, 2, \cdots, p).$$

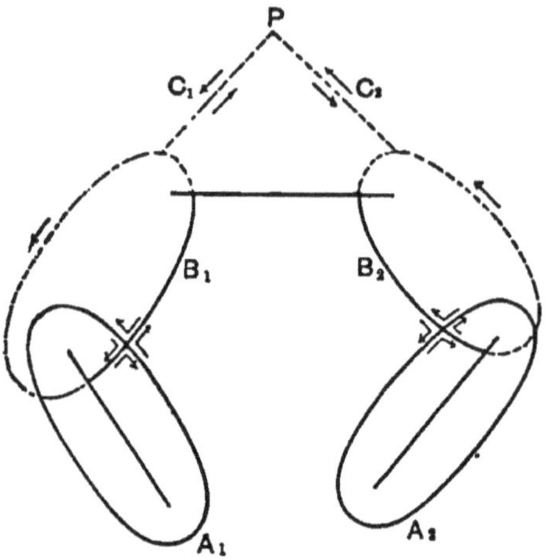

Fig. 68

In a 5-ply connected Rieman surface, four cross-cuts reduce the surface to simple connexion. In Fig. 68 the cross-cuts are C_1+B_1, A_1, C_2+B_2, A_2; C_1+B_1 and C_2+B_2 being sigma-shaped.

By following the arrows in Fig. 68, it is easy to see that the complete boundary of the dissected Riemann surface can be described by the continuous motion of a point, the two banks of a cross-cut being described in opposite directions.

§ 173. *Systems of curves which delimit regions.* The closed curve A_1 of Fig. 68 does not by itself delimit a region of the surface; in other words, it is possible to pass along a continuous curve from a point on one bank of A_1 to the point immediately opposite, without cutting the rim or passing out of the surface. B_1 is such a curve. Again, A_1, B_1 in Fig. 68 do not conjointly delimit a region of the surface. The dissection of the Riemann surface thus raises the question as to the maximum number of closed curves, which neither singly nor in sets delimit a portion of surface. Also it is evident that the cross-cuts are capable of deformation. To what extent can such deformation be effected?

In Riemann's memoir on the Abelian functions these questions are discussed and various theorems are arrived at, which are of importance in the subject known as Analysis Situs. These theorems were used by Riemann as a basis upon which to build up the theory of the dissection of a multiply connected surface. In this chapter we have accomplished the dissection by the use of theorems due to Lüroth and Clebsch; but as Riemann's theorems are interesting in themselves and throw valuable light upon the effects of cross-cuts, we shall give a brief account of his method.

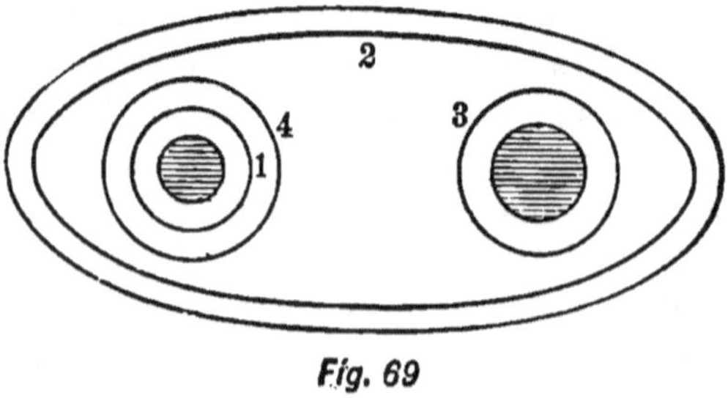

Fig. 69

I. Instead of a multiply connected Riemann surface with only one rim, we can employ the multiply connected single sheet of § 168 with several rims. The curve 1 of Fig. 69 does not by itself delimit a region; when taken in conjunction with 4 it delimits a region. It is true that it is impossible to get from one side of 1 to the other, without crossing the boundary; but nevertheless 1 does not delimit a region. To meet the case of a surface with more than one rim, we say that a curve does not delimit a region, when it is possible to reach the boundary from each of two opposite points of the curve.

II. Let $w = \sqrt{(z-a)(z-b)}$; the corresponding Riemann surface is two-sheeted, with a bridge from a to b. Let A be an oval curve round the bridge $\overline{ab}$, and let c, d be opposite points on the outer and inner banks. If the puncture be made at a place e, outside A, it is impossible to connect e with d, except by a line which crosses A. The curve A delimits a region.

III. If the puncture extend through both sheets, A does not delimit a region; if, however, a second curve A' be drawn vertically below A, it will no longer be possible to draw a curve from d to the puncture without crossing A or A'; therefore A, A' conjointly delimit a region.

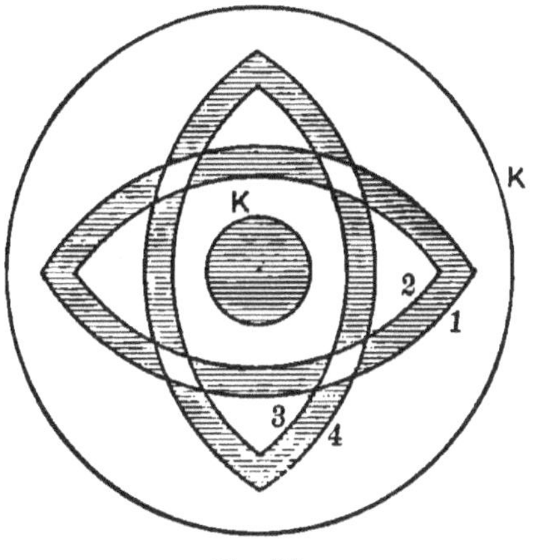

Fig. 70

IV. The systems (1, 4); (1, 2, 3); (2, 3, 4) delimit regions (Fig. 69).

V. Let K, K (Fig. 70) form the boundary of a surface. The curves 1, 2 delimit a region, and so do the curves 3, 4. The four curves divide the surface into two sets of points:—

(1) points in the shaded portions, which have the property that lines which join them to the boundary must meet the four curves in an odd number of points;

(2) points in the unshaded portions which have the property that lines which join them to the boundary must meet the curves an even number of times, or not at all. Points of the former kind are known as *interior* points; points of the latter kind as *exterior* points. The curves 1, 2, 3, 4 taken together separate the interior points from the exterior. The collective space filled by interior points is said to be 'completely bounded' or 'delimited' by the four curves, and the curves are said to be delimiting curves. With this convention as to interior and exterior points, the theorems, which we are about to prove, hold good even when the curves intersect. However, the reader will find it easier to follow the argument by drawing a figure in which the curves do not intersect.

Theorem. In any surface T, let A, B, C be systems of circuits. If A and B be delimiting curves and if A and C be delimiting curves, then B and C are delimiting curves. Formal proof is superfluous; the following considerations will suffice. Since A can be deformed into B without passing over the boundary, it must, in the process of deformation, trace out a continuous surface, which lies wholly within T. This region of T is delimited by A and B. In the same way A and C delimit a region of T. But the two regions, so obtained, adjoin each other along the lines A. Therefore, if these lines be suppressed, there is a region delimited by B and C.

Theorem. Let $A_1, A_2, \cdots, A_n$ be a system of n circuits on a surface, of which neither the whole nor a part delimits a region, but of which some or all, taken in conjunction with any other circuit, do delimit a region. Let $B_1, B_2, \cdots, B_n$ be another system of n circuits, possessing the first property of the A's; namely, that neither the whole nor a part delimits a region. Then this system, taken in conjunction with any other circuit, delimits a region. That is, the B's possess the second property of the A's.

Let us here understand that 'all of n things' is included under 'some of n things,' but that 'none of n things' is excluded. By the hypothesis, B_1, with some of the A's, delimits a region. Let A_1 be one of these A's. Take any circuit C which is not a boundary. C also, with some of the A's, delimits a region. There are two cases: —

CASE 1. A_1 is included amongst these A's.

CASE 2. A_1 is not included.

In Case 1, A_1, with some of C, $A_2, A_3, \cdots, A_n$, delimits a region;

A_1, with some of $B_1, A_2, A_3, \cdots, A_n$, delimits a region;

hence a system taken from $C, B_1, A_2, A_3, \cdots, A_n$ delimits a region.

We say that C must belong to this system; for if not, some of $B_1, A_2, \cdots, A_n$ delimit a region.

But B_1, A_1 by themselves, or with other A's, delimit a region. Hence, by the previous theorem, some of $A_1, A_2, \cdots, A_n$ delimit a region, which contradicts the hypothesis.

In Case 2, C, with some of $A_2, A_3, \cdots, A_n$, delimits a region. Hence, *a fortiori*, C, with some of $B_1, A_2, A_3, \cdots, A_n$, delimits a region. Thus, in either case, the system $B_1, A_2, \cdots, A_n$, which has the first property of the A's (that is, that neither the whole nor a part of it delimits a region), has also the second property of the A's. Continuing the reasoning, all the A's can be replaced by B's; which proves the theorem.

Theorem. Let $A_1, A_2, \cdots, A_n$ be as in the last theorem; and let $C_1, C_2, \cdots, C_n$ be another system of circuits, such that A_r and C_r $(r = 1, 2, \cdots, n)$ delimit a region; then the A's may be replaced by the C's.

To prove this, we must show that no system of C's can delimit a region. If possible, let some of the C's, say $C_1, C_2, \cdots, C_\kappa$, delimit a region, and let this system be called a delimiting system. Since A_1, C_1 form a delimiting system, it follows that $(A_1 C_2 C_3 \cdots C_\kappa)$ is a delimiting system. And since A_2, C_2 form a delimiting system, it follows that $(A_1 A_2 C_3 \cdots C_\kappa)$ is a delimiting system. Replacing in this way each C by A, we see that the original supposition demands that $A_1, A_2, \cdots, A_\kappa$ shall delimit a region; but they do not. Hence no system of the C's delimits a region, and therefore, by the last theorem, the C's may replace the A's.*

* These theorems on delimiting curves were stated by Riemann; the proofs in the text were given by Königsberger. Elliptische Functionen, pp. 47 *et seq.* We may also refer the reader to a memoir by Simart, Commentaire sur deux mémoires de Riemann; and to Lamb's Hydrodynamics, Note B.

Riemann's theorems on delimiting curves are immediately applicable to a Riemann surface. If, after a certain number of cross-cuts have been made, without severing the surface, we can draw a circuit on the surface such that it is possible to get from the interior to the exterior region without cutting the boundary, the surface is not yet simple, and more cross-cuts will be needed; but if no such circuit can be made, the surface has been dissected by the cross-cuts. Riemann defined the order of connexion of a surface by this maximum number of cross-cuts, increased by 1. We have seen that on a Riemann surface of deficiency p, $2p$ cross-cuts can be drawn, that these $2p$ curves neither singly nor collectively delimit a region, and that some of them delimit a region when taken with any other closed curve. Hence, by Riemann's definition, the order of connexion of a Riemann surface of deficiency p is $2p+1$; a result which agrees with that of § 170. For the $2p$ curves which serve the part of cross-cuts can be substituted any other $2p$ curves, of which neither the whole nor a part delimits a region. Thus the meridian and latitude curves of the normal surface can be deformed into new meridian and latitude curves, when the new meridian and latitude curves B_κ', A_κ' are such that A_κ, A_κ' and B_κ, B_κ', $(\kappa = 1, 2, \cdots, p)$ form $2p$ delimiting systems.

When a cross-cut has been made in an $(n+1)$-ply connected surface T, the new surface T′ is, as we know, only n-ply connected. It is interesting to show, graphically, that only $n-1$ curves of the type A can be drawn on T′.

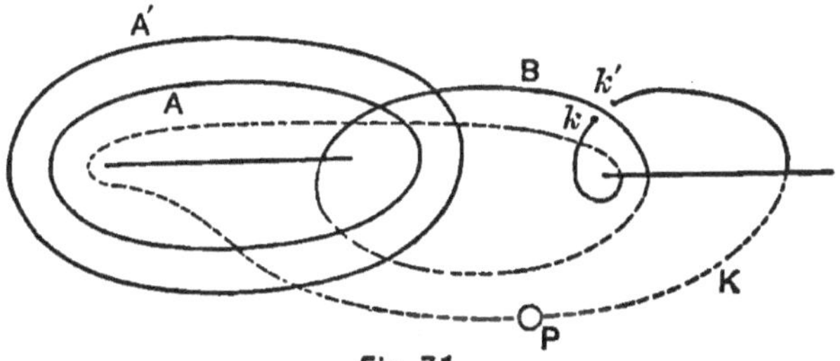

Fig. 71

The figure is that of a triply connected and two-sheeted surface, but the argument applies generally. A and B are circuits on the triply connected surface. Neither separately nor together do they delimit a region; for the line K connects the points k, k', on opposite sides of B, with the puncture P. Treat K as a cross-cut, and, when it is made, let the new surface be T′. A is a closed circuit on T′, which does not delimit a region. We wish to prove that every other circuit A', which does not of itself delimit a region, does so when taken in conjunction with A. Now A' is a circuit on the uncut surface T. Therefore A', A, B delimit a region T_1 on T. We prove, in the first place, that B must be omitted from the boundary of T_1. For if B were part of that boundary, it would be impossible to get from one side of B to the other; but the figure shows that K does pass on T from k to k'. Hence A, A' must delimit a region on T; that is, there is a region T_1 on T, from no point of which can a curve be drawn to P, without meeting A' or A. But further, A' and A also delimit a region T_1 on the surface T′; that is, it is impossible to pass from a point of the region T_1 to either P or the cross-cut K. For K does not meet either A' or A; and therefore if a curve, starting from a point of T_1, could once get to any point of K, it could be continued along K to P; but we have already seen that P cannot be reached.

§ 174. *Algebraic functions on a Riemann surface.*

It was proved in Chapter V., § 154, that when w is, throughout the z-plane, an n-valued function of z with a finite number of poles and algebraic critical points, and free from essential singularities, this function is connected with z by an equation

$$w^n + r_1(z)w^{n-1} + r_2(z)w^{n-2} + \cdots + r_n(z) = 0,$$

where $r_1(z)$, $r_2(z)$, $\cdots$, $r_n(z)$ are rational fractions. Let the denominators be removed and let the resulting equation be $F(w^n, z^m) = 0$. The corresponding Riemann surface T is n-sheeted. If the resulting surface be connected, it will be possible to pass, by properly chosen paths, from any value of w_1 to each of the $n-1$ associated values w_2, w_3, $\cdots$, w_n. That the equation $F=0$ is, under these circumstances, irreducible can be shown by the following considerations. Assume that $F(w, z) = F_1(w, z) \cdot F_2(w, z)$, where $F_1(w, z)$ is irreducible. At an ordinary point $z = a$ in the Argand diagram there must be some branch, say w_1, which makes F_1 vanish at a and throughout a region which contains a. In the neighbourhood of $z = a$ the function $F_1(w_1, z)$ can be represented as an integral series $P(z-a)$, and this integral series vanishes identically throughout its circle of convergence in virtue of the theorem of § 87. The method of continuation shows that the analytic function derived from this element is everywhere zero; but by suitable closed paths of z in the z-plane, w_1 can be changed into w_2, w_3, $\cdots$, w_n. Hence $F_1(w_1, z)$ can be made to pass into $F_1(w_2, z)$, $F_1(w_3, z)$, $\cdots$, $F_1(w_n, z)$. This proves that when $F_1(w, z)$ vanishes at $z = a$ for $w = w_1$, it also vanishes for w_2, w_3, $\cdots$, w_n; in other words, an equation $F_1(w, a) = 0$, of lower degree than n in w, is satisfied by n values. This being impossible, the original equation $F = 0$ must be irreducible.

It is evident that any rational function of w and z, say $R(w, z)$, is one-valued upon the Riemann surface T for $F(w^n, z^m) = 0$; we shall prove the converse theorem that every function one-valued upon T and continuous except at certain poles and branch-places, at which the order of infinity is a finite integer, is of the form $R(w, z)$.

Let the n values of w which correspond to a given z be w_1, w_2, $\cdots$, w_n; and let the corresponding values of R be R_1, R_2, $\cdots$, R_n. The symmetric combinations

$$s = R_1 w_1^{\kappa-1} + R_2 w_2^{\kappa-1} + \cdots + R_n w_n^{\kappa-1} \quad (\kappa = 1, 2, \cdots, n),$$

are one-valued functions of z, for they take the same values at the n places attached to z. Now a one-valued function of z which has only polar discontinuities is a rational function of z. Choosing λ_1, $\lambda_2, \cdots, \lambda_{n-1}$ so that

$$w_\mu^{n-1} + \lambda_1 w_\mu^{n-2} + \lambda_2 w_\mu^{n-3} + \cdots + \lambda_{n-1} = 0 \quad (\mu = 2, 3, \cdots, n),$$

we get

$$R_1(w_1^{n-1} + \lambda_1 w_1^{n-2} + \lambda_2 w_1^{n-3} + \cdots + \lambda_{n-1}) = s_n + \lambda_1 s_{n-1} + \cdots + \lambda_{n-1} s_1.$$

The form of the equations in w_μ shows that these quantities are the roots of an equation

$$w^{n-1} + \lambda_1 w^{n-2} + \cdots + \lambda_{n-1} = 0;$$

but they are known to be the roots of

$$F(w^n, z^m)/(w - w_1) = 0;$$

that is, of $$(f_0 w^n + f_1 w^{n-1} + \cdots + f_n)/(w - w_1) = 0,$$

or of

$$f_0 w^{n-1} + (f_0 w_1 + f_1) w^{n-2} + \cdots + (f_0 w_1^{n-1} + f_1 w_1^{n-2} + \cdots + f_{n-1}) = 0;$$

hence,

$$\lambda_1 = (f_0 w_1 + f_1)/f_0; \ \cdots; \ \lambda_{n-1} = (f_0 w_1^{n-1} + f_1 w_1^{n-2} + \cdots + f_{n-1})/f_0.$$

Therefore

$$R_1 = \frac{f_0 s_n + (f_0 w_1 + f_1) s_{n-1} + \cdots + (f_0 w_1^{n-1} + f_1 w_1^{n-2} + \cdots + f_{n-1}) s_1}{f_0 w_1^{n-1} + (f_0 w_1 + f_1) w_1^{n-2} + \cdots + (f_0 w_1^{n-1} + f_1 w_1^{n-2} + \cdots + f_{n-1})}.$$

Now w_1 is any value of w attached to the given z, and R_1 is the corresponding value of R. We may therefore omit the suffixes and write the relation between R and w in the form

$$R(w, z) = \frac{f_0 s_n + (f_0 w + f_1) s_{n-1} + \cdots + (f_0 w^{n-1} + f_1 w^{n-2} + \cdots + f_{n-1}) s_1}{\delta F/\delta w},$$

which proves the theorem. [This proof is given by Briot, Théorie des fonctions abéliennes, Appendix B; for another proof see Prym, Beweis eines Riemann'schen Satzes, Crelle, t. lxxxiii., p. 251. Also see Klein's Schrift., p. 57, and Klein-Fricke, Modulfunctionen, t. i., p. 499.]

When W is a rational function $R(w, z)$ of w, z, there corresponds to a given place (z, w) on the surface T, a single pair (z, W). It is not always true, conversely, that to a given pair (z, W) corresponds the single place (z, w). Suppose W to be a function for which the converse statement holds; then the surface T will serve for the pairs (z, W) as well as for the pairs (z, w). At an ordinary point z, the n associated pairs (z, w) are distinct. Therefore the n pairs (z, W) must be distinct. That is, W is an n-valued algebraic function of z. Starting with a surface constructed to represent a single algebraic function w we have shown that from the single n-valued algebraic function w of z there can be derived a whole system of such functions (called by Riemann *like-branched* functions). On the surface T each of these functions W is one-valued, and the surface T can be constructed by the equation in W, z, in place of $F=0$. The function w is distinguished among these functions merely by the fact that we began with it; it is itself a rational function of z and R, inasmuch as it is one-valued on the surface T. [For an algebraic proof of this by means of the process of finding the Highest Common Divisor, see Dedekind and Weber, Crelle, t. xcii., § 13.] The degrees in w of the equations are all equal to n, but the degrees in z will naturally be different. When an equation is of degree m in z, the function will be said to be m-placed on the surface; it takes any assigned value at m places, distinct or coincident (see § 108). When $w-b=P_\kappa(z-a)$, κ of the places at which $w=b$ coincide at $z=a$; but when $w-b=P_\kappa(z-a)^{1/r}$, the branch-place (a, b) is still a place at which κ values of w are equal to b. In fact, at a branch-place at which r sheets hang together $(z-a)^{1/r}$ is to be regarded as a quantity of the same order as $z-a$ at an ordinary place; or, to use a convenient mode of expression, $(z-a)^{1/r}=0^1$ at $z=a$. Similarly if, at an infinity, $1/w=P_\kappa(z-a)^{1/r}$, the place (a, ∞) counts as κ simple poles, and $(z-a)^{1/r}=\infty^1$ at a.

Integration.

§ 175.. We have seen that n sheets can be connected in such a way that an n-valued function of z is represented uniquely at each point of the n-sheeted surface, and that the branch-cuts can be determined as soon as we know the expansions at the branch-points. We have further explained how a Riemann surface can be reduced to simple connexion by the drawing of cross-cuts. The boundary of the dissected surface consists of the banks of the $2p$ cross-cuts, where each cross-cut has two banks; also this boundary can be described by the continuous motion of a point. It remains for us to examine the effect of paths of integration upon the dissected and undissected surfaces.

Suppose that w is an algebraic function of z which satisfies the equation $F(w^n, z^m)=0$. Then w is an n-valued function of z, say $w = w(z)$. The integral $I(z) = \int_{z_0}^{z} w(z)dz$ may perfectly well be infinitely many-valued for each point of the Riemann surface on which w is one-valued; in fact $I(z)$ is, in general, a function with properties distinct from those of $w(z)$. A familiar example is $\int_1^z dz/z$, or $\log z$, which belongs to a class of functions different from that of $1/z$. Whereas $1/z$ requires only a simple sheet, the number of sheets required for $\log z$ would be infinite. It is extremely convenient to be able to represent the integral of an algebraic function uniquely upon the surface T, which is associated with that function. Riemann solved this problem by the ingenious device of the cross-cut. The function of the $2p$ cross-cuts is to reduce the multiply connected surface to simple connexion; thereby permitting the application to Riemann surfaces, of Cauchy's theorems on the integration of functions holomorphic and meromorphic in simply connected plane regions. It will be seen presently that a path of integration which lies entirely within the dissected surface is subject to the same rules as a path in Cauchy's theory; but that the traversing of a cross-cut generally introduces an abrupt change into the value of the integral. While the branch-cuts are *bridges* between sheets otherwise unconnected, the cross-cuts are *barriers* which dissociate adjacent places in the same sheet.

In conformity with the definition of § 131, a function is said to be holomorphic on a simply connected region of T which includes no branch-place, when it is one-valued, finite, and continuous within that region. We have now to extend Cauchy's theorem (§ 133) to a region of the Riemann surface. The method used by Riemann depends on the transformation of a double integral into a simple integral.

§ 176. *Cauchy's theorem for a Riemann surface.* If C be the boundary of a delimited region Γ on the Riemann surface, and if $\phi(z)$ be one-valued and continuous in Γ, we have when each rim is described positively,

$$\int_C \phi(z)dz = 0.$$

Let $\phi(z) = u + iv$. Then

$$\int \phi(z)dz = \int_C (udx - vdy) + i\int_C (vdx + udy).$$

Suppose that the surface Γ is cut by two sets of vertical planes parallel to the axes of x and y. A strip made by two consecutive parallel planes may run entirely in one sheet, or it may meet a branch-cut and pass for a part of its course into another sheet; there may be several strips between the same vertical planes (in which case the strips may be said to be in the same vertical); and further, the boundary may consist of various rims. But however

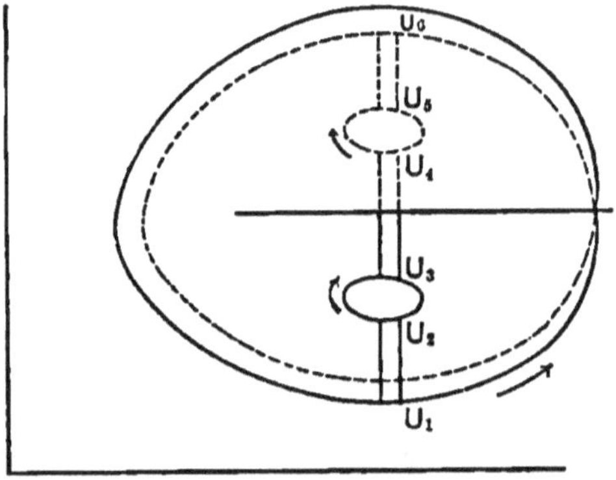

Fig. 72

many times strips in the same vertical may meet rims, and however many times they may change sheets, it will always be true that

$$\int \frac{\delta U}{\delta y} dy = U_2 - U_1 + U_4 - U_3 + \cdots + U_{2t} - U_{2t-1},$$

where U is a real function of x, y, which, with its first differential quotients, is continuous throughout Γ, and $2t =$ the number of points in which the strip meets the boundary.

Hence,

$$dx \int \frac{\delta U}{\delta y} dy = (U_2 - U_1)dx + (U_4 - U_3)dx + \cdots + (U_{2t} - U_{2t-1})dx;$$

that is, the parts of the integrals $\int\int \frac{\delta U}{\delta y} dxdy$ and $-\int U dx$, which the strip contributes, are equal. Hence for the whole region these integrals are equal. Similarly,

$$\int\int \frac{\delta U}{\delta x} dxdy = \int U dy.$$

Hence, by writing $U = u$, v successively, we get

$$\int \phi(z) dz = \int_c (udx - vdy) + i \int_c (vdx + udy)$$

$$= - \int\int \left(\frac{\delta u}{\delta y} + \frac{\delta v}{\delta x}\right) dxdy + i \int\int \left(\frac{\delta u}{\delta x} - \frac{\delta v}{\delta y}\right) dxdy.$$

But $\frac{\delta u}{\delta x}-\frac{\delta v}{\delta y}=0$, and $\frac{\delta u}{\delta y}+\frac{\delta v}{\delta x}=0$.

Hence, $\int\phi(z)dz=0$.

In this proof the restriction that $\frac{\delta U}{\delta x}, \frac{\delta U}{\delta y}$ are to be continuous throughout Γ, may be to some extent removed. For example, if $\frac{\delta U}{\delta x}, \frac{\delta U}{\delta y}$ be discontinuous at a finite number of places of Γ, while U is one-valued and continuous at these points, the integral $\int\frac{\delta U}{\delta x}dx$ is still $=U_2-U_1+U_4-U_3+\cdots$. To fix ideas, let $\int_a^b\frac{\delta U}{\delta x}dx$ become infinite for one value $x=c$ between $x=a$ and $x=b$.

Then

$$\int_a^b\frac{\delta U}{\delta x}dx=\lim_{\epsilon=0}\int_a^{c-\epsilon}\frac{\delta U}{\delta x}dx+\lim_{\eta=0}\int_{c+\eta}^b\frac{\delta U}{\delta x}dx$$

$$=\lim_{\epsilon=0,\ \eta=0}\left\{U(c-\epsilon, y)-U(a, y)+U(b, y)-U(c+\eta, y\right\}.$$

But U is one-valued and continuous, therefore

$$\lim U(c-\epsilon, y)=\lim U(c+\eta, y),$$

and the integral $=U(b, y)-U(a, y)$, a quantity independent of c. [See Königsberger's Elliptische Functionen, p. 69.]

§ 177. A function $\phi(z)$, which is one-valued and continuous at all places of a Riemann surface, is a constant. For let the function take the values $\phi_1, \phi_2, \cdots, \phi_n$ at the n places which lie in a vertical line; then each of the symmetric combinations

$$\phi_1^\kappa+\phi_2^\kappa+\cdots+\phi_n^\kappa,$$

where $\kappa=1, 2, \cdots, n$, is a one-valued and continuous function of z, for all values of z, and is therefore a constant (§ 140). Hence the equation whose roots are $\phi_1, \phi_2, \cdots, \phi_n$ has constant coefficients. Hence the roots are themselves constant for all values of z, and since they are continuous, they are equal to the same constant.

It follows that an algebraic function of the surface is defined, except as to a constant factor, when its zeros and poles are assigned. For if two such functions have the same zeros and poles, with the same orders of multiplicity, their ratio is everywhere one-valued and continuous. Compare § 145.

§ 178. The theorems which relate to the value of an integral $\int_{z_0}^{z} f(z)dz$, where $f(z)$ is a many-valued function, are greatly simplified by the use of the Riemann surface for $f(z)$, but the methods employed by Riemann are, in essence, identical with those of Cauchy. The fundamental theorem in integration is that the value of the integral $\int f(z)dz$, taken positively over the rims C of a delimited region Γ, is zero when $f(z)$ is continuous in that region.

To fix ideas let us consider the integral of a function $R(w, z)$ which is rational in w and z, where w is an algebraic function defined by $F(w^n, z^m)=0$. This integral is called an *Abelian integral.* The advantage which results from this limitation of the field of discussion mainly arises from the fact that we know the nature of the branchings and discontinuities in the case of the algebraic function and the order of connexion of the corresponding Riemann surface.

When $R(w, z)$ is continuous throughout a region Γ of finite extent upon the surface T, delimited by a system of rims C, the value of $\int Rdz$, taken positively (with regard to Γ) over all the rims C, is zero. Suppose that the region Γ is delimited by an exterior rim C and by interior rims $C_1, C_2, \cdots, C_\lambda$, we have the extension of the theorem of § 135, the difference being that a rim may pass several times round an interior point before returning to its initial point. Suppose that the region in question contains an infinity c; this infinity must be cut out by a small closed curve described r times round c, say an r-fold circle, where r is the number of turns that must be made before the curve can be closed. This curve (c) must be added to the other rims, and the theorem runs

$$\int_C Rdz = \sum_{\kappa=1}^{\lambda} \int_{C_\kappa} Rdz + \int_{(c)} Rdz,$$

where the curves $C, C_\kappa, (c)$ are described positively with regard to the region delimited by them.

When Γ is a simply connected region on T, such that R has no infinities within Γ, the value of $\int Rdz$ taken over any closed line in Γ must necessarily vanish. As an immediate consequence, it follows that when I, II are two paths which lie in Γ and go from a place s_0 to a place s, the value of the integral is the same whichever path be chosen. This theorem may be stated somewhat differently. When, on the Riemann surface T, a path I which connects a place s_0 with another place s is deformed continuously into a second path II between the same two places, the value of the integral is unaffected so long as the path does not cross a place at which R is infinite. To complete the theory, we require some way of dealing with the discontinuities of R. Let the infinities within a simply connected region of finite extent delimited by the paths I, II be $c_1, c_2, \cdots, c_\mu$; these places must be cut out from the surface T. The simply connected region has now become a multiply connected one within which R is everywhere continuous. Accordingly,

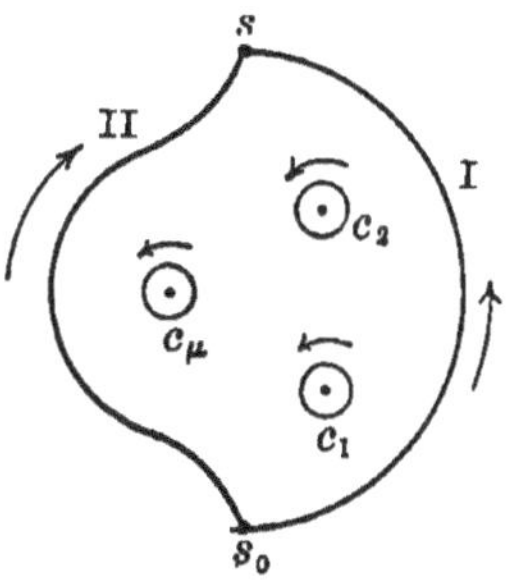

Fig. 73

$$\int_I Rdz = \int_{II} Rdz + \sum_{\kappa=1}^{\mu} \int_{(c_\kappa)} Rdz,$$

where the integration is effected in the sense indicated in Fig. 73.

On the simply connected surface T′, which results from T by the drawing of $2p$ cross-cuts, the integral $\int Rdz$ may be finite. This will be the case when the infinities of R are not infinities of the integral. In the most general case, however, the infinities of R affect the value of the integral, as has been already explained in speaking of Cauchy's theory of Residues. The surface T′ must be changed again into a multiply connected surface by cutting out those infinities which affect the value of the integral, and this new multiply connected surface must be reduced to simple connexion by fresh cross-cuts. On the surface T″ the value of the integral $\int_{s_0}^{s} Rdz$ is completely independent of the connecting path.

§ 179. When w is replaced by its values in terms of z, let the expression $R(w, z)$ become the n-valued function $\phi(z)$. The theorems which follow apply also to functions of a more extended character than $R(w, z)$; but we shall not have occasion to use these in this treatise, and consequently omit them from consideration.

When Γ is a region of T which lies entirely in one sheet, and encloses no branch-place, the same deductions can be made from Cauchy's theorem as in Chapter V. Since that branch of the many-valued function $\phi(z)$ which is represented on Γ is, *throughout* Γ, a holomorphic function of z, Cauchy's methods are immediately applicable. For example, if the infinities

$$c_1, c_2, \cdots, c_\mu,$$

be excluded by small closed curves $C', C'', \cdots, C^{(\mu)}$, then

$$\frac{1}{2\pi i}\int_C \frac{\phi(z)dz}{z-t} = \frac{1}{2\pi i}\sum_{\kappa=1}^{\mu}\int_{C^{(\kappa)}} \frac{\phi(z)dz}{z-t} + \phi(t),$$

where $\phi(t)$ is the value of ϕ at a place (t, w) of Γ. When $\phi(z)$ has no infinities in Γ, this equation becomes

$$\phi(t) = \frac{1}{2\pi i}\int_C \frac{\phi(z)dz}{z-t},$$

and the same consequences follow from this formula as in Chapter V. The most important of these is that at any place (z_1, w_1) of the region Γ, $\phi(z)$ can be represented by an integral series $P(z-z_1)$, with a circle of convergence not infinitely small. Hence it follows that $\phi'(z)$, $\phi''(z)$, etc., are holomorphic throughout the same domain.

Hitherto the supposition has been that Γ contains no branch-point, and one of the results arrived at has been that in the neighbourhood of each place (z_1, w_1), $\phi(z)$ can be represented by an *integral* series in $z-z_1$. We now make the supposition that (z_1, w_1) is a place at which r sheets hang together, and seek to find the expansions in the neighbourhood of this branch-place. Let a circle C pass r times round (z_1, w_1), so as to limit a region Γ within which there is no branch-place of $\phi(z)$ other than (z_1, w_1), and no infinity. The transformation

$$z - z_1 = \zeta^r$$

transforms Γ into a circular region Γ_1 in the ζ-plane, and the r-fold circle C into an ordinary circle C_1. Since the function $\phi(z)$ recovers its initial value after a description of C, the function $\phi(z_1+\zeta^r)$, or $\psi(\zeta)$, must recover its value after a description of C_1. That is, $\psi(\zeta)$ is holomorphic in Γ_1. Applying Cauchy's theorems to the function $\psi(\zeta)$, we have for the region Γ_1,

$$\int_{C_1}\zeta^h\psi(\zeta)d\zeta = 0 \qquad (h=1, 2, 3, \cdots),$$

and

$$\psi(\zeta') = \frac{1}{2\pi i}\int_{C_1}\frac{\psi(\zeta)d\zeta}{\zeta-\zeta'},$$

where ζ' is a point inside Γ_1. Hence

$$\int_C \phi(z)(z-z_1)^{-1+(h+1)/r}dz = 0 \qquad (h=1, 2, 3, \cdots),$$

and $$\phi(z') = \frac{1}{2r\pi i}\int_C \frac{\phi(z)}{(z-z_1)^{1-1/r}}\cdot\frac{dz}{(z-z_1)^{1/r}-(z'-z_1)^{1/r}},$$

where (z', w') is any place in Γ. By the use of the binomial theorem, the quantity under the integral sign becomes an integral series in $(z'-z_1)^{1/r}$, and therefore

$$\phi(z') = P(z'-z_1)^{1/r} = a_0 + P_\lambda(z'-z_1)^{1/r}.$$

Here the r values of ϕ attached to the r places of Γ, which lie on a vertical line, are associated with a cyclic system of r expansions. The differential quotients of $\phi(z')$ can be found from the series. For example, omitting accents,

$$\phi'(z) = (z - z_1)^{-1+\lambda/r} P_0(z - z_1)^{1/r}.$$

This result shows that $\phi'(z_1)$ is infinite when $\lambda < r$, although the corresponding value of $\phi(z_1)$ is finite. The only places at which $\phi'(z)$ can become infinite are either branch-places or infinities of $\phi(z)$.

Next let (z_1, w_1) be an infinity as well as a branch-place, but otherwise let the suppositions with regard to Γ and C be the same as before. By an application of Laurent's theorem to the function $\psi(\zeta)$, we have

$$\psi(\zeta) = \zeta^{-q} P_0(\zeta),$$

where q is *finite*, since it is assumed that $\zeta = 0$ is not an essential singularity. Hence

$$\phi(z) = (z - z_1)^{-q/r} P_0(z - z_1)^{1/r}.$$

The expansions at $z = \infty$ are found by putting $z = 1/z'$, *i.e.* by replacing $z - \infty$ by $1/z'$. If r sheets hang together at ∞, r sheets of the Riemann surface for $\phi_1(z'), = \phi(z)$, must hang together at $z' = 0$. Suppose that $z = \infty$ is not a singular point, so that $\phi_1(z')$ has no singular points within a finite neighbourhood of the origin.

Then $$\phi_1(z') = P(z'^{1/r}),$$

where z' is any point of this neighbourhood. Thus

$$\phi(z) = P(z^{-1/r}).$$

When $z = \infty$ is an infinity, but not an essential singularity, of $\phi(z)$, the expansion runs

$$\phi(z) = z^{q/r} P_0(z^{-1/r}).$$

The reader will observe that the results obtained in Chapter IV. can be derived from Cauchy's theorems. They state that if w be an algebraic function of z, the expansions for $R(w, z)$ at various points are of the forms

$$P(z-a);\; (z-a)^{-q} P_0(z-a);\; P(1/z);\; z^q P_0(1/z);$$
$$P(z-a)^{1/r};\; (z-a)^{-q/r} P_0(z-a)^{1/r};\; P(z^{-1/r});\; z^{q/r} P_0(z^{-1/r});$$

where q, r are positive integers, and $r > 1$.

§ 180. *Integrals of algebraic functions.* So long as the path of integration is not permitted to pass over a cross-cut of the surface

T″ (§ 178) the integral is one-valued. The cross-cuts serve as barriers which separate the branches of the many-valued integral. But now it is possible to remove this restriction that the barriers are not to be passed over. The cross-cuts required to make T simply connected were $2p$ in number, and further cross-cuts connected the curves round the infinities of $\int R dz$ with the boundary of T′. No curve is to be drawn round any infinity of R which yields no logarithmic term to $\int R dz$. For instance, if (z_1, w_1) be a place at which

$$R = (z - z_1)^{-q/r} P_0 (z - z_1)^{1/r}, \ (q > r),$$

the integral $\int R dz$ is logarithmically infinite when, and only when, $P_0(z - z_1)^{1/r}$ contains a term $(z - z_1)^{-1+q/r}$. When this term occurs, the value of $\frac{1}{2\pi i}\int R dz$, taken r times round z_1, is called the residue at the point (§ 146). Thus the residue is $r \times$ the coefficient of the term.

Let T″ be the simply connected surface derived from the canonically dissected surface T′, by the drawing of cross-cuts from the curves $C^{(\kappa)}$, round logarithmic infinities, to the boundary of T′. Let $I(z) = \int_{z_0, w_0}^{z, w} R dz$, where the path between (z_0, w_0), (z, w) can be drawn freely on T; $J(z) =$ the same integral when the path is restricted to T″. Then $J(z)$ is a one-valued and continuous function on the surface T″. Any difference which may exist between $I(z)$ and $J(z)$ must arise from passages over cross-cuts. In order to distinguish the two directions across a cross-cut, an arrow-head is attached to the cut. The cut may be regarded as a stream with right (or negative) and left (or positive) banks (Neumann, Abel'sche Integrale, ch. vii.). Two places a, infinitely near but on opposite banks, can be distinguished as a_+ and a_-. Now in passing from a to a' (Fig. 74) the values of dz and of R, at opposite places of the cut, are equal.

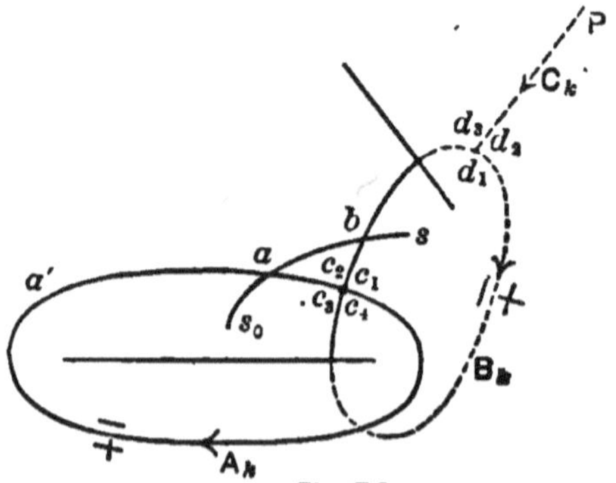

Fig. 74

Hence, $\int_{a_-}^{a'_-} R dz$, along the right bank, $= \int_{a_+}^{a'_+} R dz$, along the left bank; or $\quad J(a_-') - J(a_-) = J(a_+') - J(a_+)$.

Hence $$J(a_+')-J(a_-')=J(a_+)-J(a_-);$$

that is, the difference between the values of J at opposite banks of the cross-cut is constant. This constant is called a modulus of periodicity, or *period*, of the integral. To each cross-cut there belongs a period. The full importance of this idea will be seen later (ch. x.).

To connect $I(z)$ with $J(z)$, suppose that the path is s_0abs (Fig. 74), which crosses the cut A_κ from right to left, and the cut B_κ from left to right. Then

$$\int_{s_0a_-a_+b_+b_-s} Rdz=\int_{s_0a_-} Rdz+\int_{a_+b_+} Rdz+\int_{b_-s} Rdz,$$

the infinitely small paths a_-a_+ and b_+b_- being neglected; they contribute nothing to the integral, as R is supposed finite in their neighbourhood. Each of the three pieces s_0a_-, a_+b_+, b_-s lies on T″. Therefore

$$\int_{s_0a_-} Rdz=J(a_-);$$

$$\int_{a_+b_+} Rdz=\int_{s_0b_+} Rdz-\int_{s_0a_+} Rdz,$$

where the paths lie on T″,

$$=J(b_+)-J(a_+);$$

$$\int_{b_-s} Rdz=J(s)-J(b_-).$$

Hence $$I(s)=J(s)+J(b_+)-J(b_-)-J(a_+)+J(a_-),$$

or, if ω, ω' be the periods at the cross-cuts A, B,

$$I(s)=J(s)-\omega+\omega'.$$

So in general, when the path on T crosses over the κth cross-cut λ_κ times positively, *i.e.* from left to right, and λ_κ' times negatively, *i.e.* from right to left, and when the period attached to this cross-cut is ω_κ, the value of $I(s)$ is

$$J(s)+\sum_\kappa(\lambda_\kappa-\lambda_\kappa')\,\omega_\kappa,$$

where the summation extends over all cross-cuts traversed by the path of integration.

It has been stated that the period along a cross-cut is constant; but this has only been proved so long as the cut meets no other cut.

Applying the theorem to the cut A_κ (Fig. 74) which meets the cut B_κ and no other cut, we see that

$$J(c_2)-J(c_3)=J(a_+)-J(a_-)=J(c_1)-J(c_4);$$

therefore $$J(c_2)-J(c_1)=J(c_3)-J(c_4);$$

that is, the periods of B_κ, on both sides of the junction c, are equal.

But also
$$J(c_2)-J(c_1)=J(d_3)-J(d_1),$$

and $$J(c_3)-J(c_4)=J(d_2)-J(d_1).$$

Therefore $$J(d_2)-J(d_3)=0;$$

that is, the period across a cut C_κ is zero.

A path of integration round A_κ (along either bank), in the direction of the arrow, leads from the negative to the positive bank of B_κ. This shows that the period round A_κ is equal to the period across B_κ. But a path round B_κ, in the direction of the arrow, leads from the positive to the negative bank of A_κ, and the period round $B_\kappa=-\omega_\kappa$, where ω_κ is the period across A_κ.

The reader will readily prove that if several cross-cuts start from a place s of the surface, the sum of the periods across them is zero. Observing that a change in the direction of a cross-cut changes the sign of the corresponding period, we have the rule:—

The sum of the periods of the streams which flow to a place = the sum of the periods of the streams which flow from the place.

When the path of integration is drawn on Klein's normal surface, the theorems already enunciated for the plane Riemann surface take a form which is readily comprehended. Assuming that $\int Rdz$ has no logarithmic infinities, two contours round the same handle, or through the same handle, can be deformed continuously into each other, and the integrals round the contours are equal. On the other hand, a meridian and a latitude curve are not reconcilable, nor is a curve of either kind reconcilable with a curve of the same kind, but belonging to a different handle; and, in these cases, the integrals round the two curves may be different. Any curve which can be contracted continuously until it vanishes gives a zero integral. Thus the p integrals along p selected closed curves through the p handles, and the p integrals along p selected closed curves round the p handles afford $2p$ periods. The most general path between two places can be deformed continuously into a closed path which passes λ_κ times in one direction round the κth handle, and λ_κ' times in one direction through it, where $\kappa=1, 2, \cdots, p$, followed by some special path between the two places. Hence the most general integral is equal to the integral along the special path, together with $\sum_{\kappa=1}^{p}(\lambda_\kappa\omega_\kappa + \lambda_\kappa'\omega_\kappa')$, where ω_κ, ω_κ' are the integrals round the κth latitude and meridian curves, taken in the assigned directions. The modifications necessary in the case that $\int Rdz$ contains one or more logarithmic infinities are suggested by the theorems of the preceding articles. For example, if A, A' be two reconcilable circuits which contain between them μ places $c_1, c_2, \cdots, c_\mu$, at which the integral is logarithmically infinite, the integrals of R along A, A' are no longer equal, but

$$\int_A Rdz - \int_{A'} Rdz = \sum_{\kappa=1}^{\mu} \int_{(c_\kappa)} Rdz,$$

where the integration is performed in the same sense for all the curves A, A', (c_κ).

The reader is referred for further information to Klein's Schrift, Ueber Riemann's Theorie der Algebraischen Functionen und ihrer Integrale. Leipzig, 1882.

§ 181. We shall now give some examples of the theory just explained.

(1) $$dw = dz/z = d\rho/\rho + id\theta.$$

The critical points are $z=0$, $z=\infty$; and the residues of these points are $1, -1$. Hence both points must be cut out. The surface is now doubly connected. To reduce it to simple connexion, we make a cut from 0 to ∞. Now w is a holomorphic function of z.

Let $w=0$ when $z=1$. Then

$$w \equiv u + iv = \log\rho + i\theta,$$

log denoting the real logarithm, and

$$u = \log\rho,\ v = \theta.$$

Hence half-lines from the origin in the z-plane map into lines parallel to the real axis in the w-plane; and a circle round the origin in the z-plane maps into a part of a line parallel to the imaginary axis in the w-plane. For if z start always from the positive or left-hand bank of the cut, ending at the negative bank, the value of v increases, for any circle, by 2π. Hence if we draw the w-path C_+ which corresponds to the positive bank of the cut, the path C_- which corresponds to the other bank is obtained by displacing C_+ through $2\pi i$; and therefore the z-plane passes into an infinite strip of the w-plane, of breadth 2π. To make the strip straight, we make the z-cut pass along the real axis, and as in § 106, we shall take it along the negative half of this axis, Fig. 75 (a). To map the entire w-plane, we require infinitely many sheets in the z-plane, which

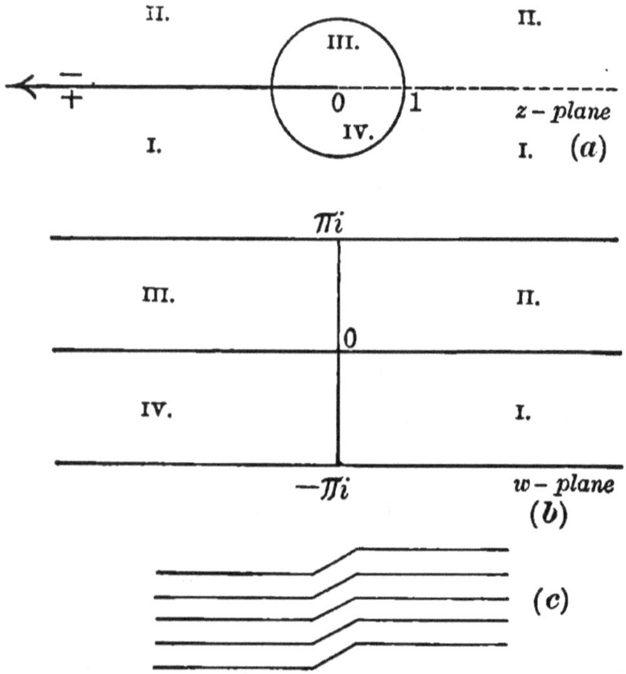

Fig. 75

hang together at 0 and ∞, and are joined by a bridge which lies along the cross-cut. Fig. 75 (c) shows a section of the bridge as it appears when one looks from 0 to $-\infty$.

(2) $$dw = dz/(z-a)(z-b).$$

Here the critical points are a, b. The transformation

$$t = (z-a)/(z-b)$$

will send them to 0 and ∞, and we have

$$dt = (a-b)dz/(z-b)^2,$$

$$dt/t = (a-b)dw.$$

Hence the w-plane is mapped on the t-plane as in the preceding example, while the t-plane maps on the z-plane by a quasi-inversion. Therefore if we take as a cross-cut any arc of a circle from a to b, the z-plane maps on a straight strip of the w-plane, whose breadth is $2\pi/|a-b|$. All other such arcs give w-lines parallel to the edges of the strip. Lines perpendicular to the edges of the strip correspond to circles whose limiting points are a, b.

(3) $$dI = dz/\sqrt{1-z^2}.$$

The critical points are $z=1, -1, \infty$. The points $1, -1$ have a zero residue; at ∞ the residue is $\pm i$. Let $w^2 = 1-z^2$, and let the z-surface on which w is one-valued be constructed. Let the branch-cut lie along the real axis from -1 to 1. When $z=i$, let the value $w=\sqrt{2}$ be assigned to the upper sheet. The places ∞ being cut out of both sheets, the cross-cut may be drawn along the imaginary axis, coinciding with the positive part of this axis in the upper sheet and with the negative part in the lower sheet (Fig. 76). We must now assign a value to I at a given place. Let $I=0$ at the place s_-, which is infinitely near $(0, 1)$ on the right of the cut. The value of I is now everywhere determinate; at the place s_+ on the left of the cross-cut, the value is 2π, as appears either from the residue at ∞ or by integrating round the branch-points as in § 138.

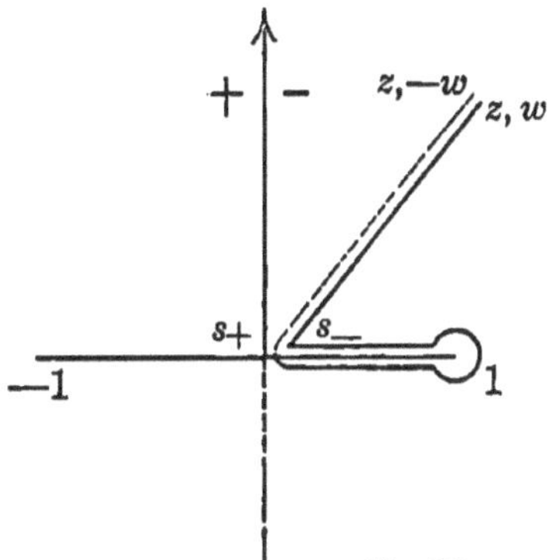

Fig. 76

If I be the value at any place (z, w), the value I' at $(z, -w)$ is $\pi - I$. For

$$I = \int_{s_-}^{z,\,w} dz/w,$$

$$I' = \int_{s_-}^{z,\,-w} dz/w,$$

$$= \int_{s_-}^{1,\,0} dx/+\sqrt{1-x^2} + \int_{1,\,0}^{0,\,-1} dx/-\sqrt{1-x^2} + \int_{0,\,-1}^{z,\,-w} dz/w,$$

$$= \pi - I.$$

In such cases as we have been considering, the z-paths which correspond to lines parallel to the axes in the I-plane may be determined directly from the differential equation. For instance, in the case

$$dI = dz/\sqrt{1-z^2},$$

let I move parallel to the real axis. Then, equating amplitudes, we have

$$\tau - 1/2(\phi_1 + \phi_2) = 0,$$

where ϕ_1, ϕ_2, τ are the amplitudes of the strokes $z-1$, $z+1$, and of the tangent of z's path. Hence the tangent bisects the angle -1, z, 1; and the path is a hyperbola. For developments of this idea, see Franklin, American Journal, t. xi.

§ 182. (4) Let $w^2 = (z-a_1)(z-a_2)\cdots(z-a_{2p+2})$.

The z-plane is in this case covered by two sheets; and the points a_κ are branch-points. The bridges may be drawn from a_1 to a_2, from

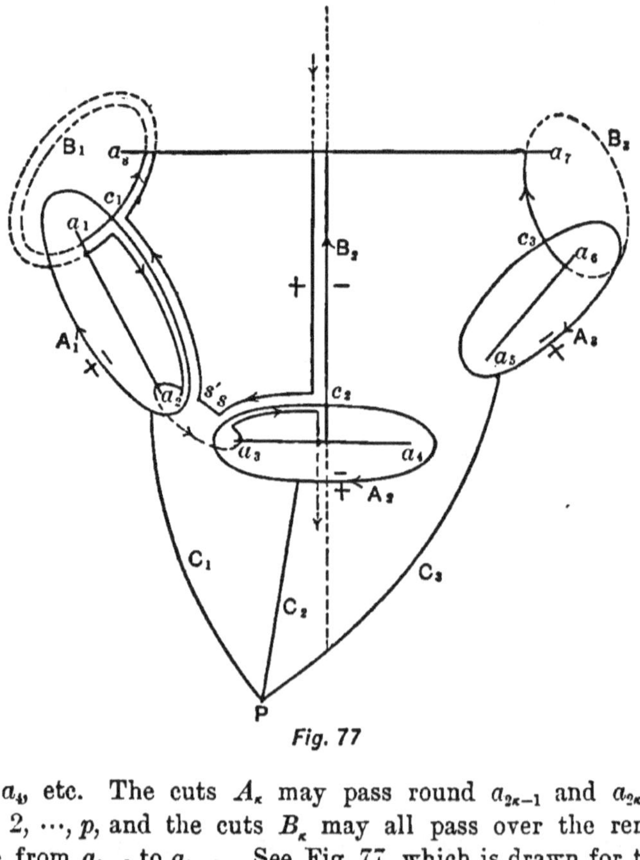

Fig. 77

a_3 to a_4, etc. The cuts A_κ may pass round $a_{2\kappa-1}$ and $a_{2\kappa}$, where $\kappa = 1, 2, \cdots, p$, and the cuts B_κ may all pass over the remaining bridge from a_{2p+1} to a_{2p+2}. See Fig. 77, which is drawn for the case $p = 3$. When the cuts C_κ are drawn, the surface becomes simple.*

* In Fig. 77 the cuts C_κ are drawn to the cuts A_κ. This is convenient in a complicated figure.

Consider the integral $I = \int z^\lambda dz/w$. It is finite when $z=\infty$ if $\lambda < p$, by § 136, and at every branch-point the residue is 0 if $\lambda \geqq 0$. There are therefore p such integrals which are finite at all places of the surface, for we may take $\lambda = 0, 1, \cdots, p-1$.

By Riemann's extension of Cauchy's theorem a cut A_κ or B_κ may be contracted till all its points lie arbitrarily near the lines which join the included branch-points. The places of the cut pair off (see Fig. 78) into places α, α' at which the values of w are arbitrarily nearly opposite; therefore, in the description of the cut, $z^\lambda dz/w$ runs twice through the series of values which it takes in passing from one branch-point to the other. It follows that the integral taken from one branch-point to the other is half the period due to the description of the cut which encloses those branch-points.

Fig. 78

Let the periods *across* A_κ be ω_κ ($\kappa = 1, 2, \cdots, p$), and those *across* B_κ be ω_κ'. Then Fig. 77 shows that the periods *round* A^κ are ω_κ', while those *round* B_κ are $-\omega_\kappa$.

It follows at once from what has been said that

$$\int_{a_2}^{a_1} dI = \tfrac{1}{2}\omega_1', \quad \int_{a_1}^{a_3} dI = \tfrac{1}{2}\omega_1, \text{ etc.}$$

But also we are able to express the value of the integral taken from any branch-point to any other by means of the periods. For instance, the path which leads from a_2 to a_3 in the lower plane (Fig. 77) is stopped, in returning to a_2, at the place s_-; it can be continued, as shown in the figure, to c_2, along the whole of B_2, and from c_2 to s_+; from s_+ direct to s_+', then to c_1, round B_1, from c_1 to s_-', and finally back to a_2. By Cauchy's fundamental theorem the integral along the whole path is zero; and this path is made up of the integral from a_2 to a_3 reckoned twice and of the integrals round B_2 and B_1; the integrals along A_2 and A_1 cancel, each being taken twice, in opposite directions and in the same sheet.

Therefore
$$2\int_{a_2}^{a_3} dI + \omega_2 - \omega_1 = 0,$$

and
$$\int_{a_2}^{a_3} dI = \tfrac{1}{2}(\omega_1 - \omega_2).$$

§ 183. *Birational transformations.*

The subject of birational transformations of curves dates back to Riemann's memoir on the Abelian functions (Werke, 1st ed., p. 111).

It was first presented in its geometric form by Clebsch in a memoir on the applications of Abelian functions to Geometry (Crelle, t. lxiii.).

Let the curve $F_1(W, Z)=0$ arise from the curve $F(w, z)=0$, by the elimination of w and z, where W and Z are rational algebraic functions of w, z. When, conversely, $F(w, z)$ arises from $F_1(W, Z)$ by the elimination of w, z, where w and z are rational algebraic functions of W, Z, the curves correspond one-to-one, and the transformation is said to be *birational.* By means of the birational transformations

$$W=R(w, z),\ w=r(W, Z),$$

$$Z=S(w, z),\ z=s(W, Z),$$

the points of the surface associated with $F=0$ correspond one-to-one with the points of the surface associated with $F_1=0$. Therefore the maximum number of retrosections must be the same for the first surface as for the second; this shows that p is the same for both surfaces. That is, p is an invariant with regard to birational transformations. It is not true, conversely, that two Riemann surfaces, with the same order of connexion, can always be connected by a birational transformation, except in the special case $p=0$.

The number p is an invariant of the surface T not merely with regard to birational transformations, but also with regard to continuous deformations. For a description of the possibilities as regards continuous deformation we refer the reader to two memoirs by Klein, Math Ann., tt. vii. and ix.; and to Klein's Schrift, p. 25. That two surfaces must have the same p, if they are to correspond point to point, can be proved as above. Jordan has proved the less evident theorem that the sufficient condition in order that two surfaces may be deformable continuously, the one into the other, is the equality of their p's (C. Jordan, Sur la déformation des surfaces, Liouville, ser. 2, t. xi., 1866; Dyck, Beiträge zur Analysis Situs, Math. Ann., tt. xxxii., xxxiii.).

Thus the number p is the only invariant of a surface in the Geometry of Situation, a theorem of importance in the ulterior Riemann theory.

Suppose that W and Z are μ-placed and ν-placed algebraic functions of T. By reason of $Z=S(w, z)$ the surface T can be represented conformly on an n_1-sheeted surface T_1 spread over the Z-plane, so that the points of T and T_1 correspond one-to-one. Since the function W is one-valued on T and ∞^1 at μ places, the function must be one-valued also on T_1, and continuous at all places other than the μ places which correspond to the μ infinities on T. At these μ places it is ∞^1. Hence to a given W correspond μ values of Z. Similar reasoning shows that to a given Z correspond ν values of W. Accordingly the new equation which connects W, Z must be $F_1(W^\nu, Z^\mu) = 0$.*

When this equation is irreducible, all functions which are one-valued on T_1 and continuous, except at isolated places at which the singularities are non-essential, can be represented as rational integral algebraic functions of z_1 and w_1, and the transformation is birational.

§ 184. Any rational fraction $G_1(w, z)/G_2(w, z)$ can be expressed in the form

$$r_0(z) + r_1(z)w + r_2(z)w^2 + \cdots + r_{n-1}(z)w^{n-1},$$

where the expressions $r(z)$ are rational fractions in z. For

$$\frac{G_1(w_\kappa, z)}{G_2(w_\kappa, z)} = \frac{G_2(w_1, z) \cdots G_2(w_{\kappa-1}, z)\, G_2(w_{\kappa+1}, z) \cdots G_2(w_n, z)}{\prod\limits_{r=1}^{n} G_2(w_r, z)} \cdot G_1(w_\kappa, z).$$

The denominator contains symmetric combinations of the w's, and therefore reduces to a function of z only. The numerator contains symmetric combinations of $w_1, \cdots, w_{\kappa-1}, w_{\kappa+1}, \cdots, w_n$, and can be expressed in terms of w_κ only. Hence, using $F(w_\kappa^n, z) = 0$,

$$\frac{G_1(w_\kappa, z)}{G_2(w_\kappa, z)} = r_0(z) + r_1(z)w_\kappa + \cdots + r_{n-1}(z)w_\kappa^{n-1}.$$

The n equations formed by giving κ the values $1, 2, \cdots, n$ can be replaced by the single equation

$$\frac{G_1(w, z)}{G_2(w, z)} = r_0(z) + r_1(z)w + \cdots + r_{n-1}(z)w^{n-1}.$$

The following pages will contain a brief account of Kronecker's methods, as developed in his memoir Ueber die Discriminante algebraischer Functionen einer Variabeln, Crelle, t. xci., p. 301. (See also the memoir by Dedekind and Weber, Theorie d. algeb. Funct. einer Veränderlichen, Crelle, t. xcii., p. 181; and Klein-Fricke, Modulfunctionen, t. ii., p. 486.)

---

* Owing to the exceptional case mentioned in § 167, it may happen that $F_1(W, Z)$ is reducible. Riemann has proved that in this case F_1 is of the form $[\Phi(W^{\nu_1}, Z^{\mu_1})]^\kappa$, where Φ is irreducible (Werke, p. 112). The surface T_1 for $F_1 = 0$ is then spread κ times over a surface of n_1/κ sheets.

When the irreducible equation $F=0$ is in its most general form it may be written

$$g_0(z)w^n + g_1(z)w^{n-1} + \cdots + g_n(z) = 0,$$

where the quantities g are integral polynomials in z. In what follows g and r are used for integral polynomials and rational fractions in z, in much the same way as $P(z)$ for an integral series in Chapter III. The functions of the system $R(w, z)$ are called algebraic functions of the surface T. That the system is a closed one is shown by the fact that the functions which result from the operations of addition, subtraction, multiplication, and division, when applied to members of the system, are themselves members of the system. Call this closed system (R). We have the theorem that every function of the system (R) is expressible uniquely in the form

$$r_0(z) + r_1(z)w + r_2(z)w^2 + \cdots + r_{n-1}(z)w^{n-1},$$

and that conversely every function, of the kind just written down, belongs to (R). Here the members of (R) are expressed linearly in terms of a system $1, w, w^2, \cdots, w^{n-1}$, and the coefficients in the linear expression are rational fractions in z. The n functions, $1, w, w^2, \cdots, w^{n-1}$ are said to form a *basis* of (R). Kronecker has pointed out that the limitation that all functions are to be expressed in terms of this basis solely is partly needless, partly injurious. We can choose any other set $\eta_1, \eta_2, \cdots, \eta_n$, which satisfies the equations

$$\eta_\kappa = r_0^{(\kappa)} + r_1^{(\kappa)}w + \cdots + r_{n-1}^{(\kappa)}w^{n-1} \quad (\kappa = 1, 2, \cdots, n),$$

where the r's are rational fractions in z, with a determinant which does not vanish identically; and every function $R(w, z)$ is then expressible in the form

$$R(w, z) = \rho_1\eta_1 + \rho_2\eta_2 + \cdots + \rho_n\eta_n,$$

where the ρ's are rational fractions in z. The necessary and sufficient condition in order that $\eta_1, \eta_2, \cdots, \eta_n$ may form a basis, is that no combination

$$\rho_1\eta_1 + \rho_2\eta_2 + \cdots + \rho_n\eta_n$$

is to vanish identically. [See Dedekind and Weber, loc. cit., p. 186.]

§ 185. *Integral algebraic functions of the surface* T.

When $g_0(z) = 1$ in the irreducible equation $F=0$, the function w is said to be an *integral algebraic function of the surface*. The function w cannot be infinite for any finite value of z; and conversely every algebraic function of the surface, which is finite for all finite values of z, is an integral algebraic function of T. It can be proved that the sum, difference, and product of two integral algebraic functions of T are themselves integral algebraic functions of T. Calling the system of the integral algebraic functions of T, for shortness, the system (G), we have the theorem that an integral algebraic function of any member of (G) is itself a member of (G). Thus the system (G) is closed.

Suppose that $\zeta_1, \zeta_2, \cdots, \zeta_n$ form a basis, the members of this basis being selected from the system (G). All functions of the system

$$g_1(z)\zeta_1 + g_2(z)\zeta_2 + \cdots + g_n(z)\zeta_n,$$

where the g's are any integral polynomials in z, belong to (G). But it is not true conversely that every member of (G) can be written in this form. Let us assume that there is a member of (G) which can be written

$$\frac{g_1\zeta_1 + g_2\zeta_2 + \cdots + g_n\zeta_n}{(z-c)},$$

where the g's are not all divisible algebraically by $z-c$. Suppose that the remainders of $g_1, \ldots, g_n$, after the division, are $a_1, a_2, \ldots, a_n$; then

$$\zeta = \frac{a_1\zeta_1 + a_2\zeta_2 + \cdots + a_n\zeta_n}{z-c}$$

is an integral algebraic function of the surface. We know that at least one of the a's is different from zero. Assume $a_1 \neq 0$. The n functions $\zeta, \zeta_2, \ldots, \zeta_n$ of the system G, form a basis of (R). Denoting the values of $\zeta, \zeta_\kappa, (\kappa = 1, 2, \cdots n,)$ at n associated places of T by $\zeta^{(1)}$, $\zeta^{(2)}, \ldots, \zeta^{(n)}$, and $\zeta_{\kappa 1}, \zeta_{\kappa 2}, \ldots, \zeta_{\kappa n}$, we have, by the theory of determinants,

$$\begin{vmatrix} \zeta^{(1)}, & \zeta_{21}, & \cdots, & \zeta_{n1} \\ \zeta^{(2)}, & \zeta_{22}, & \cdots, & \zeta_{n2} \\ \cdot & \cdot & \cdot & \cdot \\ \cdot & \cdot & \cdot & \cdot \\ \zeta^{(n)}, & \zeta_{2n}, & \cdots, & \zeta_{nn} \end{vmatrix}^2 = \frac{a_1^2}{(z-c)^2} \begin{vmatrix} \zeta_{11}, & \zeta_{21}, & \cdots, & \zeta_{n1} \\ \zeta_{12}, & \zeta_{22}, & \cdots, & \zeta_{n2} \\ \cdot & \cdot & \cdot & \cdot \\ \cdot & \cdot & \cdot & \cdot \\ \zeta_{1n}, & \zeta_{2n}, & \cdots, & \zeta_{nn} \end{vmatrix}^2,$$

say $$\Delta(\zeta, \zeta_2, \zeta_3, \cdots, \zeta_n) = \frac{a_1^2}{(z-c)^2}\Delta(\zeta_1, \zeta_2, \ldots, \zeta_n).$$

Here both determinants are integral rational functions of z; hence the determinant on the left-hand side is of lower order than the determinant on the right. By proceeding in this way, we can remove the factors $(z-c)^2$ from $\Delta(\zeta_1, \zeta_2, \ldots, \zeta_n)$ up to a certain point.

Then no further factors can be removed. Suppose that $\zeta_1', \zeta_2', \ldots, \zeta_n'$ are the ζ's when this stage has been reached. Omitting accents, these new quantities

$$\zeta_1, \zeta_2, \ldots, \zeta_n$$

form, in Klein's nomenclature, a minimal basis. [See Dedekind and Weber, loc. cit., p. 194; Klein, loc. cit., p. 493.] Since there are no longer any available denominators $z-c$, every algebraic function of the surface T can now be expressed in the form

$$g_1\zeta_1 + g_2\zeta_2 + \cdots + g_n\zeta_n,$$

where the g's are integral polynomials in z; and, conversely, every function of this form is a member of (G). The analogy to the theorem for functions of the system (R) is evident.

Let us now construct a minimal basis *ab initio*. We know (§ 184) that every function $R(w, z)$, which is at the same time an *integral algebraic function of the surface*, is expressible in the form

$$r_0(z) + r_1(z)w + \cdots + r_{n-1}(z)w^{n-1},$$

in which all the denominators are factors of $D(z)$, where

$$D(z) = F'(w_1)\cdot F'(w_2)\cdots F'(w_n),$$

$F'(w_\kappa)$ standing for $\delta F/\delta w_\kappa$. In other words, the denominators are factors of the discriminant of F. Select from the integral algebraic functions of the set

$$r_0 + r_1 w + \cdots + r_\kappa w^\kappa,$$

that which has the highest denominator in the coefficient of w^κ. Let

$$s_0 + s_1 w + s_2 w^2 + \cdots + s_\kappa w^\kappa$$

be a function which satisfies this requirement. Suppose this denominator to be $d_{\kappa+1}$, with a corresponding numerator $n_{\kappa+1}$. As the expression s_κ is in its lowest terms, it must be possible to find integral polynomials $\rho_{\kappa+1}$, $\sigma_{\kappa+1}$, such that

$$\rho_{\kappa+1}n_{\kappa+1} + \sigma_{\kappa+1}d_{\kappa+1} = 1.$$

The integral algebraic function of T,

$$\rho_{\kappa+1}\{s_0 + s_1 w + \cdots + s_\kappa w^\kappa\} + \sigma_{\kappa+1}w^\kappa,$$

begins with a term $\dfrac{1}{d_{\kappa+1}}w^\kappa$. Call this function $\theta_{\kappa+1}$.

Let $R(w, z)$ be an integral algebraic function of the surface. When it is expressed in the above form, let its order in w be κ. The denominator of the coefficient of w^κ is a factor of $d_{\kappa+1}$. For, otherwise,

$$R(w, z)+\theta_{\kappa+1}$$

is an integral algebraic function of the surface, such that the denominator of its highest term w^κ is of higher order than that of $d_{\kappa+1}$. This is contrary to supposition. We shall prove that the θ's form a minimal basis.

§ 186. *The expression for $R(w, z)$ in terms of the θ's.* The function $R(w, z)$ is expressible in the form

$$\frac{g_{\kappa+1}(z)}{d_{\kappa+1}(z)}w^\kappa + \text{terms in } w^{\kappa-1}, w^{\kappa-2}, \cdots, 1.$$

Hence $R(w, z)-g_{\kappa+1}(z)\cdot\theta_{\kappa+1}=$ an integral algebraic function of the surface, of order not higher than $\kappa-1$ in w. This expression in turn can be changed, by subtraction of $g_\kappa(z)\theta_\kappa$, into an integral algebraic function of the surface, of order $\kappa-2$ in w, and so on. Thus,

$$R(w, z)=g_{\kappa+1}(z)\theta_{\kappa+1}+g_\kappa(z)\theta_\kappa+\cdots+g_1(z)\theta_1.$$

Let the values of θ_κ at n associated places $(z, w_1), \cdots, (z, w_n)$ be $\theta_{\kappa 1}, \theta_{\kappa 2}, \cdots, \theta_{\kappa n}$, and form the determinant of the basis, viz.,

$$\begin{vmatrix} \theta_{11} & \theta_{21} & \cdots & \theta_{n1} \\ \theta_{12} & \theta_{22} & \cdots & \theta_{n2} \\ \cdot & \cdot & \cdot & \cdot \\ \cdot & \cdot & \cdot & \cdot \\ \theta_{1n} & \theta_{2n} & & \theta_{nn} \end{vmatrix}.$$

The square of this determinant is a symmetric function of $w_1, w_2, \cdots, w_n$, since the determinant itself is an alternating function. Call the square of this determinant $\Delta(z)$; we suppose that the θ's have been multiplied previously by such constants as will make the coefficient of the highest power of z in $\Delta(z)$ equal to unity. The discriminant $D(z)$ of the equation

$$F(w^n, z^m)=0,$$

in which the coefficient of w^n is 1 and the coefficients of the remaining powers of w are integral polynomials, is the square of the determinant

$$\begin{vmatrix} 1 & w_1 & w_1^2 & \cdots & w_1^{n-1} \\ 1 & w_2 & w_2^2 & \cdots & w_2^{n-1} \\ \cdot & \cdot & \cdot & \cdot & \cdot \\ \cdot & \cdot & \cdot & \cdot & \cdot \\ 1 & w_n & w_n^2 & \cdots & w_n^{n-1} \end{vmatrix}.$$

Since $$\theta_{\kappa+1} = \frac{1}{d_{\kappa+1}} w^{\kappa} + r_2(z) w^{\kappa-1} + \cdots + r_{\kappa+1}(z),$$

it follows that

$$D = \Delta \cdot (d_2 d_3 \cdots d_n)^2.$$

This shows that $\Delta \neq 0$. Since it has been proved that the determinant of the θ's does not vanish, and that all integral algebraic functions of the surface are expressible in the form $\sum_1^n g_\kappa \theta_\kappa$, it follows that the θ's constitute a *minimal basis.*

Let $R_1, R_2, \cdots, R_n$ be n members of the system (G), and let the expressions for the R's in terms of the θ's be

$$R_\kappa = f_{\kappa 1}\theta_1 + f_{\kappa 2}\theta_2 + \cdots + f_{\kappa n}\theta_n \; (\kappa = 1, 2, \cdots n),$$

where the f's are integral rational functions of z. Then, denoting by $R_{\kappa 1}, R_{\kappa 2}, \cdots, R_{\kappa n}$, the values of R_κ at n associated places, we have

$$\begin{vmatrix} R_{11} & R_{12} & \cdots & R_{1n} \\ R_{21} & R_{22} & \cdots & R_{2n} \\ \cdot & \cdot & \cdot & \cdot \\ \cdot & \cdot & \cdot & \cdot \\ R_{n1} & R_{n2} & \cdots & R_{nn} \end{vmatrix}^2 = \begin{vmatrix} f_{11} & f_{12} & \cdots & f_{1n} \\ f_{21} & f_{22} & \cdots & f_{2n} \\ \cdot & \cdot & \cdot & \cdot \\ \cdot & \cdot & \cdot & \cdot \\ f_{n1} & f_{n2} & \cdots & f_{nn} \end{vmatrix}^2 \cdot \Delta(z)$$

$$= \{G(z)\}^2 \cdot \Delta(z).$$

This is a re-statement of the theorem of § 185; it shows that the square of the determinant of the R's is at lowest $c\Delta(z)$, where c is a constant. In this case of the minimum determinant the expressions R form a minimal basis, and may replace $\theta_1, \theta_2, \cdots, \theta_n$.

Let W be a rational function of w, z, which belongs to (G); and let $R_1 = 1$, $R_2 = W$, $R_3 = W^2, \cdots, R_n = W^{n-1}$; the equation connecting the determinants becomes

$$D_1(z) = \Delta(z)\{G(z)\}^2,$$

where D_1 is the determinant of the system $1, W, W^2, \cdots, W^{n-1}$. Hence D_1, as well as D, is divisible by $\Delta(z)$. The expression $\Delta(z)$ is a divisor of the discriminants of all the irreducible equations of order n which define functions W of the system (G). The expression $\Delta(z)$ is called, by Kronecker, the *essential divisor* of the discriminant. The remarkable property of the essential divisor is that it is unaffected by a birational transformation which transforms $F(w^n, z) = 0$ into $F_1(W^n, z) = 0$, where the coefficients of w^n, W^n are 1. The remaining part of the discriminant is a complete square; it is called the *unessential divisor.* A factor of $\Delta(z)$ is an essential factor; a factor of the square root of the unessential divisor is an unessential factor. The essential divisor is equal, save as to a constant, to the square of the determinant of any minimal basis.

Corollary. The square of the determinant of the coefficients a in

$$1 = a_{11}\theta_1 + a_{21}\theta_2 + \cdots + a_{n1}\theta_n,$$
$$w = a_{12}\theta_1 + a_{22}\theta_2 + \cdots + a_{n2}\theta_n,$$
$$\cdot \quad \cdot \quad \cdot \quad \cdot \quad \cdot \quad \cdot \quad \cdot \quad \cdot \quad \cdot \quad \cdot \quad \cdot$$
$$\cdot \quad \cdot \quad \cdot \quad \cdot \quad \cdot \quad \cdot \quad \cdot \quad \cdot \quad \cdot \quad \cdot \quad \cdot$$
$$w^{n-1} = a_{1n}\theta_1 + a_{2n}\theta_2 + \cdots + a_{nn}\theta_n,$$

is equal to the unessential divisor.

Conversely, the square of the determinant of the coefficients when $\theta_1, \theta_2, \cdots, \theta_n$ are expressed in terms of $1, w, w^2, \cdots, w^{n-1}$, is the reciprocal of the unessential divisor. But we know that every integral algebraic function of the surface is equal to

$$\sum_1^n g_r\theta_r;$$

hence, there is no integral algebraic function of the surface $r_0(z) + r_1(z)w + \cdots + r_{n-1}(z)w^{n-1}$, which contains in its denominator factors distinct from the unessential factors of the discriminant.

§ 187. Kronecker has proved that the essential divisor of the discriminant of $F=0$ is the highest common divisor of the discriminants of the integral algebraic functions of T. Let

$$v = \lambda_1\theta_1 + \lambda_2\theta_2 + \cdots + \lambda_n\theta_n,$$

where the λ's are arbitrary coefficients. The function v satisfies an irreducible equation of order n; and in this equation the coefficients are integral polynomials in z, except the coefficient of v^n, which is unity. The discriminant D_v must be divisible by $\Delta(z)$, and the remaining part must be a perfect square. Suppose that

$$D_v = \Delta(z)\{\delta(z)\}^2 \cdot \{G(z, \lambda_1, \lambda_2, \cdots, \lambda_n)\}^2;$$

where G is an integral polynomial in z and the λ's, which contains no factor independent of the λ's. We have to prove that $\delta(z)$ reduces to a constant. Assume, if possible, that $\delta(z)$ contains a factor $z-c$, where c is a constant independent of z and the λ's. In the system (G) there must be a rational function $R(v, z)$ which is of the form

$$\frac{g_0(z) + g_1(z)v + g_2(z)v^2 + \cdots + g_{n-1}(z)v^{n-1}}{z-c}.$$

Suppose that the lowest order in v for such functions is s, and that one of these functions of order s is $\frac{\psi(v)}{z-c}$. Then $g_s(z)$ is not divisible by $z-c$. Let t be an undetermined quantity. Since $tv-\frac{\psi(v)}{z-c}$ is an integral algebraic function of T, it may be written

$$\mu_1\theta_1+\mu_2\theta_2+\cdots+\mu_n\theta_n,$$

where the μ's are integral rational functions of $z, \lambda_1, \lambda_2, \cdots, \lambda_n$, and symmetry shows that its discriminant must be

$$\Delta(z)\{\delta(z)\}^2\{G(z, \mu_1, \mu_2, \cdots, \mu_n)\}^2.$$

But the discriminant of $tv-\frac{\psi(v)}{z-c}$ is equal to the product

$$\Pi\left\{t(v_\alpha-v_\beta)-\frac{\psi(v_\alpha)-\psi(v_\beta)}{z-c}\right\}^2,$$

where α, β take all pairs of values, from $1, 2, \cdots, n$, such that $\alpha>\beta$. This product is equal to

$$D_v\cdot\Pi\left\{t-\frac{\psi(v_\alpha)-\psi(v_\beta)}{(z-c)(v_\alpha-v_\beta)}\right\}^2.$$

Equating the two values for the discriminant, we have

$$\pm G(z, \mu_1, \mu_2, \cdots, \mu_n)=G(z, \lambda_1, \lambda_2, \cdots, \lambda_n)\cdot\Pi\left\{t-\frac{\psi(v_\alpha)-\psi(v_\beta)}{(z-c)(v_\alpha-v_\beta)}\right\}.$$

The arbitrary quantities λ can be chosen so that the two functions G are finite and do not vanish when $z=c$. Hence the above equation cannot be true if the second factor be infinite for $z=c$. The initial supposition requires, then, that

$$\Phi(t)=\Pi\left\{t-\frac{\psi(v_\alpha)-\psi(v_\beta)}{(z-c)(v_\alpha-v_\beta)}\right\}$$

be finite for $z=c$, and therefore also for all values of z. Hence $\Phi(t)=0$ is an equation in t with coefficients which are integral polynomials in z; it defines integral algebraic functions of the surface. It follows that

$$\frac{1}{z-c}\sum_{r=2}^{n}\frac{\psi(v_1)-\psi(v_r)}{v_1-v_r}$$

must be an integral algebraic function of the surface. The following considerations prove that this is impossible. The function just written down is equal to

$$-\frac{1}{z-c}\psi'(v_1)+\frac{1}{z-c}\sum_{r=1}^{n}\frac{\psi(v_1)-\psi(v_r)}{v_1-v_r},$$

and the term which involves the highest power of v_1 in this expression is

$$\frac{1}{z-c}(n-s)g_s v_1^{s-1}.$$

This is contrary to the supposition that s is the lowest order in v for the system of expressions

$$\frac{g_0(z)+g_1(z)v+\cdots+g_{n-1}v^{n-1}}{z-c};$$

hence $\delta(z)$ must reduce to a constant.

To return to the discriminant D_v. Choose the arbitrary quantities λ so that $G(z, \lambda_1, \lambda_2, \cdots, \lambda_n)$ shall not contain any essential factor. As this is always possible, it follows that the essential divisor $\Delta(z)$ is the highest common divisor of the discriminants of all the integral algebraic functions of the surface, which satisfy irreducible equations of the nth order. The square root of the unessential divisor $\{G(z, \lambda_1, \lambda_2, \cdots, \lambda_n)\}^2$ may have, for special values of the λ's, repeated factors; or again the unessential factors may be partly the same as essential factors. Kronecker has proved the highly important theorem that it is possible to choose the λ's so that G has no factor in common with $\Delta(z)$ or with $\delta G/\delta z$.

When the quantities λ are left arbitrary, G has no divisor in common with $\delta G/\delta z$. Assume, if possible, that this is not the case. Let $H(z, \lambda_1, \lambda_2, \cdots, \lambda_n)$ be one of the common irreducible divisors; and let $G=HK$. Then the equations

$$G=HK,$$

$$G'=HK'+KH',$$

show that K must be divisible by H. When G is divided by H^2 the quotient is an integral rational function of z and the λ's. Let the quotient be Q. Since $G=H^2Q$, it follows that any divisor of G, G' is also a divisor of $\delta G/\delta\lambda_\kappa$ $(\kappa=1, 2, \cdots, n)$. But the product-expression of the discriminant shows that

$$G\sqrt{\Delta(z)}=\pm\Pi\{\lambda_1(\theta_{1\alpha}-\theta_{1\beta})+\lambda_2(\theta_{2\alpha}-\theta_{2\beta})+\cdots+\lambda_n(\theta_{n\alpha}-\theta_{n\beta})\},$$

where α, β are taken from $1, 2, \cdots, n$, and $\alpha>\beta$; hence one of these linear factors must be equal to another of these factors, multiplied by some constant. Suppose that the factor which contains suffixes α, β is equal to the factor with suffixes γ, δ, multiplied by a constant A; and use the equations

$$\theta_1=1,\ \theta_2=a+a_1w,\ \theta_3=b+b_1w+b_2w^2,\ \theta_4=c+c_1w+c_2w^2+c_3w^3,\ \cdots.$$

Then $$Aa_1(w_\alpha - w_\beta) = a_1(w_\gamma - w_\delta),$$

$$Ab_2(w_\alpha^2 - w_\beta^2) + Ab_1(w_\alpha - w_\beta) = b_2(w_\gamma^2 - w_\delta^2) + b_1(w_\gamma - w_\delta),$$

$$Ac_3(w_\alpha^3 - w_\beta^3) + Ac_2(w_\alpha^2 - w_\beta^2) + Ac_1(w_\alpha - w_\beta)$$
$$= c_3(w_\gamma^3 - w_\delta^3) + c_2(w_\gamma^2 - w_\delta^2) + c_1(w_\gamma - w_\delta),$$

. . . .

Since a_1, b_2, c_3, $\cdots$, are different from zero, it follows that

$$w_\alpha + w_\beta = w_\gamma + w_\delta,$$

$$w_\alpha^2 + w_\alpha w_\beta + w_\beta^2 = w_\gamma^2 + w_\gamma w_\delta + w_\delta^2.$$

Eliminating w_δ, we have the relation

$$(w_\alpha - w_\gamma)(w_\beta - w_\gamma) = 0.$$

This equation is not satisfied in general. Hence the initial assumption that G and $\delta G/\delta z$ have a common divisor is disproved. Here it is assumed that the λ's are arbitrary. Suppose that the λ's are integral polynomials in z with arbitrary orders and coefficients. The expression which results from the elimination of z from G, G' does not vanish identically; therefore special values can be found for the orders and coefficients, such that G and G' have still no common factor. Also values of the λ's can be found which ensure that G shall not contain any of the essential factors. Suppose that these values are λ_1', λ_2', $\cdots$, λ_n'. Hence we have found a function,

$$w' = \lambda_1'\theta_1 + \lambda_2'\theta_2 + \cdots + \lambda_n'\theta_n,$$

which belongs to the system (G), and which has a discriminant whose unessential factors are unrepeated, and distinct from the essential factors.

§ 188. Kronecker has shown that this separation of the discriminant into two divisors can be applied when the algebraic function w is not integral.

Let $F(w^n, z^m) = f_0 w^n + f_1 w^{n-1} + f_2 w^{n-2} + \cdots + f_n = 0$, where f_0, f_1, f_2, $\cdots$, f_n are at most of degree m in z. When two finite values of w become equal, $F'(w) = 0$. Hence if w_1, w_2, $\cdots$, w_n be the n branches, all the branch-points must be included in the equation

$$f_0^{n-2}F'(w_1)F'(w_2)\cdots F'(w_n) = 0,$$

say $f_0^{n-2}E(z) = 0$. The function $E(z)$ is one-valued because the roots z_1, z_2, $\cdots$, z_n enter symmetrically. When z is finite, $E(z)$ can only be made infinite by choosing $w = \infty$. When $w = \infty$, $f_0 = 0$;

but $f_0 w$ is finite, for the coefficient of w^{n-1}, namely, $f_0 w + f_1$, is zero. Since $f_0 w$ is finite, the factor f_0^{n-2} makes $E(z)$ finite, for when one of the w's is infinite, that one of the factors $F'(w)$ which becomes infinite $\propto f_0 w^{n-1}$. Let $f_0^{n-2}E(z) \equiv Q(z)$. The function $Q(z)$ is one-valued and finite for all finite values of z. Its degree is at most

$$m(n-2) + mn \text{ or } 2m(n-1).$$

Without loss of generality it may be assumed that f_n, f_0 have no factor in common; for the discriminant is unaffected when $w+b$ is written for w. Write $w' = f_0 w$. Then

$$F_1(w', z) = w'^n + g_1 w'^{n-1} + \cdots + g_n = 0$$

defines an integral algebraic function of the surface T associated with $F_1 = 0$. Since $\delta F_1/\delta w = f_0^{n-2}\delta F/\delta w$, the discriminant $D(z)$ of $F_1 = 0$ satisfies the equation

$$D(z) = f_0^{(n-1)(n-2)} \cdot Q(z).$$

Kronecker has proved that $f_0^{(n-1)(n-2)}$ enters into the unessential divisor of $D(z)$, so that

$$D(z) = f_0^{(n-1)(n-2)} \cdot \{\delta(z)\}^2 \cdot \Delta(z);$$

hence $$Q(z) = \{\delta(z)\}^2 \cdot \Delta(z).$$

The expression $\Delta(z)$ is called the essential divisor of Q, and $\{\delta(z)\}^2$ the unessential divisor.

§ 189. *The geometric aspect of the discriminant.* Let (ξ_1, ξ_2, ξ_3), (η_1, η_2, η_3), be the centres of two pencils of rays

$$w(\eta\xi) + (\eta\zeta) = 0, \quad z(\xi\eta) + (\xi\zeta) = 0,$$

where $(\xi\eta) = 0$, $(\xi\zeta) = 0$, $(\eta\zeta) = 0$, are the equations of the lines connecting ξ, η with each other and with a third point ζ whose co-ordinates are $(\zeta_1, \zeta_2, \zeta_3)$. If w and z be connected by an equation $F(w, z) = 0$, the intersection of corresponding rays of the two pencils traces out a plane curve. Starting with any plane curve C of order α and class β, it is possible to associate with it an equation $F(w, z) = 0$. The form of F will depend upon the choice of the points ξ, η. Assume that ξ, η have no specialty of position with regard to C; this means that ξ, η are not to lie on C, or on any of the singular lines associated with C, understanding by singular lines (1) the join of two singular points, (2) singular tangents, (3) tangents at singular points, (4) tangents passing through a singular point. Further, the line $(\xi\eta)$ is not to touch C, and is not to pass through a singular point (see Smith, loc. cit., § 170). Projecting the line ($\xi\eta$) to infinity, we can now treat w, z as Cartesian co-ordinates. The order a of $F=0$ is also the degree in w, z separately, the line at infinity meets the curve in a distinct points, and therefore the multiple points of the curve lie in the finite part of the plane. It should be especially noticed that when a line $z=a$ touches the curve, its contact is simple, and that it passes through no multiple point; also that no two multiple points lie on a line $z=a$. When the equation $F(w, z)=0$ has been put into this special form, it is easy to assign a meaning to the factors of the discriminant; and when the geometric significance of these factors has been found, the extensions to more complicated cases (which arise when the restrictions are removed as to tangents $z=a$ having simple contact, tangents at the multiple points being oblique to the axes, etc.,) are readily effected. The discriminant $Q(z)$ of $F(w, z)$ with regard to w is found by the elimination of w from $F=0$ and $\delta F/\delta w=0$; that is, the roots of $Q(z)=0$ correspond to the points of intersection of $F=0$ with the first polar of the point at infinity on the w-axis. Remembering that $Q(z)$ contains $\Pi(w_r-w_s)^2$, it is easy to see what factors are contributed by those branch-points which are distinct from the multiple points (*i.e.* the points of contact of the tangents from ξ to C), and by the multiple points of the curve.

(i.) If $z=a$ be an ordinary tangent at ($z=a$, $w=b$), there are two series

$$w_\kappa - b = P_1(z-a)^{1/2},\ w_\lambda - b = Q_1(z-a)^{1/2},$$

and $$(w_\kappa - w_\lambda)^2 \propto z-a.$$

(ii.) If (a, b) be a node,

$$w_\kappa - b = P_1(z-a),\ w_\lambda - b = Q_1(z-a),$$

and $$(w_\kappa - w_\lambda)^2 \propto (z-a)^2.$$

(iii.) If (a, b) be a cusp,

$$w_\kappa - b = P_3(z-a)^{1/2},\ w_\lambda - b = Q_3(z-a)^{1/2},$$

and $$(w_\kappa - w_\lambda)^2 \propto (z-a)^3.$$

(iv.) If (a, b) be a multiple point which can be resolved, by the methods of Chapter IV., into ν nodes and κ cusps, then the corresponding factor of the product $\Pi(w_r-w_s)^2 \propto (z-a)^{2\nu+3\kappa}$ * (see Chapter IV., § 121).

* The equation $F=0$ has been so arranged that no line $z=a$ cuts the curve at more than two consecutive points; in a more general case, it might happen that a tangent $z=a$ cuts in r consecutive points, the remaining $a-r$ points of intersection being in the finite part of the plane and distinct. In this case, those terms of the product $\Pi(w_r-w_s)^2$ which correspond to the point of contact contribute to Q a factor $(z-a)^{r-1}$.

Thus $Q(z)$ can be resolved into two parts. The first part consists of linear factors $(z-z_1)(z-z_2)\cdots(z-z_\beta)$ which arise from the ordinary vertical tangents; the second part consists of multiple factors $(z-z_1')^{2\nu_1+3\kappa_1}(z-z_2')^{2\nu_2+3\kappa_2}\cdots(z-z_\gamma')^{2\nu_\gamma+3\kappa_\gamma}$, where γ is the number of multiple points. Regarding the ν nodes and κ cusps at a multiple point $z=z'$ as equivalent to $\nu+\kappa$ nodes and κ branch-points, a factor $(z-z')^{2\nu+3\kappa}$ may be subdivided into two parts $(z-z')^{2\nu+2\kappa}$ and $(z-z')^{\kappa}$, where the former arises from the $\nu+\kappa$ nodes, and the latter from the κ branch-points. Thus the discriminant $Q(z)$ can be resolved into a part which arises from the $(\beta+\sum_{s=1}^{\gamma}\kappa_s)$ branch-points, and into another part which arises from the $\sum_{s=1}^{\gamma}(\nu_s+\kappa_s)$ nodes. It should be observed that the latter part is a perfect square; also that the two parts have factors in common, owing to the presence of cusps. The former part determines the branching of the corresponding Riemann surface.

In the light of these results, let us examine the essential and unessential divisors as they appear in Kronecker's theory. Starting with an equation $F(w, z)=0$, we can, by the use of birational transformations, arrive at an equation $F_1(W, z)=0$, the square root of whose unessential divisor consists of factors *which are distinct from one another and from the essential factors.* The deduction is that the final equation has no singularities higher than nodes.

In the final equation $F_1=0$, the essential divisor is precisely the same as the essential divisor for $F=0$; therefore Kronecker's transformations do not affect the branching of the surface T.

§ 190. *A special kind of Riemann surface.* When w and z are real and connected by an equation $F=0$ with real coefficients, this equation can be represented by a Cartesian curve, whereas the same equation requires for its graphic representation a Riemann surface in the case that w and z are complex. Klein has indicated a method whereby the Riemann surface can be directly associated with the curve. Let us consider, for example, the equation

$$w^2/a^2+z^2/b^2=1,$$

where a, b are real numbers; the curve is evidently an ellipse. This ellipse is enveloped by real and imaginary tangents. From every external point two real tangents can be drawn, and through every internal point pass two imaginary tangents. If each real tangent be replaced by its point of contact with the ellipse, and if each imaginary tangent be replaced by the real point through which it passes, the resulting complex of points will cover doubly the region bounded by the ellipse. These points lie on a new kind of Riemann surface whose form is a flattened ellipsoid which passes through the elliptic contour $w^2/a^2 + z^2/b^2 = 1$. As a second example consider a curve of the third class which consists of two real branches, *viz.* a tricuspidal branch lying within an oval. At points without the oval or within the tricusp the three tangents are real, and therefore the corresponding points of the surface lie on the two branches of the curve. Each point which lies between the tricusp and the oval has two imaginary tangents passing through it. Hence the new form of Riemann surface must be two sheeted, one extending from the one branch to the other, the two sheets being connected by the two branches. By a continuous deformation it can be changed into the ordinary ring-surface of deficiency 1. Incidentally this proves that the deficiency of the curve in question is 1, an illustration of the way in which Klein's surfaces can be used to prove theorems in higher plane curves. The general reasoning by which Klein was led to the construction of such surfaces is, briefly, the following: the equation $F(w, z) = 0$ with coefficients which may be complex can be transformed into a homogeneous equation $f(x_1, x_2, x_3) = 0$ in three variables. If (x_1, x_2, x_3) be tangential co-ordinates, every straight line in the plane is represented by a set of 3 co-ordinates. To obtain a geometric representation of the equation $f = 0$ such that to each system of values (x_1, x_2, x_3) corresponds a single point, the line (x_1, x_2, x_3) must be replaced by some point. For those systems which are real and satisfy $f = 0$ the point chosen is the point of contact of the line with the curve, whereas for those systems which are imaginary the point chosen is the real point through which the corresponding imaginary tangent passes. In general a real point which belongs to one imaginary tangent will belong to several others. Hence the multiple covering of the plane. Each portion of the plane is supposed to be covered with as many sheets as there are imaginary tangents from any point of it to the curve. This construction ensures, in general, a one-to-one point representation of the systems (x_1, x_2, x_3) which satisfy $f(x_1, x_2, x_3) = 0$. Usually the coefficients of $f = 0$ are real, and then conjugate imaginaries occur in pairs. The covering of the plane is then effected, in all its portions, by an even number of sheets, but the number of sheets may vary from part to part of the plane. That the point of contact of a real tangent with the curve should be chosen as the representative point for the (x_1, x_2, x_3) of the line is natural, inasmuch as the real intersection of an imaginary tangent with its conjugate tangent becomes, when the two lines coincide with a real line, the common point of contact.

Certain special cases need to be considered in order to make the theory complete. A real double tangent with real points of contact has two points attached to it; an isolated double tangent and a real inflexional tangent are represented by the totality of points which lie upon them, etc. When this surface of Klein's is compared with the ordinary Riemann surface which arises from the fundamental equation, it will be noticed that the one-to-one correspondence fails for those values (x_1, x_2, x_3) which correspond to the real inflexional tangents and to the isolated real double tangents. For instance, in the ordinary Riemann surface an isolated real double tangent is represented by two points, whereas in Klein's surface it is represented by a complete line. The two surfaces are not deducible, the one from the other, by an orthomorphic transformation, for they are not similar in respect to their smallest parts. For further details with respect to this interesting theory the student is referred to memoirs by Klein in the Mathematische Annalen, tt. vii., x., by Harnack in the same journal, t. ix., and by Haskell, Amer. Jour., t. xiii.

References. Riemann's Werke (particularly the Inaugural Dissertation and the memoir on Abelian Functions); Clebsch, Geometrie, t. i., Sechste Abtheilung; Durège, Theorie der Funktionen; Harnack's Calculus; Klein, Ueber Riemann's Theorie der Algebraischen Functionen und ihrer Integrale; Klein-Fricke, Theorie der Elliptischen Modulfunctionen, t. i.; Königsberger, Elliptische Functionen; Neumann, Abel'sche Integrale, 2d ed.; Prym, Neue Theorie der Ultraelliptischen Functionen; Thomae, Abriss einer Theorie der Complexen Functionen, 2d ed.; Casorati, Teorica delle Funzioni di Variabili Complesse; Klein, Ikosaeder.

In this chapter we have considered the Riemann surface as springing from a basis-equation $F(w^n, z^m) = 0$. But in the ulterior Riemann theory as contemplated by Riemann himself, and developed by Klein, Poincaré, and many others, a surface is taken as the starting-point, and functions are defined by the surface. The student who wishes to become acquainted with these results should consult the Mathematische Annalen, where he will find many important memoirs by Klein, Hurwitz, Dyck, etc.; in particular we may refer him to Klein's memoir, Neue Beiträge zur Riemann'schen Theorie, Math. Ann., t. xxi.

CHAPTER VII.

ELLIPTIC FUNCTIONS.

§ 191. A one-valued function $f(u)$ of the complex variable u has a primitive period ω, when, for all finite values of u and for all integral values of n,

$$f(u + n\omega) = f(u);$$

the equation ceasing to hold for fractional values of n. The quantities ω, 2ω, 3ω, $\cdots$, are all of them periods of $f(u)$, but it is usual to single out the primitive period ω as the period par excellence.

As an example of a singly periodic function, consider $\exp 2\pi i u/\omega$. Mark off in the z-plane the series of points u_0, $u_0 \pm \omega$, $u_0 \pm 2\omega$, $\cdots$, and through them draw any set of parallel lines. The plane is in this way divided into bands of equal breadths; in each band the function takes all its values.

Fourier's Theorem. Let $f(u)$ be a function with the period ω, and let it have no singular points within a strip of the plane bounded by straight rims which are parallel to the stroke ω. The transformation $z = \exp 2\pi i u/\omega$ changes the strip into a ring of the u-plane, bounded by circles whose centres are $z = 0$ (see § 181). To each point of the ring correspond infinitely many points u, but, since these are congruent to the modulus ω, to each point of the ring there corresponds only one value of $f(u)$. Further $f(u)$, regarded as a function of z, has no singular points in the ring. Hence by Laurent's Theorem (§ 141),

$$f(u) = \sum_{-\infty}^{\infty} a_m z^m,$$

where
$$a_m = \frac{1}{2\pi i}\int \frac{f(u)}{z^{m+1}}\,dz,$$

the integral being taken along a circle concentric with the ring and within it;

or
$$a_m = \frac{1}{\omega}\int_{u_0}^{u_0+\omega} f(u)du \exp\left(-2m\pi iu/\omega\right),$$

where u_0 is any point of the strip.

Therefore, replacing u by a in the integral, we have

$$f(u) = \frac{1}{\omega}\sum_{m=-\infty}^{\infty}\int_{u_0}^{u_0+\omega} f(a)da \exp\left[2m\pi i(u-a)/\omega\right].$$

[Jordan, Cours d'Analyse, t. ii., p. 312.]

§ 192. *Double Periodicity.* Let $f(u)$ be a one-valued doubly periodic function with primitive periods ω_1, ω_2. This implies that:—

(i.) $f(u+m_1\omega_1+m_2\omega_2)=f(u)$, whatever integral values be assigned to m_1, m_2;

(ii.) The above equation must not hold for any fractional values of m_1, m_2;

(iii.) If $f(u+\omega_3)=f(u)$, ω_3 forms one of the system $m_1\omega_1+m_2\omega_2$, where m_1, m_2 are integral.

The periods include all members of the system $m_1\omega_1+m_2\omega_2$; but it is usual to understand by the phrase 'a pair of periods,' not any two periods, but a primitive pair.

§ 193. Before the discussion of the general theory of double periodicity, we give two algebraic lemmas upon the possible values of $m_1\omega_1+m_2\omega_2$ and $m_1\omega_1+m_2\omega_2+m_3\omega_3$, where ω_1, ω_2, ω_3 are complex numbers, and the m's are integers.

Lemma I. The expression $m_1\omega_1+m_2\omega_2$ can be made arbitrarily small when ω_2/ω_1 is real and incommensurable; and can be made $=\omega_1/q$, when ω_2/ω_1 is real and commensurable, q being some positive integer >1.

(i.) Let λ, $=\omega_2/\omega_1$, be real and incommensurable. Convert it into a continued fraction, and let n_1/n_2, n_1'/n_2' be consecutive convergents. Then λ lies between n_1/n_2 and n_1'/n_2', and therefore differs from either by less than $1/n_2n_2'$. Hence

$$\frac{\omega_2}{\omega_1}-\frac{n_1}{n_2}=\pm\frac{\theta}{n_2n_2'},$$

where θ is a proper fraction, and

$$|n_1\omega_1-n_2\omega_2|<\frac{|\omega_1|}{n_2'};$$

thus $|n_1\omega_1-n_2\omega_2|$ can be made as small as we please.

(ii.) Let ω_2/ω_1 be commensurable and $=p/q$, where p, q are integers. Because $\omega_2/\omega_1=p/q$, the expression

$$m_1\omega_1+m_2\omega_2=\omega_1(m_1+m_2p/q)=\omega_1(m_1q+m_2p)/q.$$

But p, q are relatively prime, hence it is possible to choose m_1, m_2 so as to make $m_1q+m_2p=1$.

Lemma II. If ω_1, ω_2, ω_3 be any triangle of points, it is always possible to find integers m_1, m_2, m_3, such that

$$|m_1\omega_1+m_2\omega_2+m_3\omega_3|<\epsilon.$$

Let $\overline{0\,\omega_1}$ meet $\overline{\omega_2\omega_3}$ at u_1. It is evident from the preceding lemma that a point u_1' can be found which divides $\overline{\omega_2\omega_3}$ in the ratio of two integers m_3, m_2, and which is such that $|u_1'-u_1|<\epsilon_1$. Draw $\overline{0\,u}$ parallel to $\overline{\omega_2\omega_3}$, meeting $\overline{\omega_1u_1'}$ at u. Let $\omega_1=\lambda(\omega_1-u_1)$; then also

$$|u|=\lambda|u_1'-u_1|<\lambda\epsilon_1.$$

Again, a point u' can be found which divides $\overline{\omega_1u_1'}$ commensurably, say in the ratio $(m_2+m_3)/m_1$, and is such that $|u'-u|<\epsilon_2$.

Since $\quad u_1'=(m_2\omega_2+m_3\omega_3)/(m_2+m_3)$

and $\quad u'=\{m_1\omega_1+(m_2+m_3)u_1'\}/(m_1+m_2+m_3)$,

therefore $\quad u'=(m_1\omega_1+m_2\omega_2+m_3\omega_3)/(m_1+m_2+m_3)$;

and since $\quad |u'-u|<\epsilon_2$

and $\quad |u|<\lambda\epsilon_1$,

therefore $\quad |u'|<\lambda\epsilon_1+\epsilon_2$.

Hence $\quad |m_1\omega_1+m_2\omega_2+m_3\omega_3|<|m_1+m_2+m_3|(\lambda\epsilon_1+\epsilon_2)$,

which proves the lemma, since m_1, m_2, m_3, λ are finite. It is assumed that the zero-point is not on the join of ω_2 and ω_3; if it were, the preceding lemma would apply.

The first lemma shows that if $f(u)$ be a one-valued doubly periodic function, with the periods ω_1, ω_2, the ratio ω_2/ω_1 must be complex. A real commensurable ratio would involve the existence of a period ω_1/q, contrary to supposition. And a real incommensurable ratio would involve the existence of arbitrarily small numbers amongst the numbers $m_1\omega_1+m_2\omega_2$, and therefore also the existence of a one-valued function $f(u)$, which can be made to assume the same value, as often as we please, along a line of finite length. Such a function must reduce to a constant. Hence the ratio ω_2/ω_1 is complex.

Theorem. There cannot be a triply periodic one-valued function of a single variable.

For, if ω_1, ω_2, ω_3 be the three periods, which are unconnected by a homogeneous linear relation, the corresponding points form a triangle. By the second lemma it is possible to choose m_1, m_2, m_3, so as to make $m_1\omega_1 + m_2\omega_2 + m_3\omega_3$ arbitrarily small. Thus within any region of the plane, however small, the function assumes the value $f(u)$ an infinite number of times. Such a function reduces to a constant. The same conclusion holds for a function with a finite number of values for each value of the variable. For the behaviour of a function with an infinite number of values, see a paper by Casorati (Acta Math., t. viii.). Throughout this chapter we shall consider solely *one-valued* doubly periodic functions which have no essential singularities in the *finite* part of the plane. It will (lemma I.) always be supposed that ω_2/ω_1 is complex. It is an evident consequence of the theorem, that there can be no one-valued function with more than two periods.

The inversion problem in Abelian integrals leads to Abelian functions of p variables, which are one-valued in those variables and $2p$-tuply periodic. By a $2p$-tuply periodic function $f(u_1, u_2, \cdots, u_p)$ is to be understood one which is unaltered by the addition to $u_1, u_2, \cdots, u_p$ of the simultaneous periods $\omega_{\kappa 1}, \omega_{\kappa 2}, \cdots, \omega_{\kappa p} (\kappa = 1, 2, \cdots, 2p)$. It can be proved that such a function, if one-valued, cannot have more than $2p$ sets of simultaneous periods (Riemann, Werke, p. 276). To return to functions of a single variable, some misconceptions on the subject of multiple periodicity are widely spread. Giulio Vivanti has pointed out that Prym, Clebsch and Gordan (Theorie der Abelschen Functionen) and Tychomandritsky (Treatise on the Inversion of the Hyperelliptic Integrals) have assumed that the points in a region at which a κ-tuply periodic function $f(u)$ takes a given value fill that region. Let u be one of these points; the other points are given by $u + m_1\omega_1 + \cdots + m_\kappa\omega_\kappa$. Now it can be proved that these values of u can be arranged in a determinate order $u_1, u_2, u_3, \cdots$, so that each value of u occurs once, and only once, with a determinate suffix. Hence although the number of points $u_1, u_2, u_3, \cdots$ within the region is infinitely great, the known theorem that the points of the region cannot be numbered in the way just described, shows that the region contains points which do not belong to the system $u_1, u_2, u_3, \cdots$. See Lüroth and Schepp's translation of Dini's work: Grundlagen für eine Theorie der Functionen einer veränderlichen reellen Grösse, p. 99. Göpel (Crelle, t. xxxv.) appears to have been the first to point out that Jacobi's theorem on the non-existence of triply periodic one-valued functions of a single variable, does not cover the case of infinitely many-valued functions. To judge by his reply to Göpel, Jacobi would seem to have denied the possibility of the occurrence of such functions. Vivanti's paper is in the Rendiconti del Circolo Matematico di Palermo, Sulle funzioni ad infiniti valori; there is an accompanying note by Poincaré.

§ 194. *Primitive pairs of periods.* If (ω_1, ω_2) be a primitive pair, $m_1\omega_1 + m_2\omega_2$ is a period, provided m_1, m_2 be positive or negative integers. Also *all* the periods of the function are included in the system $m_1\omega_1 + m_2\omega_2$; in other words, a primitive pair of periods is one from which all other periods can be built up by addition or subtraction.

Taking any initial point u_0, mark off in the u-plane the points $u_0 + m_1\omega_1 + m_2\omega_2$, where m_1, m_2 are integers.

We shall speak of the system of points as a network. All the points for which m_1 is constant lie on a line; and all these lines are parallel. Similarly, m_2 = a constant, gives another system of parallel lines. The two systems of parallel lines divide the plane into an infinite number of equal parallelograms. Any such parallelogram is called a parallelogram of periods. A point $u' = u_0 + m_1\omega_1 + m_2\omega_2$ is said to be congruent to u_0, and the equation is written

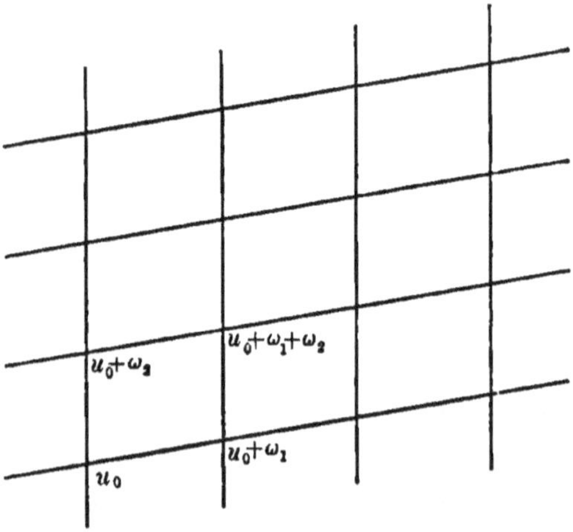

Fig. 79

$$u' \equiv u_0 \text{ (mod. } \omega_1, \omega_2),$$

or shortly

$$u' \equiv u_0.$$

It is evident that the number of primitive pairs of periods is infinite. For instance, ω_1 and ω_2 being one primitive pair, ω_1 and $\omega_1 + \omega_2$ form another primitive pair. But it must not be inferred that every pair of primitive periods is a primitive pair. For instance, from $\omega_1 + \omega_2$ and $\omega_1 - \omega_2$ we cannot, by addition and subtraction, recover the original pair (ω_1, ω_2). Taking, generally,

$$\text{(i.)} \qquad \left.\begin{aligned} \omega_1' &= m_1\omega_1 + m_2\omega_2 \\ \omega_2' &= n_1\omega_1 + n_2\omega_2 \end{aligned}\right\},$$

where the coefficients are integers, the pair (ω_1', ω_2') will be primitive, if

$$\text{(ii.)} \qquad \left.\begin{aligned} \omega_1 &= \mu_1\omega_1' + \mu_2\omega_2' \\ \omega_2 &= \nu_1\omega_1' + \nu_2\omega_2' \end{aligned}\right\},$$

where the coefficients are integers.

Now, solving the first system of equations for ω_1 and ω_2, and comparing the results with the second system, we see that

$$\text{(iii.)} \qquad \begin{vmatrix} m_1 & m_2 \\ n_1 & n_2 \end{vmatrix} = \pm 1.$$

This condition is sufficient, as well as necessary; for if we can recover the original pair (ω_1, ω_2), we can form all possible periods by integral combinations.

The ratio ω_2/ω_1 is of the form $\alpha + i\beta$, where β is not zero. The ratio ω_1/ω_2 is $(\alpha - i\beta)/(\alpha^2 + \beta^2)$. The coefficients of i in the two ratios have opposite signs; we shall suppose always that ω_1 and ω_2 are so chosen that in ω_2/ω_1 the coefficient of i is positive. This is required for the convergence of the series that arise in the theory of the theta-functions.

Consider the original parallelogram (ω_1, ω_2) orthogonally projected into a square. Taking, in the new plane, u_0 as origin, the direction of ω_1 as that of the real axis, and the absolute value of ω_1 as the unit of length, the numbers m_1, m_2 and n_1, n_2 are the rectangular co-ordinates of the projections of $u_0 + \omega_1'$ and $u_0 + \omega_2'$. Equation (iii.) states that the area of the projection of the new parallelogram (ω_1', ω_2') is equal to that of the square. Hence *all the primitive parallelograms of periods are of equal area.*

The simplest change from one primitive pair to another is that in which a period is common to both pairs. The formulae (i.) are then of the type

$$\omega_1' = \omega_1, \; \omega_2' = n\omega_1 \pm \omega_2, \cdots \qquad \text{(iv.)}.$$

By a series of such changes any new parallelogram may be obtained. For starting with the equations (i.),

$$\left.\begin{aligned} \omega_1' &= m_1\omega_1 + m_2\omega_2 \\ \omega_2' &= n_1\omega_1 + n_2\omega_2 \end{aligned}\right\}, \; m_1n_2 - m_2n_1 = \pm 1,$$

let m_1/n_1 be converted into a continued fraction. First let the convergent immediately preceding m_1/n_1 be m_2/n_2. In this case (iii.) is satisfied. Let the convergents, in reverse order, be

$$m_1/n_1, \; m_2/n_2, \; m_3/n_3, \cdots, \; m_{r+1}/n_{r+1},$$

and the corresponding quotients $q_1, q_2, q_3, \cdots, q_{r+1}$, so that

$$\left.\begin{aligned} m_1 &= q_1m_2 + m_3, \; m_2 = q_2m_3 + m_4, \cdots \\ n_1 &= q_1n_2 + n_3, \; n_2 = q_2n_3 + n_4, \cdots \end{aligned}\right\} \qquad \text{(v.)}.$$

Then

$$\begin{aligned} \omega_1' &= m_1\omega_1 + m_2\omega_2 = m_2(q_1\omega_1 + \omega_2) + m_3\omega_1, \\ \omega_2' &= n_1\omega_1 + n_2\omega_2 = n_2(q_1\omega_1 + \omega_2) + n_3\omega_1. \end{aligned}$$

For the first transformation let us take

$$\left.\begin{aligned} \omega_1 &= \omega_1 \\ \omega_3 &= q_1\omega_1 + \omega_2 \end{aligned}\right\},$$

so that
$$\left.\begin{aligned} \omega_1' &= m_2\omega_3 + m_3\omega_1 \\ \omega_2' &= n_2\omega_3 + n_3\omega_1 \end{aligned}\right\}.$$

We have now smaller numbers to deal with. Proceeding in the same way, we get, if $\omega_4 = q_2\omega_3 + \omega_1$,

$$\left.\begin{aligned} \omega_1' &= m_3\omega_4 + m_4\omega_3 \\ \omega_2' &= n_3\omega_4 + n_4\omega_3 \end{aligned}\right\},$$

and so on. Now, in the last convergent, either the numerator or the denominator is 1.

If $n_{r+1} = 1$, $m_{r+1} = q_{r+1}$; the final equations in (v.) are $m_r = q_r m_{r+1} + 1$, $n_r = q_r$, and the final change of periods is

$$\left.\begin{aligned} \omega_1' &= m_{r+1}\,\omega_{r+2} + \omega_{r+1} \\ \omega_2' &= \omega_{r+2} \end{aligned}\right\}.$$

Similarly, if $m_{r+1} = 1$; but if $m_1 \geqq n_1$, this case cannot occur.

A simple example will make the process clearer. We wish to change the periods from ω_1, ω_2, to $67\,\omega_1 + 30\,\omega_2$, $29\,\omega_1 + 13\,\omega_2$, by changing only one period at a time.

Now $67/29 = 2 + \frac{1}{3+}\frac{1}{4+}\frac{1}{2}$. Here $q_1=2$, $q_2=4$, $q_3=3$, $q_4=2$, $\omega_3 = 2\,\omega_1 + \omega_2$, $\omega_4 = 4\,\omega_3 + \omega_1$, $\omega_5 = 3\,\omega_4 + \omega_3$, $\omega_6 = 2\,\omega_5 + \omega_4$, and the pair (ω_5, ω_4) is primitive.

Secondly, if m_2/n_2 be not the convergent preceding m_1/n_1, let μ/ν be that convergent. Then the equations

$$\left.\begin{aligned} m_1 n_2 - m_2 n_1 &= \pm 1 \\ m_1 \nu - n_1 \mu &= \pm 1 \end{aligned}\right\}$$

require that

either $$m_2 = \mu + m_1 t, \quad n_2 = \nu + n_1 t,$$

or $$m_2 = m_1 t - \mu, \quad n_2 = n_1 t - \nu,$$

t being an integer (see Todhunter's Algebra, p. 389). We have, then,

$$\left.\begin{aligned} \omega_1' &= m_1\omega_1 + m_2\omega_2 = m_1(\omega_1 + t\omega_2) \pm \mu\omega_2 \\ \omega_2' &= n_1\omega_1 + n_2\omega_2 = n_1(\omega_1 + t\omega_2) \pm \nu\omega_2 \end{aligned}\right\}.$$

Thus the single change from ω_1, ω_2, to $\omega_1 + t\omega_2$, $\pm\,\omega_2$, reduces this case to the preceding one. It may be remarked that in a change of the kind exemplified by (iv.), the new parallelogram and the old have a side in common, and lie between the same parallels, so that they are equal in area. Hence we see again that all fundamental parallelograms are equal in area.

Corresponding to the parallelogram of periods, there is for the $2p$-tuply periodic functions of p variables a parallelopiped of periods in $2p$-dimensional space.

X § 195. A one-valued doubly periodic function, whose only essential singular point is at ∞, is called an *elliptic function*. In the present paragraph we shall build up a fundamental elliptic function. A primitive pair of periods will be denoted by $2\,\omega_1$, $2\,\omega_2$, although in the general treatment of periodicity we shall continue to denote a period by ω.

Consider the series

$$\Sigma'\left\{\frac{1}{(u-w)^2}-\frac{1}{w^2}\right\} \quad . \quad . \quad . \quad . \quad . \quad . \quad . \quad . \quad (A),$$

where $w=2m_1\omega_1+2m_2\omega_2$, $u\neq w$, and $m_1, m_2=0, \pm 1, \pm 2, \cdots$, the accent signifying, as in Chapter V., that the combination $m_1=m_2=0$ is excluded. Here the remark may be made that an accent attached to Σ or Π always signifies the exclusion of an infinite term or factor.

We shall denote the series obtained from (A) by omitting all terms in which $|w|<|u|$ by $\phi(u)$, and the series of excluded terms by $\psi(u)$. It is first to be proved that the series $\phi(u)$ is absolutely convergent. The general term is $\dfrac{2u(1-u/2w)}{w^3(1-u/w)^2}$. Comparing this with $2u/w^3$, we see that $\phi(u)$ is absolutely convergent if $\Sigma_1 2u/w^3$ be absolutely convergent, where Σ_1 denotes summation for all values of w such that $|w|>|u|$. It was proved in § 77 that the series $\Sigma' 2u/w^3$, of which $\Sigma_1 2u/w^3$ forms a part, is absolutely convergent. Hence $\phi(u)$ is absolutely convergent. The terms in $\psi(u)$ are finite in number, and yield an ordinary rational function of u. Hence (A) is absolutely convergent.

Expanding each term of $\phi(u)$, we see that

$$\frac{1}{(u-w)^2}-\frac{1}{w^2}=\frac{2u}{w^3}+\frac{3u^2}{w^4}+\cdots;$$

and that the series for $\phi(u)$,

$$\Sigma_1\left\{\frac{2u}{w^3}+\frac{3u^2}{w^4}+\cdots\right\},$$

is still absolutely convergent. For if accents denote absolute values,

$$\begin{aligned}\Sigma_1&\left\{\frac{2u'}{w'^3}+\frac{3u'^2}{w'^4}+\cdots\right\}\\ &=\Sigma_1\frac{2u'(1-u'/2w')}{w'^3(1-u'/w')^2}\\ &=\text{a convergent series.}\end{aligned}$$

Rearranging the series for $\phi(u)$ according to ascending powers of u, it is evident that

$$\phi(u)=c_1u+c_2u^2+c_3u^3+\cdots,$$

where $c_n = \Sigma_1 \frac{n+1}{w^{n+2}}$. This equation holds for all values of $|u|$ less than the minimum $|w|$ in Σ_1, say $<\rho$. Hence $\phi(u)+\psi(u)=$ an integral series + a rational fraction, when $|u|<\rho$. That is, the series (A) represents an analytic function of u, and the convergence is uniform, since $\phi(u)+\psi(u)$ does not depend on ρ. Hence the differential quotient of the series (A)

$$=\psi'(u)+\Sigma_1 c_n \cdot nu^{n-1}=\psi'(u)+\Sigma_1 \frac{-2}{(u-w)^3}.$$

Because $\psi(u)$ contains only a finite number of terms,

$$\psi'(u)=\Sigma \frac{-2}{(u-w)^3},$$

where the summation extends over all the values of w, which enter into $\psi(u)$. Hence

$$\psi'(u)+\Sigma_1 \frac{-2}{(u-w)^3}=\Sigma' \frac{-2}{(u-w)^3},$$

where w takes all possible values, except $w=0$. We have proved that the series (A) gives on differentiation $\Sigma' \frac{-2}{(u-w)^3}$; therefore the series

$$\wp(u)=\frac{1}{u^2}+\Sigma'\left\{\frac{1}{(u-w)^2}-\frac{1}{w^2}\right\} \quad . \quad . \quad . \quad . \quad (1)$$

gives $$\wp'(u)=-2\,\Sigma \frac{1}{(u-w)^3} \quad . \quad . \quad . \quad . \quad . \quad . \quad . \quad . \quad (2).$$

From the composition of $\wp(u)$ it is evident that the function has infinities at *all* points of the network

$$2m_1\omega_1+2m_2\omega_2 \quad (m_1, m_2, =0, \pm 1, \pm 2, \cdots),$$

and that each infinity is of order two. There are no other infinities; further the function is even, because the change of u into $-u$ does not affect the value of the right-hand side. Similarly $\wp'(u)$ is an odd function of u.

We have in the function $\wp(u)$, obtained as above, an example of Mittag-Leffler's theorem (§ 150). The series $\Sigma' \frac{1}{(u-w)^2}$ is conditionally convergent, inasmuch as $\Sigma' \frac{1}{w^2}$ is conditionally convergent. But the series $\Sigma'\left\{\frac{1}{(u-w)^2}-\frac{1}{w^2}\right\}$ is unconditionally convergent.

The integral series for $\Sigma'\left\{\frac{1}{(u-w)^2}-\frac{1}{w^2}\right\}$, namely

$$2u\Sigma'\frac{1}{w^3}+3u^2\Sigma'\frac{1}{w^4}+\cdots,$$

contains only even powers of u, because $\Sigma'\frac{1}{w^{2\kappa+1}}=0$, the terms cancelling each other in pairs. To prove the double periodicity of $\wp(u)$, we observe that the right-hand side of (2) is manifestly unaltered by the substitution of $u+2\omega_1$, or of $u+2\omega_2$ for u, since these substitutions merely interchange the terms of an absolutely convergent series. Hence

$$\wp'(u+2\omega_1)=\wp'(u),$$
$$\wp'(u+2\omega_2)=\wp'(u).$$

Integration gives

$$\wp(u+2\omega_1)=\wp(u)+C.$$

Write $u=-\omega_1$; therefore

$$\wp(\omega_1)=\wp(\omega_1)+C,$$

and $C=0$. Hence

and similarly $$\left.\begin{aligned}\wp(u+2\omega_1)&=\wp(u)\\ \wp(u+2\omega_2)&=\wp(u)\end{aligned}\right\} \quad \text{(3)}.$$

Thus $\wp(u)$ is a doubly periodic function, with periods $2\omega_1$, $2\omega_2$. The relations

$$\wp'(u+2\omega_1)=\wp'(u),\ \wp'(u+2\omega_2)=\wp'(u),\ \wp'(u+2\omega_1+2\omega_2)=\wp'(u),$$

give $$\wp'(\omega_1)=0,\ \wp'(\omega_2)=0,\ \wp'(\omega_1+\omega_2)=0 \quad \text{(4)}.$$

The doubly periodic function $\wp(u)$ is the simplest doubly periodic function. Its introduction was due to Weierstrass. While Jacobi's functions $sn(u)$, $cn(u)$, $dn(u)$ are recognized as valuable for numerical work, it is now generally conceded that Weierstrass's functions $\wp(u)$ and $\sigma(u)$ — the latter function will be defined later — form the proper basis for the theory of elliptic functions.

It will be noticed that in one sense four infinities 0, $2\omega_1$, $2\omega_2$, $2\omega_1+2\omega_2$, occur in the parallelogram $(2\omega_1, 2\omega_2)$, one at each corner, but three of these must be regarded as belonging to adjacent parallelograms.

§ 196. The best insight into the nature of the properties of doubly periodic functions, with periods ω_1, ω_2, is probably obtained by the use of Liouville's methods. Liouville's theorems are referred to by their author in the Comptes Rendus, t. xix., and were communicated to Borchardt and Joachimsthal in 1847 (see Crelle, t. lxxxviii.). Cauchy anticipated Liouville in part. Our account of this theory is based on Briot and Bouquet.

We denote by A the region bounded by a parallelogram of periods, whose angular points are u_0, $u_0+\omega_1$, $u_0+\omega_1+\omega_2$, $u_0+\omega_2$, and denote by $f(u)$ a function with the periods ω_1, ω_2. The sides of the parallelogram need not be straight; so that, if required, poles of $f(u)$ can be removed from the rim by small deformations of the sides of the parallelogram, which conserve the parallelogram shape. It is evident that $f(u)$ is fully determined throughout the u-plane, when its values are known at every point within the parallelogram, and at every point along two of the sides. It will be shown presently that fewer conditions than these suffice to fix $f(u)$.

Suppose that c is a pole, whose order of multiplicity is κ; then, in the neighbourhood of c,

$$f(u)=\frac{a_\kappa}{(u-c)^\kappa}+\frac{a_{\kappa-1}}{(u-c)^{\kappa-1}}+\cdots+\frac{a_1}{u-c}+P(u-c)$$

$$=(u-c)^{-\kappa}P_0(u-c).$$

Here c counts as κ simple infinities, and $-\kappa$ is the residue of $\frac{f'(u)}{f(u)}$ at $u=c$, since

$$\frac{f'(u)}{f(u)}=-\frac{\kappa}{u-c}+P(u-c).$$

If $c'\equiv c(\text{mod. }\omega_1,\omega_2)$, the order of multiplicity of the pole c' is also κ. For, if h be sufficiently small,

$$f(c'+h)=f(c+h)=h^{-\kappa}P_0(h).$$

Suppose next that c is a zero, whose order of multiplicity is κ; then, in the neighbourhood of c,

$$f(u)=(u-c)^\kappa P_0(u-c),$$

$$\frac{f'(u)}{f(u)}=\frac{\kappa}{u-c}+P(u-c),$$

and κ is the residue of $\frac{f'(u)}{f(u)}$ at $u=c$. Similar reasoning to that already used shows that κ is the order of multiplicity of the zero at every point c' congruent to c.

Theorem I. A doubly periodic function, which is holomorphic in a parallelogram A, reduces to a constant.

For within a parallelogram A in the finite part of the plane, the absolute value of $f(u)$ is finite; therefore within all the congruent parallelograms it is finite. This shows that the function is finite throughout the plane, and is consequently a constant (§ 140). Thus every one-valued doubly periodic function $f(u)$ has *at least* one pole in the parallelogram A.

Comparison of two doubly periodic functions $f_1(u)$, $f_2(u)$, with the same periods, whose zeros and poles are common, and occur to the same orders of multiplicity.

Since $\dfrac{f_2(u)}{f_1(u)}$ is a doubly periodic function, and since the zeros and poles of the numerator are neutralized by the zeros and poles of the denominator, the quotient in question must have no infinities (and no zeros) in the parallelogram of periods. Therefore, by Theorem I., $\dfrac{f_2(u)}{f_1(u)}=$ a constant. Thus doubly periodic functions can be determined by their zeros and poles within the parallelogram. This theorem is remarkable as an instance of the determination of a function when certain properties are given.

Again, if $f_1(u)$, $f_2(u)$ be two doubly periodic functions with the same poles and with the same coefficients $a_1, a_2, \cdots, a_\kappa$ at the poles c, then $f_2(u)-f_1(u)$ is a doubly periodic function, which is everywhere finite, and

$$f_2(u)=f_1(u)+C.$$

Theorem II. The sum of the residues of $f(u)$, with respect to points inside A, is zero.

Let $\int f(u)du$ be taken round the rim of A. Its value is 0, because the function assumes the same values at congruent points of opposite sides, while those sides are described in opposite directions. Hence the sum of the residues is zero. An important inference is that there are *at least* two infinities of $f(u)$ in A. We have already proved that there must be at least one simple pole. If there were only one simple pole c, we should have $f(u)=a_1/(u-c)+g(u)$, where $g(u)$ is holomorphic in A; but we have proved that the sum of the residues in A is equal to 0, therefore $a_1=0$, contrary to supposition. Hence we have the following theorem:—

Theorem III. If a doubly periodic function $f(u)$ be not a constant, it has at least two poles of order 1, or one pole of order 2, within the parallelogram of periods.

The function $\wp(u)$ is an example of a doubly periodic function, with one pole of order 2. The number of infinities of $f(u)$ in A, or the sum of the orders of multiplicity of the poles, is called the order of the function $f(u)$. When a pole, situated on a side of A, is regarded as belonging to A, the congruent pole on the opposite side must be regarded as belonging to the adjacent parallelogram. If the pole occur at a vertex and be counted as belonging to A, the poles at the other vertices must be excluded.

An immediate deduction from the foregoing theorems is that every doubly periodic function of the *second* order is expressible in the form

$$f(u) = \frac{a}{u-\alpha} - \frac{a}{u-\beta} + g(u),$$

where α, β are the two poles, and $g(u)$ is holomorphic throughout A. Hence, if $f_1(u)$, $f_2(u)$ be functions of the second order with the same poles α, β, and residues a_1, a_2, so that

$$f_1(u) = \frac{a_1}{u-\alpha} - \frac{a_1}{u-\beta} + g_1(u),$$

$$f_2(u) = \frac{a_2}{u-\alpha} - \frac{a_2}{u-\beta} + g_2(u),$$

it follows that $\frac{1}{a_1}f_1(u) - \frac{1}{a_2}f_2(u)$ is a doubly periodic function holomorphic throughout A; it therefore reduces to a constant.

§ 197. *Theorem IV.* The order of $f(u)$ is equal to the number of zeros of $f(u)-k$, whatever be the value of the constant k.

Since $\quad f(u) = f(u+\omega_1) = f(u+\omega_2)$, it follows that

$$f'(u) = f'(u+\omega_1) = f'(u+\omega_2).$$

Hence $f'(u)/\{f(u)-k\}$ is a doubly periodic function. By Theorem II., the sum of its residues is 0. But the function $=\frac{d}{du}\log\{f(u)-k\}$; therefore the sum of its residues is equal to the number of the zeros of $f(u)-k$ diminished by the number of infinities of $f(u)-k$ (see § 146). But each infinity of $f(u)-k$ is an infinity of $f(u)$, and conversely. Therefore the order of $f(u)$ is the number of zeros of $f(u)-k$. The theorem states that a doubly periodic function $f(u)$, of order m, takes every value m times within A. In particular, the number of zeros of $f(u)$ is the same as the number of infinities.

Theorem V. If c_r' be a zero and c_r an infinity of $f(u)$, both situated within A,

$$\Sigma c_r - \Sigma c_r' \equiv 0.$$

More generally, if d_r be one of the points at which the function takes a given value k,

$$\Sigma c_r - \Sigma d_r \equiv 0.$$

To prove this, we make use of a theorem concerning residues. If $g(u)$ be one-valued throughout the finite part of the plane,

$$\int_A \frac{g(u)f'(u)du}{f(u)-k} = \Sigma \int_{(d_r)} \frac{g(u)f'(u)du}{f(u)-k} + \Sigma \int_{(c_r)} \frac{g(u)f'(u)du}{f(u)-k}$$

$$= \Sigma g(d_r)\int_{(d_r)} \frac{f'(u)du}{f(u)-k} + \Sigma g(c_r)\int_{(c_r)} \frac{f'(u)du}{f(u)-k}$$

$$= 2\pi i \Sigma\{g(d_r) - g(c_r)\}.$$

Let $g(u) = u$, and let u_0 be a corner of the parallelogram.

Then $$\int_A \frac{uf'(u)du}{f(u)-k} = \int_{u_0}^{u_0+\omega_1} \left\{\frac{uf'(u)du}{f(u)-k} - \frac{(u+\omega_2)f'(u+\omega_2)du}{f(u+\omega_2)-k}\right\}$$

$$+ \int_{u_0}^{u_0+\omega_2} \left\{\frac{(u+\omega_1)f'(u+\omega_1)du}{f(u+\omega_1)-k} - \frac{uf'(u)du}{f(u)-k}\right\},$$

or, by virtue of the periodicity of $f(u)$,

$$= -\omega_2 \int_{u_0}^{u_0+\omega_1} d\log[f(u)-k] + \omega_1 \int_{u_0}^{u_0+\omega_2} d\log[f(u)-k]$$

$$= -\omega_2 \log\frac{f(u_0+\omega_1)-k}{f(u_0)-k} + \omega_1 \log\frac{f(u_0+\omega_2)-k}{f(u_0)-k}$$

$$= 2\pi i(m_1\omega_1 + m_2\omega_2), \text{ where } m_1, m_2 \text{ are integers.}$$

Hence $\Sigma d_r - \Sigma c_r = m_1\omega_1 + m_2\omega_2$. This important result may be restated in the following way: —

The sum of the points at which $f(u)$ takes any assigned value is congruent to a constant.

Theorem VI. If c_1, c_2 be the poles, in A, of a function $f(u)$ *of the second order,*

$$f(u) = f(c_1 + c_2 - u).$$

To prove this, we observe that the points u_1, u_2, at which the function takes one and the same value, are connected by the relation

$$u_1 + u_2 \equiv c_1 + c_2;$$

therefore $$f(u_1) = f(c_1 + c_2 - u_1),$$

where u_1 can take all values.

From this equation it follows that

$$f'(c_1 + c_2 - u) = -f'(u).$$

Hence, when $u \equiv \frac{c_1 + c_2}{2}$, the derived function $f'(u)$ is either zero or infinite. The infinities of $f'(u)$ are the points c_1, c_2, each counted twice (§ 196). Thus $\frac{c_1 + c_2}{2}$ is a zero. The other zeros

$$\equiv \frac{c_1 + c_2 + \omega_1}{2}, \quad \frac{c_1 + c_2 + \omega_2}{2}, \quad \frac{c_1 + c_2 + \omega_1 + \omega_2}{2}.$$

If c_1, c_2 coincide, so that $f(u)$ has now a pole of the second order in A, $f'(u)$ becomes a function of the third order, with zeros congruent to $c_1 + \frac{\omega_1}{2}$, $c_1 + \frac{\omega_2}{2}$, $c_1 + \frac{\omega_1 + \omega_2}{2}$. If the origin be taken at a pole, the zeros are congruent to $\frac{\omega_1}{2}$, $\frac{\omega_2}{2}$, $\frac{\omega_1 + \omega_2}{2}$. The function $f'(u)$ now bears to $\wp'(u)$ a constant ratio, if the periods be the same.

The theorem shows that when the centroid of two non-congruent poles of an elliptic function of the second order is taken as origin, $f(u) = f(-u)$, and the function is even.

§ 198. *Theorem VII.* Two doubly periodic functions f_1, f_2, whose networks of periods have a common network, are connected by an algebraic equation.

Let m_1, m_2 be the orders of the functions in the smallest parallelogram formed by a pair of common periods ω_1, ω_2. Let

$$\phi(u) = \sum_{n_1=0}^{\mu_1} \sum_{n_2=0}^{\mu_2} a_{n_1 n_2} f_1^{n_1} f_2^{n_2} \quad . \quad . \quad . \quad . \quad . \quad . \quad . \quad . \quad . \quad . \quad \text{(i.)}$$

where μ_1, μ_2 are positive integers at our disposal, and the a's are constants. The function $\phi(u)$ is doubly periodic, with periods ω_1, ω_2, and with the same poles as those of f_1, f_2. If it be possible to choose μ_1, μ_2 so as to make all the infinities disappear, without reducing all the coefficients a, other than a_{00}, to zero, then $\phi(u)$ must be a constant, and an algebraic relation connects f_1. f_2.

Let c be a pole of $\phi(u)$; then, in the neighbourhood of c,

$$\phi(u)=\frac{\lambda_\kappa}{(u-c)^\kappa}+\frac{\lambda_{\kappa-1}}{(u-c)^{\kappa-1}}+\cdots+\frac{\lambda_1}{u-c}+P(u-c).$$

The coefficients $\lambda_\kappa, \lambda_{\kappa-1}, \cdots, \lambda_1$ are linear and homogeneous expressions in the coefficients a, as can be seen by substituting in (i.) the expansions of f_1, f_2, in powers of $u-c$. Equating to zero all the coefficients λ, we get as many equations as there are infinities of $\phi(u)$ in the parallelogram ω_1, ω_2; *i.e.* $\mu_1 m_1+\mu_2 m_2$. Now the number of a's is $(\mu_1+1)(\mu_2+1)$, and μ_1, μ_2 can always be chosen so as to make

$$(\mu_1+1)(\mu_2+1)>\mu_1 m_1+\mu_2 m_2;$$

therefore μ_1, μ_2 can always be chosen so as to make the number of the equations in the a's less than the number of the a's. This proves that there exist functions $\phi(u)$ which are nowhere infinite, and consequently reduce to constants. The number of these algebraic equations in f_1, f_2 is infinite, but the corresponding expressions must be composed of powers of a single irreducible expression and of irrelevant factors. In this unique irreducible expression $\mu_1 \leqq m_2$, $\mu_2 \leqq m_1$. For when f_2 is given, u has m_2 values, and therefore f_1 has, in general, m_2 values. But it may happen, for instance, that the m_2 values of u determine only $m_2/2$ values of f_1; the degree of the irreducible equation is then diminished. One or other of the equations

$$\mu_1=m_2, \ \mu_2=m_1,$$

must, however, hold good. [The above proof is Jordan's; see his Cours d'Analyse, t. ii., p. 335.]

Corollary. Let $f(u)$ be a doubly periodic function of order m, and let $f'(u)$ be its derivative. There exists an irreducible algebraic relation

$$\phi(f, f')=0$$

between $f(u)$ and $f'(u)$, whose degrees in f, f' are not greater than $2m, m$.

Since f' is an algebraic function of f, we must first determine the number of its branches, and thus discover the highest order to which f' occurs in the equation $\phi=0$. Suppose that $u=c$ is a pole of multiplicity κ. Then, in the neighbourhood of c, f and f' are of the forms

$$f(u)=(u-c)^{-\kappa}P_0(u-c),$$

$$f'(u)=(u-c)^{-(\kappa+1)}Q_0(u-c),$$

and all the poles of $f'(u)$ are coincident, in position, with the poles of $f(u)$. From this it follows that when f is finite, so is f'; whence the coefficient of the highest power of f' in ϕ must be a constant; that is, f' is an integral algebraic function of f in the sense of § 185.

Corresponding to that infinite value of f, which arises from $u=c$, there are κ expansions for f' in descending powers of $f^{1/\kappa}$, the initial term being $f^{(\kappa+1)/\kappa}$. To the $\kappa_1+\kappa_2+\cdots,=m$, infinities of $f(u)$ correspond $\kappa_1+\kappa_2+\cdots,=m$, branches of f'. Thus f' occurs to the order m in the equation $\phi(f,f')=0$.

To a given value of $f(u)$ correspond m values of u within A, say $u_1, u_2, \cdots, u_m$, and all the remaining values are congruent to $u_1, u_2, \cdots, u_m$. Hence, when f is given, the doubly periodic function f' takes m values $f'(u_1), f'(u_2), \cdots, f'(u_m)$. Any symmetric combination of the m branches $f'(u_1), \cdots, f'(u_m)$ of $\phi(f,f')=0$ is an integral function of f. Therefore

$$S_1 f'(u_r)=g_1(f),\ S_2 f'(u_r)=g_2(f),\ \cdots,\ S_m f'(u_r)=g_m(f),$$

where $S_\lambda f'(u_r)$ means the sum of the products of the m branches, λ at a time, and $g_1, g_2, \cdots, g_m$ denote polynomials. We have still to determine the orders of $g_1(f), g_2(f), \cdots, g_m(f)$ in f.

Since, when $f=\infty$, the order of each branch f' of f is of the form

$(\kappa+1)/\kappa,\ \leqq 2,\quad S_1 f'(u_r)$ is at most ∞^2 when $f=\infty$,

$S_2 f'(u_r)$ is at most ∞^4 when $f=\infty$,

. . . .,

$S_m f'(u_r)$ is at most ∞^{2m} when $f=\infty$.

Hence the equation $\phi(f,f')=0$ may be written

$$f'^m - g_1(f)f'^{m-1}+\cdots+(-1)^{m-1}g_{m-1}(f)f'+(-1)^m g_m(f)=0 \cdots \text{(ii.)},$$

where $g_\iota(f)$ contains f, at most, to the power $f^{2\iota}$.

The coefficient $g_{m-1}(f)$ vanishes identically; for

$$u_1+u_2+\cdots+u_m=\text{a constant},$$

therefore $\qquad du_1/df+du_2/df+\cdots+du_m/df=0,$

or $\qquad \Sigma\, 1/f'(u_\iota)=0,$

so that $\qquad g_{m-1}(f)/g_m(f)=0.$

§ 199. A special case of great importance is that in which $f(u)$ is of the second order. Then the equation (ii.) becomes

$$f'^2=\text{a quartic in } f.$$

The importance of this formula renders a more direct proof desirable. Let the poles c_1, c_2 of $f(u)$ be distinct; and let $(c_1+c_2)/2=c$. Consider the expression ψ, where

$$\psi(u)=[f(u)-f(c)][f(u)-f(c+\omega_1/2)]$$
$$[f(u)-f(c+\omega_2/2)][f(u)-f(c+\overline{\omega_1+\omega_2}/2)].$$

Evidently $\psi(u)$ is doubly periodic, with periods ω_1, ω_2; also the poles c_1, c_2 of f are poles of ψ, each of order 4. The four factors have zeros $u=c$, $c+\omega_1/2$, $c+\omega_2/2$, $c+(\omega_1+\omega_2)/2$, and each zero is of order 2, since it is a simple zero of the common derivative $f'(u)$ of the four factors. Hence ψ is an elliptic function with 2 fourfold poles and 4 double zeros. The function $\{f'(u)\}^2$ has the same periods, the same zeros, and the same poles. Hence $\psi(u)/\{f'(u)\}^2$ is a constant, and therefore $\{f'(u)\}^2=$ a quartic in $f(u)$.

When c_1 and c_2 coincide, we consider

$$\psi(u)=[f(u)-f(c+\omega_1/2)][f(u)-f(c+\omega_2/2)][f(u)-f(c+\overline{\omega_1+\omega_2}/2)],$$

and the quartic reduces to a cubic.

§ 200. Let $f(u)$ be an elliptic function of the second order, and let $F(u)$ be any elliptic function of order m, with the same periods. Then

$$F(u)=(\phi_m+\psi_{m-2}f')/\chi_m,$$

where $\phi_m, \psi_{m-2}, \chi_m$ are polynomials in f of degrees m, $m-2$, m.

Let us take as origin the centroid of the two poles of $f(u)$, which lie in A. Let the poles, in A, of $F(u)$ be $c_1, c_2, \cdots, c_m$. Let $F(u)$ be separated into an even and an odd part by means of the equations

$$F(u)+F(-u)=2F_1(u),$$
$$F(u)-F(-u)=2F_2(u).$$

The poles of $F_1(u)$ are both those of $F(u)$ and those of $F(-u)$; hence $F_1(u)$ is an elliptic function of order $2m$, with poles congruent to $\pm c_\kappa$ where $\kappa=1, 2, \cdots, m$. From § 197, $F_1(u)$ has also $2m$ zeros, say $\pm c_1', \pm c_2', \cdots, \pm c_m'$. The function

$$\prod_{\kappa=1}^{m}\{f(u)-f(c_\kappa')\}/\{f(u)-f(c_\kappa)\}$$

has the same periods, the same poles, and the same zeros as $F_1(u)$. Hence the ratio of these functions is constant, and we can write

$$F_1(u)=\phi_m(f)/\chi_m(f):$$

The function $F_2(u)$ has the same poles as $F_1(u)$, and has therefore $2m$ zeros. Four of these are $0, \omega_1/2, \omega_2/2, (\omega_1+\omega_2)/2$; and are the zeros of $f'(u)$; the remaining $2(m-2)$ zeros may be written $\pm c_1'', \pm c_2'', \cdots, \pm c_{m-2}''$. As before the function

$$\prod_{\kappa=1}^{m-2}\{f(u)-f(c_\kappa'')\}/\prod_{\kappa=1}^{m}\{f(u)-f(c_\kappa)\}$$

is in a constant ratio to $F_2(u)/f'(u)$, and we can write

$$F_2(u)/f'(u)=\psi_{m-2}(f)/\chi_m(f).$$

Therefore, finally,

$$F(u)=F_1(u)+F_2(u)=(\phi_m+\psi_{m-2}f')/\chi_m.$$

This theorem of Liouville's has been extended by Weierstrass to one-valued functions of p variables, which are $2p$-tuply periodic and have no essential singularity for finite values of the variables. Weierstrass proves that $p+1$ functions of this nature, which have a common parallelopiped of periods, are connected by an algebraic relation; and as a special case, that if $f(u_1, u_2, \cdots, u_p)$ and $\phi(u_1, u_2, \cdots, u_p)$ be two such functions, both functions are connected with $\delta f/\delta u_1, \delta f/\delta u_2, \cdots, \delta f/\delta u_p$ by algebraic relations, whence ϕ is a rational function of f and its p first derivatives (Crelle, t. lxxxix.; translated by Molk in Bulletin des Sc. Math., ser. 2, t. vi.).

§ 201. We now recur to the function $\wp u$, and the notation of § 195.*

In the neighbourhood of the origin we have

$$\frac{1}{(u-w)^2}=\frac{1}{w^2}\left(1-\frac{u}{w}\right)^{-2}=\frac{1}{w^2}+2\frac{u}{w^3}+3\frac{u^2}{w^4}+\cdots.$$

Now the absolutely convergent series $\Sigma'\frac{1}{w^{2n+1}}$ consists of pairs of opposite terms, and is therefore zero for all positive integers $n>1$. Hence, if $s_{2n}=\Sigma'\frac{1}{w^{2n}}$, we have from equations (1) and (2) of § 195,

$$\left.\begin{aligned}\wp u &= 1/u^2+3s_4u^2+5s_6u^4+\cdots\\ \wp' u &= -2/u^3+2\cdot 3s_4u+4\cdot 5s_6u^3+\cdots\end{aligned}\right\}\quad . \quad . \quad . \quad . \quad . \quad (5).$$

Now if we assume, from § 199, the relation

$$\wp'^2u-(4b\wp^3u+6c\wp^2u+4d\wp u+e)=0,$$

and determine the coefficients from equations (5), so that the relation holds when $u=0$, then the equation with these coefficients

*It will be convenient to write $\wp u$, $\wp^2 u$, etc., instead of $\wp(u)$, $[\wp(u)]^2$, etc., where no ambiguity arises. This is the custom in the case of such familiar functions as $\sin x$. Further, where only one argument is in question, we shall on occasion write $\wp$ for $\wp u$. The formulæ in Weierstrass's theory are from Schwarz, Formeln und Lehrsätze, and for the proofs we are largely indebted to Halphen, Fonctions Elliptiques.

will hold for all values of u. For since the expression on the left-hand side of the equation is a doubly periodic function, which has no infinities within the parallelogram of periods, it must reduce to a constant. This constant must be 0, otherwise the expression could not vanish for $u=0$.

Now $\wp'^2 u$ begins with $4/u^6$, therefore $b=1$. Equating to zero the coefficients of $1/u^4$, $1/u^2$, and the absolute term, we have

$$6c=0,$$

$$4\cdot 3^2 s_4 \quad +4d+2^3\cdot 3\cdot s_4=0,$$

$$4\cdot 3\cdot 5\, s_6 \ +e+2^4\cdot 5\cdot s_6=0,$$

so that $c=0$, $4d=-2^2\cdot 3\cdot 5\, s_4$, $e=-2^2\cdot 5\cdot 7\, s_6$. Instead of $4d$ and e, we shall write $-g_2$, $-g_3$, so that

$$\left.\begin{aligned} g_2&=2^2\cdot 3\cdot 5\,\Sigma'\frac{1}{w^4} \\ g_3&=2^2\cdot 5\cdot 7\,\Sigma'\frac{1}{w^6} \end{aligned}\right\} \quad \ldots \ldots \ldots \quad (6).$$

These are called the invariants of $\wp u$. The relation between $\wp u$ and its derivative is now

$$\wp'^2 u=4\wp^3 u-g_2\wp u-g_3 \quad \ldots \ldots \ldots \quad (7).$$

We know that the zeros of $\wp' u$ are ω_1, ω_2, $\omega_1+\omega_2$. Let the corresponding values of $\wp u$ be e_1, e_2, e_3. Thus

$$\left.\begin{aligned} &4\wp^3 u-g_2\wp u-g_3=4(\wp u-e_1)(\wp u-e_2)(\wp u-e_3) \\ \text{where}\quad &\wp(\omega_1)=e_1,\ \wp(\omega_2)=e_2,\ \wp(\omega_1+\omega_2)=e_3 \end{aligned}\right\} \quad . \quad (8);$$

and, by comparing the two sides of the equation, we get

$$\left.\begin{aligned} e_1+e_2+e_3&=0, \\ e_2e_3+e_3e_1+e_1e_2&=-g_2/4, \\ e_1e_2e_3&=g_3/4. \end{aligned}\right\} \quad \ldots \ldots \ldots \quad (9).$$

By the use of the series (5) we can express any coefficient in terms of g_2, g_3. From the relation

$$\wp'^2=4\wp^3-g_2\wp-g_3,$$

we derive, by differentiation,

$$2\wp''=12\wp^2-g_2,$$

$$\wp'''=12\wp\wp'.$$

This last relation is the best for our purpose. We have

$$\begin{cases} \wp u = 1/u^2 + 3\,s_4 u^2 + \cdots + (2\,n-1)s_{2n}u^{2n-2} + \cdots, \\ \wp' u = -2/u^3 + 2\cdot 3\,s_4 u + \cdots + (2\,n-1)(2\,n-2)\,s_{2n}u^{2n-3} + \cdots, \\ \wp''' u = -24/u^5 + 2\cdot 3\cdot 4\cdot 5\,s_6 u + \cdots + (2\,n-4)\,(2\,n-3)(2\,n-2) \\ \qquad\qquad\qquad\qquad\qquad (2\,n-1)\,s_{2n}u^{2n-5} + \cdots. \end{cases}$$

Hence, substituting in $\wp''' = 12\,\wp\wp'$ and equating coefficients,

$$\begin{aligned} &(2\,n-4)\,(2\,n-3)\,(2\,n-2)\,(2\,n-1)\,s_{2n} \\ &\quad = 12\{-2(2\,n-1)\,s_{2n} + 2\cdot 3(2\,n-5)\,s_4 s_{2n-4} + \cdots \\ &\quad + (2\,n-2)\,(2\,n-1)s_{2n} + 3(2\,n-6)\,(2\,n-5)\,s_4 s_{2n-4} + \cdots\} \\ &\quad = 12(2\,n-4)\,\{(2\,n-1)\,s_{2n} + 3(2\,n-5)\,s_4 s_{2n-4} + \cdots\}. \end{aligned}$$

Therefore

$$\begin{aligned} (n-3)(2\,n-1)\,(2\,n+1)\,s_{2n} = 3\{3(2\,n-5)\,s_4 s_{2n-4} \\ + \cdots + (2\,n-5)\cdot 3 s_{2n-4}s_4\}. \end{aligned}$$

[The right-hand expression is written in this form to avoid separate statements for n odd and n even.]

From this formula it is at once evident that all the expressions s_8, s_{10}, $\cdots$ are rational functions of s_4, s_6, and therefore, by (6), of g_2 and g_3.

Homogeneity. If by the notations $\wp(u\,|\,\omega_1, \omega_2)$ and $\wp(u\,;\,g_2, g_3)$ we call attention to the periods and invariants, then from (1) and (6)

$$\wp(u\,|\,\omega_1, \omega_2) = \wp(u\,;\,g_2, g_3) = \frac{1}{\mu^2}\wp\left(\frac{u}{\mu}\middle|\frac{\omega_1}{\mu}, \frac{\omega_2}{\mu}\right) = \frac{1}{\mu^2}\wp\left(\frac{u}{\mu}\,;\,\mu^4 g_2, \mu^6 g_3\right).$$

For, from its definition, $\wp(u\,|\,\omega_1, \omega_2)$ is a homogeneous function of degree -2 in u, ω_1, ω_2 (§ 195). Also, by equation (6), g_2 and g_3 are homogeneous in ω_1, ω_2, and of degrees -4, -6. Hence the equation

$$\wp'^2 = 4\,\wp^3 - g_2\wp - g_3$$

shows that $\wp'^2$ is homogeneous and of degree -6 in u, ω_1, ω_2, so that $\wp'$ is homogeneous and of degree -3 in u, ω_1, ω_2. This was to be expected from equation (2), § 195.

If in the expression of s_{2n} by means of g_2, g_3, a term $g_2^{\alpha}g_3^{\beta}$ enter, we have

$$g_2^{\alpha}g_3^{\beta} = \frac{1}{\mu^{2n}}\cdot\mu^{4\alpha+6\beta}g_2^{\alpha}g_3^{\beta},$$

so that $2\alpha + 3\beta = n$. This shows at once what terms can enter. In particular, when n is even, β must be even; and when n is a multiple of 3, so also is α. The series for $\wp u$ begins

$$\wp u = 1/u^2 + g_2 u^2/2^2 \cdot 5 + g_3 u^4/2^2 \cdot 7 + g_2^2 u^6/2^4 \cdot 3 \cdot 5^2 \\ + 3 g_2 g_3 u^8/2^4 \cdot 5 \cdot 7 \cdot 11 + \cdots \quad . \quad . \quad . \quad . \quad (10).$$

The vanishing of an invariant. There are special values of ω_2/ω_1 for which one of the invariants vanishes.

When the parallelogram of periods is a square, $\omega_2 = i\omega_1$. Hence, if w be a period, iw, $-w$, $-iw$ are also periods. When n is odd, $\frac{1}{w^{2n}} + \frac{1}{(iw)^{2n}}$ is zero, and all the series s_{4m+2} vanish. In particular s_6, and therefore g_3, is zero. The relation of homogeneity gives, on putting $\mu = i$,

$$\wp(u;\ g_2,\ 0) = -\wp\left(\frac{u}{i};\ g_2,\ 0\right),$$

or

$$\wp(u) = -\wp(iu).$$

Next let $\omega_2 = v\omega_1$, where v is an imaginary cube root of unity. The periods can be arranged in triads w, vw, v^2w. The sum

$$\frac{1}{w^{2n}} + \frac{1}{(vw)^{2n}} + \frac{1}{(v^2w)^{2n}}$$

vanishes, if n be not a multiple of 3. Thus $s_{6m\pm 2}$ is zero. In particular s_4, and therefore g_2, is zero. The relation of homogeneity gives, putting $\mu = v$,

$$\wp(u;\ 0,\ g_3) = \frac{1}{v^2}\wp\left(\frac{u}{v};\ 0,\ g_3\right)$$

or

$$\wp(vu) = v\wp(u).$$

This is the simplest case of complex multiplication of the argument.

§ 202. *The addition theorem.*

Suppose that the relation

$$\wp' u = a\wp u + b \quad . \quad . \quad . \quad . \quad . \quad . \quad . \quad . \quad . \quad . \quad (11)$$

holds when a and b are constants. We derive, from the differential equation (7), the relation

$$4\wp^3 - g_2\wp - g_3 = (a\wp + b)^2,$$

which gives three values for $\wp u$. Each of these will give two values of u in the parallelogram of periods, but of the six values so obtained only three satisfy (11). We shall find the sum of these three values, by using a theorem which is a very special case of Abel's Theorem.

Let $$f(\wp)=4\wp^3-g_2\wp-g_3-(a\wp+b)^2=0,$$

and let the roots of this equation be $\wp_1, \wp_2, \wp_3$, the corresponding values of u being u_1, u_2, u_3. When a, b undergo variations, the quantities u_1, u_2, u_3 and the roots $\wp_1, \wp_2, \wp_3$ also vary. Let $da, db, d\wp_r, du_r (r=1, 2, 3)$ be the corresponding variations. Then

$$f'(\wp_r)d\wp_r-2(a\wp_r+b)(\wp_r da+db)=0,$$

or $$\frac{d\wp_r}{a\wp_r+b}=\frac{2\wp_r da}{f'(\wp_r)}+\frac{2\,db}{f'(\wp_r)}.$$

Now, generally, if $f(\wp)$ be a polynomial of degree n, with zeros $\wp_1, \wp_2, \cdots, \wp_n$, and if $\phi(\wp)$ be another polynomial of degree m,

$$\frac{\phi(\wp)}{f(\wp)}=\sum_{r=1}^{n}\frac{\phi(\wp_r)}{f'(\wp_r)}\cdot\frac{1}{\wp-\wp_r}.$$

Multiplying by $\wp$ and then making $\wp$ infinite, we have

$$\Sigma\frac{\phi(\wp_r)}{f'(\wp_r)}=\lim\frac{\wp\phi(\wp)}{f(\wp)},$$

so that $$\Sigma\frac{\phi(\wp_r)}{f'(\wp_r)}=0,$$

when $m+1<n$. In the case under consideration $n=3$, and

$$\Sigma\frac{\wp_r}{f'(\wp_r)}=0,\quad \Sigma\frac{1}{f'(\wp_r)}=0;$$

therefore $$\Sigma\frac{d\wp_r}{a\wp_r+b}=0,$$

or $$\Sigma du_r=0,$$

or $$\Sigma u_r=\text{a constant}.$$

Suppose that initially $a=b=0$, and therefore that

$$\wp_1=e_1,\quad \wp_2=e_2,\quad \wp_3=e_3.$$

We can take for the corresponding values of u

$$u_1=\omega_1,\quad u_2=\omega_2,\quad u_3=\omega_1+\omega_2,$$

or any values congruent to these. Therefore the constant reduces to a period, and

$$u_1+u_2+u_3\equiv 0.$$

Eliminating a, b between the relations

$$\wp_r' = a\wp_r + b, \quad (r = 1, 2, 3)$$

we get

$$\begin{vmatrix} 1 & \wp_1 & \wp_1' \\ 1 & \wp_2 & \wp_2' \\ 1 & \wp_3 & \wp_3' \end{vmatrix} = 0 \quad \cdots \quad (12).$$

Again, $(\wp_2' - \wp_3')/(\wp_2 - \wp_3) = a$; and since

$$\wp_1 + \wp_2 + \wp_3 = a^2/4,$$

we get

$$\wp_1 + \wp_2 + \wp_3 = (\wp_2' - \wp_3')^2/4(\wp_2 - \wp_3)^2,$$

or, writing $u_2 = u$, $u_3 = v$, $u_1 = -(u+v)$, and noticing that $\wp u$ is an even function,

$$\wp(u+v) = \frac{1}{4}\left(\frac{\wp'u - \wp'v}{\wp u - \wp v}\right)^2 - \wp u - \wp v \quad \cdots \quad (13).$$

The relation (13) is known as the addition theorem of $\wp u$. Changing the sign of v, we have

$$\wp(u-v) = \frac{1}{4}\left(\frac{\wp'u + \wp'v}{\wp u - \wp v}\right)^2 - \wp u - \wp v \quad \cdots \quad (14).$$

The addition of this to the previous equation gives

$$\wp(u+v) + \wp(u-v) = \frac{1}{2}\frac{\wp'^2u + \wp'^2v}{(\wp u - \wp v)^2} - 2(\wp u + \wp v);$$

but $\quad \wp'^2u + \wp'^2v = 4(\wp^3u + \wp^3v) - g_2(\wp u + \wp v) - 2g_3,$

hence $\quad \wp(u+v) + \wp(u-v) = \dfrac{2(\wp u + \wp v)(\wp u\wp v - \frac{1}{4}g_2) - g_3}{(\wp u - \wp v)^2} \quad \cdot \quad (15).$

Again, the subtraction of (14) from (13) gives

$$\wp(u+v) - \wp(u-v) = -\frac{\wp'u\wp'v}{(\wp u - \wp v)^2} \quad \cdots \quad (16).$$

The equations (15) and (16) give

$$\wp(u+v) = \frac{(\wp u + \wp v)(2\wp u\wp v - \frac{1}{2}g_2) - g_3 - \wp'u\wp'v}{2(\wp u - \wp v)^2} \quad \cdot \quad (17).$$

Notation of periods. To give symmetry to our formulæ we shall usually denote by $2\omega_1$, $2\omega_2$, $2\omega_3$ three periods whose sum is zero, and which are such that one of the three pairs is a primitive pair.

It follows at once that all three pairs are primitive pairs. We shall use λ, μ, ν to denote the numbers 1, 2, 3 in any order. We have always

$$\omega_\lambda + \omega_\mu + \omega_\nu = 0.*$$

§ 203. In (13) put $\quad v = \omega_\lambda.$

Then $\quad \wp v = e_\lambda, \ \wp' v = 0,$

and
$$\wp(u+\omega_\lambda) = \frac{1}{4}\left(\frac{\wp' u}{\wp u - e_\lambda}\right)^2 - \wp u - e_\lambda;$$

or, since
$$\wp'^2 u = 4(\wp u - e_1)(\wp u - e_2)(\wp u - e_3),$$

$$\wp(u+\omega_\lambda) = \frac{(\wp u - e_\mu)(\wp u - e_\nu)}{\wp u - e_\lambda} - \wp u - e_\lambda$$

$$= \frac{e_\lambda \wp u + e_\mu e_\nu + e_\lambda^2}{\wp u - e_\lambda} = e_\lambda + \frac{2e_\lambda^2 + e_\mu e_\nu}{\wp u - e_\lambda}.$$

Therefore

$$\{\wp(u+\omega_\lambda) - e_\lambda\}\{\wp u - e_\lambda\} = (e_\mu - e_\lambda)(e_\nu - e_\lambda) \quad . \quad . \quad . \quad . \quad (18)$$

Again, in (13), let $\quad v = u.$

Then
$$\wp 2u + 2\wp u = \lim_{v=u} \frac{1}{4}\left(\frac{\wp' u - \wp' v}{\wp u - \wp v}\right)^2,$$

$$= \frac{1}{4}\left(\lim_{v=u} \frac{\wp' u - \wp' v}{\wp u - \wp v}\right)^2,$$

$$= \frac{1}{4}\left(\frac{\wp'' u}{\wp' u}\right)^2;$$

therefore
$$\wp 2u = \frac{(3\wp^2 - g_2/4)^2 - 2\wp(4\wp^3 - g_2\wp - g_3)}{4\wp^3 - g_2\wp - g_3}$$

$$= \frac{\wp^4 + \frac{1}{2}g_2\wp^2 + 2g_3\wp + \frac{1}{16}g_2^2}{4\wp^3 - g_2\wp - g_3}.$$

Now if we regard $4\wp^3 - g_2\wp - g_3$ as a quartic in $\wp$ whose leading coefficient is zero, so that one zero is at infinity and the others at e_1, e_2, e_3, the invariants of this quartic are g_2, g_3 (§ 42). Also the Hessian is (§ 41)

$$H = -(\wp^4 + \tfrac{1}{2}g_2\wp^2 + 2g_3\wp + \tfrac{1}{16}g_2^2);$$

hence
$$H = -\wp 2u \cdot \wp'^2 u \quad . \quad . \quad . \quad . \quad . \quad . \quad . \quad . \quad . \quad . \quad . \quad (19).$$

---

* When it is desirable to specify a primitive pair, we shall select $2\omega_1$, $2\omega_2$, in conformity with our previous notation. As Weierstrass selects $2\omega_1$, $2\omega_3$, the formulæ will occasionally diverge from those in the Formeln und Lehrsätze.

It is now proper to see how the Jacobian J is expressed in terms of $\wp$. Writing, for an instant, the quartic as a homogeneous function of two variables p, q, we have

$$J = \tfrac{1}{8}\begin{vmatrix} U_p & U_q \\ H_p & H_q \end{vmatrix},$$

or, since $\quad pU_p + qU_q = 4U$ and $pH_p + qH_q = 4H$,

$$J = \tfrac{1}{2}\begin{vmatrix} U_p & U \\ H_p & H \end{vmatrix}.$$

In our case $U = \wp'^2 u$, $U_p = \dfrac{\delta}{\delta\wp}(4\wp^3 - g_2\wp - g_3) = 2\wp''$,

$$H = -\wp 2u\wp'^2 u,\ H_p = \frac{\delta H}{\delta\wp} = -2\wp' 2u\wp' u - 2\wp 2u\wp'' u;$$

hence $$J = \begin{vmatrix} \wp'' u & \wp'^2 u \\ -\wp' 2u\cdot\wp' u - \wp 2u\cdot\wp'' u, & -\wp 2u\cdot\wp'^2 u \end{vmatrix} = \wp'^3 u\cdot\wp' 2u \qquad (20).$$

The well-known relation (Salmon, Higher Algebra, 4th ed., § 204)

$$J^2 = -4H^3 + g_2 HU^2 - g_3 U^3,$$

which connects the invariants and covariants of the quartic, follows at once from the identity

$$\wp'^2 2u = 4\wp^3 2u - g_2\wp 2u - g_3.$$

The zeros of the Jacobian.

Formula (20) shows that the Jacobian vanishes when $2u$ is half a primitive period; *i.e.* when $u = \omega_1/2,\ 3\omega_1/2,\ \omega_1/2 + \omega_2,\ 3\omega_1/2 + \omega_2,$ $\omega_2/2, 3\omega_2/2,\ \omega_2/2 + \omega_1,\ 3\omega_2/2 + \omega_1,\ (\omega_1 + \omega_2)/2,\ 3(\omega_1 + \omega_2)/2,\ (3\omega_1 + \omega_2)/2,$ $(\omega_1 + 3\omega_2)/2$. The corresponding six values of $\wp$ are obtained at once from (18), and are all included in the single formula $e_\lambda \pm \sqrt{(e_\mu - e_\lambda)(e_\nu - e_\lambda)}$, which agrees with § 39. We may add that the comparison of (18) with equations (iii.) of § 38 shows that to the points u, $u + \omega_\lambda$, where $\lambda = 1, 2, 3$, corresponds a quartic of the form $mU + H$ in the $\wp$-plane, and that the comparison of (19) with § 41 shows that the centroid of the quartic is $\wp 2u$.

✓ § 204. *The function ζu.*

The integration of the general term in (1) gives

$$-1/(u - w) - u/w^2.$$

The series composed of these terms becomes unconditionally convergent for every value of $u(\neq w)$, upon the addition of the constant $-1/w$ to the general term. For

$$1/(u-w) = -(1/w + u/w^2 + u^2/w^3 + \cdots),$$

so that

$$\Sigma'\{1/(u-w) + 1/w + u/w^2\}$$

is unconditionally convergent.

Adding the term $1/u$ in order to make the origin an infinity, we have an odd function $\zeta(u)$ or $\zeta(u \mid \omega_1, \omega_2)$:—

$$\left.\begin{aligned} \zeta(u) &= 1/u + \Sigma'(1/(u-w) + 1/w + u/w^2) \\ &= 1/u + \int_0^u (1/u^2 - \wp u)du, \\ \text{and} \quad d\zeta(u)/du &= -\wp u. \end{aligned}\right\} \quad (21).$$

By $\int_0^u (1/u^2 - \wp u)du$ is to be understood

$$\int_0^u - \left\{ \frac{g_2}{2^2 \cdot 5} u^2 + \frac{g_3}{2^2 \cdot 7} u^4 + \frac{g_2^2}{2^4 \cdot 3 \cdot 5^2} u^6 + \cdots \right\} du,$$

i.e.

$$-\left\{ \frac{g_2}{2^2 \cdot 5} \frac{u^3}{3} + \frac{g_3}{2^2 \cdot 7} \frac{u^5}{5} + \frac{g_2^2}{2^4 \cdot 3 \cdot 5^2} \frac{u^7}{7} + \cdots \right\},$$

a series with a finite circle of convergence; within that circle

$$\zeta(u) = 1/u - \frac{g_2}{2^2 \cdot 5} \cdot \frac{u^3}{3} - \frac{g_3}{2^2 \cdot 7} \frac{u^5}{5} - \cdots.$$

The first of the formulæ included in (21) defines $\zeta(u)$ for the whole u-plane as a one-valued function of u, which is ∞^1 at the points w and nowhere else. The second of the formulæ gives an element of $\zeta(u) - 1/u$; the extension to external points u can be effected by the addition theorem for $\zeta(u)$.

From the equation $\wp(u + 2\omega_\lambda) = \wp u$, we deduce

$$\frac{d}{du}\{\zeta(u + 2\omega_\lambda) - \zeta u\} = 0,$$

and, therefore, $\zeta(u + 2\omega_\lambda) - \zeta u =$ a constant.

Write $u = -\omega_\lambda$; therefore the constant $= 2\zeta\omega_\lambda$. If $\zeta\omega_\lambda$ be denoted by η_λ,

$$\left.\begin{aligned} \zeta(u + 2\omega_\lambda) &= \zeta u + 2\eta_\lambda, \\ \zeta\omega_\lambda &= \eta_\lambda. \end{aligned}\right\} \quad . \quad . \quad . \quad . \quad . \quad . \quad . \quad . \quad (22).$$

Equation (16), when integrated with respect to u, gives

$$\zeta(u+v)-\zeta(u-v)=-\wp'v/(\wp u-\wp v)+c.$$

To determine the constant, put $u=0$, $\wp u=\infty$. Then the left-hand side becomes

$$\zeta v-\zeta(-v) \text{ or } 2\zeta v.$$

Hence

$$\zeta(u+v)-\zeta(u-v)-2\zeta v=-\wp'v/(\wp u-\wp v) \quad \cdot \quad (23).$$

The interchange of u and v gives

$$\zeta(u+v)-\zeta(v-u)-2\zeta u=-\wp'u/(\wp v-\wp u).$$

Hence, by addition, we arrive at the *addition theorem*

$$\zeta(u+v)-\zeta u-\zeta v=(\wp'u-\wp'v)/2(\wp u-\wp v).$$

Example. Writing $u_1+u_2+u_3=0$, we may put the addition theorem in the form

$$\zeta u_1+\zeta u_2+\zeta u_3=\sqrt{\wp u_1+\wp u_2+\wp u_3}.$$

§ 205. *The expression of an elliptic function $f(u)$, of order m, in terms of the function ζ.*

Let $2\omega_1$, $2\omega_2$ be the periods of $f(u)$. Let c_κ be a pole of order m_κ; in the neighbourhood of c_κ

$$f(u)=\frac{a_{m_\kappa}^{(\kappa)}}{(u-c_\kappa)^{m_\kappa}}+\frac{a_{m_\kappa-1}^{(\kappa)}}{(u-c_\kappa)^{m_\kappa-1}}+\cdots+\frac{a_1^{(\kappa)}}{u-c_\kappa}+P(u-c_\kappa);$$

also (§ 196) $\Sigma a_1^{(\kappa)}=0$.

Now ζu has only one pole in the parallelogram, namely $u=0$; and since $\zeta u=1/u+P(u)$, the residue at this pole is 1. Hence $\zeta(u-c_\kappa)$, $\zeta'(u-c_\kappa)$, $\cdots$, $\zeta^{(m_\kappa-1)}(u-c_\kappa)$ are infinite at c_κ, like

$$1/(u-c_\kappa),\ -1/(u-c_\kappa)^2,\ \cdots,\ (-1)^{m_\kappa-1}(m_\kappa-1)!/(u-c_\kappa)^{m_\kappa}.$$

The function

$$f(u)-[a_1^{(\kappa)}\zeta(u-c_\kappa)-a_2^{(\kappa)}\zeta'(u-c_\kappa)$$
$$+\cdots+(-1)^{m_\kappa-1}a_{m_\kappa}^{(\kappa)}\zeta^{(m_\kappa-1)}(u-c_\kappa)/(m_\kappa-1)!]$$

has therefore no pole at c_κ, but it is not an elliptic function, since $\zeta(u-c_\kappa)$ is not elliptic. But if we remove in this way all the irreducible poles of $f(u)$ (*i.e.* all the poles within the parallelogram of periods), we arrive at the expression

$$f(u) - \sum_\kappa a_1^{(\kappa)}\zeta(u-c_\kappa) + \sum_\kappa a_2^{(\kappa)}\zeta'(u-c_\kappa) - \cdots,$$

which *is* an elliptic function; for $\zeta'(u-c_\kappa)$, $\zeta''(u-c_\kappa)$, ⋯, are elliptic functions, and $\Sigma a_1^{(\kappa)}\zeta(u-c_\kappa)$ is also an elliptic function, because

$$\begin{aligned}\Sigma a_1^{(\kappa)}\zeta(u-c_\kappa+2\omega_\lambda) &= \Sigma a_1^{(\kappa)}\{\zeta(u-c_\kappa)+2\eta_\lambda\}\\ &= \Sigma a_1^{(\kappa)}\zeta(u-c_\kappa),\end{aligned}$$

where $\lambda = 1, 2$.

Further, it has no infinities within the parallelogram of periods; hence it is a constant. Therefore

$$\begin{aligned}f(u) = a_0 + \Sigma a_1^{(\kappa)}\zeta(u-c_\kappa) - \Sigma a_2^{(\kappa)}\zeta'(u-c_\kappa) + \cdots\\ + \Sigma(-1)^{m_\kappa-1}a_{m_\kappa}^{(\kappa)}\zeta^{(m_\kappa-1)}(u-c_\kappa)/(m_\kappa-1)! \quad . \quad . \quad . \quad (24).\end{aligned}$$

Instead of ζ', ζ'', ⋯, we can of course write $-\wp$, $-\wp'$, ⋯. This theorem is assigned by Briot and Bouquet to Hermite. [Fonctions Elliptiques, p. 257; Halphen, t. i., p. 461.]

§ 206. *The function* σu.

The general term in (21) gives, on integration,

$$\begin{aligned}&\log(1-u/w) + u/w + u^2/2w^2,\\ &= -u^3/3w^3 - u^4/4w^4 - \cdots.\end{aligned}$$

The series just written down is convergent if $|u| < |w|$. Now however large be the finite quantity u, it only needs the exclusion of a *finite* number of values of w, to secure that, for every remaining value of w, the condition of convergence is satisfied. Hence the complete series $\Sigma'\{\log(1-u/w) + u/w + u^2/2w^2\}$ is convergent. By the use of exponentials, and by the introduction of u, we get the function $\sigma(u)$:—

$$\sigma u = u\Pi'(1-u/w)e^{u/w+u^2/2w^2} \quad . \quad . \quad . \quad . \quad . \quad . \quad . \quad . \quad (25),$$

where $\quad w = 2m_1\omega_1 + 2m_2\omega_2 \quad (m_1, m_2 = 0, \pm 1, \pm 2, \cdots)$.

The function $\sigma(u \mid \omega_1, \omega_2)$, or shortly σu, is connected with the functions $\zeta(u \mid \omega_1, \omega_2)$, $\wp(u \mid \omega_1, \omega_2)$, by the relations

$$\left.\begin{aligned}\frac{d}{du}\log\sigma u = \sigma'u/\sigma u = \zeta u\\ -\frac{d^2}{du^2}\log\sigma u = \wp u\end{aligned}\right\} \quad . \quad . \quad . \quad . \quad . \quad . \quad (26).$$

The function σu can be defined ab initio by means of formula (25), which affords an example of the theorem of § 151. The unconditional convergence of the product on the right-hand side of (25) follows from the theorem referred to.

The factor $\exp\,(u/w + u^2/2w^2)$ removes from the series for $\log\,(1 - u/w)$ the coefficients $\Sigma 1/w$ and $\Sigma 1/w^2$, which are not unconditionally convergent.

The function σu, defined in this way, is the one from which Weierstrass evolves the theory of elliptic functions. It is the simplest transcendental integral function with simple zeros situated at all points of the network. The most general function of the same nature, with the same zeros, is of the form $e^{G(u)} \cdot \sigma u$. See § 151. The reader will at once notice the resemblance of formula (25) to

$$\sin u = u \prod_{-\infty}^{\infty}{}' (1 - u/\kappa\pi) e^{u/\kappa\pi},$$

which was proved in § 152.

In obtaining the function σu by successive integrations of the function $\wp u$, we have followed Halphen. From the point of view of the elliptic integral, it seems that the function $\wp u$ is the one which presents itself most naturally.

§ 207. From the relation

$$\zeta(u + 2\,\omega_\lambda) - \zeta u = 2\,\eta_\lambda,$$

we deduce

$$\log \sigma(u + 2\,\omega_\lambda)/\sigma u = 2\,\eta_\lambda u + C,$$

$$\sigma(u + 2\,\omega_\lambda) = e^{2\eta_\lambda u + C} \cdot \sigma u.$$

To find e^C, write $u = -\,\omega_\lambda$;

therefore $$e^{-2\eta_\lambda \omega_\lambda + C} = -\,1,$$

and $$\sigma(u + 2\,\omega_\lambda) = -\,e^{2\eta_\lambda(u + \omega_\lambda)} \sigma u \quad \cdot \quad \cdot \quad \cdot \quad \cdot \quad \cdot \quad \cdot \quad (27).$$

Take a triad of periods $2\,\omega_\lambda$, $2\,\omega_\mu$, $2\,\omega_\nu$, such that

$$\omega_\lambda + \omega_\mu + \omega_\nu = 0.$$

By a repeated use of (27) we get

$$\sigma(u + 2\,\omega_\lambda + 2\,\omega_\mu + 2\,\omega_\nu) = -\,e^{2\eta_\lambda(u + \omega_\lambda + 2\omega_\mu + 2\omega_\nu) + 2\eta_\mu(u + \omega_\mu + 2\omega_\nu) + 2\eta_\nu(u + \omega_\nu)} \cdot \sigma u,$$

or $$\sigma u = -\,e^{2(\eta_\lambda + \eta_\mu + \eta_\nu)(u + \omega_\nu) + 2(\eta_\lambda \omega_\mu - \eta_\mu \omega_\lambda)} \cdot \sigma u.$$

Hence $$\eta_\lambda + \eta_\mu + \eta_\nu = 0$$

(a relation which can also be deduced from § 204);

and $$\eta_\lambda \omega_\mu - \eta_\mu \omega_\lambda = \begin{vmatrix} \eta_\lambda & \eta_\mu \\ \omega_\lambda & \omega_\mu \end{vmatrix} = (2r-1)\pi i/2 \quad \cdot \quad \cdot \quad (28),$$

when r is án integer as yet undetermined (see p. 320).

Again, from (27),

$$\sigma(u + 2m_1\omega_1) = -e^{2\eta_1(u+2m_1\omega_1-\omega_1)}\sigma(u + 2m_1\omega_1 - 2\omega_1)$$
$$= (-1)^{m_1} e^{2m_1\eta_1(u+m_1\omega_1)}\sigma u.$$

Therefore, if $\hat{\omega} = m_1\omega_1 + m_2\omega_2$, and $\hat{\eta} = m_1\eta_1 + m_2\eta_2$,

$$\sigma(u + 2\hat{\omega}) = (-1)^{m_1} e^{2m_1\eta_1(u+2m_2\omega_2+m_1\omega_1)}\sigma(u + 2m_2\omega_2)$$
$$= (-1)^{m_1+m_2} e^{2m_1\eta_1(u+\hat{\omega}+m_2\omega_2)+2m_2\eta_2(u+\hat{\omega}-m_1\omega_1)}\sigma u$$
$$= (-1)^{m_1m_2+m_1+m_2} e^{2\hat{\eta}(u+\hat{\omega})}\sigma u \quad \cdot \quad \cdot \quad \cdot \quad \cdot \quad (29).$$

From (23), by integration with regard to v,

$$\log\sigma(u+v) + \log\sigma(u-v) - 2\log\sigma v = \log(\wp u - \wp v) + \text{const},$$

so that

$$\sigma(u+v)\sigma(u-v)/\sigma^2 v = c(\wp v - \wp u),$$

where c is independent of v.

Let v tend to zero; the principal value of σv is v (formula 25), and that of $\wp v$ is $\frac{1}{v^2}$ (formula 1).

Hence $$\sigma^2(u)/v^2 = c/v^2,\ c = \sigma^2 u,$$

and $$\wp v - \wp u = \frac{\sigma(u+v)\sigma(u-v)}{\sigma^2 u \sigma^2 v} \quad \cdot \quad \cdot \quad \cdot \quad \cdot \quad \cdot \quad \cdot \quad \cdot \quad (30),$$

a formula of cardinal importance in Elliptic Functions.

As a special case we have, when $v = u$,

$$\sigma 2u/\sigma^4 u = -\wp' u \quad \cdot \quad \cdot \quad \cdot \quad \cdot \quad \cdot \quad \cdot \quad \cdot \quad \cdot \quad \cdot \quad \cdot \quad \cdot \quad (31).$$

Integral series for σu.

From the formula

$$\zeta u = 1/u - \frac{g_2}{2^2\cdot 5}\frac{u^3}{3} - \frac{g_3}{2^2\cdot 7}\frac{u^5}{5}\cdots,$$

combined with $\zeta u = \frac{d}{du}\log\sigma u$, it can easily be shown that $\sigma u = uP_0(u^2)$.

The actual values of the coefficients are best determined by the differential equation

$$\frac{\delta^2\sigma(u\,;\,g_2,g_3)}{\delta u^2}=12\,g_3\frac{\delta\sigma(u\,;\,g_2,g_3)}{\delta g_2}+\frac{2}{3}g_2{}^2\frac{\delta\sigma(u\,;\,g_2,g_3)}{\delta g_3}$$
$$-\tfrac{1}{12}\,g_2u^2\sigma(u\,;\,g_2,g_3).$$

[Other forms of this differential equation are given in a paper by Forsyth, Quar. Jour. t. xxii., p. 1, Halphen, t. i., p. 299; the formula is given in Schwarz's Formeln und Lehrsätze, p. 6.]

The series for σu is

$$\sigma u=u\left\{1-\frac{g_2u^4}{2\cdot 5!}-6\,g_3\frac{u^6}{7!}-\frac{9}{4}g_2{}^2\frac{u^8}{9!}-\frac{18\,g_2g_3}{11!}u^{10}\ldots\right\}\quad .\quad .\quad .\quad (32).$$

This series is convergent throughout the finite part of the plane.

For $$\sigma 2u=-\sigma^4u\wp' u=\sigma^4u\frac{d^3}{du^3}\log\sigma u$$

$$=\sigma^3u\sigma'''u-3\,\sigma^2u\sigma' u\sigma'' u+2\,\sigma u\sigma'^3u.$$

Assume that the series for σu converges throughout a circle (δ), centre 0 and radius δ. Since σu is, by the above formula, an integral function of $\sigma\frac{u}{2}$ and its derivatives, and since $\sigma\frac{u}{2}$ and its derivatives can be expressed as integral series throughout the domain given by $\left|\frac{u}{2}\right|<\delta$, it follows that σu can be expressed as an integral series *within a circle of radius* 2δ. By continued repetitions of this process it can be shown that the radius of convergence is infinitely great.

§ 208. *The functions* $\sigma_\lambda u$.

In $$\sigma(u+2\,\omega_\lambda)=-\,e^{2\eta_\lambda(u+\omega_\lambda)}\sigma u,$$

write u for $u+\omega_\lambda$; the resulting formula is

$$\sigma(u+\omega_\lambda)=-\,e^{2\eta_\lambda u}\sigma(u-\omega_\lambda),$$

whence $$e^{-\eta_\lambda u}\sigma(u+\omega_\lambda)=e^{\eta_\lambda u}\sigma(\omega_\lambda-u).$$

Dividing by $\sigma\omega_\lambda$, we arrive at an even function $\sigma_\lambda u$, allied to σu, which is defined by

$$\sigma_\lambda u=\frac{e^{-\eta_\lambda u}\sigma(u+\omega_\lambda)}{\sigma\omega_\lambda}=\frac{e^{\eta_\lambda u}\sigma(\omega_\lambda-u)}{\sigma\omega_\lambda}\quad .\quad .\quad .\quad (33).$$

The zeros of $\sigma_\lambda u$ are of the first order, and are situated at all the points congruent to ω_λ. The value of $\sigma_\lambda(0)$ is 1, whereas $\sigma'(0)=1$.

From (29) it follows that

$$\sigma(u+\omega_\lambda+2\bar{\omega})=(-1)^{m_1m_2+m_1+m_2}\sigma(u+\omega_\lambda)e^{2\bar{\eta}(u+\omega_\lambda+\bar{\omega})};$$

this equation when expressed in terms of σ_λ is

$$\sigma_\lambda(u+2\bar{\omega})=(-1)^{m_1m_2+m_1+m_2}\sigma_\lambda u\cdot e^{2\bar{\eta}(u+\bar{\omega})+2(\bar{\eta}\omega_\lambda-\bar{\omega}\eta_\lambda)}\quad . \quad (34).$$

Since $\bar{\eta}\omega_\lambda-\bar{\omega}\eta_\lambda=m_1(\eta_1\omega_\lambda-\eta_\lambda\omega_1)+m_2(\eta_2\omega_\lambda-\eta_\lambda\omega_2)$,

the factor $e^{2(\bar{\eta}\omega_\lambda-\bar{\omega}\eta_\lambda)}$ is, from (28), $(-1)^{m_2}$ when $\lambda=1$, $(-1)^{m_1}$ when $\lambda=2$, and $(-1)^{m_1+m_2}$ when $\lambda=3$.

In (30), write $v=\omega_\lambda$.

Then $$\wp u-e_\lambda=-\frac{\sigma(u+\omega_\lambda)\sigma(u-\omega_\lambda)}{\sigma^2u\sigma^2\omega_\lambda}=\left(\frac{\sigma_\lambda u}{\sigma u}\right)^2\quad . \quad . \quad . \quad (35).$$

Writing $\dfrac{\sigma_\lambda u}{\sigma u}=\xi_\lambda$, we have from (35) the differential equations

$$(d\xi_\lambda/du)^2=(\xi_\lambda^2+e_\lambda-e_\mu)(\xi_\lambda^2+e_\lambda-e_\nu)\quad . \quad . \quad . \quad (36).^*$$

Also from (35), $\sigma_\mu^2-\sigma_\nu^2=(e_\nu-e_\mu)\sigma^2$,

and $(e_\mu-e_\nu)\sigma_\lambda^2+(e_\nu-e_\lambda)\sigma_\mu^2+(e_\lambda-e_\mu)\sigma_\nu^2=0$ } . (37).

Multiplying together the three equations of the set (35), we have

$$\wp'^2u=4(\sigma_1u\sigma_2u\sigma_3u/\sigma^3u)^2.$$

In taking the square root, the sign is determined by supposing u small. The principal values of $\wp'u$, σu, $\sigma_\lambda u$, being $-2/u^3$, u, 1, the value of $\wp'u$ must be

$$\wp'u=-2\sigma_1u\sigma_2u\sigma_3u/\sigma^3u,$$

which, on comparison with (31), gives

$$\sigma(2u)=2\sigma u\sigma_1u\sigma_2u\sigma_3u\quad . \quad . \quad . \quad . \quad . \quad . \quad . \quad . \quad . \quad . \quad (38).$$

§ 209. The property of the σ-function that it has one irreducible zero and no infinity, combined with its one-valuedness, makes it a suitable element for the construction of the general elliptic function. For instance,

---

* These equations form a means of passage from Weierstrass's function $\wp u$ to Jacobi's functions snu, cnu, dnu. We find

$$\wp u-e_\lambda=(e_\nu-e_\lambda)/sn^2(\sqrt{e_\nu-e_\lambda}u,\kappa),$$

where $\kappa^2=(e_\mu-e_\lambda)/(e_\nu-e_\lambda)$;

so that if e_λ, e_μ, e_ν be real and e_μ lie between e_λ and e_ν, $\kappa^2<1$. The reader is referred to Weierstrass-Schwarz, p. 30.

$$f(u) = k \prod_{\kappa=1}^{m} \frac{\sigma(u - c_\kappa')}{\sigma(u - c_\kappa)} \exp 2(\mu_1\eta_1 + \mu_2\eta_2)u \quad . \quad . \quad (39),$$

where $\sum_{\kappa=1}^{m} (c_\kappa' - c_\kappa) = 2(\mu_1\omega_1 + \mu_2\omega_2)$, and μ_1, μ_2 are integers,

is the most general form for an elliptic function of the mth order, with zeros c_κ' and poles c_κ.

For from formula (27), when u becomes $u+2\omega_\lambda$, $\sigma(u-c_\kappa')/\sigma(u-c_\kappa)$ acquires the factor $\exp 2\eta_\lambda(c_\kappa - c_\kappa')$; therefore $f(u)$ acquires the

factor $\exp [2\eta_\lambda \Sigma(c_\kappa - c_\kappa') + 4(\mu_1\eta_1 + \mu_2\eta_2)\omega_\lambda]$

or $\exp [4\mu_1(\eta_1\omega_\lambda - \eta_\lambda\omega_1) + 4\mu_2(\eta_2\omega_\lambda - \eta_\lambda\omega_2)]$

or 1, by formula (28).

We can now generalize formula (30) as follows: —

Let Δ_n denote the determinant of n rows

$$\begin{vmatrix} 1, & \wp u_1, & \wp' u_1 \cdots \wp^{(n-2)} u_1 \\ 1, & \wp u_2, & \wp' u_2 \cdots \wp^{(n-2)} u_2 \\ \cdot & \cdot & \cdot \quad \cdot \quad \cdot \\ 1, & \wp u_n, & \wp' u_n \cdots \wp^{(n-2)} u_n \end{vmatrix},$$

where $u_1, u_2, \cdots, u_n$ lie within the parallelogram of periods A. This expression, regarded as a function of u_1, is doubly periodic, and infinite at $u = 0$ in the same way as $\wp^{(n-2)}u_1$; that is, in the same way as $(-)^n(n-1)!/u_1^n$. There are no other poles within A, and therefore Δ_n is an elliptic function of u_1, of the order n. Therefore it has n zeros whose sum = the sum of the poles, = 0. The determinant vanishes when $u_\iota = u_\kappa$; and therefore the zeros are $u_2, u_3, \cdots, u_n$, and $-(u_2 + u_3 + \cdots + u_n)$. Accordingly we can write

$$\Delta_n = k \frac{\sigma(u_1 - u_2)\sigma(u_1 - u_3)\cdots\sigma(u_1 - u_n)\sigma(u_1 + u_2 + \cdots + u_n)}{\sigma^n u_1},$$

where k is independent of u_1. Applying the same reasoning to the other arguments, we can write

$$\Delta_n / k_n = \frac{\sigma(\Sigma u_\kappa)\Pi\sigma(u_\iota - u_\kappa)}{\Pi\sigma^n u_\kappa},$$

where $\iota, \kappa = 1, 2, \cdots, n$, $\iota < \kappa$, and k_n is independent of all the arguments.

To determine k_n let both sides of this equation be multiplied by u_n^n, and then let $u_n = 0$. The effect on the right side is to change n into $n-1$, while the left side becomes $(-)^n(n-1)!\,\Delta_{n-1}/k_n$, where Δ_{n-1} is the determinant obtained by changing n into $n-1$. We have therefore

$$(-)^n(n-1)!\,\Delta_{n-1}/k_n = \Delta_{n-1}/k_{n-1},$$

or

$$k_n = (-)^n(n-1)!\,k_{n-1}.$$

Therefore $k_n = (-)^{n+n-1+\cdots+3}(n-1)!\,(n-2)!\cdots 2!\,k_2.$

From formula (30) $k_2 = 1$; therefore

$$\Delta_n = (-)^{(n-1)(n-2)/2}(n-1)!(n-2)!\cdots 2!\,\frac{\sigma(\Sigma u_\kappa)\Pi\sigma(u_\iota - u_\kappa)}{\Pi\sigma^n u_\kappa} \quad . \quad (40).$$

Example. Deduce the theorem (12) from this result.

§ 210. *The addition of half-periods to the arguments.*

Write, in $\sqrt{\wp u - e_\nu} = \dfrac{\sigma_\nu u}{\sigma u}$, $u = \omega_\mu$;

we get $\sqrt{e_\mu - e_\nu} = \sigma_\nu\omega_\mu/\sigma\omega_\mu.$

But $\sigma_\nu\omega_\mu = e^{-\eta_\nu\omega_\mu}\sigma(-\omega_\lambda)/\sigma\omega_\nu$;

hence $\sqrt{e_\mu - e_\nu} = -\,e^{-\eta_\nu\omega_\mu}\sigma\omega_\lambda/\sigma\omega_\mu\sigma\omega_\nu \quad . \quad . \quad . \quad . \quad . \quad (41).$

From the expressions for $\sqrt{e_\mu - e_\lambda}$ and $\sqrt{e_\nu - e_\lambda}$ we derive, by multiplication,

$$\sqrt{e_\mu - e_\lambda \cdot e_\nu - e_\lambda} = e^{-\eta_\lambda(\omega_\mu+\omega_\nu)}/\sigma^2\omega_\lambda = e^{\eta_\lambda\omega_\lambda}/\sigma^2\omega_\lambda.$$

There is no ambiguity of signs, because all functions such as $\sigma_\nu u/\sigma u$ are one-valued. The formulæ for the addition of a half-period are

$$\sigma(u+\omega_\lambda) = e^{\eta_\lambda u}\sigma\omega_\lambda\sigma_\lambda u,$$

$$\sigma_\lambda(u+\omega_\lambda) = -\,e^{\eta_\lambda(u+\omega_\lambda)}\sigma u/\sigma\omega_\lambda = -\sqrt{e_\mu - e_\lambda}\cdot\sqrt{e_\nu - e_\lambda}\cdot e^{\eta_\lambda u}\sigma\omega_\lambda\sigma u,$$

$$\sigma_\mu(u+\omega_\nu) = e^{-\eta_\mu(u+\omega_\nu)}\sigma(u-\omega_\lambda)/\sigma\omega_\mu = -\,e^{-\eta_\mu(u+\omega_\nu)}e^{-\eta_\lambda u}\sigma\omega_\lambda\sigma_\lambda u/\sigma\omega_\mu,$$

$$= \sqrt{e_\nu - e_\mu}\,e^{\eta_\nu u}\cdot\sigma\omega_\nu\sigma_\lambda u.$$

When u is replaced by $-u$ in these formulæ, σ changes into $-\sigma$ and σ_λ is unaffected. Hence

$$\left.\begin{aligned}\sigma(u\pm\omega_\lambda) &= \pm\,e^{\pm\eta_\lambda u}\sigma\omega_\lambda\sigma_\lambda u,\\ \sigma_\lambda(u\pm\omega_\lambda) &= \mp\sqrt{(e_\mu - e_\lambda)(e_\nu - e_\lambda)}\cdot e^{\pm\eta_\lambda u}\sigma\omega_\lambda\sigma u,\\ \sigma_\mu(u\pm\omega_\nu) &= \sqrt{e_\nu - e_\mu}\cdot e^{\pm\eta_\nu u}\sigma\omega_\nu\sigma_\lambda u\end{aligned}\right\} \quad . \quad . \quad (42).$$

From these, by division,

$$\frac{\sigma_\lambda(u+\omega_\lambda)}{\sigma_\lambda(u-\omega_\lambda)} = -e^{2\eta_\lambda u},$$

$$\frac{\sigma_\mu(u+\omega_\nu)}{\sigma_\mu(u-\omega_\nu)} = e^{2\eta_\nu u};$$

these formulæ are equivalent to special cases of (34), namely,

$$\left.\begin{aligned} \sigma_\lambda(u+2\omega_\lambda) &= -e^{2\eta_\lambda(u+\omega_\lambda)}\sigma_\lambda u, \\ \sigma_\mu(u+2\omega_\nu) &= e^{2\eta_\nu(u+\omega_\nu)}\sigma_\mu u \end{aligned}\right\} \quad . \quad . \quad . \quad . \quad . \quad (43).$$

We see from these equations and (27) that $\sigma_\lambda u/\sigma u$ and $\sigma_\mu u/\sigma_\nu u$ are doubly periodic functions, and that $2\omega_\lambda$, $4\omega_\mu$, $4\omega_\nu$ are primitive periods. These facts can be also deduced from (35).

Example. Prove that

$$\sigma(u+\omega_\mu-\omega_\nu) = (e_\mu - e_\nu)e^{(\eta_\mu-\eta_\nu)u}\sigma^2\omega_\mu\sigma^2\omega_\nu\sigma_\lambda u/\sigma\omega_\lambda \quad . \quad . \quad . \quad (44).$$

§ 211. *Addition theorems.*

Starting from the algebraic identity

$$(\wp u_2-\wp u_3)(\wp u-\wp u_1)+(\wp u_3-\wp u_1)(\wp u-\wp u_2)+(\wp u_1-\wp u_2)(\wp u-\wp u_3)=0,$$

introduce the σ-functions by means of the addition-theorem. If the common denominator be suppressed, there remains the remarkable "Equation with three terms": —

$$\left.\begin{aligned} &\sigma(u_2+u_3)\sigma(u_2-u_3)\sigma(u+u_1)\sigma(u-u_1) \\ &+\sigma(u_3+u_1)\sigma(u_3-u_1)\sigma(u+u_2)\sigma(u-u_2) \\ &+\sigma(u_1+u_2)\sigma(u_1-u_2)\sigma(u+u_3)\sigma(u-u_3)=0 \end{aligned}\right\} \quad . \quad . \quad . \quad (45).$$

This equation, with the condition $\sigma'0=1$, is in itself sufficient to define σu. This remark was made by Weierstrass (Sitzungsberichte der Berl. Akad., 1882) and verified by Delisle (Math. Ann., t. xxx.). See also Halphen, t. i., p. 188.

We proceed to some important transformations, given in Weierstrass-Schwarz, § 38. The results were discovered by Jacobi, as relations between theta-functions. The reader is referred to Enneper-Müller, § 19.

Write

$$u+u_1=a,\quad u-u_1=b,\quad u_2+u_3=c,\quad u_2-u_3=d,$$

$$u+u_2=a',\quad u-u_2=b',\quad u_3+u_1=c',\quad u_3-u_1=d',$$

$$u+u_3=a'',\quad u-u_3=b'',\quad u_1+u_2=c'',\quad u_1-u_2=d'',$$

so that

$$\left.\begin{array}{lll} a' = \tfrac{1}{2}(a+b+c+d), & a''= \tfrac{1}{2}(a'+b'+c'+d'), & a = \tfrac{1}{2}(a''+b''+c''+d'') \\ b' = \tfrac{1}{2}(a+b-c-d), & b''= \tfrac{1}{2}(a'+b'-c'-d'), & b = \tfrac{1}{2}(a''+b''-c''-d'') \\ c' = \tfrac{1}{2}(a-b+c-d), & c''= \tfrac{1}{2}(a'-b'+c'-d'), & c = \tfrac{1}{2}(a''-b''+c''-d'') \\ d' = \tfrac{1}{2}(-a+b+c-d), & d''= \tfrac{1}{2}(-a'+b'+c'-d'), & d = \tfrac{1}{2}(-a''+b''+c''-d'') \end{array}\right\} \quad (46),$$

and

$$\left.\begin{array}{lll} a''=\tfrac{1}{2}(a+b+c-d), & a=\tfrac{1}{2}(a'+b'+c'-d'), & a'=\tfrac{1}{2}(a''+b''+c''-d'') \\ b''=\tfrac{1}{2}(a+b-c+d), & b=\tfrac{1}{2}(a'+b'-c'+d'), & b'=\tfrac{1}{2}(a''+b''-c''+d'') \\ c''=\tfrac{1}{2}(a-b+c+d), & c=\tfrac{1}{2}(a'-b'+c'+d'), & c'=\tfrac{1}{2}(a''-b''+c''+d'') \\ d''=\tfrac{1}{2}(a-b-c-d), & d=\tfrac{1}{2}(a'-b'-c'-d'), & d'=\tfrac{1}{2}(a''-b''-c''-d'') \end{array}\right\} \cdot (47),$$

and also

$$a^2+b^2+c^2+d^2 = a'^2+b'^2+c'^2+d'^2 = a''^2+b''^2+c''^2+d''^2$$
$$= 2(u^2+u_1^2+u_2^2+u_3^2).$$

The equation (45) becomes

$$\sigma a\sigma b\sigma c\sigma d + \sigma a'\sigma b'\sigma c'\sigma d' + \sigma a''\sigma b''\sigma c''\sigma d'' = 0 \quad . \quad . \quad . \quad . \quad (48).$$

For a, b write $a+\omega_\lambda$, $b+\omega_\lambda$ $(\lambda = 1, 2, 3)$; after the removal of the factor $e^{2\eta_\lambda u}\cdot\sigma^2\omega_\lambda$, there remains

$$\sigma_\lambda a\sigma_\lambda b\sigma c\sigma d + \sigma_\lambda a'\sigma_\lambda b'\sigma c'\sigma d' + \sigma_\lambda a''\sigma_\lambda b''\sigma c''\sigma d'' = 0.$$

Write in (48), $a+\omega_\lambda$, $b-\omega_\mu$, $c-\omega_\nu$ for a, b, c, where λ, μ, ν are 1, 2, 3 in any order. After the removal of a factor there remains

$$\sigma_\lambda a\cdot\sigma_\mu b\cdot\sigma_\nu c\cdot\sigma d + \sigma_\lambda a'\cdot\sigma_\mu b'\cdot\sigma_\nu c'\cdot\sigma d' + \sigma_\lambda a''\cdot\sigma_\mu b''\cdot\sigma_\nu c''\cdot\sigma d'' = 0.$$

Write in (48), $a-\omega_\mu$, $b-\omega_\mu$, $c+\omega_\nu$, $d-\omega_\nu$ for a, b, c, d.

After the removal of a factor, there remains

$$\sigma_\mu a\sigma_\mu b\sigma_\nu c\sigma_\nu d - \sigma_\mu a'\sigma_\mu b'\sigma_\nu c'\sigma_\nu d' + (e_\mu - e_\nu)\sigma_\lambda a''\sigma_\lambda b''\sigma c''\sigma d'' = 0.$$

Write in (48), $a+\omega_\lambda+2\omega_\nu$, $b+\omega_\lambda$, $c+\omega_\lambda$, $d-\omega_\lambda$ for a, b, c, d. After removal of a factor, there remains

$$(e_\mu - e_\nu)\sigma_\lambda a\sigma_\lambda b\sigma_\lambda c\sigma_\lambda d + (e_\nu - e_\lambda)\sigma_\mu a'\sigma_\mu b'\sigma_\mu c'\sigma_\mu d'$$
$$+ (e_\lambda - e_\mu)\sigma_\nu a''\sigma_\nu b''\sigma_\nu c''\sigma_\nu d'' = 0.$$

Lastly, write in (48), $a+\omega_\lambda$, $b+\omega_\lambda$, $c+\omega_\lambda$, $d-\omega_\lambda$ for a, b, c, d. After removal of the factor $-e^{2\eta_\lambda a''}\sigma^4\omega_\lambda$, there remains

$$\sigma_\lambda a\sigma_\lambda b\sigma_\lambda c\sigma_\lambda d-\sigma_\lambda a'\sigma_\lambda b'\sigma_\lambda c'\sigma_\lambda d'+(e_\lambda-e_\mu)(e_\lambda-e_\nu)\sigma a''\sigma b''\sigma c''\sigma d''=0.$$

Collecting these, we have the table

$$\left\{\begin{array}{lr}
\sigma a\sigma b\sigma c\sigma d+\sigma a'\sigma b'\sigma c'\sigma d'+\sigma a''\sigma b''\sigma c''\sigma d''=0 \quad \cdots & (48),\\
\sigma_\lambda a\sigma_\lambda b\sigma c\sigma d+\sigma_\lambda a'\sigma_\lambda b'\sigma c'\sigma d'+\sigma_\lambda a''\sigma_\lambda b''\sigma c''\sigma d''=0 \quad \cdots & (49),\\
\sigma_\lambda a\sigma_\mu b\sigma_\nu c\sigma d+\sigma_\lambda a'\sigma_\mu b'\sigma_\nu c'\sigma d'+\sigma_\lambda a''\sigma_\mu b''\sigma_\nu c''\sigma d''=0 \quad \cdots & (50),\\
\sigma_\mu a\sigma_\mu b\sigma_\nu c\sigma_\nu d-\sigma_\mu a'\sigma_\mu b'\sigma_\nu c'\sigma_\nu d'+(e_\mu-e_\nu)\sigma_\lambda a''\sigma_\lambda b''\sigma c''\sigma d''=0 & (51),\\
(e_\mu-e_\nu)\sigma_\lambda a\sigma_\lambda b\sigma_\lambda c\sigma_\lambda d+(e_\nu-e_\lambda)\sigma_\mu a'\sigma_\mu b'\sigma_\mu c'\sigma_\mu d' & \\
\qquad +(e_\lambda-e_\mu)\sigma_\nu a''\sigma_\nu b''\sigma_\nu c''\sigma_\nu d''=0 \quad \cdots & (52),\\
\sigma_\lambda a\sigma_\lambda b\sigma_\lambda c\sigma_\lambda d-\sigma_\lambda a'\sigma_\lambda b'\sigma_\lambda c'\sigma_\lambda d'+(e_\lambda-e_\mu)(e_\lambda-e_\nu)\sigma a''\sigma b''\sigma c''\sigma d''=0 & (53).
\end{array}\right.$$

If in these we put

$$a=0,\quad b=0,\quad c=u+v,\quad d=u-v,$$

so that $a'=u,\quad b'=-u,\quad c'=v,\quad d'=v,$

$$a''=v,\quad b''=-v,\quad c''=u,\quad d''=-u,$$

then we have

$$\sigma(u+v)\sigma(u-v)+\sigma_\lambda{}^2u\sigma^2v-\sigma_\lambda{}^2v\sigma^2u=0,$$

$$\sigma_\nu(u+v)\sigma(u-v)+\sigma_\lambda u\sigma_\mu u\sigma_\nu v\sigma v-\sigma_\lambda v\sigma_\mu v\sigma_\nu u\sigma u=0,$$

$$\sigma_\nu(u+v)\sigma_\nu(u-v)-\sigma_\mu{}^2u\sigma_\nu{}^2v-(e_\mu-e_\nu)\sigma_\lambda{}^2v\sigma^2u=0,$$

$$(e_\mu-e_\nu)\sigma_\lambda(u+v)\sigma_\lambda(u-v)+(e_\nu-e_\lambda)\sigma_\mu{}^2u\sigma_\mu{}^2v+(e_\lambda-e_\mu)\sigma_\nu{}^2u\sigma_\nu{}^2v=0,$$

$$\sigma_\lambda(u+v)\sigma_\lambda(u-v)-\sigma_\lambda{}^2u\sigma_\lambda{}^2v+(e_\lambda-e_\mu)(e_\lambda-e_\nu)\sigma^2u\sigma^2v=0.$$

Again, if we put

$$a''=0,\quad b''=0,\quad c''=u+v,\quad d''=u-v,$$

so that $a=u,\quad b=-u,\quad c=v,\quad d=v,$

$$a'=v,\quad b'=-v,\quad c'=u,\quad d'=-u,$$

we obtain only one new formula, from (51),

$$\sigma_\mu{}^2u\sigma_\nu{}^2v-\sigma_\mu{}^2v\sigma_\nu{}^2u+(e_\mu-e_\nu)\sigma(u+v)\sigma(u-v)=0.$$

If we put $a=0,\quad b=u+v,\quad c=u-v,\quad d=0,$

so that $a'=u,\quad b'=v,\quad c'=-v,\quad d'=u,$

$$a''=u,\quad b''=v,\quad c''=-v,\quad d''=-u,$$

we obtain only one new formula, again from (51),

$$\sigma_\mu(u+v)\sigma_\nu(u-v)-\sigma_\mu u\sigma_\mu v\sigma_\nu u\sigma_\nu v+(e_\mu-e_\nu)\sigma_\lambda u\sigma_\lambda v\sigma u\sigma v=0.$$

We have, then, the special table:

$$\sigma(u+v)\sigma(u-v)=\sigma^2u\sigma_\lambda{}^2v-\sigma^2v\sigma_\lambda{}^2u \quad (54),$$

$$(e_\nu-e_\mu)\sigma(u+v)\sigma(u-v)=\sigma_\mu{}^2u\sigma_\nu{}^2v-\sigma_\mu{}^2v\sigma_\nu{}^2u \quad (55),$$

$$\sigma_\lambda(u+v)\sigma_\lambda(u-v)=\sigma_\lambda{}^2u\sigma_\lambda{}^2v-(e_\lambda-e_\mu)(e_\lambda-e_\nu)\sigma^2u\sigma^2v \quad (56),$$

$$(e_\nu-e_\mu)\sigma_\lambda(u+v)\sigma_\lambda(u-v)=(e_\nu-e_\lambda)\sigma_\mu{}^2u\sigma_\mu{}^2v+(e_\lambda-e_\mu)\sigma_\nu{}^2u\sigma_\nu{}^2v \quad (57),$$

$$\sigma_\lambda(u+v)\sigma_\lambda(u-v)=\sigma_\mu{}^2u\sigma_\lambda{}^2v-(e_\lambda-e_\mu)\sigma^2u\sigma_\nu{}^2v \quad (58),$$

$$\sigma_\lambda(u+v)\sigma_\lambda(u-v)=\sigma_\lambda{}^2u\sigma_\mu{}^2v-(e_\lambda-e_\mu)\sigma_\nu{}^2u\sigma^2v \quad (59),$$

$$\sigma_\lambda(u+v)\sigma(u-v)=\sigma_\lambda u\sigma u\sigma_\mu v\sigma_\nu v-\sigma_\mu u\sigma_\nu u\sigma_\lambda v\sigma v \quad (60),$$

$$\sigma(u+v)\sigma_\lambda(u-v)=\sigma_\lambda u\sigma u\sigma_\mu v\sigma_\nu v+\sigma_\mu u\sigma_\nu u\sigma_\lambda v\sigma v \quad (61),$$

$$\sigma_\mu(u+v)\sigma_\nu(u-v)=\sigma_\mu u\sigma_\nu u\sigma_\mu v\sigma_\nu v-(e_\mu-e_\nu)\sigma_\lambda u\sigma u\sigma_\lambda v\sigma v \quad (62).$$

Here (59) is obtained from (58) by an interchange of u and v; and (61) is obtained from (60) by writing $-v$ for v.

§ 212. *Multiplication of the arguments of* $\wp u$, σu.

It is evident, *a priori*, that $\wp nu$, where n is an integer, is a rational algebraic function of $\wp u$. For since $\wp nu$ is an elliptic function with the same periods as $\wp u$, it is rationally expressible in terms of $\wp u$ and $\wp' u$. And it cannot contain $\wp'$ to odd powers, since $\wp nu$ is an even function. Hence $\wp nu$ is rationally expressible in $\wp u$ alone.

It is easy to express σnu in terms of σu, $\sigma 2u$, $\sigma 3u$, $\sigma 4u$. In the equation with three terms, put $u=0$, $u_1=u$, $u_2=mu$, $u_3=nu$, where m and n are integers. Then

$$-\sigma(m+n)u\sigma(m-n)u\sigma^2u-\sigma(n+1)u\sigma(n-1)u\sigma^2mu$$
$$+\sigma(m+1)u\sigma(m-1)u\sigma^2nu=0 \quad (63).$$

Let $m=n+1$. Then

$$\sigma(2n+1)u\sigma^3u=\sigma(n+2)u\sigma^3nu-\sigma(n-1)u\sigma^3(n+1)u \quad (64).$$

Replace m, n by $n+1$, $n-1$. Then

$$\sigma 2nu\sigma 2u\sigma^2u=\sigma nu\{\sigma(n+2)u\sigma^2(n-1)u-\sigma(n-2)u\sigma^2(n+1)u\} \quad (65).$$

These equations give $\sigma 5u$, $\sigma 7u$, $\cdots$, and $\sigma 6u$, $\sigma 8u$, $\cdots$, in terms of σu, $\sigma 2u$, $\sigma 3u$, $\sigma 4u$. We have already proved that

$$\sigma 2u=2\sigma\sigma_1\sigma_2\sigma_3.$$

Write $v=u$ in (57), and use $e_\nu-e_\mu=(\sigma_\mu{}^2-\sigma_\nu{}^2)/\sigma^2$; the resulting equation is

$$(\sigma_\mu{}^2-\sigma_\nu{}^2)\sigma_\lambda 2u=(\sigma_\lambda{}^2-\sigma_\nu{}^2)\sigma_\mu{}^4+(\sigma_\mu{}^2-\sigma_\lambda{}^2)\sigma_\nu{}^4.$$

Therefore $\sigma_\lambda 2u = -\sigma_\mu{}^2\sigma_\nu{}^2 + \sigma_\nu{}^2\sigma_\lambda{}^2 + \sigma_\lambda{}^2\sigma_\mu{}^2.$

In (54), write $2u$, u for u, v. Then

$$\sigma 3u \cdot \sigma = \sigma^2 2u\sigma_\lambda{}^2 u - \sigma_\lambda{}^2 2u\sigma^2 u$$
$$= \sigma^2\{4\sigma_\lambda{}^4\sigma_\mu{}^2\sigma_\nu{}^2 - (-\sigma_\mu{}^2\sigma_\nu{}^2 + \sigma_\nu{}^2\sigma_\lambda{}^2 + \sigma_\lambda{}^2\sigma_\mu{}^2)^2\};$$

therefore

$$\sigma 3u = \sigma(\sigma_2\sigma_3 + \sigma_3\sigma_1 + \sigma_1\sigma_2)(-\sigma_2\sigma_3 + \sigma_3\sigma_1 + \sigma_1\sigma_2)(\sigma_2\sigma_3 - \sigma_3\sigma_1 + \sigma_1\sigma_2)(\sigma_2\sigma_3 + \sigma_3\sigma_1 - \sigma_1\sigma_2).$$

The formula for $\sigma 4u$ in terms of σ, σ_1, σ_2, σ_3 can be obtained from $\sigma 4u = 2\sigma 2u\sigma_1 2u\sigma_2 2u\sigma_3 2u.$

Example. Prove that, if $\tau_\lambda = \sigma_\lambda 2u$,

$$\sigma 3u = \sigma(\tau_2\tau_3 + \tau_3\tau_1 + \tau_1\tau_2),$$

$$\sigma_\lambda 3u = \sigma_\lambda(-\tau_\mu\tau_\nu + \tau_\nu\tau_\lambda + \tau_\lambda\tau_\mu).$$

To express $\wp nu$ in terms of $\wp u$, we have

$$\wp nu - \wp u = -\frac{\sigma(n+1)u \cdot \sigma(n-1)u}{\sigma^2 nu \cdot \sigma^2 u}.$$

If we write $\psi_r = \sigma ru/\sigma^{r^2}u$,

then
$$\wp nu - \wp u = -\frac{\psi_{n+1}\psi_{n-1}}{\psi_n{}^2} \quad \ldots\ldots\ldots \quad (66).$$

Again, the equation (63) becomes

$$\psi_{m+n}\psi_{m-n} = \psi_{m+1}\psi_{m-1}\psi_n{}^2 - \psi_{n+1}\psi_{n-1}\psi_m{}^2,$$

the factor $\sigma^{2(m^2+n^2+1)}$ dividing out.

Noting that $\psi_1 = 1$, we infer, as on the preceding page, that

$$\left.\begin{aligned} \psi_{2n+1} &= \psi_{n+2}\psi_n{}^3 - \psi_{n-1}\psi_{n+1}{}^3, \\ \psi_{2n}\psi_2 &= \psi_n(\psi_{n+2}\psi_{n-1}{}^2 - \psi_{n-2}\psi_{n+1}{}^2) \end{aligned}\right\} \quad \ldots \quad (67).$$

We have now to express ψ_2, ψ_3, ψ_4 in terms of $\wp$.

We have already $\psi_2 = -\wp'$, from (31).

Again, $\wp 2u - \wp u = -\psi_3/\wp'^2,$

and from (19) $\wp 2u = -H/\wp'^2;$

therefore
$$\psi_3 = H + \wp\wp'^2,$$
$$= 3\wp^4 - \tfrac{3}{2}g_2\wp^2 - 3g_3\wp - \tfrac{1}{16}g_2{}^2.$$

If in formula (16) we replace u and v by $2u$ and u, we obtain

$$\wp 3u - \wp u = -\frac{\wp' 2u \cdot \wp' u}{(\wp 2u - \wp u)^2}.$$

The left-hand expression is $\psi_4 \wp' / \psi_3^2$, and the right-hand one is $-J\wp'^2/\psi_3^2$, by formula (20). Hence

$$\psi_4 = -J\wp'.$$

The equations determine ψ_n, when $n > 4$, in terms of $\psi_1, \psi_2, \psi_3, \psi_4$ (Halphen, t. i., p. 100).

§ 213. *Expression of σu as a singly infinite product.*

The doubly infinite product in (25) is absolutely convergent, and the factors may be taken in any order. The factors for which $m_2 = 0$ give

$$u\Pi'(1 - u/2m_1\omega_1) \exp(u/2m_1\omega_1 + u^2/8m_1^2\omega_1^2),\ m_1 = \pm 1, \pm 2, \cdots.$$

Now $$\prod_{\mu=-\infty}^{\infty}{}' (1 - x/\mu) \exp(x/\mu) = (\sin \pi x)/\pi x;$$

therefore $$u\Pi'(1 - u/2m_1\omega_1) \exp(u/2m_1\omega_1) = \frac{2\omega_1}{\pi} \sin \frac{\pi u}{2\omega_1};$$

also $$\Pi' e^{u^2/8m_1^2\omega_1^2} = e^{u^2/4\omega_1^2 \cdot (\sum_1^{\infty} 1/m^2)} = e^{\pi^2 u^2/24\omega_1^2}.$$

Hence the selected factors give

$$2\omega_1/\pi \cdot \sin \pi u/2\omega_1 \cdot \exp(\pi^2 u^2/24\omega_1^2) \quad \cdot \quad \cdot \quad \cdot \quad \cdot \quad (A).$$

Take next all the factors for which m_2 has a given value other than zero. These give

$$\exp \tfrac{1}{2} \sum_{m_1=-\infty}^{\infty} u^2/4(m_1\omega_1 + m_2\omega_2)^2 \cdot$$
$$\prod_{m_1=-\infty}^{\infty} [\{1 - u/2(m_1\omega_1 + m_2\omega_2)\} \exp u/2(m_1\omega_1 + m_2\omega_2)] \quad \cdot \quad (B).$$

Now $$\sum_{\mu=-\infty}^{\infty} 1/(\mu + x)^2 = \pi^2 \csc^2 \pi x,$$

so that $$\sum_{m_1=-\infty}^{\infty} 1/(m_1\omega_1 + m_2\omega_2)^2 = \frac{\pi^2}{\omega_1^2} \csc^2 \pi m_2\omega_2/\omega_1.$$

Thus (B) begins with the factor

$$\exp\{\pi^2 u^2/8\omega_1^2 \cdot \csc^2 \pi m_2\omega_2/\omega_1\}.$$

We can bring the remaining factors of (B) into the domain of the sine-formula, by writing

$$1 - u/2(m_1\omega_1 + m_2\omega_2) = \frac{1 + (2m_2\omega_2 - u)/2m_1\omega_1}{1 + m_2\omega_2/m_1\omega_1},$$

and by introducing, for each factor of the form $(1+x/m_1)$, the exponential $\exp(-x/m_1)$. Thus we write

$$\{1-u/2(m_1\omega_1+m_2\omega_2)\}\exp\{u/2(m_1\omega_1+m_2\omega_2)\}$$

$$=\frac{\{1+(2m_2\omega_2-u)/2m_1\omega_1\}\exp\{-(2m_2\omega_2-u)/2m_1\omega_1\}}{\{1+m_2\omega_2/m_1\omega_1\}\cdot\exp\{-m_2\omega_2/m_1\omega_1\}}e^{\frac{u}{2(m_1\omega_1+m_2\omega_2)}-\frac{u}{2m_1\omega_1}}.$$

This being done for all values of m_1, except 0, we have for the factor series in (B)

$$(1-u/2m_2\omega_2)\cdot\exp(u/2m_2\omega_2)\cdot\frac{\sin\pi(2m_2\omega_2-u)/2\omega_1}{\pi(2m_2\omega_2-u)/2\omega_1}\cdot\frac{\pi m_2\omega_2/\omega_1}{\sin\pi m_2\omega_2/\omega_1}$$
$$\cdot\exp\left[\frac{u}{2}\sum_{-\infty}^{\infty}{}'\left\{\frac{1}{m_1\omega_1+m_2\omega_2}-\frac{1}{m_1\omega_1}\right\}\right]\quad\cdot\quad\cdot\quad\cdot\quad\text{(C)}.$$

But
$$\sum_{-\infty}^{\infty}{}'\left(\frac{1}{\mu+x}-\frac{1}{\mu}\right)=\pi\cot\pi x-1/x;$$

therefore
$$\sum_{-\infty}^{\infty}{}'\left(\frac{1}{m_1\omega_1+m_2\omega_2}-\frac{1}{m_1\omega_1}\right)=\frac{1}{\omega_1}\{\pi\cot\pi m_2\omega_2/\omega_1-\omega_1/m_2\omega_2\}.$$

The expression (C) is

$$\frac{\sin\pi(2m_2\omega_2-u)/2\omega_1}{\sin\pi m_2\omega_2/\omega_1}\cdot\exp\left\{\frac{\pi u}{2\omega_1}\cdot\cot\frac{\pi m_2\omega_2}{\omega_1}\right\}.$$

By compounding (A) and (B) we get

$$\sigma u=\frac{2\omega_1}{\pi}\cdot\sin\frac{\pi u}{2\omega_1}\exp\left[\frac{\pi^2u^2}{4\omega_1^2}\left(\frac{1}{6}+\frac{1}{2}\sum_{-\infty}^{\infty}{}'\csc^2\frac{\pi m_2\omega_2}{\omega_1}\right)\right]$$
$$\times\prod_{-\infty}^{\infty}{}'\left[\frac{\sin\pi(2m_2\omega_2-u)/2\omega_1}{\sin\pi m_2\omega_2/\omega_1}\exp\left\{\frac{\pi u}{2\omega_1}\cot\frac{\pi m_2\omega_2}{\omega_1}\right\}\right];$$

or pairing off opposite values of m_2, and using the elementary formula

$$\sin(x+y)\sin(x-y)=\sin^2x-\sin^2y,$$

$$\sigma u=\frac{2\omega_1}{\pi}\cdot\sin\frac{\pi u}{2\omega_1}\cdot e^{\frac{\eta_1u^2}{2\omega_1}}\prod_{m_2=1}^{\infty}\left[1-\frac{\sin^2\frac{\pi u}{2\omega_1}}{\sin^2\frac{\pi m_2\omega_2}{\omega_1}}\right]\quad\cdot\quad\text{(68)},$$

where
$$\eta_1=\frac{\pi^2}{2\omega_1}\left(\frac{1}{6}+\sum_{1}^{\infty}\csc^2\pi m_2\omega_2/\omega_1\right)\quad\cdot\quad\cdot\quad\cdot\quad\cdot\quad\cdot\quad\cdot\quad\text{(69)}.$$

The constant η_1 is the same as that introduced in § 204.

For the relation

$$\sigma(u+2\omega_1)=-e^{2\eta_1(u+\omega_1)}\sigma u,$$

which is immediately deducible from (68), gives by logarithmic differentiation

$$\zeta(u+2\omega_1)=\zeta u+2\eta_1.$$

In the formula

$$\zeta u=\frac{\eta_1 u}{\omega_1}+\frac{\pi}{2\omega_1}\cot\frac{\pi u}{2\omega_1}+\frac{\pi}{2\omega_1}\sum_1^\infty\left[\cot\pi\left(\frac{m_2\omega_2}{\omega_1}+\frac{u}{2\omega_1}\right)-\cot\pi\left(\frac{m_2\omega_2}{\omega_1}-\frac{u}{2\omega_1}\right)\right],$$

derived from (68) by logarithmic differentiation, write $u=\omega_2$.

Hence

$$\eta_2=\frac{\eta_1\omega_2}{\omega_1}+\frac{\pi}{2\omega_1}\cot\frac{\pi\omega_2}{2\omega_1}+\frac{\pi}{2\omega_1}\sum_1^\infty\left[\cot\pi\frac{(2m_2+1)\omega_2}{2\omega_1}-\cot\pi\frac{(2m_2-1)\omega_2}{2\omega_1}\right],$$

and $\omega_1\eta_2-\omega_2\eta_1=\lim_{m_2=+\infty}\frac{\pi}{2}\cot\pi\frac{(2m_2+1)\omega_2}{2\omega_1}=\pm\frac{\pi}{2}i;$

$$\left.\begin{aligned}\eta_1\omega_2-\eta_2\omega_1&=\frac{\pi}{2}i,\text{ when the imaginary part of }\frac{\omega_2}{\omega_1}\text{ is }+\\&=-\frac{\pi}{2}i,\text{ when the imaginary part of }\frac{\omega_2}{\omega_1}\text{ is }-\end{aligned}\right\}\text{(70)}.$$

The former assumption will be made (see § 194).

This formula is the Weierstrassian analogue to Legendre's relation $EK'+E'K-KK'=\frac{\pi}{2}$ (Cayley's Elliptic Functions, p. 49).

The formula (68) can be written in a form which is better adapted for applications. Let $\omega_2/\omega_1=\omega$, $m_2=m$. Then

$$\sigma u=\frac{2\omega_1}{\pi}\sin\frac{\pi u}{2\omega_1}e^{\eta_1\frac{u^2}{2\omega_1}}\prod_{m=1}^\infty\left(1-\frac{\sin^2\frac{\pi u}{2\omega_1}}{\sin^2\pi m\omega}\right)$$

$$=\frac{2\omega_1}{\pi}\sin\frac{\pi u}{2\omega_1}e^{\eta_1\frac{u^2}{2\omega_1}}\prod_{m=1}^\infty\frac{\sin\left(m\omega-\frac{u}{2\omega_1}\right)\pi}{\sin\pi m\omega}\cdot e^{-\pi iu/2\omega_1}$$

$$\cdot\prod_{m=1}^\infty\frac{\sin\left(m\omega+\frac{u}{2\omega_1}\right)\pi}{\sin\pi m\omega}\cdot e^{\pi iu/2\omega_1}.$$

Writing* $e^{\pi iu/2\omega_1}=z$, $e^{\pi i\omega}=q$, we get

$$\frac{\sin\pi\left(m\omega-\frac{u}{2\omega_1}\right)}{\sin\pi m\omega}e^{-\pi iu/2\omega_1}=\frac{q^{2m}z^{-2}-1}{q^{2m}-1}.$$

* Weierstrass uses h for Jacobi's q, and Klein uses r for q^2.

Hence $$\sigma u = \frac{2\omega_1}{\pi}\sin\frac{\pi u}{2\omega_1}\, e^{\eta_1\frac{u^2}{2\omega_1}} \prod_{m=1}^{\infty}\frac{1-q^{2m}z^{-2}}{1-q^{2m}} \prod_{m=1}^{\infty}\frac{1-q^{2m}z^{2}}{1-q^{2m}}$$

$$= \frac{2\omega_1}{\pi}\sin\frac{\pi u}{2\omega_1}\, e^{\eta_1\frac{u^2}{2\omega_1}} \prod_{m=1}^{\infty}\frac{1-2q^{2m}\cos\frac{\pi u}{\omega_1}+q^{4m}}{(1-q^{2m})^2} \quad . \quad (71).$$

Logarithmic differentiation gives

$$\zeta u = \frac{\pi}{2\omega_1}\cot\frac{\pi u}{2\omega_1} + \eta_1\frac{u}{\omega_1} + \frac{2\pi}{\omega_1}\sum_1^{\infty}\frac{q^{2m}\sin\frac{\pi u}{\omega_1}}{1-2q^{2m}\cos\frac{\pi u}{\omega_1}+q^{4m}}.$$

The series on the right-hand side can be put into a more useful shape. We have when $|x|<1$,

$$1/(1-x) = 1 + x + x^2 + \cdots + x^n + \cdots.$$

Let $x = r(\cos\theta \pm i\sin\theta)$, and subtract the two series thus obtained. The result is that when $|r|<1$

$$\frac{2r\sin\theta}{1-2r\cos\theta+r^2} = \sum_1^{\infty} 2r^n \sin n\theta.$$

This is an identity which holds for complex values of r as well as for real values. Let $r = q^{2m}$, where $|q|<1$; then

$$\sum_1^{\infty}\frac{q^{2m}\sin\theta}{1-2q^{2m}\cos\theta+q^{4m}} = \sum_{n=1}^{\infty}\sum_{m=1}^{\infty} q^{2mn}\sin n\theta.$$

The condition $|q|<1$ implies that the imaginary part of ω_2/ω_1 is positive. Now let the summation be performed for m, and we have the result

$$\sum_1^{\infty}\frac{q^{2n}}{1-q^{2n}}\sin n\theta.$$

Thus $$\zeta u = \frac{\pi}{2\omega_1}\cot\frac{\pi u}{2\omega_1} + \frac{\eta_1 u}{\omega_1} + \frac{2\pi}{\omega_1}\sum_1^{\infty}\frac{q^{2n}}{1-q^{2n}}\sin\frac{n\pi u}{\omega}. \quad (72).$$

Differentiating with regard to u, we have

$$\wp u = \frac{\pi^2}{4\omega_1{}^2}\csc^2\frac{\pi u}{2\omega_1} - \frac{\eta_1}{\omega_1} - \frac{2\pi^2}{\omega_1{}^2}\sum_1^{\infty}\frac{nq^{2n}\cos\frac{n\pi u}{\omega_1}}{1-q^{2n}} \quad . \quad (73).$$

Expand in powers of u: we have

$$\wp u = \frac{1}{u^2} + \frac{g_2}{20}u^2 + \frac{g_3}{28}u^4 + \cdots.$$

The development of $\csc^2 \pi u/2\omega_1$ to the first few powers can be found by the following considerations: — Let $\omega_1 = \pi/2$, and let ω_2 be a pure imaginary whose absolute value is ∞. Then q becomes 0,

$$\begin{aligned} \sigma u &= e^{u^2/6} \sin u, \\ \zeta u &= u/3 + \cot u, \\ \wp u &= -1/3 + \csc^2 u, \\ g_2 &= 4/3, \quad g_3 = 8/27. \end{aligned}$$

Hence, from the expansion for $\wp u$,

$$\csc^2 u = \frac{1}{u^2} + \frac{1}{3} + \frac{4}{3 \cdot 20} u^2 + \frac{8}{27 \cdot 28} u^4 + \cdots,$$

and the right-hand side of (73) becomes

$$\frac{\pi^2}{4\,\omega_1^2} \left\{ \frac{4\,\omega_1^2}{\pi^2 u^2} + \frac{1}{3} + \frac{1}{15} \frac{\pi^2 u^2}{4\,\omega_1^2} + \frac{2}{27 \cdot 7} \frac{\pi^4 u^4}{16\,\omega_1^4} + \cdots \right\}$$

$$-\frac{\eta_1}{\omega_1} - \frac{2\pi^2}{\omega_1^2} \sum_1^\infty \frac{n q^{2n}}{1 - q^{2n}} \left\{ 1 - \frac{\pi^2 n^2 u^2}{2\,\omega_1^2} + \frac{\pi^4 n^4 u^4}{4!\,\omega_1^4} - \cdots \right\}.$$

Equating the coefficients of the powers of u, we have

$$\eta_1 = \frac{\pi^2}{12\,\omega_1} - \frac{2\pi^2}{\omega_1} \sum_1^\infty \frac{n q^{2n}}{1 - q^{2n}} \quad \cdots \cdots \quad (74);$$

$$\left(\frac{\omega_1}{\pi}\right)^4 g_2 = \frac{1}{12} + 20 \sum_1^\infty \frac{n^3 q^{2n}}{1 - q^{2n}} \quad \cdots \cdots \quad (75),$$

$$\left(\frac{\omega_1}{\pi}\right)^6 g_3 = \frac{1}{216} - \frac{7}{3} \sum_1^\infty \frac{n^5 q^{2n}}{1 - q^{2n}} \quad \cdots \cdots \quad (76).$$

Many other formulæ of this kind will be found in Halphen, t. i., ch. xiii.

§ 214. *Expression of $\sigma_\lambda u$ as a singly infinite product.*
From formula (71),

$$\sigma u = \frac{2\,\omega_1}{\pi} e^{\eta_1 u^2/2\omega_1} \frac{z - z^{-1}}{2i} \prod_1^\infty \frac{1 - q^{2m} z^{-2}}{1 - q^{2m}} \prod_1^\infty \frac{1 - q^{2m} z^2}{1 - q^{2m}}.$$

From this can be readily deduced analogous expressions for the allied functions $\sigma_1 u$, $\sigma_2 u$, $\sigma_3 u$. When u changes into $u + \omega_1$, z becomes iz, and the exponent $\eta_1 u^2/2\,\omega_1$ becomes $\eta_1 u^2/2\,\omega_1 + \eta_1 u + \frac{1}{2}\eta_1\omega_1$. When u changes into $u + \omega_2$, z becomes $q^{1/2} z$, and $\eta_1 u^2/2\,\omega_1$ becomes

$$\begin{aligned} \eta_1 u^2/2\,\omega_1 + \eta_1 \omega (u + \omega_2/2) &= \eta_1 u^2/2\,\omega_1 + (\eta_2 + \pi i/2\,\omega_1)(u + \omega_2/2) \\ &= \eta_1 u^2/2\,\omega_1 + \eta_2 u + \tfrac{1}{2}\eta_2\omega_2 + \pi i u/2\,\omega_1 + \tfrac{1}{4}\pi i \omega. \end{aligned}$$

When u changes into $u+\omega_3$, z becomes $-iq^{-1/2}z$, and $\eta_1 u^2/2\omega_1$ becomes

$$\eta_1 u^2/2\omega_1+\eta_1\omega_3/\omega_1\cdot(u+\omega_3/2)=\eta_1 u^2/2\omega_1+(\eta_3-\pi i/2\omega_1)(u+\omega_3/2),$$

or
$$\eta_1 u^2/2\omega_1+\eta_3 u-\pi i u/2\omega_1+\tfrac{1}{2}\eta_3\omega_3+\tfrac{1}{4}\pi i(1+\omega).$$

Hence

$$\sigma(u+\omega_1)=\frac{2\omega_1}{\pi}e^{\eta_1\frac{u^2}{2\omega_1}+\eta_1 u+\frac{1}{2}\eta_1\omega_1}\cdot\frac{z+z^{-1}}{2}\cdot\prod_1^\infty\frac{(1+q^{2m}z^{-2})(1+q^{2m}z^2)}{(1-q^{2m})^2}.$$

Put $u=0$, and therefore $z=1$. Then

$$\sigma\omega_1=\frac{2\omega_1}{\pi}\cdot e^{\frac{1}{2}\eta_1\omega_1}\prod_1^\infty\left(\frac{1+q^{2m}}{1-q^{2m}}\right)^2.$$

Hence

$$\sigma_1 u=\frac{e^{-\eta_1 u}\cdot\sigma(u+\omega_1)}{\sigma\omega_1}=e^{\frac{\eta_1 u^2}{2\omega_1}}\frac{z+z^{-1}}{2}\prod_1^\infty\frac{(1+q^{2m}z^{-2})(1+q^{2m}z^2)}{(1+q^{2m})^2}\quad\cdot\ (77).$$

Again

$$\sigma(u+\omega_2)=\frac{2\omega_1}{\pi}\cdot e^{\frac{\eta_1 u^2}{2\omega_1}+\eta_2 u+\frac{1}{2}\eta_2\omega_2}\cdot z\cdot q^{\frac{1}{4}}\cdot\frac{q^{\frac{1}{2}}z-q^{-\frac{1}{2}}z^{-1}}{2i}\prod_1^\infty\frac{(1-q^{2m-1}z^{-2})(1-q^{2m-1}z^2)}{(1-q^{2m})^2}$$

$$=\frac{\omega_1 i}{\pi q^{1/4}}\cdot e^{\frac{\eta_1 u^2}{2\omega_1}+\eta_2 u+\frac{1}{2}\eta_2\omega_2}\cdot\prod_1^\infty\frac{(1-q^{2m-1}z^{-2})(1-q^{2m-1}z^2)}{(1-q^{2m})^2}.$$

Putting $u=0,\ z=1$,

$$\sigma\omega_2=\frac{\omega_1 i}{\pi q^{1/4}}e^{\frac{1}{2}\eta_2\omega_2}\prod_1^\infty\frac{(1-q^{2m-1})^2}{(1-q^{2m})^2}.$$

Therefore

$$\sigma_2 u=\frac{e^{-\eta_2 u}\sigma(u+\omega_2)}{\sigma\omega_2}=e^{\eta_1 u^2/2\omega_1}\prod_1^\infty\frac{(1-q^{2m-1}z^{-2})(1-q^{2m-1}z^2)}{(1-q^{2m-1})^2}\quad\cdot\quad(78).$$

Lastly,

$$\sigma(u+\omega_3)=\frac{2\omega_1}{\pi}e^{\frac{\eta_1 u^2}{2\omega_1}+\eta_3 u+\frac{1}{2}\eta_3\omega_3}\cdot z^{-1}\cdot\sqrt{i}q^{1/4}$$

$$\cdot\left(-\frac{q^{-1/2}z+q^{1/2}z^{-1}}{2}\right)\cdot\prod_1^\infty\frac{(1+q^{2m+1}z^{-2})(1+q^{2m-1}z^2)}{(1-q^{2m})^2}$$

$$=-\frac{\sqrt{i}\omega_1}{\pi q^{1/4}}e^{\frac{\eta_1 u^2}{2\omega_1}+\eta_3 u+\frac{1}{2}\eta_3\omega_3}\prod_1^\infty\frac{(1+q^{2m-1}z^{-2})(1+q^{2m-1}z^2)}{(1-q^{2m})^2},$$

so that, putting $u=0,\ z=1,$

$$\sigma\omega_3 = -\frac{\sqrt{i\omega_1}}{\pi q^{1/4}} e^{\frac{1}{2}\eta_3\omega_3} \prod_1^\infty \frac{(1+q^{2m-1})^2}{(1-q^{2m})^2},$$

and

$$\sigma_3 u = \frac{e^{-\eta_3 u}\cdot\sigma(u+\omega_3)}{\sigma\omega_3} = e^{\frac{\eta_1 u^2}{2\omega_1}} \prod_1^\infty \frac{(1+q^{2m-1}z^{-2})(1+q^{2m-1}z^2)}{(1+q^{2m-1})^2} \quad . \quad (79).$$

If we write

$$\left.\begin{aligned} \prod_{m=1}^\infty (1-q^{2m}) &= Q_0 \\ \prod_{m=1}^\infty (1+q^{2m}) &= Q_1 \\ \prod_{m=1}^\infty (1-q^{2m-1}) &= Q_2 \\ \prod_{m=1}^\infty (1+q^{2m-1}) &= Q_3 \end{aligned}\right\} \quad . \quad . \quad . \quad . \quad . \quad . \quad . \quad (80),$$

we have, from what precedes,

$$\sigma\omega_1 = \frac{2\,\omega_1}{\pi} e^{\frac{1}{2}\eta_1\omega_1} \frac{Q_1^2}{Q_0^2},$$

$$\sigma\omega_2 = \frac{2\,i\omega_1}{\pi} e^{\frac{1}{2}\eta_2\omega_2} \frac{Q_2^2}{2\,q^{1/4}Q_0^2},$$

$$\sigma\omega_3 = -\frac{2\sqrt{i\omega_1}}{\pi} e^{\frac{1}{2}\eta_3\omega_3} \frac{Q_3^2}{2\,q^{1/4}Q_0^2}.$$

Comparing these with § 210, we have

$$\left.\begin{aligned} \sqrt{e_2-e_1\cdot e_3-e_1} &= \left(\frac{\pi}{2\,\omega_1}\right)^2 \frac{Q_0^4}{Q_1^4} \\ \sqrt{e_3-e_2\cdot e_1-e_2} &= -\left(\frac{\pi}{2\,\omega_1}\right)^2 \frac{4\,q^{1/2}Q_0^4}{Q_2^4} \\ \sqrt{e_1-e_3\cdot e_2-e_3} &= -\left(\frac{\pi}{2\,\omega_1}\right)^2 \frac{4\,iq^{1/2}Q_0^4}{Q_3^4} \end{aligned}\right\} \quad . \quad . \quad . \quad (81).$$

Let the discriminant of the cubic $4\wp^3 - g_2\wp - g_3$ be Δ; that is, let

$$\Delta = g_2^3 - 27\,g_3^2 = 16(e_2-e_3)^2(e_3-e_1)^2(e_1-e_2)^2.$$

From the preceding equations

$$\Delta = \left(\frac{\pi}{\omega_1}\right)^{12} q^2 \frac{Q_0^{24}}{Q_1^8 Q_2^8 Q_3^8}.$$

But $\quad Q_0Q_1Q_2Q_3 = \Pi(1-q^{4m})\Pi(1-q^{4m-2}) = Q_0.$

Hence $\quad Q_1Q_2Q_3=1,$

and
$$\Delta \doteq \left(\frac{\pi}{\omega_1}\right)^{12} q^2\Pi(1-q^{2m})^{24} \quad . \quad . \quad . \quad . \quad . \quad . \quad . \quad . \quad . \quad (82).$$

In this formula take the logarithms of both sides and differentiate with regard to q. This gives

$$\frac{d\log(\Delta\omega_1^{12})}{dq}=\frac{2}{q}-48\sum_{m=1}^{\infty}\frac{mq^{2m-1}}{1-q^{2m}},$$

or
$$\frac{d\log(\Delta\omega_1^{12})}{d\log q}=2-48\sum_{m=1}^{\infty}\frac{mq^{2m}}{1-q^{2m}}.$$

Comparing with (74), we have*

$$\eta_1\omega_1=\frac{\pi^2}{24}\frac{d\log(\Delta\omega_1^{12})}{d\log q},$$

or, since $\log q=\pi i\omega$,

$$\eta_1\omega_1=\frac{\pi}{24i}\frac{d\log(\Delta\omega_1^{12})}{d\omega} \quad . \quad . \quad . \quad . \quad . \quad . \quad . \quad . \quad (83).$$

If ω_2 vary, while ω_1 is constant, this gives

$$\eta_1=-\frac{\pi i}{24}\frac{\delta\log\Delta}{\delta\omega_2} \quad . \quad . \quad . \quad . \quad . \quad . \quad . \quad . \quad . \quad . \quad . \quad . \quad (84).$$

If we had interchanged ω_1 and ω_2, η_1 and η_2, throughout, we should have obtained the equation

$$\eta_2=\frac{\pi i}{24}\frac{\delta\log\Delta}{\delta\omega_1} \quad . \quad . \quad . \quad . \quad . \quad . \quad . \quad . \quad . \quad . \quad . \quad (85).$$

Substituting these values of η_1 and η_2 in the relation

$$\begin{vmatrix}\eta_1 & \eta_2\\ \omega_1 & \omega_2\end{vmatrix}=\pi i/2,$$

we obtain
$$\omega_1\frac{\delta\log\Delta}{\delta\omega_1}+\omega_2\frac{\delta\log\Delta}{\delta\omega_2}=-12,$$

which can be written down at once from the fact that Δ is a form of degree -12 in ω_1, ω_2.

For a full discussion of the relations between the quantities ω_1, ω_2, η_1, η_2, g_2, g_3, $\log\Delta$, we refer the reader to Halphen, t. i., ch. 9, and Klein-Fricke, Modulfunctionen, t. i., pp. 116–123.

* For other proofs of this formula, see Halphen, t. i., pp. 259, 321.

§ 215. *The q-series for σu and $\sigma_\lambda u$.*

Let $$\phi_n(s) = \prod_{m=1}^{n} (1 + q^{2m-1}s)(1 + q^{2m-1}s^{-1})$$

$$= a_0 + a_1(s + s^{-1}) + \cdots + a_n(s^n + s^{-n}),$$

where $$a_n = q^{\Sigma(2m-1)} = q^{n^2}.$$

For s write q^2s. Then

$$\phi_n(q^2s)/\phi_n(s) = \frac{(1 + q^{2n+1}s)(1 + q^{-1}s^{-1})}{(1 + qs)(1 + q^{2n-1}s^{-1})}$$

$$= \frac{1 + q^{2n+1}s}{qs + q^{2n}}.$$

Multiplying up, and equating in the two series the coefficients of s^{m+1}, we have

$$a_{m+1}q^{2(m+n+1)} + a_m q^{2m+1} = a_{m+1} + q^{2n+1}a_m,$$

or $$a_m = a_{m+1} \cdot \frac{1 - q^{2(m+n+1)}}{q^{2m+1} - q^{2n+1}},$$

whence $$a_{n-1} = a_n \cdot \frac{1 - q^{4n}}{q^{2n-1} - q^{2n+1}} = q^{(n-1)^2} \cdot \frac{1 - q^{4n}}{1 - q^2},$$

$$a_{n-2} = a_{n-1} \cdot \frac{1 - q^{4n-2}}{q^{2n-3} - q^{2n+1}} = q^{(n-2)^2} \cdot \frac{(1 - q^{4n})(1 - q^{4n-2})}{(1 - q^2)(1 - q^4)},$$

$$a_m = q^{m^2} \frac{(1 - q^{4n})(1 - q^{4n-2}) \cdots (1 - q^{2(m+n+1)})}{(1 - q^2)(1 - q^4) \cdots (1 - q^{2n-2m})}.$$

When n is made infinitely great, and $m \leqq$ an assigned integer κ,

$$\lim a_m = q^{m^2}/Q_0;$$

and when $m > \kappa$, the absolute value of the denominator lies between fixed finite limits, namely, the least and greatest terms of the sequence

$$|1 - q^2|,\ |1 - q^2|\,|1 - q^4|, \cdots.$$

Therefore

$$\phi_\infty(s) - \{1 + q(s + s^{-1}) + \cdots + q^{\kappa^2}(s^\kappa + s^{-\kappa})\}/Q_0$$

$$= \sum_{r=1}^{\infty} q^{(\kappa+r)^2} b_{\kappa+r}(s^{\kappa+r} + s^{-\kappa-r}),$$

where $|b_{\kappa+r}|$ also lies between fixed finite limits.

The series $\Sigma q^{m^2}(s^m + s^{-m})$ is absolutely convergent, for it separates into two absolutely convergent series; hence the right-hand side of the last equation can be made as small as we please by taking κ large enough. Therefore

$$Q_0 \prod_{m=1}^{\infty} (1+q^{2m-1}s)(1+q^{2m-1}s^{-1}) = 1 + \sum_{m=1}^{\infty} q^{m^2}(s^m + s^{-m}) \quad . \quad . \quad (86).$$

This important algebraic identity appears in Gauss's Works, t. iii., p. 434; and in Jacobi's Fundamenta Nova, §§ 63, 64. Cayley's statement of Jacobi's proof will be found in his Elliptic Functions, § 385. See also Briot and Bouquet, p. 315; Clifford, Works, p. 459. The above method is due to Biehler, Sur les Fonctions doublement périodiques considérées comme des limites de fonctions entières, Crelle, t. lxxxviii. See Enneper-Müller, pp. 97, 113; Jordan's Cours d'Analyse, t. i., p. 148.

Let $s = z^2 = e^{\pi i u/\omega_1}$; the right-hand side of the above identity becomes

$$1 + 2q\cos\frac{\pi u}{\omega_1} + 2q^4 \cos\frac{2\pi u}{\omega_1} + \cdots,$$

or, in Jacobi's notation, $\theta_3 \dfrac{\pi u}{2\omega_1}$.

Therefore, from (79),

$$Q_0 Q_3^2 \sigma_3 u = e^{\eta_1 u^2/2\omega_1} \theta_3 \frac{\pi u}{2\omega_1}.$$

When $u = 0$, this equation becomes

$$Q_0 Q_3^2 = \theta_3(0),$$

and therefore

$$\sigma_3 u = e^{\eta_1 u^2/2\omega_1} \theta_3\left(\frac{\pi u}{2\omega_1}\right) \Big/ \theta_3(0) \quad . \quad . \quad . \quad . \quad . \quad . \quad . \quad (87).$$

If in the preceding work we write iz instead of z, and therefore $u + \omega_1$ instead of u, we obtain in the same way

$$\sigma_2 u = e^{\eta_1 u^2/2\omega_1} \theta\left(\frac{\pi u}{2\omega_1}\right) \Big/ \theta(0) \quad . \quad . \quad . \quad . \quad . \quad . \quad . \quad (88),$$

where $\qquad \theta v = 1 - 2q\cos 2v + 2q^4 \cos 4v - \cdots.$

Again, in (86) let $s = qz^2$.

Then

$$Q_0(1+z^{-2}) \prod_{m=1}^{\infty} (1+q^{2m}z^2)(1+q^{2m}z^{-2}) = 1 + \sum_{m=1}^{\infty} (q^{m^2+m}z^{2m} + q^{m^2-m}z^{2-m}),$$

or

$$Q_0(z+z^{-1})\Pi(1+q^{2m}z^2)(1+q^{2m}z^{-2}) = z + z^{-1} + \sum_{m=1}^{\infty} q^{m(m+1)}(z^{2m+1} + z^{-2m-1})$$

$$= q^{-1/4}\left[2q^{1/4}\cos\frac{\pi u}{2\omega_1} + 2q^{9/4}\cos\frac{3\pi u}{2\omega_1} + 2q^{25/4}\cos\frac{5\pi u}{2\omega_1} + \cdots\right].$$

The series in brackets is Jacobi's function $\theta_2 \frac{\pi u}{2\omega_1}$.

From (77)

$$Q_0 Q_1^2 \cdot \sigma_1 u = \frac{1}{2} e^{\eta_1 u^2/2\omega_1} q^{-1/4} \theta_2 \frac{\pi u}{2\omega_1},$$

whence it follows that

$$\sigma_1 u = e^{\eta_1 u^2/2\omega_1} \theta_2 \left(\frac{\pi u}{2\omega_1}\right) \Big/ \theta_2(0) \quad \cdot \quad \cdot \quad \cdot \quad \cdot \quad \cdot \quad \cdot \quad \cdot \quad (89).$$

If in the preceding work we write iz instead of z, we obtain from (71)

$$Q_0^3 \sigma u = \frac{\omega_1}{\pi} q^{-1/4} e^{\eta_1 u^2/2\omega_1} \theta_1 \frac{\pi u}{2\omega_1},$$

where $\theta_1 v = 2q^{1/4} \sin v - 2q^{9/4} \sin 3v + 2q^{25/4} \sin 5v - \cdots,$

an odd function of v, which vanishes when $v = 0$.

Let $\theta_1' v$ stand for $\frac{\delta \theta_1 v}{\delta v}$; after differentiation with respect to u let $u = 0$; then since $\sigma'(0) = 1$,

$$Q_0^3 = \tfrac{1}{2} q^{-1/4} \theta_1'(0),$$

and

$$\sigma u = \frac{2\omega_1}{\pi} e^{\eta_1 u^2/2\omega_1} \theta_1 \left(\frac{\pi u}{2\omega_1}\right) \Big/ \theta_1'(0) \quad \cdot \quad \cdot \quad \cdot \quad \cdot \quad \cdot \quad (90).$$

Examples. Prove Jacobi's relations

$$\theta(0)\theta_2(0)\theta_3(0) = \theta_1'(0),$$

$$\theta^4(0) + \theta_2^4(0) = \theta_3^4(0).$$

There is considerable diversity of notation for Jacobi's θ-functions. The above notation follows Jacobi's final usage, except that Jacobi writes $\theta_\lambda(v, q)$ for $\theta_\lambda v$, where $\lambda = 1, 2, 3$; and $\theta(v, q)$ for θv. Weierstrass writes $\theta_\lambda v$ or $\theta_\lambda(v, q)$ for what in the above notation would be $\theta_\lambda(\pi v)$, and writes $\theta_0 v$ instead of $\theta(\pi v)$. Halphen (t. i., p. 253) follows Weierstrass. Cayley, in his Elliptic Functions, adopts Jacobi's earlier notation, — that of the Fundamenta Nova. The subject of q-series has been very fully entered into, from the standpoint of the real variable, by Glaisher in papers which have appeared in the Quarterly Journal, the Proceedings of the London Mathematical Society, and the Messenger of Mathematics.

The Elliptic Integrals.

§ 216. If y be connected with x by an irreducible algebraic equation $F(x, y)=0$, of deficiency p, $\int R(x, y)dx$, when R is a rational function of x and y, is the general Abelian integral. When $F(x, y)=0$ takes the simpler form $y^2=c_0(x-a_1)(x-a_2)\cdots(x-a_{2p+1})$, or $y^2=c_0(x-a_1)(x-a_2)\cdots(x-a_{2p+2})$, the integral is the general hyperelliptic integral of order p. When $p=1$, the Abelian integral becomes elliptic, the general elliptic integral being

$$\int R(x, \sqrt{\text{a quartic in } x})dx;$$

when $p=2$, the Abelian integral becomes hyperelliptic of order 1, but when $p>2$, the hyperelliptic integrals form only a portion of the complete system of Abelian integrals. The values (x, y) which belong to $F(x, y)=0$, and all rational functions $R(x, y)$, form what is termed by Weierstrass the algebraic formation or configuration ('Gebilde'). The properties of the Abelian integral and of every function of the integral must be ultimately dependent upon the properties of the algebraic configuration. The method which we have adopted in this chapter starts with the elliptic functions and leaves out of account the connexion with the configuration. In the next few paragraphs we shall follow another order, and endeavour to indicate the connexion of the elliptic function with the elliptic integral.

§ 217. If $F(x, y)=0$ be of deficiency 1, the corresponding Riemann surface is triply connected and is reducible to simple connexion by two cross-cuts. Is the equation reducible in all cases, by birational transformations, to one of the form

$$y^2=c_0(x-a_1)(x-a_2)(x-a_3)(x-a_4)?$$

In other words, can there be found transformations $x_1=R_1(x, y)$, $y_1=R_2(x, y)$, which give $x=R_1'(x_1, y_1)$, $y=R_2'(x_1, y_1)$, and transform $F(x, y)=0$ into $y_1^2=$ a quartic in x_1? Before showing that this can be done we must explain briefly how the theory of the algebraic Riemann surface can be connected with the theory of the algebraic plane curve.

Among the rational functions on the surface, we are entitled to select two, x and y, such that the equation between x and y, when regarded as a plane curve, presents no other multiple points than simple nodes (§ 189). We can further assume that the nodes and branch-points lie at a finite distance. This curve will be called the basis-curve.

Let any rational function of x and y be $\frac{\phi_1(x, y)}{\phi_2(x, y)}$, where ϕ_1 and ϕ_2 are rational polynomials. Instead of the places on the surface at which this function takes an assigned value w, we can speak of the points on the curve F, which are cut out by the curve $\phi_1 - w\phi_2 = 0$. When w varies, we obtain different groups of points, in which the curve F is cut by the curves of the pencil $\phi_1 - w\phi_2 = 0$. Any two of these groups of points are said to be *equivalent* or coresidual. Through any point of the curve F there passes generally but one curve of the pencil, and the rational function has a unique value. But when a basis-point of the pencil is on the curve F, ϕ_1/ϕ_2 takes at this point the form $0/0$; and its true value must be determined as the limit of the values of ϕ_1/ϕ_2 at adjacent points of F. That is, at a basis-point on F, the value of ϕ_1/ϕ_2 is obtained by taking that curve of the pencil which touches F at the point. When a basis-point coincides with a node of F, there will be two curves of the pencil which touch F at the node, and two distinct values of ϕ_1/ϕ_2.

Particular importance attaches to those pencils of curves which pass through all the nodes of F. These are called adjoint curves.

It will be noticed that we are not concerned with the pencil $\phi_1 - w\phi_2 = 0$ as a whole, but only with its intersections with F; these alone correspond to places on the Riemann surface. And also that imaginary intersections are to be included as much as real ones. It must not be supposed by readers acquainted with the theory of curves that the connexion with the theory of Riemann surfaces is solely in the interest of the latter; the curve is by far the greater gainer.

§ 218. Let the curve $F(x, y) = 0$ be of order κ and deficiency 1. The number of nodes is $\frac{1}{2}\kappa(\kappa - 3)$; and the number of points through which a curve of order $\kappa - 2$ can be drawn is $\frac{1}{2}(\kappa - 2)(\kappa + 1)$. Hence, since $\frac{1}{2}(\kappa - 2)(\kappa + 1) - \frac{1}{2}\kappa(\kappa - 3) = \kappa - 1$, a pencil of adjoint curves $\phi_1 - w\phi_2 = 0$ can be drawn through the nodes and through $\kappa - 2$ other assigned points of F. Each curve of the pencil has, with F, $\kappa(\kappa - 2) - \kappa(\kappa - 3) - (\kappa - 2)$ remaining intersections. Therefore the function w or ϕ_1/ϕ_2 takes an assigned value twice; that is, we have a function which is one-valued and two-placed on the Riemann surface which represents y as a function of x. Let this surface have m sheets; as y is to be rational in x and w, it is *necessary* that to the m places in a vertical line be attached m distinct values of w. Thus the equation between w and x must be of order m in w; and it is of order 2 in x. Let it be $F_1(x^2, w^m)=0$. The following considerations show that y is now a rational function of x, w. Given a pair of values x, w, there is only one corresponding y. Hence the two equations $w=\frac{\phi_1}{\phi_2}$, $F(x, y)=0$, when treated as equations in y only, have a single common root. They therefore give on elimination y as a rational function of the coefficients, *i.e.* y as a rational function of x and w. Thus the necessary condition for passing birationally from $F(x, y)=0$ to $F_1(x, w)=0$ is also the sufficient condition.

The new equation $F_1(x^2, w^m)=0$ can be transformed, in a similar manner, into $F_2(z^2, w^2)=0$, by the use of a function z which is two-placed on the surface which represents x as a one-valued function of w. The new equation is

$$f_0w^2+2f_1w+f_2=0,$$

where f_0, f_1, f_2 are polynomials in z of orders $\gtreqless 2$, and leads to

$$w=\frac{-f_1\pm\sqrt{Q}}{f_0},$$

where Q is a quartic in z. Let $y_1=\sqrt{Q}$, $x_1=z$; then the equation between w and z passes into

$$y_1^2=Q(x_1).$$

Here the last transformation is birational, for y_1 satisfies the necessary and sufficient condition; namely, it takes two values when z is given.

This equation, which is the standard form for the case $p=1$ can be written

$$y_1^2=c_0(x_1-a_1)(x_1-a_2)(x_1-a_3)(x_1-a_4);$$

and this can be brought into the form used by Weierstrass by the birational transformations

$$x_1-a_4=-1/z_1,$$
$$y_1=cw_1/z_1^2;$$

the new branch-points in the z_1-plane are e_1, e_2, e_3, ∞, where

$$a_\lambda-a_4=-1/e_\lambda,\quad \lambda=1, 2, 3;$$

and the new equation is

$$-c^2w_1^2=c_0(z_1-e_1)(z_1-e_2)(z_1-e_3)/e_1e_2e_3,$$

or, if

$$-4c^2=c_0/e_1e_2e_3,$$
$$w_1^2=4(z_1-e_1)(z_1-e_2)(z_1-e_3).$$

We can suppose the origin in the z_1-plane transferred to the centroid of the points e_λ, so that $e_1 + e_2 + e_3 = 0$.

The transformations of x_1 and z_1 have been bilinear. Hence if the quartic $Q(x_1)$ be

$$c_0x_1^4 + 4\,c_1x_1^3 + 6\,c_2x_1^2 + 4\,c_3x_1 + c_4,$$

whose invariants are

$$g_2 = c_0c_4 - 4\,c_1c_3 + 3\,c_2^2,$$

$$g_3 = c_0c_2c_4 + 2\,c_1c_2c_3 - c_0c_3^2 - c_4c_1^2 - c_2^3,$$

the invariants of the transformed expression will be of the forms $g_2' = \mu^2g_2$, $g_3' = \mu^3g_3$. But in the transformed expression $c_0' = 0$, $c_1' = 1$, $c_2' = 0$, and therefore $g_2' = -4c_3'$, $g_3' = -c_4'$.

Hence the new equation is

$$w^2 = 4\,z^3 - g_2'z - g_3',$$

where $$g_2' = \mu^2g_2,\ g_3' = \mu^3g_3,$$

or, if we replace z by μz, and w by $\sqrt{\mu^3}w$,

$$w^2 = 4\,z^3 - g_2z - g_3.$$

The conclusion of this paragraph is that the values of x and y which satisfy an equation of deficiency 1 can be expressed rationally in terms of z and $\sqrt{4\,z^3 - g_2z - g_3}$, and therefore also they can be expressed rationally in terms of $\wp u$ and $\wp' u$. [Klein, Einleitung in die Geometrische Funktionentheorie, pp. 208, 209; see also Thomae, Abriss einer Theorie der Funktionen, p. 122; Riemann, Werke, p. 111; Clebsch and Gordan, § 20; Salmon, Higher Plane Curves, § 366 and § 195.]

§ 219. *Elliptic Integrals.*

The consideration of integrals of the form $\int R(x, y)dx$, where x and y belong to the elliptic configuration, is now reduced to the consideration of $\int R(z, w)dz$, where

$$w^2 = 4\,z^3 - g_2z - g_3.$$

The simplest integral, setting aside those which are expressible as rational functions of z and w, is $u = \int \frac{dz}{w}$. If we suppose that $u = 0$ when $z = \infty$, we have, as implied in the last paragraph,

$$z = \wp u,$$

$$w = \wp' u.$$

Hence $$\int R(z, w)\,dz = \int f(u)\,du,$$

where $$f(u) = R(\wp u, \wp' u)\,\wp' u.$$

Since this last expression is an elliptic function of u, the elliptic integral is also the integral of an elliptic function with regard to the argument.

In § 205, any elliptic function was expressed by means of $\zeta(u - c_\kappa)$ and its derivatives, where c_κ was a pole of the function. Therefore in the integral of such a function there will occur:—

(1) a term $a_0 u$;

(2) the terms $\sum\limits_\kappa a_1^{(\kappa)} \log \sigma(u - c_\kappa)$;

(3) terms of the form $\zeta(u - c_\kappa)$, that is, by the addition theorem of § 204, of the form $\zeta(u) +$ an elliptic function of u;

(4) elliptic functions of u, which arise from the higher derivatives of $\zeta(u - c_\kappa)$.

Since (§ 205) $\Sigma a_1^{(\kappa)} = 0$, the terms in (2) can be written

$$\Sigma a_1^{(\kappa)} \log \frac{\sigma(c_\kappa - u)}{\sigma u \sigma c_\kappa},$$

in which a constant taken from the constant of integration is incorporated.

Accordingly,

$$\int f(u)\,du = a_0 u + \Sigma a_1^{(\kappa)} \log \frac{\sigma(c_\kappa - u)}{\sigma u \sigma c_\kappa} - \Sigma a_2^{(\kappa)} \zeta u + f_1(u),$$

where $f_1(u)$ is an elliptic function with the same periods as $f(u)$.

The integral u or $\int_{z, w}^{\infty} dz/w$ is an integral of the first kind. Its distinguishing characteristic is that on the surface T, which represents w as a one-valued function of z, it is everywhere finite. The function ζu or $-\int \wp u\,du$ takes on the surface T the form $+\int^{z, w} z\,dz/w$, and is an elementary integral of the second kind. It has the single infinity $z = \infty$, and if we write $z = 1/t^2$, it behaves near $t = 0$ like $-1/t$.

By § 204, we have

$$\zeta(u - c) - \zeta u + \zeta c = (\wp' u + \wp' c)/2(\wp u - \wp c).$$

On integration,

$$\log \frac{\sigma(c-u)}{\sigma u \sigma c} + u\zeta c = \int^u \frac{\wp' u + \wp' c}{2(\wp u - \wp c)}\, du$$

$$= \int^{z,\,w} \frac{w + w_0}{2(z - z_0)} \frac{dz}{w},$$

an integral which is finite on T except at (z_0, w_0) and at ∞; near these places it behaves like $\log(z - z_0)$ and $\log z^{1/2}$ respectively. Such an integral is an elementary integral of the third kind.

Hence any elliptic integral can be expressed by means of an integral of the first kind, an elementary integral of the second kind, a finite number of elementary integrals of the third kind, and a rational function of z and w. [Weierstrass-Schwarz, § 56. The resolution of the general elliptic integral into integrals of the three kinds was effected by Legendre, Fonctions Elliptiques, t. i., ch. 4.]

§ 220. We shall now show how some properties of the function $z = \wp u$ can be deduced from the consideration of the integral $u = \int dz/w$ on the surface T. This surface is constructed from the equation

$$w^2 = 4(z - e_1)(z - e_2)(z - e_3).$$

The branch-points are e_1, e_2, e_3, ∞.

Let the branch-cuts be along the straight lines $e_1 \cdots e_3$, $e_2 \cdots \infty$. Let the dissected surface T′ be formed from T by drawing a cross-cut A round e_1, e_3, and a cross-cut B round e_3, e_2 (Fig. 80a). In accordance with the notation of this chapter, let $2\omega_1$ and $2\omega_2$ be the periods across A and B (§ 180).

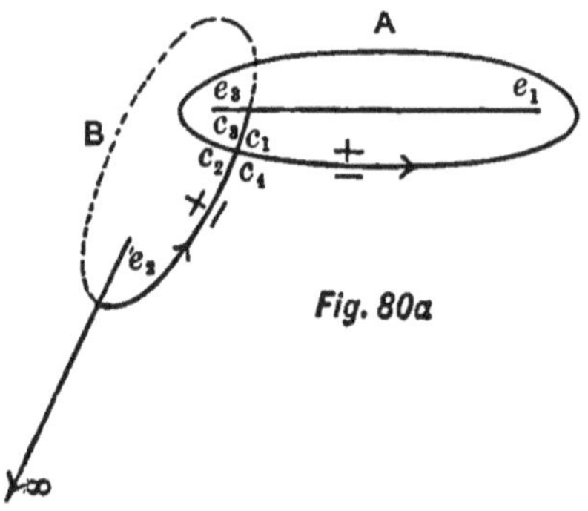

Fig. 80a

The integral $u = \int dz/w$ is one-valued on T′, and moreover it is everywhere finite (§ 182). From § 182 it follows further that

$$\int_{e_1}^{e_3} dz/w = \omega_2, \quad \int_{e_3}^{e_2} dz/w = -\omega_1,$$

$$\int_{e_2}^{\infty} dz/w = -\omega_2.$$

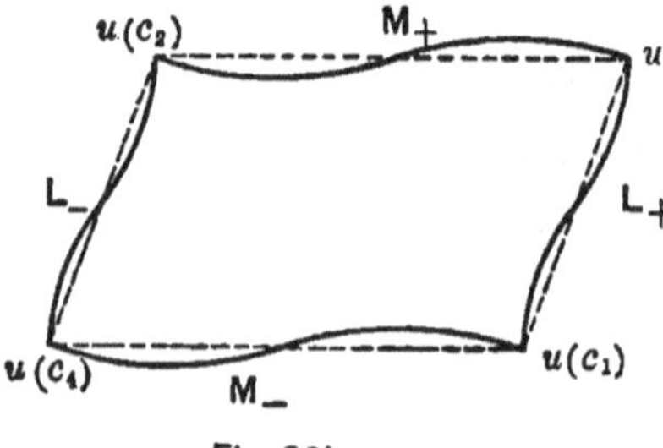

Fig. 80b

Let the value $u = 0$ be assigned to the point $z = \infty$. Then the values of u at e_1, e_2, e_3, are $\omega_1, \omega_2, \omega_1 + \omega_2$.

Let s and $\bar{s}$ be any two places of the surface T′ which lie in the same vertical. The relation between $u(s)$ and $u(\bar{s})$ can be written readily. Consider two paths which start from $\bar{\infty}$ and lie always in the same vertical. The integrands at s and $\bar{s}$ are dz/w and $-dz/w$; therefore

$$u(s)+u(\bar{s})=0.$$

This equation must be modified when either path of integration crosses a cut. For instance, at points near e_3, within both A and B, the equation becomes

$$u(s)+u(\bar{s})=2\omega_1+2\omega_2.$$

To map the z-surface on the u-plane.

From § 157, z is a one-valued function of u, and it has just been seen that at two places of T′ which have the same z, the values of u are in general different. Therefore to a given u corresponds only one place of T′, and therefore only one sheet is required in the u-plane. In general du/dz, $=1/w$, is neither 0 nor ∞, and there is isogonality. But near e_λ we have $du/dz=\dfrac{1}{\sqrt{z-e_\lambda}}P_0(\sqrt{z-e_\lambda})$, $\lambda=1, 2, 3$, and therefore $u-\omega_\lambda=P_1(\sqrt{z-e_\lambda})$, where $\omega_3=\omega_1+\omega_2$; whence it follows that while z turns round e_λ, u turns half as fast, in the same sense, round ω_λ. Near $z=\infty$, $u=P_1(1/z^{1/2})$, and again the isogonality breaks down.

The region of the u-plane into which the surface T′ maps is bounded by that curve into which the boundary of T′ maps. Let (z, w) start from the junction at c_3 and describe the positive bank of A upstream; and let L_+ be the corresponding path of u. When (z, w) continues from c_1 along the negative bank of B downstream, let M_- be the path of u. While (z, w) describes the negative bank of A downstream, u describes a path L_-, which is obtained by displacing L_+ through $-2\omega_1$; and while (z, w) describes the positive bank of B upstream, u describes a path M_+, which is obtained by displacing M_- through $2\omega_2$. Since (z, w) has now reached its starting-place, the whole u-path is closed. Thus the boundary of T′ maps into a curvilinear *parallelogram*.

Since u is finite for all values of z, the whole of the surface T′ is mapped on the *interior* of the parallelogram.

By giving a suitable form to the cross-cuts, the parallelogram can be made rectilinear. It is convenient to suppose that the bridge from e_1 to e_3 is shifted with the cross-cut A, so as to remain within it.

Now let us remove the restriction that z is not to cross A or B. Suppose that z crosses A positively; then the integral has a new range of values, which are obtained by increasing the old range of values by $2\omega_1$. To represent these we require a new parallelogram in the u-plane, obtained by shifting the old parallelogram through $2\omega_1$. In this way we require, to represent all possible values of the integral on T, a network of parallelograms, which cover the u-plane completely and without overlapping. A path on T which starts from a place (z_0, w_0) with a value u_0, and returns to this place after crossing A m_+ times positively and m_- times negatively, and B m_+' times positively and m_-' times negatively, gives as the final value of the integral

$$u_0 + 2m\omega_1 + 2m'\omega_2,$$

where
$$m_+ - m_- = m, \quad m_+' - m_-' = m';$$

and therefore the pair (z_0, w_0) is represented by infinitely many points of the u-plane, all congruent to u_0. Thus z and w are one-valued, doubly-periodic functions of u. When (z, w) is restricted to T′, the correspondence of u and (z, w) is $(1, 1)$, and the correspondence of z and w is $(3, 2)$; therefore the correspondence of u and z is $(2, 1)$, and that of u and w is $(3, 1)$. In other words, the orders of the elliptic functions z and w are respectively 2 and 3.

In passing, it should be pointed out that a new idea makes its appearance here. Since u is an infinitely many-valued function of z, the ordinary Riemann surface which represents u as a function of z would cover infinitely many sheets of the z-plane; but the infinitely many parallelograms serve equally well to show the connexion between z and u. The importance of this idea, which consists in the replacing of sheets lying one below the other, by regions of a plane which lie side by side and cover the plane without gaps, has been shown by Klein in his memoirs on Modular Functions in the Mathematische Annalen. See Klein-Fricke, Modulfunctionen, *passim*, and Poincaré's papers on Fuchsian and Kleinian Functions in the Acta Mathematica.

In this paragraph it has been assumed that the ratio ω_2/ω_1 is complex. This fact follows from a general theorem on the periods of Abelian Integrals, which will be proved in Chapter X. It will appear, in the same chapter, that on a surface of deficiency p there are p linearly independent integrals of the first kind; whence it will follow that in the present case every integral of the first kind is of the form $au + b$.

§ 221. It follows from the equation

$$du = dz/2\sqrt{(z - e_1)(z - e_2)(z - e_3)},$$

as in § 181, that to straight lines in the u-plane there correspond in the z-plane curves which obey the relation

$$\tau - \tfrac{1}{2}\Sigma\phi_\lambda = \text{constant},$$

where τ is the inclination of the tangent at any point z, and ϕ_λ is the amplitude of $z - e_\lambda$. These are the curves along which the cross-cuts must be drawn, when the parallelograms in the u-plane are to be rectilinear. In general, these curves are transcendental, but they contain a system of algebraic curves when g_2 and g_3 are real.

In this case, either the branch-points e_1, e_2, e_3 are all real, or one is real and two are complex and conjugate.

When the discriminant > 0, the branch-points are real. Let $e_2 < e_3 < e_1$. The integral $\int dz/2\sqrt{(z - e_1)(z - e_2)(z - e_3)}$, when taken along the real axis from $-\infty$ to e_2, is imaginary, and the integral from e_2 to e_3 is real; therefore ω_2 is imaginary and ω_1 is real, so that the parallelogram of periods is a rectangle.

Let u describe a line parallel to the axis of imaginaries; the equation of such a line is, in circular co-ordinates,

$$u + \bar{u} = 2\alpha,$$

where u and $\bar{u}$ are conjugate, and α is real.

Since g_2 and g_3 are real, the series for z, or $\wp u$, given in formula (10), § 201, has real coefficients, and therefore $\wp\bar{u}$ is conjugate to $\wp u$. Let $\bar{z} = \wp\bar{u}$, and let $\gamma = \wp 2\alpha$. The relation between z, $\bar{z}$, γ which results from $u + \bar{u} = 2\alpha$ is obtained symmetrically as follows:—

From § 202, z, $\bar{z}$, γ are roots of

$$\wp'u = a\wp u + b,$$

or

$$4\wp^3 - g_2\wp - g_3 - (a\wp + b)^2 = 0.$$

Therefore

$$z + \bar{z} + \gamma = a^2/4,$$

$$z\bar{z} + \gamma(z + \bar{z}) = -(2ab + g_2)/4,$$

$$\gamma z\bar{z} = (b^2 + g_3)/4.$$

Eliminating a and b, we get

$$\{z\bar{z} + \gamma(z + \bar{z}) + g_2/4\}^2 = 4(z + \bar{z} + \gamma)(\gamma z\bar{z} - g_3/4).$$

The equation is that of a Cartesian; and from the theory of curves, the single foci are e_1, e_2, e_3, and the triple focus is γ. As the complete curve depends on the single constant γ or $\wp 2\alpha$, it includes the representations of four lines in the rectangle; namely, those which meet the real axis at the points $\alpha, \omega_1 \pm \alpha, 2\omega_1 - \alpha$. To the first and fourth lines correspond the outer oval in both sheets of the Riemann surface; to the second and third lines, the inner oval in both sheets.

It may be remarked that the leading properties of the Cartesian, which are given in Salmon's Plane Curves and Williamson's Differential Calculus, are direct interpretations of standard formulæ for the functions $\wp$ and σ. Thus formula (18), coupled with the symmetry of the curve about the real axis, states that the curve is its own inverse with regard to any focus.

Again, a focal distance ρ_λ can be expressed as follows: Let $u = \alpha + v$; then

$$\rho_\lambda = \sqrt{\{\wp(\alpha+v) - e_\lambda\}\{\wp(\alpha - v) - e_\lambda\}},$$

or, from formula (35),

$$\rho_\lambda = \sigma_\lambda(\alpha+v)\sigma_\lambda(\alpha-v)/\sigma(\alpha+v)\sigma(\alpha-v).$$

Similarly, $\quad \sqrt{a_\lambda} = \sigma_\lambda(2\alpha)/\sigma(2\alpha),$

where a_λ is the distance from the triple focus to a single focus. Again, if ρ be the distance from the point whose parameter is v to the triple focus,

$$\rho^2 = \{\wp(\alpha+v) - \wp 2\alpha\}\{\wp(\alpha-v) - \wp 2\alpha\},$$

or, from (30),

$$= \sigma(3\alpha+v)\sigma(3\alpha-v)/\sigma(\alpha+v)\sigma(\alpha-v)\sigma^4 2\alpha.$$

In § 211, let

$$a,\ b,\ c,\ d\ = 2\alpha,\ 0,\ 3\alpha+v,\ 3\alpha - v;$$

then $\quad a',\ b',\ c',\ d'\ = 4\alpha,\ -2\alpha,\ \alpha+v,\ v-\alpha,$

and $\quad a'',\ b'',\ c'',\ d'' = \alpha+v,\ \alpha-v,\ 4\alpha,\ -2\alpha.$

Then (49) gives

$$\sigma_\lambda 2\alpha\sigma(3\alpha+v)\sigma(3\alpha-v) - \sigma_\lambda 4\alpha\sigma_\lambda 2\alpha\sigma(\alpha+v)\sigma(\alpha-v)$$
$$- \sigma_\lambda(\alpha+v)\sigma_\lambda(\alpha-v)\sigma 4\alpha\sigma 2\alpha = 0;$$

therefore $\quad \rho^2 = \dfrac{\sigma 4\alpha}{\sigma_\lambda 2\alpha\sigma^3 2\alpha}\rho_\lambda + \dfrac{\sigma_\lambda 4\alpha}{\sigma^4 2\alpha};$

or, since $\quad \sigma 2u = 2\sigma u\sigma_1 u\sigma_2 u\sigma_3 u,$

and $$\sigma_\lambda 2u = -\sigma_\mu{}^2u\sigma_\nu{}^2u + \sigma_\nu{}^2u\sigma_\lambda{}^2u + \sigma_\lambda{}^2u\sigma_\mu{}^2u,$$

$$\rho^2 = 2\sqrt{\alpha_\mu\alpha_\nu}\rho_\lambda + \alpha_\nu\alpha_\lambda + \alpha_\lambda\alpha_\mu - \alpha_\mu\alpha_\nu;$$

that is, $$\sqrt{\alpha_2\alpha_3}\rho_1 - \alpha_2\alpha_3 = \sqrt{\alpha_3\alpha_1}\rho_2 - \alpha_3\alpha_1 = \sqrt{\alpha_1\alpha_2}\rho_3 - \alpha_1\alpha_2$$
$$= \tfrac{1}{2}(\rho^2 - \alpha_2\alpha_3 - \alpha_3\alpha_1 - \alpha_1\alpha_2).$$

Very similar treatment applies to the orthogonal system of Cartesians, with the same foci, which correspond to lines parallel to the real axis in the u-plane.

Example. When the discriminant is negative, and e_1, e_2 are conjugate, show that ω_1, ω_2 are conjugate, and that lines parallel to the axes map into unipartite Cartesians.

§ 222. *Reduction to the canonical form.*

Let $f(z)$ be a quartic; the integral $u = \int_{y,\sqrt{f(y)}}^{x,\sqrt{f(x)}} dz/\sqrt{f(z)}$ is reduced to the canonical form $\int_s^\infty \frac{ds}{\sqrt{4s^3 - g_2s - g_3}}$ by the formula

$$s = \frac{\sqrt{f(x)}\sqrt{f(y)} + \alpha_x{}^2\alpha_y{}^2}{2(x-y)^2},$$

where $\alpha_x{}^2\alpha_y{}^2$ is the second polar of y with regard to $f(x)$; namely, if

$$f(x) = c_0x^4 + 4c_1x^3 + 6c_2x^2 + 4c_3x + c_4,$$

$$\alpha_x{}^2\alpha_y{}^2 = c_0x^2y^2 + 2c_1xy(x+y) + c_2(x^2 + 4xy + y^2) + 2c_3(x+y) + c_4.$$

The following proof of this theorem of Weierstrass's is due to Professor Klein, and is reproduced by his courteous permission from a manuscript course of lectures.

Introduce homogeneous co-ordinates by writing z_1/z_2 for z, and let $(x\,y)$ stand for $y_1x_2 - y_2x_1$.

Given $u = \int_{y,\sqrt{fy}}^{x,\sqrt{fx}} \frac{(zdz)}{\sqrt{f(z_1, z_2)}}$, the functions $\wp u$, $\wp' u$ belong to the algebraic configuration; *i.e.* are rational on the Riemann surface. Let us consider the parallelograms in the u-plane which are maps of the cut Riemann surface. Starting with the definition

$$\wp u = \frac{1}{u^2} + \Sigma' \left\{ \frac{1}{(u-w)^2} - \frac{1}{w^2} \right\},$$

we know that $\wp u$ is a one-valued analytic function of u throughout the finite part of the u-plane, and therefore also over the whole of the surface T. The function $\wp u$ must consequently be rational upon T, and must therefore be a rational function of x, $\sqrt{f(x)}$, y, $\sqrt{f(y)}$. Since $\wp u$ assumes a given value *twice* in the parallelogram of periods, it also takes an assumed value *twice* on the surface T. But the number of constants contained in a rational function which repeats its value twice on T is 2. Therefore the most general value of such a function is

$$c\wp u + c'.$$

We have to find the most general rational function of x, $\sqrt{f(x)}$, y, $\sqrt{f(y)}$ which is ∞^2 at $(y, \sqrt{f(y)})$ *only*.

Let $\dfrac{\phi_1(x, y)}{\phi_2(x, y)}$ be the function. Since this function is to be ∞^2 for $(x, \sqrt{f(x)}) = (y, \sqrt{f(y)})$, assume

$$\phi_2(x, y) = (xy)^2.$$

The denominator is ∞^2 for $(x, \sqrt{f(x)}) = (y, -\sqrt{f(y)})$, as well as for $(x, \sqrt{f(x)}) = (y, \sqrt{f(y)})$. Therefore the numerator must be 0^2 for $(x, \sqrt{f(x)}) = (y, -\sqrt{f(y)})$.

Now

$\sqrt{f(x)}\ \sqrt{f(y)} + a_x^2 a_y^2$ is 0^2, when $(x, \sqrt{f(x)}) = (y, -\sqrt{f(y)})$.

Therefore

$$\frac{\sqrt{f(x)}\sqrt{f(y)} + a_x^2 a_y^2}{(xy)^2} \text{ is of the form } c\wp u + c'.$$

Now the functions $\wp u$, $\wp' u$ are covariants with regard to x_1, x_2 and y_1, y_2, of weights 2 and 3. This covariantive character cán be proved as follows: let u be transformed by means of

$$z_1 = a_1 z_1' + a_2 z_2', \quad z_2 = b_1 z_1' + b_2 z_2',$$

into
$$\mu u' = \int_{y', \sqrt{f_1(y')}}^{x', \sqrt{f_1(x')}} (z'dz') / \sqrt{f_1(z')},$$

where
$$\mu = a_1 b_2 - a_2 b_1,$$

and let g_2', g_3' be the invariants of $f_1(z)$.

Then $\quad \wp(u'; g_2', g_3') = \wp(u/\mu; \mu^4 g_2, \mu^6 g_3) = \mu^2 \wp(u; g_2, g_3),$

and
$$\wp'(u'; g_2', g_3') = \mu^3 \wp'(u; g_2, g_3).$$

Thus $\wp$, $\wp'$ are also covariants with regard to the two sets of variables x_1, x_2 and y_1, y_2. They are of orders 0, 0, and of weights

2, 3. It will be remarked that u itself has equally a covariantive character, and is of weight -1.

Since the expressions $\dfrac{\sqrt{f(x)}\sqrt{f(y)}+a_x^2a_y^2}{(xy)^2}$, $\wp u$, are covariants of weight 2, c and c' must be invariants of weights 0, 2. When f is written in Weierstrass's normal form

$$u=\int_{y_1=1,\, y_2=0}^{x_1,\, x_2}\frac{(zdz)}{\sqrt{z_2(4z_1^3-g_2z_1z_2^2-g_3z_2^3)}},\quad x_1/x_2=\wp u,$$

we have $\sqrt{f(y)}=0,\ a_x^2a_y^2=2x_1x_2$.

Hence

$$\frac{2x_1x_2}{x_2^2}=c\frac{x_1}{x_2}+c'.$$

Therefore $c=2$, $c'=0$, and these are the values in the general case, since c and c' are invariant.

Therefore

$$\wp u=\frac{\sqrt{f(x)}\sqrt{f(y)}+a_x^2a_y^2}{2(xy)^2}.$$

For other proofs of this formula see Halphen, t. ii., p. 358; Enneper-Müller, p. 27; Klein, Math. Ann., t. xxvii., p. 455.

References. In this chapter we have made constant use of the work of Schwarz, Formeln und Lehrsätze nach Vorlesungen und Aufzeichnungen des Herrn K. Weierstrass; and of Halphen's Fonctions Elliptiques. We are also indebted to Müller's edition of Enneper's Elliptische Functionen, Theorie und Geschichte (where also an abundant bibliography will be found), and to Briot and Bouquet's Fonctions Elliptiques. On the important subject of modular functions the reader will not fail to consult Klein-Fricke, Theorie der Elliptischen Modulfunctionen.

Weierstrass's theory was first made accessible to the English-speaking public by Daniels (Amer. Jour., tt. vi. and vii.) and Forsyth (Quar. Jour., t. xxii). In Greenhill's recent work on the Applications of Elliptic Functions this theory is developed side by side with the older theory. Among other books whose main object is the theory of elliptic functions we may mention those of Bobek, Cayley, Durège, Königsberger, Thomae (Theorie der Functionen einer complexen veränderlichen Grösse), and Weber. Cayley's work has been translated into Italian by Brioschi. Reference must also be made to the great originators of the subject, —Euler (Institutiones Calculi Integralis), Legendre (Fonctions Elliptiques), Abel (Works), Jacobi (Works), and Gauss (Works); to whom we may now add Hermite and Weierstrass. For information on the early history of the subject the reader may consult Königsberger's valuable work, Zur Geschichte der Theorie der Elliptischen Transcendenten in den Jahren 1826–9, and Enneper-Müller's book already mentioned.

CHAPTER VIII.

Double Theta-Functions.

§ 223. Consider the double series

$$\sum_{n_1=-\infty}^{\infty}\sum_{n_2=-\infty}^{\infty} \exp \pi i[\tau_{11}(n_1+g_1/2)^2+2\tau_{12}(n_1+g_1/2)(n_2+g_2/2)$$
$$+\tau_{22}(n_2+g_2/2)^2+2(n_1+g_1/2)(v_1+h_1/2)$$
$$+2(n_2+g_2/2)(v_2+h_2/2)] \quad \cdot \cdot \cdot \cdot \cdot \cdot \cdot \cdot \quad (1),$$

where g_1, g_2, h_1, h_2 are integers, and n_1, n_2 are supposed to take all integral values from $-\infty$ to ∞ independently of each other. This series converges absolutely when the quantities $\tau_{rs}=\alpha_{rs}+i\beta_{rs}$ are chosen so that

$$\beta_{11}>0, \quad \beta_{22}>0, \quad \beta_{12}^2<\beta_{11}\beta_{22}.$$

For the absolute value of the general term is

$$\exp -\pi[\beta_{11}(n_1+g_1/2)^2+2\beta_{12}(n_1+g_1/2)(n_2+g_2/2)+\beta_{22}(n_2+g_2/2)^2$$
$$+2y_1(n_1+g_1/2)+2y_2(n_2+g_2/2)],$$

where iy_1, iy_2 are the imaginary parts of v_1, v_2. When n_1 (or n_2) becomes infinitely great, the terms

$$\beta_{11}(n_1+g_1/2)^2+2\beta_{12}(n_1+g_1/2)(n_2+g_2/2)+\beta_{22}(n_2+g_2/2)^2$$

tend to $+\infty$; and since their sum is arbitrarily great as compared with the remaining terms in the brackets, the absolute value of the general term tends to the limit 0. Since

$$\lim_{t=\infty}\frac{e^{-t}}{t^{-\kappa}}=0, \quad (\kappa=1, 2, 3, \cdots),$$

where t is a real positive quantity, it follows that the sum of the series composed of the absolute values of the terms of (1) is less than that of the series

$$(-\pi)^{-\kappa}\Sigma\Sigma[\beta_{11}(n_1+g_1/2)^2+2\beta_{12}(n_1+g_1/2)(n_2+g_2/2)+\beta_{22}(n_2+g_2/2)^2$$
$$+2y_1(n_1+g_1/2)+2y_2(n_2+g_2/2)]^{-\kappa}.$$

It can be proved that this series is convergent when $\kappa > 1$, by the extension of Cauchy's integral test which was employed in § 77. The terms of the series are the ordinates of the surface

$$\xi_3 = (\beta_{11}\xi_1^2 + 2\beta_{12}\xi_1\xi_2 + \beta_{22}\xi_2^2 + 2y_1\xi_1 + 2y_2\xi_2)^{-\kappa}$$

at the points of a network formed by unit squares in the plane $\xi_1\xi_2$. The ordinate ξ_3 is infinite only at points in the plane $\xi_1\xi_2$ which lie on a conic; and the condition $\beta_{12}^2 < \beta_{11}\beta_{22}$ ensures that this conic is an ellipse. When this ellipse is referred to its axes, the surface takes the form

$$\xi_3 = (\gamma_{11}\eta_1^2 + \gamma_{22}\eta_2^2 + \gamma_0)^{-\kappa},$$

where γ_{11} and γ_{22} have the same sign.

Let a contour be drawn in the plane $\xi_1\xi_2$, which encloses this ellipse, and conceive a cylinder drawn through the contour, parallel to the axis of ξ_3. The volume contained between the surface and the plane, exterior to the cylinder, is

$$V = \int\int \xi_3 d\eta_1 d\eta_2;$$

or if

$$\sqrt{\gamma_{11}}\eta_1 = \rho \cos\theta, \quad \sqrt{\gamma_{22}}\eta_2 = \rho \sin\theta,$$

$$V = \int\int \frac{\rho d\rho d\theta}{\sqrt{\gamma_{11}\gamma_{22}}(\rho^2 + \gamma_0)^\kappa};$$

V is finite if the integral remain finite when $\rho = \infty$; that is, if $\kappa > 1$.

Now the sum of the ordinates is equally the sum of parallelopipeds whose bases are the unit squares of the network; and if we assign to each square that ordinate which is furthest from the origin, these parallelopipeds lie within the volume V; and their sum is convergent when V is finite. Within the cylinder there is only a finite number of squares, and therefore when those ordinates are excluded which stand on the ellipse, the sum of all the ordinates is convergent. In the case when the ellipse is imaginary, the cylinder may be dispensed with.

The excluded ordinates annul the real part of the exponent in (1), but do not affect its convergence. Accordingly, the series (1) is absolutely convergent.

[Picard, Traité d'Analyse, t. i., p. 265; Riemann, Werke, p. 452; Jordan, Cours d'Analyse, first edition, t. i., p. 169.]

§ 224. Subject to the inequalities already stated, the series (1) defines a function of v_1, v_2, which is one-valued and continuous for all finite values of the variables. This function is called a *Double Theta-Function*, and is written in one or other of the forms

$$\theta\begin{bmatrix} g_1 & g_2 \\ h_1 & h_2 \end{bmatrix}(v_1, v_2); \; \theta\begin{bmatrix} g_1 & g_2 \\ h_1 & h_2 \end{bmatrix}(v_1, v_2, \tau_{11}, \tau_{12}, \tau_{22});$$

the latter form indicates that the *moduli*, or parameters, are $\tau_{11}, \tau_{12}, \tau_{22}$. The symbol $\begin{bmatrix} g_1 & g_2 \\ h_1 & h_2 \end{bmatrix}$ is called the *mark*, or characteristic, of the function; and the quantities g_1, g_2, h_1, h_2 are called the *constituents*, or elements, of the mark. We shall often write the theta-function more shortly

$$\theta\left|\begin{matrix} g_r \\ h_r \end{matrix}\right|(v_r).$$

§ 225. *Differential equations.* When the general term in (1) is differentiated twice with regard to v_r, it is multiplied by a factor $\{2\pi i(n_r + g_r/2)\}^2$; and when it is differentiated once with regard to τ_{rr}, it is multiplied by $\pi i(n_r + g_r/2)^2$. The ratio of these factors is $4\pi i$, and is therefore independent of n_1, n_2. Hence

$$\delta^2\theta/\delta v_r^2 = 4\pi i\delta\theta/\delta\tau_{rr}, \quad (r = 1, 2);$$

and similarly,

$$\delta^2\theta/\delta v_1\delta v_2 = 2\pi i\delta\theta/\delta\tau_{12}.$$

§ 226. *The addition of even integers to the constituents of a mark.* In the series (1) replace g_1 by $g_1 \pm 2$. The effect is the same as that produced by writing $n_1 \pm 1$ for n_1. But $n_1 \pm 1$, like n_1, takes all integral values from $-\infty$ to $+\infty$. Hence the series is not altered, and we have the theorem that

$$\theta\begin{bmatrix} g_1 \pm 2 & g_2 \\ h_1 & h_2 \end{bmatrix}(v_1, v_2) = \theta\begin{bmatrix} g_1 & g_2 \\ h_1 & h_2 \end{bmatrix}(v_1, v_2).$$

Similarly, g_2 may be increased or diminished by 2, without affecting the value of the function. Next replace h_1 by $h_1 \pm 2$. The general term of (1) now acquires a factor $\exp \pm 2\pi i(n_1 + g_1/2)$, whose value is $(-1)^{g_1}$. Similarly, when h_2 is replaced by $h_2 \pm 2$, the general term acquires the factor $(-1)^{g_2}$. By a combination of the preceding results, it results immediately that

$$\theta\left|\begin{matrix} g_r + 2\lambda_r \\ h_r + 2\mu_r \end{matrix}\right|(v_r) = (-1)^{\Sigma g_r\mu_r}\,\theta\left|\begin{matrix} g_r \\ h_r \end{matrix}\right|(v_r) \quad . \quad . \quad . \quad . \quad (2).$$

With the help of this formula it is always possible to replace the integers g_r, h_r by 0 or 1; this leaves the original theta-function unchanged, except perhaps as to sign. The general formula for such a reduction shows, for example, that

$$\theta\begin{bmatrix} 3 & 8 \\ 4 & 15 \end{bmatrix}(v_1, v_2) = \theta\begin{bmatrix} 1 & 0 \\ 0 & 1 \end{bmatrix}(v_1, v_2).$$

§ 227. *Even and odd functions.* Changing the signs of v_1, v_2, the general term in $\theta\left|\begin{matrix} g_r \\ h_r \end{matrix}\right|(-v_r)$ is

$$\exp \pi i[(\tau_{11}, \tau_{12}, \tau_{22})(n_1+g_1/2, n_2+g_2/2)^2+2\sum_{r=1}^{2}(n_r+g_r/2)(-v_r+h_r/2)].$$

This may be written

$$\exp \pi i[(\tau_{11}, \tau_{12}, \tau_{22})(-n_1-g_1/2, -n_2-g_2/2)^2$$
$$+2\sum_{r=1}^{2}(-n_r-g_r/2)(v_r+h_r/2)+2\sum_{r=1}^{2}(n_r+g_r/2)h_r].$$

Now $\exp \pi i \sum_{r=1}^{2}(2n_rh_r+g_rh_r) = (-1)^{\Sigma g_rh_r}$, and the remaining factor differs from the general term of (1) only as regards the signs of $n_1+g_1/2$, $n_2+g_2/2$. But $-(n_r+g_r/2)$ runs through the same set of values as $n_r+g_r/2$, of course in the reverse order. Hence we have

$$\theta\left|\begin{matrix} g_r \\ h_r \end{matrix}\right|(-v_r) = (-1)^{\Sigma g_rh_r}\,\theta\left|\begin{matrix} g_r \\ h_r \end{matrix}\right|(v_r).$$

This theorem shows that a theta-function is even or odd, according as $\sum_{r=1}^{2} g_rh_r$ is even or odd; it leads to the division of marks into even or odd, according as the corresponding functions are even or odd.

§ 228. *The theory of marks.** The symbol $\begin{bmatrix} g_1 & g_2 \cdots g_p \\ h_1 & h_2 \cdots h_p \end{bmatrix}$, composed of the $2p$ real numbers g_r, h_r, is the mark associated with a p-tuple theta-function. We shall be concerned solely with marks whose constituents are integers. When the constituents of one mark differ from those of another mark by multiples of 2, the marks are said to be congruent to each other. Thus

$$\begin{bmatrix} g_1' & g_2' \cdots g_p' \\ h_1' & h_2' \cdots h_p' \end{bmatrix} \equiv \begin{bmatrix} g_1 & g_2 \cdots g_p \\ h_1 & h_2 \cdots h_p \end{bmatrix}$$

* Clifford, Mathematical Papers, p. 356.

when $\qquad g_r \equiv g_r' \pmod 2$, and $h_r \equiv h_r' \pmod 2$,

$$r = 1, 2, \cdots, p.$$

Every mark is congruent to a *reduced mark*, whose constituents are 0, 1. The mark is said to be even or odd, according as $\sum_1^p g_r h_r$ is even or odd.

The number of even and odd marks. The number of reduced marks which can be formed with $2p$ integers is evidently 2^{2p}. Let E_p be the number of even marks, F_p the number of odd marks. The first column is (0, 0), (1, 0), (0, 1), or (1, 1). The first three, when placed before a given mark, leave its character (as even or odd) unaltered; the fourth alters the character. Hence

$$E_p = 3E_{p-1} + F_{p-1},$$

$$F_p = 3F_{p-1} + E_{p-1}.$$

Therefore

$$E_p - F_p = 2(E_{p-1} - F_{p-1}) = \cdots = 2^r(E_{p-r} - F_{p-r}) = \cdots = 2^{p-1}(E_1 - F_1);$$

that is, $\qquad E_p - F_p = 2^p$. But $E_p + F_p = 2^{2p}$;

hence $\qquad E_p = 2^{2p-1} + 2^{p-1},\ F_p = 2^{2p-1} - 2^{p-1}.$

In this chapter we consider the case $p = 2$. The number of reduced marks is 16, and of these 6 are odd. These 16 marks may be arranged in the order

00	10	01	11	00	10	01	11	00	10	01	11	00	10	01	11
00	00	00	00	10	10	10	10	01	01	01	01	11	11	11	11 . (A),
(00)	(12)	(56)	(34)	(23)	(13)	(14)	(24)	(45)	(36)	(46)	(35)	(16)	(26)	(15)	(25)

and, for this arrangement of marks, $\theta\begin{bmatrix}0\,0\\1\,0\end{bmatrix}(v_1, v_2)$ can be replaced, in what is called the current-number notation, by $\theta_5(v_1, v_2)$, and similarly for the other functions. It will be observed that $\theta_6, \theta_8, \theta_{11}, \theta_{12}, \theta_{14}, \theta_{15}$ are the 6 odd functions. The meaning of the third row in the table will be seen in the next paragraph.

§ 229. The sum or difference of two marks is found by adding or subtracting the corresponding constituents. Thus

$$\begin{bmatrix}1\,3\\2\,4\end{bmatrix} + \begin{bmatrix}5\,9\\6\,2\end{bmatrix} = \begin{bmatrix}6\,12\\8\,6\end{bmatrix} \equiv \begin{bmatrix}0\,0\\0\,0\end{bmatrix}.$$

Let us denote the reduced odd marks, namely,

$$\begin{matrix} 11 & 01 & 01 & 10 & 10 & 11 \\ 01 & 01 & 11 & 11 & 10 & 10 \end{matrix},$$

by $(1) \quad (2) \quad (3) \quad (4) \quad (5) \quad (6),$

and the even mark $\begin{matrix} 0\,0 \\ 0\,0 \end{matrix}$ by (0). From the definitions

$$(1)+(3)+(5)\equiv(0), \quad (2)+(4)+(6)\equiv(0),$$

and therefore $\quad (1)+(2)+(3)+(4)+(5)+(6)\equiv(0).$

This shows that the sum of any odd marks $\equiv$ the sum of the remaining odd marks. Further, $(1)+(3)\equiv(5)$, $(2)+(4)\equiv(6)$, etc.

Excluding (0), the other 15 marks are congruent to the sums of the 6 odd marks two at a time. We have just proved this for the odd marks, and have only to show that, of the 15 duads formed from the 6 marks, no two can give congruent marks. First $(1)+(2)$ cannot be congruent to $(1)+(3)$, since $(2)\not\equiv(3)$; and, secondly, $(1)+(2)$ cannot be congruent to $(3)+(4)$;

for $\quad (5)\not\equiv(6),$

and $\quad (1)+(2)+(5)\equiv(3)+(4)+(6);$

therefore $\quad (1)+(2)\not\equiv(3)+(4).$

Example. Prove that any even mark is congruent to the sum of three odd marks, in exactly two ways.

We shall name each of the fifteen reduced marks, (0) excluded, by that duad to which it is congruent. For example, since

$$\begin{bmatrix} 1\,1 \\ 0\,0 \end{bmatrix} \equiv \begin{bmatrix} 0\,1 \\ 1\,1 \end{bmatrix} + \begin{bmatrix} 1\,0 \\ 1\,1 \end{bmatrix} \equiv (3)+(4),$$

we shall denote $\begin{bmatrix} 1\,1 \\ 0\,0 \end{bmatrix}$ by $(3\,4)$. The corresponding theta-function will then be $\theta_{34}(v_r)$; and the excluded function will be written $\theta_{00}(v_r)$. It may be remarked that for an even mark, $(0\,0)$ excepted, the sum of the numbers in the duad is odd, and for an odd mark the sum is even. The table (A) of the preceding paragraph gives the connexion between the marks and the duads.

As in the case of the single thetas, the notations used for the double thetas are numerous. Tables for comparison are given in Cayley's memoir (Phil. Trans., t. 171, 1880), in Forsyth's memoir (ib., t. 173, 1882), and in Krause, Die Transformation der Hyperelliptischen Funktionen erster Ordnung (1886). Cayley's current number notation is from 0 to 15, instead of from 1 to 16; and he uses the letters A, B, C, D, E, F instead of $1, 2, \cdots, 6$ in forming the duads. The notation of duads employed here is based on Cayley's, and agrees with that of Staude if we replace 6 by 0 in the duads (Staude, Math. Ann., t. xxiv.).

§ 230. *The periods.*

Referring to (1), the change of v_r into $v_r + \mu_r$ is equivalent to the change of h_r into $h_r + 2\mu_r$; and, therefore, if $2\mu_r$ be an integer,

$$\theta \begin{vmatrix} g_r \\ h_r \end{vmatrix} (v_r + \mu_r) = \theta \begin{vmatrix} g_r \\ h_r + 2\mu_r \end{vmatrix} (v_r) \quad \cdot \quad \cdot \quad \cdot \quad \cdot \quad \cdot \quad (3).$$

From (2) and (3), if μ_r be an integer,

$$\theta \begin{vmatrix} g_r \\ h_r \end{vmatrix} (v_r + \mu_r) = (-1)^{\Sigma g_r \mu_r} \theta \begin{vmatrix} g_r \\ h_r \end{vmatrix} (v_r).$$

Let $\mu_1 = 1$, $\mu_2 = 0$. When $g_1 \equiv 0 \pmod 2$,

$$\theta \begin{vmatrix} g_r \\ h_r \end{vmatrix} (v_1 + 1, v_2) = \theta \begin{vmatrix} g_r \\ h_r \end{vmatrix} (v_1, v_2),$$

and we say that 1, 0 are a pair of periods of the function; but, when $g_1 \equiv 1 \pmod 2$, 2 and 0 are a pair of periods. Another pair of periods is 0, 1 when $g_2 \equiv 0 \pmod 2$; 0, 2 when $g_2 \equiv 1 \pmod 2$.

Write for shortness

$$\phi(x_1, x_2) = \tau_{11} x_1^2 + 2\tau_{12} x_1 x_2 + \tau_{22} x_2^2.$$

Then the typical term in (1) is

$$\exp \pi i \left[\phi(n_1 + g_1/2, n_2 + g_2/2) + 2\Sigma(n_r + g_r/2)(v_r + h_r/2)\right].$$

Consider the effect of changing $n_r + g_r/2$ into $n_r + g_r/2 + \lambda_r$. The expression in square brackets is increased by

$$\Sigma(n_r + g_r/2) \frac{\delta\phi(\lambda_1, \lambda_2)}{\delta\lambda_r} + \phi(\lambda_1, \lambda_2) + 2\Sigma\lambda_r(v_r + h_r/2),$$

and therefore it becomes

$$\phi(n_1 + g_1/2, n_2 + g_2/2) + 2\Sigma(n_r + g_r/2)\left(v_r + \frac{1}{2}\frac{\delta\phi(\lambda_1, \lambda_2)}{\delta\lambda_r} + h_r/2\right)$$
$$+ \phi(\lambda_1, \lambda_2) + 2\Sigma\lambda_r(v_r + h_r/2).$$

The theta-function accordingly becomes

$$\theta\begin{vmatrix} g_r \\ h_r \end{vmatrix}\left(v_r+\frac{1}{2}\frac{\delta\phi}{\delta\lambda_r}\right)\exp \pi i\,[\phi+2\,\Sigma\lambda_r(v_r+h_r/2)],$$

where $\phi=\phi(\lambda_1, \lambda_2)$.

But, as the change contemplated was merely the change of g_r into $g_r+2\lambda_r$, the function also becomes

$$\theta\begin{vmatrix} g_r+2\lambda_r \\ h_r \end{vmatrix}(v_r),$$

on the supposition that $2\lambda_r$ is an integer. Therefore

$$\theta\begin{vmatrix} g_r \\ h_r \end{vmatrix}\left(v_r+\frac{1}{2}\frac{\delta\phi}{\delta\lambda_r}\right)=\theta\begin{vmatrix} g_r+2\lambda_r \\ h_r \end{vmatrix}\cdot\exp -\pi i[\phi(\lambda_1, \lambda_2) + 2\,\Sigma\lambda_r v_r+\Sigma\lambda_r h_r] \quad . \quad . \quad (4).$$

If λ_r be an integer, the function reproduces itself except as to an exponential factor. When $\lambda_1=1$, $\lambda_2=0$, $\frac{1}{2}\frac{\delta\phi}{\delta\lambda_r}$ becomes τ_{1r}, and we have one pair of *quasi-periods*, τ_{11} and τ_{12}; when $\lambda_1=0$, $\lambda_2=1$, $\frac{1}{2}\frac{\delta\phi}{\delta\lambda_r}$ becomes τ_{r2}, and we have another pair, τ_{12} and τ_{22}. All other periods and quasi-periods are compounded from the two pairs of quasi-periods and the two pairs of periods by addition and subtraction, the general form being

$$\frac{1}{2}\frac{\delta\phi}{\delta\lambda_1}+\mu_1\epsilon_1+\mu_2\cdot 0,\quad \frac{1}{2}\frac{\delta\phi}{\delta\lambda_2}+\mu_1\cdot 0+\mu_2\cdot\epsilon_2,$$

where ϵ_1, 0 and 0, ϵ_2 are pairs of periods. Where there is no danger of misapprehension, we shall use the term periods for both the true periods and the quasi-periods.

§ 231. The theta-function satisfies the relations

$$\theta\begin{vmatrix} g_r \\ h_r \end{vmatrix}(v_1+1, v_2) \qquad =(-1)^{g_1}\theta\begin{vmatrix} g_r \\ h_r \end{vmatrix}(v_r),$$

$$\theta\begin{vmatrix} g_r \\ h_r \end{vmatrix}(v_1, v_2+1) \qquad =(-1)^{g_2}\theta\begin{vmatrix} g_r \\ h_r \end{vmatrix}(v_r)$$

$$\theta\begin{vmatrix} g_r \\ h_r \end{vmatrix}(v_1+\tau_{11}, v_2+\tau_{12})=(-1)^{h_1}\theta\begin{vmatrix} g_r \\ h_r \end{vmatrix}(v_r)\cdot\exp-\pi i(2v_1+\tau_{11}),$$

$$\theta\begin{vmatrix} g_r \\ h_r \end{vmatrix}(v_1+\tau_{12}, v_2+\tau_{22})=(-1)^{h_2}\theta\begin{vmatrix} g_r \\ h_r \end{vmatrix}(v_r)\cdot\exp-\pi i(2v_2+\tau_{22}),$$

and also satisfies three differential equations. Conversely, every function which satisfies these conditions differs from $\theta\begin{vmatrix} g_r \\ h_r \end{vmatrix}(v_r)$ only by a factor which is independent of the variables and moduli. The following is the proof generally given for this important theorem (see, for instance, Krause, Die Transformation der Hyperelliptischen Funktionen, p. 5): —

Assume, for simplicity, $g_1 = g_2 = h_1 = h_2 = 0$; and let $f(v_1, v_2)$ be a function which obeys the above equations. Because $f(v_1, v_2)$ is simply periodic as regards v_1, v_2 separately, it must be expressible as a sum of exponentials

$$\sum_{n_1=-\infty}^{\infty} \sum_{n_2=-\infty}^{\infty} a_{n_1 n_2} \exp 2\pi i(n_1 v_1 + n_2 v_2).$$

The third condition,

$$f(v_1 + \tau_{11}, v_2 + \tau_{12}) = f(v_1, v_2) \exp -\pi i(2v_1 + \tau_{11}),$$

gives

$$\Sigma \Sigma a_{n_1 n_2} \exp 2\pi i[n_1(v_1 + \tau_{11}) + n_2(v_2 + \tau_{12})]$$
$$= \Sigma \Sigma a_{n_1 n_2} \exp 2\pi i(n_1 v_1 + n_2 v_2 - v_1 - \tfrac{1}{2}\tau_{11}).$$

By equating the corresponding terms on the two sides, we get

$$a_{n_1-1, n_2} \exp \pi i[(2n_1 - 1)\tau_{11} + 2n_2\tau_{12}] = a_{n_1 n_2},$$

and repeated use of this relation gives

$$a_{n_1 n_2} = a_{0 n_2} \exp \pi i(n_1^2 \tau_{11} + 2n_1 n_2 \tau_{12}).$$

By similar reasoning based on the fourth condition, it is easy to show that

$$a_{0n_2} = a_{00} \exp \pi i n_2^2 \tau_{22}.$$

Therefore

$$f(v_1, v_2) = a_{00} \Sigma \Sigma \exp \pi i(\tau_{11} n_1^2 + 2\tau_{12} n_1 n_2 + \tau_{22} n_2^2 + 2n_1 v_1 + 2n_2 v_2)$$
$$= a_{00}\theta_{00}(v_r).$$

The quantity a_{00} is independent of the parameters, in virtue of the differential equations. For

$$\frac{\delta^2 f}{\delta v_1^2} = 4\pi i \frac{\delta f}{\delta \tau_{11}};$$

that is,

$$a_{00}\frac{\delta^2 \theta_{00}}{\delta v_1^2} = 4\pi i \left[\theta_{00}\frac{\delta a_{00}}{\delta \tau_{11}} + a_{00}\frac{\delta \theta_{00}}{\delta \tau_{11}}\right];$$

therefore $\frac{\delta a_{00}}{\delta \tau_{11}} = 0$, which shows that a_{00} is independent of τ_{11}. Similarly, a_{00} is independent of τ_{12}, τ_{22}.

§ 232. *General formulæ for the addition of periods and half-periods.*

From (3) and (4),

$$\theta\begin{vmatrix} g_r \\ h_r \end{vmatrix}\left(v_r + \mu_r + \tfrac{1}{2}\frac{\delta\phi}{\delta\lambda_r}\right) = \theta\begin{vmatrix} g_r \\ h_r + 2\,\mu_r \end{vmatrix}\left(v_r + \tfrac{1}{2}\frac{\delta\phi}{\delta\lambda_r}\right)$$

$$= \theta\begin{vmatrix} g_r + 2\,\lambda_r \\ h_r + 2\,\mu_r \end{vmatrix}(v_r)\cdot\exp -\pi i\,[\phi + 2\,\Sigma\lambda_r v_r + \Sigma\lambda_r(h_r + 2\,\mu_r)].$$

If λ_r and μ_r be integers, this is

$$\theta\begin{vmatrix} g_r \\ h_r \end{vmatrix}(v_r)\cdot\exp -\pi i[\phi + 2\,\Sigma\lambda_r v_r + \Sigma(g_r\mu_r + h_r\lambda_r)] \quad . \quad . \quad (5),$$

the formula for the addition of periods.

Let $m = \begin{vmatrix} g_r \\ h_r \end{vmatrix}$, $m_0 = \begin{vmatrix} \lambda_r \\ \mu_r \end{vmatrix}$; then for $\Sigma(g_r\mu_r + h_r\lambda_r)$ the symbol $m \mid m_0$ is used.

If λ_r and μ_r be multiples of 1/2, write $\lambda_r/2$, $\mu_r/2$, in place of λ_r, μ_r. Then

$$\theta\begin{vmatrix} g_r \\ h_r \end{vmatrix}\left(v_r + \mu_r/2 + \frac{1}{4}\delta\phi/\delta\lambda_r\right)$$

$$= \theta\begin{vmatrix} g_r + \lambda_r \\ h_r + \mu_r \end{vmatrix}(v_r)\cdot\exp -\pi i\left[\frac{\phi}{4} + \Sigma\lambda_r v_r + \frac{1}{2}\Sigma\lambda_r(h_r + \mu_r)\right] \quad . \quad . \quad (6),$$

the formula for the addition of half-periods. If, in this equation, $\begin{vmatrix} g_r \\ h_r \end{vmatrix}$ be a reduced mark, and if λ_r, μ_r receive only the values 0, 1, the reduced form of the new mark is

$$\begin{vmatrix} g_r + \lambda_r - 2\,g_r\lambda_r \\ h_r + \mu_r - 2\,h_r\mu_r \end{vmatrix},$$

since each constituent is now 0 or 1; and the formula becomes *

$$\theta\begin{vmatrix} g_r \\ h_r \end{vmatrix}\left(v_r + \mu_r/2 + \frac{1}{4}\delta\phi/\delta\lambda_r\right)$$

$$= \theta\begin{vmatrix} g_r + \lambda_r - 2g_r\lambda_r \\ h_r + \mu_r - 2h_r\mu_r \end{vmatrix}(v_r)\cdot\exp -\pi i\left[\frac{\phi}{4} + \Sigma\lambda_r v_r + \frac{1}{2}\Sigma\lambda_r(h_r + \mu_r) + \Sigma h_r\mu_r(g_r + \lambda_r)\right] \quad . \quad . \quad (7).$$

---

* It is easy to remove the restriction on λ_r, μ_r in (7), but the result is seldom needed. Most writers on multiple θ-functions leave the marks in the general formula unreduced.

Example.

$$\theta\begin{vmatrix} g_r \\ h_r \end{vmatrix}(v_r+\mu_r/2+\tfrac{1}{4}\delta\phi/\delta\lambda_r)\cdot\theta\begin{vmatrix} g_r \\ h_r \end{vmatrix}(v_r-\mu_r/2-\tfrac{1}{4}\delta\phi/\delta\lambda_r)$$
$$=(-1)^{\Sigma(g_r+\lambda_r)\mu_r}\cdot\theta^2\begin{vmatrix} g_r+\lambda_r \\ h_r+\mu_r \end{vmatrix}(v_r)\cdot\exp[-\pi i\phi/2].$$

From formula (7) it is clear that each of the 16 theta-functions can be changed into any of the other 15, save as to an exponential factor, by the addition of some pair of half-periods; the number of pairs of values of λ_r, μ_r, given by $\lambda_r=0, 1$ and $\mu_r=0, 1$ being just 15, when the case $\lambda_1=\lambda_2=\mu_1=\mu_2=0$ is excluded.

In the current number notation the result of adding two marks is not immediately evident, but the case is otherwise with the notation by means of duads. After the addition of half-periods the suffix of a theta-function is altered. To obtain the new suffix from the old, when any half-periods are added, let (α), (β), (γ), (δ), (ϵ), (ζ), be the six odd marks (1), (2), (3), (4), (5), (6), in any order, and observe that we may regard $\begin{matrix}\lambda_1 & \lambda_2 \\ \mu_1 & \mu_2\end{matrix}$ as a mark. Let it be expressed in the form $(\alpha)+(\beta)$ or $(\alpha\beta)$. Four cases arise: according as $\begin{vmatrix} g_r \\ h_r \end{vmatrix}$ is of the form $(\gamma\delta)$, $(\alpha\gamma)$, $(\alpha\beta)$, (00),

$$\begin{vmatrix} g_r+\lambda_r \\ h_r+\mu_r \end{vmatrix} \text{ is of the form } \begin{aligned}&(\alpha)+(\beta)+(\gamma)+(\delta),\\ &(\alpha)+(\beta)+(\alpha)+(\gamma),\\ &(\alpha)+(\beta)+(\alpha)+(\beta),\\ &(\alpha)+(\beta);\end{aligned}$$

and therefore $\equiv(\epsilon\zeta)$, $(\beta\gamma)$, (00), $(\alpha\beta)$. We can express this shortly by the symbolic equations

$$(\alpha\beta)\theta_{\gamma\delta}=\theta_{\epsilon\zeta},\ (\alpha\beta)\theta_{\alpha\gamma}=\theta_{\beta\gamma},\ (\alpha\beta)\theta_{\alpha\beta}=\theta_{00},$$
$$(\alpha\beta)\theta_{00}=\theta_{\alpha\beta}.$$

No attention is here paid to the exponential factor, which is readily filled in from formula (6). As an illustration of the above, let $\begin{vmatrix} g \\ h \end{vmatrix}=\begin{bmatrix}1 & 0 \\ 0 & 1\end{bmatrix}$, $\begin{vmatrix} \lambda \\ \mu \end{vmatrix}=\begin{bmatrix}0 & 1 \\ 0 & 0\end{bmatrix}$. These two marks are (36), (56); hence the new theta-function is θ_{35}; also the multiplier is

$$\exp -\pi i\left(\frac{\tau_{22}}{4}+v_2+\frac{1}{2}\right).$$

Hence

$$\theta_{36}(v_1+\tau_{12}/2,\ v_2+\tau_{22}/2)=\theta_{35}(v_1,\ v_2)\exp-\pi i(\tau_{22}/4+v_2+1/2).$$

§ 233. It is well to notice that the symbol $m \mid m_0$ of § 232 $\equiv 1$ or 0 (mod 2), according as m and m_0 do or do not contain a common letter. To prove this, we write the odd marks as follows:—

$$1, 3, 5 \quad \begin{matrix} 11 & 01 & 10 \\ 01' & 11' & 10' \end{matrix}$$

$$2, 4, 6 \quad \begin{matrix} 01 & 10 & 11 \\ 01' & 11' & 10' \end{matrix}$$

and observe that if m and m_0 be different marks, selected from the same row, $m \mid m_0 \equiv 1$; if m, m_0 be selected from different rows, $m \mid m_0 \equiv 0$, while $m \mid m \equiv 0$. Now

$$\begin{aligned} \alpha\beta \mid \alpha\gamma &= \alpha \mid \alpha + \alpha \mid \beta + \alpha \mid \gamma + \beta \mid \gamma \\ &\equiv \alpha \mid \beta + \alpha \mid \gamma + \beta \mid \gamma; \end{aligned}$$

either α, β, γ are all from the same row, or two of them are from the same row. In either case,

$$\alpha \mid \beta + \alpha \mid \gamma + \beta \mid \gamma \equiv 1.$$

Again, $$\alpha\beta \mid \gamma\delta = \alpha \mid \gamma + \alpha \mid \delta + \beta \mid \gamma + \beta \mid \delta;$$

if γ, δ be from the same row, $\alpha \mid \gamma + \alpha \mid \delta \equiv 0$, and $\beta \mid \gamma + \beta \mid \delta \equiv 0$; but if γ, δ be from different rows, $\alpha \mid \gamma + \alpha \mid \delta \equiv 1$, and $\beta \mid \gamma + \beta \mid \delta \equiv 1$. In either case

$$\alpha\beta \mid \gamma\delta \equiv 0.$$

Lastly, it is evident that $00 \mid \alpha\beta \equiv 0$.

It will be useful in the sequel to remark that, given $m_0 = (\alpha\beta)$, there are eight marks m such that $m \mid m_0 \equiv 0$; and that, since $m \mid m_0 \equiv 0$ implies $(m + m_0) \mid m_0 \equiv 0$, these eight arrange themselves in four pairs:—

$$\left.\begin{matrix} m & m+m_0=mm_0 \\ 00 & \alpha\beta \\ \gamma\delta & \epsilon\zeta \\ \gamma\epsilon & \zeta\delta \\ \gamma\zeta & \delta\epsilon \end{matrix}\right\} \quad \ldots\ldots\ldots \quad (B);$$

and that, similarly, when $m \mid m_0 \equiv 1$, there are eight marks m which may be arranged in the four pairs

$$\left.\begin{matrix} m & mm_0 \\ \alpha\gamma & \beta\gamma \\ \alpha\delta & \beta\delta \\ \alpha\epsilon & \beta\epsilon \\ \alpha\zeta & \beta\zeta \end{matrix}\right\} \quad \ldots\ldots\ldots \quad (B').$$

Since the mark m_0 can be chosen in 15 ways, (00) being excluded, there are 30 such sets of four pairs, into which fall the $16 \cdot 15/2$, or 120, pairs which can be formed from the 16 marks.

§ 234. *The Rosenhain hexads.*

The six odd marks form a Rosenhain hexad, or, simply, a hexad; and any six marks obtained from the odd ones, by the addition of the same pair of half-periods, also form a hexad. Thus there are 16 hexads, as follows: —

$$\left.\begin{array}{r|cccccc}
 & \theta_{13} & \theta_{35} & \theta_{15} & \theta_{24} & \theta_{46} & \theta_{26} \\
(12) & \theta_{23} & \theta_{46} & \theta_{25} & \theta_{14} & \theta_{35} & \theta_{16} \\
(13) & \theta_{00} & \theta_{15} & \theta_{35} & \theta_{56} & \theta_{25} & \theta_{45} \\
(14) & \theta_{34} & \theta_{26} & \theta_{45} & \theta_{12} & \theta_{16} & \theta_{35} \\
(15) & \theta_{35} & \theta_{13} & \theta_{00} & \theta_{36} & \theta_{23} & \theta_{34} \\
(16) & \theta_{36} & \theta_{24} & \theta_{56} & \theta_{35} & \theta_{14} & \theta_{12} \\
(23) & \theta_{12} & \theta_{25} & \theta_{46} & \theta_{34} & \theta_{15} & \theta_{36} \\
(24) & \theta_{56} & \theta_{16} & \theta_{36} & \theta_{00} & \theta_{26} & \theta_{46} \\
(25) & \theta_{46} & \theta_{23} & \theta_{12} & \theta_{45} & \theta_{13} & \theta_{56} \\
(26) & \theta_{45} & \theta_{14} & \theta_{34} & \theta_{46} & \theta_{24} & \theta_{00} \\
(34) & \theta_{14} & \theta_{45} & \theta_{26} & \theta_{23} & \theta_{36} & \theta_{15} \\
(35) & \theta_{15} & \theta_{00} & \theta_{13} & \theta_{16} & \theta_{12} & \theta_{14} \\
(36) & \theta_{16} & \theta_{56} & \theta_{24} & \theta_{15} & \theta_{34} & \theta_{23} \\
(45) & \theta_{26} & \theta_{34} & \theta_{14} & \theta_{25} & \theta_{56} & \theta_{13} \\
(46) & \theta_{25} & \theta_{12} & \theta_{23} & \theta_{26} & \theta_{00} & \theta_{24} \\
(56) & \theta_{24} & \theta_{36} & \theta_{16} & \theta_{13} & \theta_{45} & \theta_{25}
\end{array}\right\} \quad \cdot \quad \cdot \quad (C).$$

The hexads will be named by the pairs in the column to the left. For example, the hexad [15, 00, 13, 16, 12, 14] will be called the hexad (35). In each hexad, except the first, there are two odd functions; any two hexads have a common pair of functions; and any function occurs in six hexads. The hexads are of two types, which may be called pentagonal and triangular. The pentagonal ones are of the form

$$00,\ \alpha\beta,\ \alpha\gamma,\ \alpha\delta,\ \alpha\epsilon,\ \alpha\zeta,$$

and are obtained from the original hexad by the addition of an odd mark. The triangular ones are of the form

$$\alpha\beta,\ \beta\gamma,\ \gamma\alpha,\ \delta\epsilon,\ \epsilon\zeta,\ \zeta\delta,$$

and are obtained from the original hexad by the addition of an even mark. Any pair of functions belongs to two hexads. Thus $\theta_{\alpha\beta}$, $\theta_{\gamma\delta}$, belong to the two triangular hexads

$$(\alpha\beta\epsilon,\ \gamma\delta\zeta) \text{ and } (\alpha\beta\zeta,\ \gamma\delta\epsilon).$$

The functions $\theta_{\alpha\beta}$, $\theta_{\alpha\gamma}$ belong to the pentagonal system (α) and to the triangular system $(\alpha\beta\gamma,\ \delta\epsilon\zeta)$. The functions θ_{00}, $\theta_{\alpha\beta}$, belong to the two pentagonal systems α and β. The name of each hexad is inferred from the marks in it very simply. In the triangular hexad $(\alpha\beta\gamma,\ \delta\epsilon\zeta)$ the figures in *each* triad are all like (all odd or all even), or two are like and one unlike. If all are like, the hexad is (00); if not, the two unlike ones name the hexad. For instance, the two unlike ones in (126, 345), which are 1 and 4, name the hexad (14). The hexad is, in point of fact, derived from (246, 135) by the interchange of 1 and 4. In the pentagonal hexad (α) the hexad is named by the two figures congruent to α; for instance, we speak of the hexad (35) instead of the hexad (1).

§ 235. *The Göpel tetrads.*

Any two marks belong to two hexads; a third mark can be chosen in six ways, so that the three do not belong to any hexad. If the two be, for instance, 00 and $\alpha\beta$, the third is chosen from $\gamma\delta \cdot \epsilon\zeta$, $\gamma\epsilon \cdot \delta\zeta$, $\gamma\zeta \cdot \delta\epsilon$; for it must belong to neither of the pentagonal hexads (α), (β). Taking any of these six marks, say $\gamma\delta$, the marks 00, $\alpha\beta$, $\gamma\delta$ do not belong to any hexad, and further, no three of the four marks 00, $\alpha\beta$, $\gamma\delta$, $\epsilon\zeta$ belong to any hexad.

Such a system of four marks or functions is called a Göpel tetrad. It is evident that a Göpel tetrad may be constructed by taking any two of a set of four pairs (§ 233). From the 30 sets of pairs, two of a set can be chosen in 30×6 ways; and since a tetrad is compounded of pairs in three ways, as $00 \cdot \alpha\beta$, $\gamma\delta \cdot \epsilon\zeta$; $00 \cdot \gamma\delta$, $\alpha\beta \cdot \epsilon\zeta$; $00 \cdot \epsilon\zeta$, $\alpha\beta \cdot \gamma\delta$, the total number of Göpel tetrads is $30 \times 6/3$ or 60. There are 15 of the type $00 \cdot \alpha\beta \cdot \gamma\delta \cdot \epsilon\zeta$, and 45 of the type $\alpha\gamma \cdot \beta\gamma \cdot \alpha\delta \cdot \beta\delta$.

§ 236. *The zeros.*

It is evident that an odd θ-function vanishes when v_1 and v_2 vanish. We express this by saying that 0, 0 are a common pair of zeros of the first hexad. The second hexad is obtained by the addition of the half-periods, which have the mark (12), to v_1, v_2; that is, by writing $v_r + \tau_{1r}/2$ for v_r. Therefore $\tau_{11}/2$, $\tau_{12}/2$ are a common pair of zeros for the second hexad; and so on. Thus six pairs of zeros, which are incongruent as regards the periods, are determined for each function. For instance, referring to the position of θ_{00} in the columns of the table of hexads, we see that θ_{00} vanishes, when v_1, v_2, are equal to the half-periods (13), (35), (15), (24), (46), (26); that is, when

$$v_1 = (1+\tau_{11})/2,\ (\tau_{11}+\tau_{12})/2,\ (1+\tau_{12})/2,\ (1+\tau_{11}+\tau_{12})/2,\ \tau_{12}/2,\ (1+\tau_{11})/2,$$

$$v_2 = \tau_{12}/2,\ (1+\tau_{12}+\tau_{22})/2,\ (1+\tau_{22})/2,\ (\tau_{12}+\tau_{22})/2,\ (1+\tau_{22})/2,\ (1+\tau_{12})/2.$$

§ 237. *Approximate values of the functions.*

In the double series (1) consider the lowest powers of $\exp \pi i\tau_{11}$, $\exp \pi i\tau_{12}$, $\exp \pi i\tau_{22}$. If $g_1=0$, $g_2=0$, the lowest power occurs when $n_1=n_2=0$, and is then merely 1. If $g_1=1$, $g_2=0$, the lowest power occurs when $n_1=n_2=0$ and when $n_1=-1$, $n_2=0$; and, combining these, it is

$$2\cos\pi(v_1+h_1/2)\exp\tfrac{1}{4}\pi i\tau_{11}.$$

Similarly, if $g_1=0$, $g_2=1$, the lowest power is

$$2\cos\pi\,(v_2+h_2/2)\exp\tfrac{1}{4}\pi i\tau_{22}.$$

If $g_1=1$, $g_2=1$, the lowest powers are contributed by

$$n_1,\ n_2=0,\ 0;\ -1,\ 0;\ 0,\ -1;\ -1,\ -1,$$

and are

$$2\cos\pi\left(v_1+v_2+\frac{h_1+h_2}{2}\right)\exp\frac{\pi i}{4}(\tau_{11}+2\tau_{12}+\tau_{22}) + 2\cos\pi\left(v_1-v_2+\frac{h_1-h_2}{2}\right)\exp\frac{\pi i}{4}(\tau_{11}-2\tau_{12}+\tau_{22}).$$

These values are sometimes useful for the verification of formulæ. If we put $v_1=v_2=0$, we obtain approximate values of $\theta_{\alpha\beta}(0, 0)$ for small values of

$$\exp\pi i\tau_{11},\ \exp\pi i\tau_{12},\ \exp\pi i\tau_{22}.$$

When $(\alpha\beta)$ is any even mark, the value of $\theta_{\alpha\beta}(0, 0)$ will be denoted by $c_{\alpha\beta}$.

§ 238. *Riemann's theta-formula.*

Let

$$\left.\begin{aligned} a_r+b_r+c_r+d_r&=2a_r' \\ a_r+b_r-c_r-d_r&=2b_r' \\ a_r-b_r+c_r-d_r&=2c_r' \\ a_r-b_r-c_r+d_r&=2d_r' \end{aligned}\right\} \quad \ldots\ldots\ldots \quad \text{(i.)}.$$

$$(r=1,\ 2).$$

Solving these equations for a_r, b_r, c_r, d_r, we find that the equations (i.) are still true when unaccented letters are replaced by accented, and accented by unaccented. Hence accented and unaccented letters may be interchanged in any equations we may deduce.

When a_r becomes $$a_r+\mu_r+\frac{1}{2}\frac{\delta\phi}{\delta\lambda_r},\ \lambda_r=0,\,1;\ \mu_r=0,\,1,$$

a_r' becomes $$a_r'+\frac{1}{2}\mu_r+\frac{1}{4}\frac{\delta\phi}{\delta\lambda_r},$$

and the same change is made in b_r', c_r', d_r'. By such additions of periods and half periods, $\theta_m a_r$ becomes, from formula (5),

$$\theta_m a_r \exp -\pi i[\phi+2\Sigma\lambda_r a_r + m\mid m_0],$$

and $\theta_{m'}a_r'\theta_{m'}b_r'\theta_{m'}c_r'\theta_{m'}d_r'$ becomes, from formula (7),

$$\theta_{m''}a_r'\cdot\theta_{m''}b_r'\cdot\theta_{m''}c_r'\cdot\theta_{m''}d_r'\cdot$$
$$\exp -\pi i[\phi+\Sigma\lambda_r(a_r'+b_r'+c_r'+d_r')+\text{an even integer}],$$

where $$m'=\begin{vmatrix} g_r' \\ h_r' \end{vmatrix},\ m''=\begin{vmatrix} g_r'+\lambda_r-2g_r'\lambda_r \\ h_r'+\mu_r-2h_r'\mu_r \end{vmatrix},$$

or $$\theta_{m''}a_r'\cdot\theta_{m''}b_r'\cdot\theta_{m''}c_r'\cdot\theta_{m''}d_r'\cdot\exp -\pi i(\phi+2\Sigma\lambda_r a_r).$$

Now when m' runs through all the 16 possible values, m'' does the same.

Hence $$\sum_{m'}(-1)^{m\mid m'}\theta_{m'}a_r'\cdot\theta_{m'}b_r'\cdot\theta_{m'}c_r'\cdot\theta_{m'}d_r',$$

summed for all reduced marks, obeys the fundamental relations of § 231, when regarded as a function of a_r; that is, it reproduces itself, save as to the exponential factor $\exp -\pi i[\phi+2\Sigma\lambda_r a_r+m\mid m_0]$. Therefore it differs from $\theta_m a_r$ only by a factor independent of a_r. Similar reasoning holds for b_r, c_r, d_r, so that

$$\sum_{m'}(-1)^{m\mid m'}\theta_{m'}a_r'\cdot\theta_{m'}b_r'\cdot\theta_{m'}c_r'\cdot\theta_{m'}d_r' = k\theta_m a_r\cdot\theta_m b_r\cdot\theta_m c_r\cdot\theta_m d_r \quad\cdot\quad\cdot\quad (8),$$

where k is independent of a_r, b_r, c_r, d_r, and therefore either a numerical constant or a function of the moduli τ_{11}, τ_{12}, τ_{22}.

There is a more general form of the theta-formula which has been used by Prym and Krazer as a basis for their researches on the theta-functions. This extended formula, in the case $p=2$, can be found by changing b_r, c_r, d_r into

$$b_r+\frac{1}{2}\mu_r'+\frac{1}{4}\frac{\delta\phi'}{\delta\lambda_r'},\ c_r+\frac{1}{2}\mu_r''+\frac{1}{4}\frac{\delta\phi''}{\delta\lambda_r''},\ d_r+\frac{1}{2}\mu_r'''+\frac{1}{4}\frac{\delta\phi'''}{\delta\lambda_r'''},$$

where $\lambda_r'+\lambda_r''+\lambda_r'''=\mu_r'+\mu_r''+\mu_r'''=0$, and $\phi^{(\kappa)}$ means $\phi(\lambda_1^{(\kappa)}, \lambda_2^{(\kappa)})$. For then a_r' is unchanged, by (i.), and b_r, c_r, d_r are changed by precisely the same amounts as b_r', c_r', d_r'. By these additions of half-periods any mark m is changed into mm_1, mm_2, mm_3, where

$$m_1 = \begin{vmatrix} \lambda_r' \\ \mu_r' \end{vmatrix}, \quad m_2 = \begin{vmatrix} \lambda_r'' \\ \mu_r'' \end{vmatrix}, \quad m_3 = \begin{vmatrix} \lambda_r''' \\ \mu_r''' \end{vmatrix},$$

and the marks mm_1, mm_2, mm_3 are unreduced. The right-hand side of the formula (8), already obtained, changes by a three-fold use of formula (6) into

$$k\theta_m a_r \cdot \theta_{mm_1} b_r \cdot \theta_{mm_2} c_r \cdot \theta_{mm_3} d_r \cdot$$
$$\exp - \pi i \{ \tfrac{1}{4}(\phi' + \phi'' + \phi''') + \Sigma(\lambda_r' b_r + \lambda_r'' c_r + \lambda_r''' d_r)$$
$$+ \tfrac{1}{2}\Sigma(\lambda_r'\mu_r' + \lambda_r''\mu_r'' + \lambda_r'''\mu_r''')\}.$$

On the left-hand side of (8) there enters the same exponential, since

$$\lambda_r' b_r' + \lambda_r'' c_r' + \lambda_r''' d_r' = \lambda_r' b_r + \lambda_r'' c_r + \lambda_r''' d_r.$$

Hence, $\sum_{m'}(-1)^{m|m'}\theta_{m'} a_r' \cdot \theta_{m'm_1} b_r' \cdot \theta_{m'm_2} c_r' \cdot \theta_{m'm_3} d_r'$

$$= k\theta_m a_r . \theta_{mm_1} b_r \cdot \theta_{mm_2} c_r \cdot \theta_{mm_3} d_r \quad . \quad . \quad . \quad . \quad (9),$$

k being the same as before.

§ 239. Changing d_r' into $-d_r'$, write as in § 211,

$$\begin{array}{l|l|l}
2a_r' = a_r+b_r+c_r+d_r & 2a_r'' = a_r'+b_r'+c_r'+d_r' & 2a_r = a_r''+b_r''+c_r''+d_r'' \\
2b_r' = a_r+b_r-c_r-d_r & 2b_r'' = a_r'+b_r'-c_r'-d_r' & 2b_r = a_r''+b_r''-c_r''-d_r'' \\
2c_r' = a_r-b_r+c_r-d_r & 2c_r'' = a_r'-b_r'+c_r'-d_r' & 2c_r = a_r''-b_r''+c_r''-d_r'' \\
2d_r' = -a_r+b_r+c_r-d_r & 2d_r'' = -a_r'+b_r'+c_r'-d_r' & 2d_r = -a_r''+b_r''+c_r''-d_r''.
\end{array}$$

The formula (8) is now

$$k\theta_m a_r \cdot \theta_m b_r \cdot \theta_m c_r \cdot \theta_m d_r = \sum_{m'}(-1)^{m|m'}\theta_{m'} a_r' \cdot \theta_{m'} b_r' \cdot \theta_{m'} c_r' \cdot \theta_{m'}(-d_r');$$

and similarly, by virtue of the remark about interchange of accents,

$$k\theta_m a_r'' \cdot \theta_m b_r'' \cdot \theta_m c_r'' \cdot \theta_m(-d_r'') = \sum_{m'}(-1)^{m|m'}\theta_{m'} a_r' \cdot \theta_{m'} b_r' \cdot \theta_{m'} c_r' \cdot \theta_{m'} d_r'.$$

Let m be any odd mark. Then, if we write

$$\chi_m^{(\kappa)} \text{ for } \theta_m a_r^{(\kappa)} \cdot \theta_m b_r^{(\kappa)} \cdot \theta_m c_r^{(\kappa)} \cdot \theta_m d_r^{(\kappa)},$$

we have, after subtraction,

$$k(\chi_m + \chi_m'') = -2\sum_{m'}(-1)^{m|m'}\chi_{m'}',$$

where the summation applies to all odd marks.

It has been stated (in § 233) that when m, m' are like odd marks, $m \mid m' \equiv 1$; when m, m' are unlike, $m \mid m' \equiv 0$; and when m, m' are identical, $m \mid m' \equiv 0$. Therefore, putting for example $m = (13)$, the last equation becomes

$$k(\chi_{13} + \chi_{13}'') = -2(\chi_{13}' - \chi_{15}' - \chi_{35}' + \chi_{24}' + \chi_{26}' + \chi_{46}'),$$

or $$k(\chi_{13} + \chi_{13}'') + 4\chi_{13}' = 2(\chi_{13}' + \chi_{15}' + \chi_{35}' - \chi_{24}' - \chi_{26}' - \chi_{46}').$$

Similarly, $k(\chi_{15} + \chi_{15}'') + 4\chi_{15}' =$ the same expression. Therefore

$$k(\chi_{13} + \chi_{13}'') + 4\chi_{13}' = k(\chi_{15} + \chi_{15}'') + 4\chi_{15}'.$$

The same is true for any interchange of accents, so that

$$k(\chi_{13} + \chi_{13}') + 4\chi_{13}'' = k(\chi_{15} + \chi_{15}') + 4\chi_{15}''.$$

Therefore by subtraction

$$(k-4)(\chi_{13}' - \chi_{13}'') = (k-4)(\chi_{15}' - \chi_{15}''),$$

an equation which cannot be true for all values of a_r', a_r'', etc., except when $k = 4$. For instance, if $a_r' = a_r'' =$ the half period (15), χ_{13}' and χ_{13}'' vanish, while the other terms need not. Therefore for certain values of the variables $k = 4$, and since k is independent of the variables, its value is always 4. Writing ψ_m for $\chi_m + \chi_m' + \chi_m''$, we have

$$2\psi_{13} = \chi_{13}' + \chi_{15}' + \chi_{35}' - \chi_{24}' - \chi_{26}' - \chi_{46}',$$

and therefore

$$\psi_m + (-1)^{m|m'}\psi_{m'} = 0 \quad \cdot \cdot \cdot \cdot \cdot \cdot \cdot \cdot \cdot \quad (10),$$

where m and m' are different odd marks.

§ 240. Dealing similarly with the general formula (9), we have

$$4\theta_m a_r \cdot \theta_{mm_1} b_r \cdot \theta_{mm_2} c_r \cdot \theta_{mm_3} d_r = \Sigma(-1)^{m|m'}\theta_{m'} a_r' \cdot \theta_{m'm_1} b_r' \cdot \theta_{m'm_2} c_r' \cdot \theta_{m'm_3}(-d_r'),$$

$$4\theta_m a_r'' \cdot \theta_{mm_1} b_r'' \cdot \theta_{mm_2} c_r'' \cdot \theta_{mm_3}(-d_r'')$$
$$= \Sigma(-1)^{m|m'}\theta_{m'} a_r' \cdot \theta_{m'm_1} b_r' \cdot \theta_{m'm_2} c_r' \cdot \theta_{m'm_3} d_r'.$$

Case i. Let mm_3 be an odd mark. Then, by subtraction,

$$2(\chi_m + \chi_m'') = -\Sigma(-1)^{m|m'}\chi_{m'}',$$

where $\chi_m^{(\kappa)}$ now denotes $\theta_m a_r^{(\kappa)} \cdot \theta_{mm_1} b_r^{(\kappa)} \cdot \theta_{mm_2} c_r^{(\kappa)} \cdot \theta_{mm_3} d_r^{(\kappa)}$, and $m'm_3$ is any odd mark. Let, for example, $m_3 = (12)$. Then, because mm_3 and $m'm_3$ are odd marks, m and m' belong to the hexad

14, 16, 23, 25, 35, 46,

derived from the hexad of odd marks by the operation (12), as in § 234.

The symbol $m \mid m' \equiv 1$, when m and m' have one common figure. To fix ideas, suppose $m = (14)$; then

$$2(\chi_{14} + \chi_{14}'') = -\chi_{14}' + \chi_{16}' + \chi_{46}' - \chi_{23}' - \chi_{25}' - \chi_{35}';$$

or, if $$\psi_m = \sum_{\kappa} \chi_m^{(\kappa)},$$

$$2\psi_{14} = \chi_{14}' + \chi_{16}' + \chi_{46}' - \chi_{23}' - \chi_{25}' - \chi_{35}',$$

and therefore $$\psi_m + (-1)^{m \mid m'} \psi_{m'} = 0 \quad . \quad . \quad . \quad . \quad . \quad . \quad . \quad . \quad . \quad (11),$$

where m and m' belong to the hexad (12). And similarly for any hexad.

§ 241. *Case* ii. Adding the same equations, we have

$$2(\chi_m - \chi_m'') = \Sigma(-1)^{m \mid m} \chi_m',$$

where $m'm_3$ is any even mark, mm_3 any odd mark. Interchanging the accents, we have in all three such equations, which give on addition

$$\Sigma(-1)^{m \mid m'} \psi_{m'} = 0.$$

To reduce this equation, which contains ten terms, let us take two different values of m. First let these be of the form $(\alpha\beta)$, $(\gamma\delta)$, as for instance (12) and (34). These belong to two hexads, (23) and (14). If both values of m make mm_3 odd, m_3 must be (23) or (14). Selecting the value $m_3 = (23)$, m' is any mark not contained in the hexad (23), since $m'm_3$ is even. That is, m' is obtained from any even mark η by the addition of (23). The even marks are

00, 12, 14, 16, 23, 25, 34, 36, 45, 56,

and therefore m' is one of

23, 13, 56, 45, 00, 35, 24, 26, 16, 14.

When m takes in succession the values (12) and (34), $\Sigma(-1)^{m \mid m'} \psi_{m'}$ becomes respectively

$$-\psi_{23} - \psi_{13} + \psi_{56} + \psi_{45} + \psi_{00} + \psi_{35} - \psi_{24} - \psi_{26} - \psi_{16} - \psi_{14} = 0,$$

and $$-\psi_{23} - \psi_{13} + \psi_{56} - \psi_{45} + \psi_{00} - \psi_{35} - \psi_{24} + \psi_{26} + \psi_{16} - \psi_{14} = 0.$$

Therefore, by subtraction,

$$\psi_{35} + \psi_{45} = \psi_{16} + \psi_{26}.$$

But this equation may be obtained directly from the consideration of the ten even marks; for

$$m' = \eta + (23);$$

therefore $$m \mid m' = m \mid \eta + m \mid (23),$$

and consequently $\Sigma(-1)^{m\,|\,\eta}\psi_{m'}=0$. Writing out $\Sigma(-1)^{m\,|\,\eta}\psi_\eta$ when $m=(12)$ and (34), we may combine the expressions, and replace each η that remains by the corresponding m'. Thus, from the even marks 00, 12, 14, 16, 23, 25, 34, 36, 45, 56, $m=(12)$ gives the sequence $\psi_{00}+\psi_{12}-\psi_{14}-\psi_{16}-\psi_{23}-\psi_{25}+\psi_{34}+\psi_{36}+\psi_{45}+\psi_{56}$;

$m=(34)$ gives

$$\psi_{00}+\psi_{12}-\psi_{14}+\psi_{16}-\psi_{23}+\psi_{25}+\psi_{34}-\psi_{36}-\psi_{45}+\psi_{56};$$

the half-difference is

$$-\psi_{16}-\psi_{25}+\psi_{36}+\psi_{45} \quad . \quad . \quad . \quad . \quad . \quad . \quad . \quad . \quad (12),$$

and, when (23) is added to each mark, there results as before the equation

$$\psi_{35}+\psi_{45}=\psi_{26}+\psi_{16}.$$

The other value for m_3 is (14). This, in the second method of reduction, has only to be added to each mark in formula (12), giving

$$\psi_{46}+\psi_{36}=\psi_{25}+\psi_{15}.$$

The argument is clearly independent of the choice of $(\alpha\beta)$, $(\gamma\delta)$, when both are even; and it is left to the reader to verify that, whatever be the marks m, provided their form is $(\alpha\beta)$, $(\gamma\delta)$, the reduced equation is

$$\psi_{\alpha\zeta}+\psi_{\beta\zeta}=\psi_{\gamma\epsilon}+\psi_{\delta\epsilon}.$$

Next let the marks m be $(\alpha\beta)$, $(\alpha\gamma)$. For the sake of definiteness suppose them to be (12) and (13). The values of m_3 are the names of the hexads in which (12) and (13) occur, namely (35) and (25). When $m=(12)$, $\Sigma(-1)^{m\,|\,\eta}\psi_\eta$ is, as before,

$$\psi_{00}+\psi_{12}-\psi_{14}-\psi_{16}-\psi_{23}-\psi_{25}+\psi_{34}+\psi_{36}+\psi_{45}+\psi_{56};$$

when $m=(13)$ it is

$$\psi_{00}-\psi_{12}-\psi_{14}-\psi_{16}-\psi_{23}+\psi_{25}-\psi_{34}-\psi_{36}+\psi_{45}+\psi_{56}.$$

The half-difference is

$$\psi_{12}-\psi_{25}+\psi_{34}+\psi_{36}.$$

Therefore, when $m_3=(35)$,

$$\psi_{23}=\psi_{46}+\psi_{45}+\psi_{56},$$

and, when $m_3=(25)$,

$$\psi_{00}=\psi_{15}+\psi_{16}+\psi_{14}.$$

This proves that for two marks of the form $(\alpha\beta)$, $(\alpha\gamma)$, of which one is even and the other odd, we have results of the forms

$$\psi_{\beta\gamma}=\psi_{\epsilon\zeta}+\psi_{\zeta\delta}+\psi_{\delta\epsilon},$$

$$\psi_{00}=\psi_{\alpha\delta}+\psi_{\alpha\epsilon}+\psi_{\alpha\zeta},$$

and it is left to the reader to show, by working out a typical case, that the results hold for any pair of the form $(\alpha\beta)$, $(\alpha\gamma)$.

Lastly let the marks m be $(\alpha\beta)$, (00); say (12), (00). Then $m_3=(35)$ or (46) and $\Sigma(-1)^{m|\eta}\psi_\eta$ is

$$\psi_{00}+\psi_{12}+\psi_{14}+\psi_{16}+\psi_{23}+\psi_{25}+\psi_{34}+\psi_{36}+\psi_{45}+\psi_{56},$$

or $$\psi_{00}+\psi_{12}-\psi_{14}-\psi_{16}-\psi_{23}-\psi_{25}+\psi_{34}+\psi_{36}+\psi_{45}+\psi_{56}.$$

The half-difference is

$$\psi_{14}+\psi_{16}+\psi_{23}+\psi_{25}.$$

Therefore, when $m_3=(35)$,

$$\psi_{26}+\psi_{24}+\psi_{25}+\psi_{23}=0,$$

and, when $m_3=(46)$,

$$\psi_{16}+\psi_{14}+\psi_{15}+\psi_{13}=0.$$

This proves that for the form $(\alpha\beta)$, (00), when $(\alpha\beta)$ is even, we have results of the form

$$\psi_{\alpha\gamma}+\psi_{\alpha\delta}+\psi_{\alpha\epsilon}+\psi_{\alpha\zeta}=0;$$

and the same result holds when $(\alpha\beta)$ is odd.

Combining these facts, we have the result:—

Of the ten marks which are not in the hexad m_3, four are in any other hexad. If this other hexad be triangular, then according as the four marks are of the form $\alpha\zeta$, $\beta\zeta$, $\gamma\epsilon$, $\delta\epsilon$ or $\beta\gamma$, $\epsilon\zeta$, $\zeta\delta$, $\delta\epsilon$, there is the relation

$$\psi_{\alpha\zeta}+\psi_{\beta\zeta}=\psi_{\gamma\epsilon}+\psi_{\delta\epsilon},$$

or $$\psi_{\beta\gamma}=\psi_{\epsilon\zeta}+\psi_{\zeta\delta}+\psi_{\delta\epsilon}.$$

But if the other hexad be pentagonal, then according as the system of four marks does or does not contain (00), there is the relation

$$\psi_{00}=\psi_{\alpha\delta}+\psi_{\alpha\epsilon}+\psi_{\alpha\zeta},$$

or $$\psi_{\alpha\gamma}+\psi_{\alpha\delta}+\psi_{\alpha\epsilon}+\psi_{\alpha\zeta}=0.$$

All these come under the formula

$$\psi_m=\Sigma(-1)^{m|m'}\psi_{m'} \quad . \quad . \quad . \quad . \quad . \quad . \quad . \quad (13),$$

where m is any one of the four marks, m' is in turn the other three.

When the four marks are given, m_3 is determined as follows. To the hexad belong two other marks, which appear in the hexad (m_3). For instance, let the four be $12 \cdot 34 \cdot 45 \cdot 53$, belonging to the hexad (14). The remaining marks $16 \cdot 26$ also belong to (24). And (24), when combined with any of the four, gives an even mark.

§ 242. *Formula for the reduction of the marks in* ψ_m.

Let m_1 and m_2 be reduced marks; we shall determine the formula which will enable us to reduce the marks in such a product as $\theta_m \theta_{m+m_1} \theta_{m+m_2} \theta_{m-m_1-m_2}$.

By § 232, θ_{m+m_1} or $\theta \begin{vmatrix} g_r + \lambda_r' \\ h_r + \mu_r' \end{vmatrix}$ is reduced by the factor

$$\exp \pi i \, \Sigma h_r \mu_r' (g_r + \lambda_r').$$

Similarly for $m + m_2$; and for $m - m_1 - m_2$ we have

$$\theta \begin{vmatrix} g_r - \lambda_r' - \lambda_r'' \\ h_r - \mu_r' - \mu_r'' \end{vmatrix} = \theta \begin{vmatrix} g_r + \lambda_r' + \lambda_r'' \\ h_r + \mu_r' + \mu_r'' \end{vmatrix} \exp \pi i \Sigma (\mu_r' + \mu_r'')(g_r + \lambda_r' + \lambda_r'').$$

The last mark will be reduced by a factor of the form

$$\exp \pi i \Sigma \kappa_r (g_r + \lambda_r' + \lambda_r''),$$

where $\kappa_r \equiv 0$ when

$$h_r + \mu_r' + \mu_r'' = 0 \text{ or } 1,$$

and $\equiv 1$ when

$$h_r + \mu_r' + \mu_r'' = 2 \text{ or } 3;$$

these are all the permissible values of $h_r + \mu_r' + \mu_r''$, inasmuch as m, m_1, m_2 are reduced.

The conditions are satisfied by $\kappa_r = h_r(\mu_r' + \mu_r'') + \mu_r'\mu_r''$; the reducing factor for $m - m_1 - m_2$ is

$$\exp \pi i \, \Sigma (g_r + \lambda_r' + \lambda_r'')(\kappa_r + \mu_r' + \mu_r''),$$

and the whole reducing factor is

$$\exp \pi i \, \Sigma [2 g_r h_r (\mu_r' + \mu_r'') + h_r(\lambda_r'\mu_r' + \lambda_r''\mu_r'' + \lambda_r'''\mu_r''') \\ + (g_r - \lambda_r''')(\mu_r'\mu_r'' + \mu_r' + \mu_r'')],$$

or, omitting a factor which is independent of m,

$$\exp \pi i [\Sigma h_r(\lambda_r'\mu_r'' + \lambda_r''\mu_r') + g_r(\mu_r'\mu_r'' + \mu_r' + \mu_r'')] \quad . \quad . \quad (14).$$

This result is not altered by writing λ_r''', μ_r''' for λ_r', μ_r' or for λ_r'', μ_r''.

§ 243. *Addition theorems.*

I. When $m'm_3$ is odd, let us put

$$a_r,\ b_r,\ c_r,\ d_r = v_r + v_r',\ 0,\ 0,\ v_r - v_r'.$$

Then $$a_r',\ b_r',\ c_r',\ d_r' = v_r,\ v_r',\ v_r',\ -v_r,$$

and $$a_r'',\ b_r'',\ c_r'',\ d_r'' = v_r',\ v_r,\ v_r,\ v_r'.$$

Let $m=(13)$, $m'=(15)$. We seek to express the relations between $\theta_{13}(v \pm v')$, $\theta_{15}(v \pm v')$ and θ-functions of v, v', using for that purpose formula (11),

$$\psi_m + (-1)^{m\,m'}\psi_{m'} = 0,$$

in which mm_3, $m'm_3$ are odd marks.

Since mm_3 and $m'm_3$ are odd,

$$m_3 \equiv (00) \text{ or } (35).$$

Taking $m_3 \equiv (35)$, we have

$$m_1 + m_2 \equiv (35).$$

Now in order that the functions θ_m, $\theta_{m'}$, with the arguments $v_r \pm v_r'$, may not disappear, it is necessary that the marks

$$m_1 + (13),\ m_1 + (15),\ m_2 + (13),\ m_2 + (15)$$

be even. Therefore m_1 and m_2 must be selected from (12), (14), (16), (24), (26), (46). The condition $m_1 + m_2 \equiv (35)$ further restricts m_1, m_2 to the three pairs

$$12,\ 46;\quad 14,\ 26;\quad 16,\ 24.$$

Take $$m_1 = (12) = \begin{vmatrix} 1 & 0 \\ 0 & 0 \end{vmatrix},\quad m_2 = (46) = \begin{vmatrix} 0 & 1 \\ 0 & 1 \end{vmatrix}.$$

The reducing factor of § 242 is $(-1)^{g_2}$; that is, $+1$ for $m=(13)$, -1 for $m'=(15)$. Therefore, from formula (11), since

$$m,\ mm_1,\ mm_2,\ mm_3 = (13),\ (23),\ (25),\ (15),$$

and $$m',\ m'm_1,\ m'm_2,\ m'm_3 = (15),\ (25),\ (23),\ (13),$$

$$c_{23}c_{25}\{\theta_{13}(v+v')\theta_{15}(v-v') + \theta_{15}(v+v')\theta_{13}(v-v')\}$$
$$-2\theta_{13}\theta_{23}'\theta_{25}'\theta_{15} + 2\theta_{13}'\theta_{23}\theta_{25}\theta_{15}' = 0,$$

where the arguments of θ are v_1, v_2, and those of θ' are v_1', v_2'.

II. In the second case, when $m'm_3$ is even, we use formula (13),

$$\psi_m = \Sigma(-1)^{m|m'}\psi_{m'}.$$

Let $\quad a_r,\ b_r,\ c_r,\ d_r\ = v_r + v_r',\ v_r - v_r',\ 0,\ 0;$

then $\quad a_r',\ b_r',\ c_r',\ d_r'\ = v_r,\ v_r,\ v_r',\ -v_r';$

and $\quad a_r'',\ b_r'',\ c_r'',\ d_r'' = v_r,\ v_r,\ v_r',\ v_r'.$

Let two of the four marks be (13), (15), and $m_1 = (35)$. That $\theta_{13}(v \pm v')$ and $\theta_{15}(v \pm v')$ may not disappear, we may choose $m_2 = (12)$, $m_3 = (46)$. From either of the two hexads which contain (13) and (15) two other marks must now be selected. But, in order that for these $\theta_m(v \pm v')$ may not enter, we impose the condition that mm_2 be odd. This restricts the choice to (14), (16) from the hexad (35), or to (35), (46) from the hexad (00). Choosing

$$m = (13),\ (15),\ (14),\ (16),$$

we have

$$mm_1 \equiv (15),\ (13),\ (26),\ (24),$$

$$mm_2 \equiv (23),\ (25),\ (24),\ (26),$$

$$mm_3 \equiv (25),\ (23),\ (16),\ (14),$$

and the reducing factor $= \quad 1,\ -1,\ -1,\ \ 1;$

the reducing factor of § 242 being, as before, $(-1)^{g_2}$.

The equation

$$\psi_{13} + \psi_{15} + \psi_{14} + \psi_{16} = 0$$

becomes

$$\theta_{13}(v+v')\theta_{15}(v-v')c_{23}c_{25} + 2\theta_{13}\theta_{15}\theta_{23}'\theta_{25}' - \theta_{15}(v+v')\theta_{13}(v-v')c_{25}c_{23}$$
$$- 2\theta_{15}\theta_{13}\theta_{25}'\theta_{23}' - 2\theta_{14}\theta_{26}\theta_{24}'\theta_{16}' + 2\theta_{16}\theta_{24}\theta_{26}'\theta_{14}' = 0,$$

or $\quad c_{23}c_{25}\{\theta_{13}(v+v')\theta_{15}(v-v') - \theta_{15}(v+v')\theta_{13}(v-v')\}$
$$- 2\theta_{14}\theta_{26}\theta_{24}'\theta_{16}' + 2\theta_{24}\theta_{16}\theta_{14}'\theta_{26}' = 0.$$

Similarly, if we replace

$$a_r,\ b_r,\ c_r,\ d_r \text{ by } 0,\ v_r + v_r',\ v_r - v_r',\ 0,$$

the rest of the work being unaltered, we have

$$c_{14}c_{16}\{\theta_{24}(v+v')\theta_{26}(v-v') - \theta_{26}(v+v')\theta_{24}(v-v')\}$$
$$+ 2\theta_{13}\theta_{15}'\theta_{23}'\theta_{25} - 2\theta_{13}'\theta_{15}\theta_{23}\theta_{25}' = 0.$$

From the two equations already obtained for $\theta_{13}(v+v')\theta_{15}(v-v')$ and $\theta_{15}(v+v')\theta_{13}(v-v')$, it is evident that each of these quantities can be expressed in terms of θ-functions of v and v'.

Example. Prove

$$c_{00}c_{12}\{\theta_{35}(v+v')\theta_{46}(v-v')+\theta_{46}(v+v')\theta_{35}(v-v')\}$$
$$-2\,\theta_{35}\theta_{46}\theta_{00}'\theta_{12}'+2\,\theta_{00}\theta_{12}\theta_{35}'\theta_{46}'=0.$$

§ 244. *Relations between the squares of four theta-functions of a hexad.*

Let $a_r,\ b_r,\ c_r,\ d_r\ = v_r,\ v_r,\ 0,\ 0;$

then $a_r',\ b_r',\ c_r',\ d_r' = v_r,\ v_r,\ 0,\ 0,$

and $a_r'',\ b_r'',\ c_r'',\ d_r'' = v_r,\ v_r,\ 0,\ 0.$

Further let $m_1=(00)$. Then since $m_1+m_2+m_3=0$, $m_2=-m_3$; and ψ_m takes the form

$$3\,\theta_m^2(v_r)\cdot c_{m-m_3}\cdot c_{m+m_3}.$$

Here the reducing factor is $(-1)^{\Sigma g r \mu r'''}$, when mm_3 is reduced. Hence, dropping the factor 3, we may put

$$\psi_m=(-1)^{\Sigma g r \mu r'''}\cdot\theta_m^2(v_r)c_{mm_3}^2,$$

in formula (13). We thus have a linear relation between the squares of any four functions of a hexad.

For example, let $m_3=(00)$. Then we have the six independent relations

$$\left.\begin{aligned}c_{00}^2\theta_{00}^2&=c_{12}^2\theta_{12}^2+c_{14}^2\theta_{14}^2+c_{16}^2\theta_{16}^2\\&=c_{12}^2\theta_{12}^2+c_{23}^2\theta_{23}^2+c_{25}^2\theta_{25}^2\\&=c_{23}^2\theta_{23}^2+c_{34}^2\theta_{34}^2+c_{36}^2\theta_{36}^2\\&=c_{14}^2\theta_{14}^2+c_{34}^2\theta_{34}^2+c_{45}^2\theta_{45}^2\\&=c_{25}^2\theta_{25}^2+c_{45}^2\theta_{45}^2+c_{56}^2\theta_{56}^2\\&=c_{16}^2\theta_{16}^2+c_{36}^2\theta_{36}^2+c_{56}^2\theta_{56}^2\end{aligned}\right\}\quad.\ .\ .\ .\ .\quad(15),$$

and nine others which are readily deduced from these by subtraction. Again, if $m_3=(13)=\begin{vmatrix}10\\10\end{vmatrix}$, and

$$m=(13),\ (12),\ (14),\ (16),$$

$$=\begin{vmatrix}10\\10\end{vmatrix},\ \begin{vmatrix}10\\00\end{vmatrix},\ \begin{vmatrix}01\\10\end{vmatrix},\ \begin{vmatrix}00\\11\end{vmatrix},$$

we have, attending to the factor $(-1)^{\Sigma g_r \mu_r'''}$ or $(-1)^{g_1}$,

$$\left.\begin{array}{ll} -\theta_{13}{}^2 c_{00}{}^2 - \theta_{12}{}^2 c_{32}{}^2 + \theta_{14}{}^2 c_{34}{}^2 + \theta_{16}{}^2 c_{36}{}^2 = 0. & \\ \text{Similarly,} \quad -\theta_{35}{}^2 c_{00}{}^2 - \theta_{34}{}^2 c_{54}{}^2 + \theta_{32}{}^2 c_{52}{}^2 + \theta_{36}{}^2 c_{56}{}^2 = 0, & m_3 = (35), \\ \theta_{51}{}^2 c_{00}{}^2 + \theta_{56}{}^2 c_{16}{}^2 - \theta_{52}{}^2 c_{12}{}^2 - \theta_{54}{}^2 c_{14}{}^2 = 0, & m_3 = (15), \\ \theta_{24}{}^2 c_{00}{}^2 + \theta_{12}{}^2 c_{14}{}^2 - \theta_{32}{}^2 c_{34}{}^2 + \theta_{52}{}^2 c_{54}{}^2 = 0, & m_3 = (24), \\ \theta_{46}{}^2 c_{00}{}^2 + \theta_{14}{}^2 c_{16}{}^2 + \theta_{34}{}^2 c_{36}{}^2 - \theta_{54}{}^2 c_{56}{}^2 = 0, & m_3 = (46), \\ -\theta_{62}{}^2 c_{00}{}^2 + \theta_{16}{}^2 c_{12}{}^2 - \theta_{36}{}^2 c_{32}{}^2 - \theta_{56}{}^2 c_{52}{}^2 = 0, & m_3 = (26). \end{array}\right\} \quad . \quad (16).$$

The total number of such relations is the number of sets of four which can be chosen from the hexads; namely, $\frac{1}{2} \cdot 6 \cdot 5 \cdot 16$ or 240.

Example. Prove the relation

$$\theta_{12}{}^2 c_{14}{}^2 + \theta_{13}{}^2 c_{56}{}^2 = \theta_{14}{}^2 c_{12}{}^2 + \theta_{15}{}^2 c_{36}{}^2.$$

§ 245. *Relations between the c's.*

If in the square relation we put $v_r = 0$, we derive six relations of the form

$$c_{00}{}^4 = c_{12}{}^4 + c_{14}{}^4 + c_{16}{}^4 \quad . \quad . \quad . \quad . \quad . \quad . \quad . \quad . \quad . \quad . \quad (17),$$

and six relations

$$\left.\begin{array}{r} -c_{12}{}^2 c_{23}{}^2 + c_{14}{}^2 c_{34}{}^2 + c_{16}{}^2 c_{36}{}^2 = 0 \\ c_{12}{}^2 c_{25}{}^2 + c_{14}{}^2 c_{45}{}^2 - c_{16}{}^2 c_{56}{}^2 = 0 \\ c_{23}{}^2 c_{25}{}^2 - c_{34}{}^2 c_{45}{}^2 + c_{36}{}^2 c_{56}{}^2 = 0 \\ c_{12}{}^2 c_{14}{}^2 - c_{23}{}^2 c_{34}{}^2 + c_{25}{}^2 c_{45}{}^2 = 0 \\ -c_{12}{}^2 c_{16}{}^2 + c_{23}{}^2 c_{36}{}^2 + c_{25}{}^2 c_{56}{}^2 = 0 \\ c_{14}{}^2 c_{16}{}^2 + c_{34}{}^2 c_{36}{}^2 - c_{45}{}^2 c_{56}{}^2 = 0 \end{array}\right\} \quad . \quad . \quad . \quad . \quad . \quad . \quad (18).$$

This may be stated as follows :

The ratios of $c_{12}{}^2, c_{14}{}^2, c_{16}{}^2;\ -c_{23}{}^2, c_{34}{}^2, c_{36}{}^2;\ -c_{25}{}^2, -c_{45}{}^2, c_{56}{}^2$ to $c_{00}{}^2$ are equal to the coefficients in an orthogonal transformation of co-ordinates. (See Caspary, Crelle, t. xciv.) For if we write for these ratios

$$l_{11}, l_{12}, l_{13};\ l_{21}, l_{22}, l_{23};\ l_{31}, l_{32}, l_{33},$$

we have the relations

$$\sum_{r=1}^{3} l_{rs}{}^2 \qquad (s = 1, 2, \text{ or } 3),$$

$$\sum_{s=1}^{3} l_{rs}{}^2 \qquad (r = 1, 2, \text{ or } 3),$$

$$\sum_{r=1}^{3} l_{rs} l_{rs'} = 0 \qquad (s = 1, 2, \text{ or } 3;\ s' = 1, 2, \text{ or } 3;\ s \neq s'),$$

$$\sum_{s=1}^{3} l_{rs} l_{r's} = 0 \qquad (r = 1, 2, \text{ or } 3;\ r' = 1, 2, \text{ or } 3;\ r \neq r').$$

Interpreted in this way, the table

$$\left|\begin{matrix} c_{12}^2 & c_{14}^2 & c_{16}^2 \\ -c_{32}^2 & c_{34}^2 & c_{36}^2 \\ -c_{52}^2 & -c_{54}^2 & c_{56}^2 \end{matrix}\right| / c_{00}^2 \quad . \quad . \quad . \quad . \quad . \quad . \quad . \quad . \quad . \quad . \qquad \text{(D)}$$

affords a compendium of the relations in question. It may be observed that the sign is minus when the figures in the suffix are displaced from their natural order.

§ 246. *Relations between the squares of five theta-functions.*

Excluding θ_{00}, take any five functions $\theta_1, \theta_2, \theta_3, \theta_4, \theta_5$, no four of which belong to a hexad. Referring to the table of hexads (§ 234), we see that in any case two triads,

$$\theta_1, \theta_2, \theta_3 \text{ and } \theta_1, \theta_4, \theta_5,$$

can be chosen, which belong to different hexads. The two hexads have θ_1 and some other function θ_6 in common.

Therefore $\quad \theta_1^2, \theta_2^2, \theta_3^2, \theta_6^2$ and $\theta_1^2, \theta_4^2, \theta_5^2, \theta_6^2$,

are linearly related. Eliminating θ_6^2, we have a linear relation between $\theta_1^2, \theta_2^2, \theta_3^2, \theta_4^2, \theta_5^2$. There cannot be two such independent relations. For if there be a second linear relation between $\theta_1^2, \theta_2^2, \theta_3^2, \theta_4^2, \theta_5^2$, there is a linear relation between $\theta_1^2, \theta_2^2, \theta_3^2, \theta_4^2$, of which three belong to a hexad, and therefore have a common pair of half-periods as zeros. If then

$$\sum_{\kappa=1}^{4} k_\kappa \theta_\kappa^2 = 0,$$

$\theta_1, \theta_2, \theta_3$ vanish for values which do not make θ_4 vanish, and consequently $k_4 = 0$. The relation reduces to $\sum_{\kappa=1}^{3} k_\kappa \theta_\kappa^2 = 0$; but θ_1 and θ_2 belong to some other hexad, and therefore have a common pair of half-periods as zeros, which are not zeros of θ_3. Therefore $k_3 = 0$, and $k_1\theta_1^2 + k_2\theta_2^2 = 0$, a relation which is impossible, since all the pairs of zeros of θ_1^2, θ_2^2 are not the same.

If one of the originally selected functions be θ_{00}, the addition of suitable half-periods gives five functions not including θ_{00}; and from the relation between these five follows a similar relation between the original five.

There is then a linear relation between the squares of any five functions, no four of which belong to a hexad.

The proof supposes that there is no linear relation between four squares which do not belong to a hexad. This is evident if any three belong to a hexad, and the only outstanding case is that of a Göpel tetrad. Suppose that

$$k_{12}\theta_{12}^2 + k_{34}\theta_{34}^2 + k_{13}\theta_{13}^2 + k_{24}\theta_{24}^2 = 0;$$

then, when $v_r = 0$,

$$k_{12}c_{12}^2 + k_{34}c_{34}^2 = 0;$$

and when $v_r = \mu_r/2 + \frac{1}{4}\delta\phi/\delta\lambda_r$, and $\begin{vmatrix}\lambda_r\\ \mu_r\end{vmatrix} = (56) = \begin{vmatrix}01\\ 00\end{vmatrix}$,

it follows from (7) that

$$k_{12}c_{34}^2 + k_{34}c_{12}^2 = 0.$$

Therefore, either $k_{12} = k_{34} = 0$, or

$$c_{12}^2 = c_{34}^2.$$

The former supposition is impossible, because it would imply a linear relation between two squared theta-functions; the latter supposition is untenable, because there are three independent parameters, $\tau_{11}, \tau_{12}, \tau_{22}$, of which the c's are functions, and the approximate values for the c's in terms of the τ's disprove the equations

$$\left.\begin{matrix} c_{12} = c_{34} \\ c_{12} = -c_{34} \end{matrix}\right\}.$$

The total number of sets of five is $\frac{16 \cdot 15 \cdot 14 \cdot 13 \cdot 12}{1 \cdot 2 \cdot 3 \cdot 4 \cdot 5}$; the number of sets of which all are in a hexad is $6 \cdot 16$, and the number of sets of which four are in a hexad is $\frac{6 \cdot 5}{2} \cdot 10 \cdot 16$. Hence the number of linear relations between five squares is 1872.

The general theorem at which we have arrived is the following: Four squared theta-functions are connected by linear equations when, and only when, their marks belong to a Rosenhain hexad; but there is always a linear relation between five squared theta-functions, no four of which belong to a hexad.

Hence every θ^2 is linearly expressible in terms of four selected squares, if these four do not belong to a hexad. For instance, the four theta-functions may form a Göpel tetrad.

§ 247. *Product theorems.*

Let $\qquad a_r,\ b_r,\ c_r,\ d_r\ = v_r,\ v_r,\ 0,\ 0;$

then $\qquad a_r',\ b_r',\ c_r',\ d_r' = v_r,\ v_r,\ 0,\ 0,$

and $\qquad a_r'',\ b_r'',\ c_r'',\ d_r'' = v_r,\ v_r,\ 0,\ 0.$

Therefore $\psi_m = 3^{\Omega}{}_m\, \theta_{mm_1} c_{mm_2} c_{mm_3},$

and, from § 241,

$$\theta_m \theta_{mm_1} c_{mm_2} c_{mm_3} = \Sigma (-1)^{m|m'} \theta_{m'} \theta_{m'm_1} c_{m'm_2} c_{m'm_3},$$

the summation being for any three marks m' which, like m itself, give an even mark when combined with m_3.

Now ψ_m exists only when mm_2 and mm_3 are even, and since, regarding m_2 and m_3 as given, there are six values of m for which mm_2 is odd, six values for which mm_3 is odd, and two values for which both mm_2 and mm_3 are odd, there must be six different values for which both mm_2 and mm_3 are even. But $mm_1m_2 \equiv mm_3$, $mm_1m_3 \equiv mm_2$, since $m_1 + m_2 + m_3 = 0$. Therefore, if m be one of the values, mm_1 is another; and corresponding to any two marks m_2, m_3, there are three products of pairs of functions, $\theta_m \theta_{mm_1}$, between which there is a linear relation. The number of such relations is the number of ways in which m_2 and m_3 can be chosen; that is,

$$\tfrac{1}{2} \cdot 16 \cdot 15 \text{ or } 120.$$

Since mm_2 and mm_3 are even,

$$\Sigma (g_r + \lambda_r'') (h_r + \mu_r'') \equiv 0,$$

$$\Sigma (g_r + \lambda_r''') (h_r + \mu_r''') \equiv 0,$$

and, by addition, $m \mid m_2 + m \mid m_3 + \Sigma \lambda_r'' \mu_r'' + \Sigma \lambda_r''' \mu_r''' \equiv 0,$

therefore $m \mid m_1$ is given. This shows that the three linearly related pairs belong to one of the 30 sets of four pairs of § 233. Four relations can be formed from each set by leaving out each pair in turn, the total number of relations being $4 \times 30 = 120$.

Taking the set determined by $m_1 = (12)$, $m \mid m_1 \equiv 0$, we see that, since $m_1 = \begin{vmatrix} 1 & 0 \\ 0 & 0 \end{vmatrix}$, it must be

$$00,\ 12;\ 34,\ 56;\ 35,\ 46;\ 36,\ 45$$

The determination of m_2, m_3 is effected by the consideration that

$$m_2 + m_3 + (12) = 0,$$

$$\Sigma \lambda_r'' \mu_r'' + \Sigma \lambda_r''' \mu_r''' \equiv 0,$$

therefore $m_2,\ m_3 \equiv 00,\ 12;\ 34,\ 56;\ 35,\ 46;\ 36,\ 45.$

The reducing factors are respectively

$$1,\ 1,\ (-1)^{g_2},\ (-1)^{g_1}.$$

Therefore, taking

$$m=(00), \quad m'=(34),\ (35),\ (36),$$

and observing that $g_2=0, 1, 1, 0$ respectively, we have from

$$\psi_{00}=\psi_{34}+\psi_{35}+\psi_{36},$$

when $m_1, m_2, m_3 \equiv 12, 00, 12,$

$$\left.\begin{aligned}
&\theta_{00}\theta_{12}c_{00}c_{12}=\theta_{34}\theta_{56}c_{34}c_{56}+\theta_{36}\theta_{45}c_{36}c_{45};\\
\text{when}\quad &m_1, m_2, m_3\equiv 12, 34, 56,\\
&\theta_{00}\theta_{12}c_{34}c_{56}=\theta_{34}\theta_{56}c_{00}c_{12}+\theta_{35}\theta_{46}c_{45}c_{36};\\
\text{when}\quad &m_1, m_2, m_3\equiv 12, 35, 46,\\
&-\theta_{34}\theta_{56}c_{45}c_{36}-\theta_{35}\theta_{46}c_{00}c_{12}+\theta_{36}\theta_{45}c_{56}c_{34}=0;\\
\text{when}\quad &m_1, m_2, m_3\equiv 12, 36, 45,\\
&\theta_{00}\theta_{12}c_{36}c_{45}=-\theta_{35}\theta_{46}c_{56}c_{34}+\theta_{36}\theta_{45}c_{00}c_{12}.
\end{aligned}\right\}\quad \cdots \quad (19).$$

Example. Eliminating from these four equations the four quantities

$$\theta_{00}\theta_{12},\ \theta_{34}\theta_{56},\ \theta_{35}\theta_{46},\ \theta_{36}\theta_{45},$$

we have a determinant in the c's. Verify from the relations of the c's that this determinant vanishes.

§ 248. *The number of independent relations.*

The relations between theta-functions of the same variables, which can be derived by the preceding processes, are very numerous, but they are all reducible to thirteen independent relations, of course in very many ways. For instance, the equations (15) express all the other even functions in terms of four, say of θ_{00}, θ_{12}, θ_{34}, θ_{56}; then equations (16) express the odd functions in terms of these four, and finally the first of equations (19) connects these four. If the sixteen functions be selected as co-ordinates of a point in space of fifteen dimensions, the thirteen independent relations amongst the θ's show that the point belongs to a two-dimensional locus in this space.

Between any four theta-functions there is a relation. We have seen of what nature this relation is, when the functions belong to a hexad; we proceed to examine the nature of the relation when the four functions belong to a Göpel tetrad.

§ 249. *Göpel's biquadratic relation.*

To determine the relation between a Göpel tetrad, say θ_{34}, θ_{56}, θ_{36}, θ_{45}, selected from the pairs

$$00\cdot 12;\ 34\cdot 56;\ 35\cdot 46;\ 36\cdot 45,$$

let us connect θ_{35} with the tetrad. To do this we observe that $34 \cdot 45 \cdot 35$ belong to the hexad $(345 \cdot 126)$, and $56 \cdot 36 \cdot 35$ to $(356 \cdot 124)$. The remaining mark common to the hexads is 12. Hence the relations

$$c_{14}^2\theta_{35}^2 - c_{32}^2\theta_{36}^2 - c_{52}^2\theta_{56}^2 + c_{16}^2\theta_{12}^2 = 0,$$

$$c_{14}^2\theta_{12}^2 - c_{32}^2\theta_{34}^2 + c_{52}^2\theta_{54}^2 - c_{16}^2\theta_{35}^2 = 0,$$

are easily written down by the method already explained. Eliminating θ_{12}^2,

$$(c_{14}^4 + c_{16}^4)\theta_{35}^2 = c_{14}^2c_{32}^2\theta_{36}^2 + c_{14}^2c_{52}^2\theta_{56}^2 - c_{32}^2c_{16}^2\vartheta_{34}^2 + c_{16}^2c_{52}^2\theta_{54}^2 = U \text{ (say)}.$$

To express θ_{46} by means of the tetrad, we add the half-periods (12) to each argument, and obtain

$$(c_{14}^4 + c_{16}^4)\theta_{46}^2 = c_{14}^2c_{32}^2\theta_{54}^2 + c_{14}^2c_{52}^2\theta_{34}^2 - c_{32}^2c_{16}^2\theta_{56}^2 + c_{16}^2c_{52}\theta_{36}^2 = V \text{ (say)}.$$

Also, from formula (19),

$$c_{00}c_{12}\theta_{35}\theta_{46} = -c_{45}c_{36}\theta_{34}\theta_{56} + c_{34}c_{56}\vartheta_{36}\theta_{45} = W \text{ (say)}.$$

Eliminating θ_{35}, θ_{46}, we have a relation of the fourth order between the functions of the tetrad,

$$(c_{14}^4 + c_{16}^4)^2 W^2 = c_{00}^2c_{12}^2 UV.$$

In this equation, when expanded, the coefficients of

$$\theta_{34}^4,\ \theta_{56}^4,\ -\theta_{36}^4,\ -\theta_{54}^4$$

are equal. But also the coefficients of $\theta_{34}^2\theta_{56}^2$ and $\theta_{36}^2\theta_{45}^2$ are equal and opposite: for these coefficients are

$$c_{00}^2c_{12}^2(c_{14}^4c_{52}^4 + c_{32}^4c_{16}^4) - c_{54}^2c_{36}^2(c_{14}^4 + c_{16}^4)^2$$

and $$c_{00}^2c_{12}^2(c_{14}^4c_{32}^4 + c_{52}^4c_{16}^4) - c_{34}^2c_{56}^2(c_{14}^4 + c_{16}^4)^2,$$

and their sum is

$$c_{00}^2c_{12}^2(c_{14}^4 + c_{16}^4)(c_{32}^4 + c_{52}^4) - (c_{34}^2c_{56}^2 + c_{54}^2c_{36}^2)(c_{14}^4 + c_{16}^4)^2,$$

which vanishes, since

$$c_{32}^4 + c_{52}^4 = c_{14}^4 + c_{16}^4 \quad \text{from (17)},$$

and $$c_{00}^2c_{12}^2 = c_{34}^2c_{56}^2 + c_{54}^2c_{36}^2 \quad \text{from (19), when } v_r = 0.$$

Similarly the coefficients of $\theta_{34}^2\theta_{36}^2$ and $\theta_{54}^2\theta_{56}^2$, and those of $\theta_{34}^2\theta_{54}^2$ and $\theta_{36}^2\theta_{56}^2$ are equal. Accordingly the biquadratic relation is of the form

$$\begin{aligned}\theta_{34}^4 + \theta_{56}^4 - \theta_{36}^4 - \theta_{54}^4 &+ A(\theta_{34}^2\theta_{56}^2 - \theta_{36}^2\theta_{54}^2) \\ &+ B(\theta_{34}^2\theta_{36}^2 + \theta_{54}^2\theta_{56}^2) \\ &+ C(\theta_{34}^2\theta_{54}^2 + \theta_{36}^2\theta_{56}^2) \\ &+ D\theta_{34}\theta_{56}\theta_{36}\theta_{54} = 0 \quad . \quad . \quad . \quad (20).\end{aligned}$$

§ 250. *Kummer's 16-nodal quartic surface.*

Regarding the four functions as homogeneous co-ordinates, the equation is that of a quartic surface; and the abridged form shows that the intersections of the quadrics U, V, W are nodes. But U, V, W vanish when θ_{35} and θ_{46} vanish; that is, when $v_1, v_2 = 0, 0$, and when $v_1, v_2 =$ the half-periods (12). Hence the points $c_{34}, c_{50}, c_{36}, c_{54}$ and $c_{56}, c_{34}, c_{54}, c_{30}$ are nodes.

The symmetry of the expanded equation shows that it is unaltered by the interchange of 3 and 5, or 4 and 6; that is, by the addition to u of the half-periods (35) or (46). Hence also

$$c_{54}, c_{40}, c_{50}, c_{34} \text{ and } c_{30}, c_{54}, c_{34}, c_{56}$$

are also nodes. Further the equation is unaltered when the signs of two of the co-ordinates are changed.

Hence there are 16 nodes, the scheme of which is

$$\theta_{34} = c_{34}, c_{56}, c_{54}, c_{30},$$

$$\theta_{56} = \pm c_{50}, \pm c_{34}, \pm c_{40}, \pm c_{54},$$

$$\theta_{30} = \pm c_{36}, \pm c_{54}, \pm c_{50}, \pm c_{34},$$

$$\theta_{54} = \pm c_{54}, \pm c_{30}, \pm c_{34}, \pm c_{56},$$

where one of the signs in each column of ambiguous signs is to be $+$. The surface has therefore the maximum number of nodes (Salmon, Geometry of Three Dimensions, § 573).

The connexion between the double theta-functions and Kummer's surface was first pointed out by Cayley, Crelle, t. lxxxiii., p. 210, who was followed by Borchardt (ib., t. lxxxiii.) and Weber (ib., t. lxxxiv.). Other important memoirs in this connexion are Rohn, Math. Ann., t. xv., in which the subject of quadratic transformations is taken up; and Klein, Math. Ann., t. xxvii. A detailed account of the subject is given by Reichardt, Ueber die Darstellung der Kummer'schen Fläche durch hyperelliptische Functionen, Halle, 1887.

§ 251. *The p-tuple theta-functions.*

The series

$$\sum_{n_1=-\infty}^{\infty} \cdots \sum_{n_p=-\infty}^{\infty} \exp \pi i \left[\sum_{r=1}^{p}\sum_{s=1}^{p} \tau_{rs}(n_r + g_r/2)(n_s + g_s/2) + 2\sum_{r=1}^{p}(n_r + g_r/2)(v_r + h_r/2)\right],$$

where $\tau_{rs} = \tau_{sr}$, and g_r, h_r are integers, is absolutely convergent when the coefficient of i in $\Sigma\Sigma\tau_{rs}n_rn_s$ is expressible as the sum of p positive squares. This can be proved by an extension of the method of § 223, and we refer the reader to the authorities quoted in that paragraph. Under the condition stated, the series defines a function which is one-valued for all values of the p arguments v_r. The function is called a p-tuple theta-function, and is denoted by $\theta\begin{vmatrix} g_r \\ h_r \end{vmatrix}(v_r)$ or $\theta_m v_r$. The function is even or odd according as $\sum_{r=1}^{p} g_r h_r \equiv 0$ or 1 (see § 228), and its properties as regards periods are precisely analogous to those of the double theta-function.

Formula (9), § 238, takes the form

$$\sum_{m'}(-1)^{m|m'}\theta_{m'}a_r' \cdot \theta_{m'm_1}b_r' \cdot \theta_{m'm_2}c_r' \cdot \theta_{m'm_3}d_r'$$

$$= 2^p\theta_m a_r \cdot \theta_m b_r \cdot \theta_m c_r \cdot \theta_m d_r,$$

where in equations (i.) $r = 1, 2, \cdots, p$, and where m' is any mark $\begin{vmatrix} g_r' \\ h_r' \end{vmatrix}$, m is a given mark $\begin{vmatrix} g_r \\ h_r \end{vmatrix}$, and $m \mid m' = \sum_{r=1}^{p}(g_r h_r' + g_r' h_r)$.

References. The following books deal with the theory of theta-functions from the algebraic standpoint: Krause, Die Transformation der Hyperelliptischen Funktionen erster Ordnung; Krazer, Theorie der zweifach unendlichen Thetareihen auf Grund der Riemann'schen Thetaformel; Krazer and Prym, Neue Grundlagen einer Theorie der allgemeinen Thetafunctionen; Prym, Untersuchungen ueber die Riemann'sche Thetaformel. All these works are published by Teubner.

The theta-functions form an essential part of Riemann's theory of Abelian Integrals, and therein lies their present importance. Works on Riemann's theory contain more or less of the Algebra of the theta-functions; accordingly to the above references can be added those at the end of Chapter X.

The founders of the theory of this chapter were Göpel (Theoriæ transcendentium Abelianarum primi ordinis adumbratio levis. Crelle, t. xxxv.) and Rosenhain (Mémoire sur les fonctions de deux variables et à quatre périodes. Mém. prés. à l'Académie des Sciences de Paris, t. xi.). A bibliography will be found in Krause's book; but special mention ought to be made here of a memoir by Weber (Math. Ann., t. xiv.) and of Cayley's memoir in the Phil. Trans. (1880). The method used in § 239 and following paragraphs for the discussion of formula (9), which is due to Prym, appears to be new.

CHAPTER IX.

DIRICHLET'S PROBLEM.

§ 252. Hitherto the parts of a function $f(z)$, $= u + iv$, have been considered in the combination $u + iv$, and x, y in the combination $x + iy$. Let the functions u, v be one-valued and continuous, as also their first differential quotients, throughout a connected region of finite dimensions which includes the point z; in other words, throughout some neighbourhood of z. By § 20 of Chapter I., $f(z)$ has at this point z a differential quotient $f'(z)$ which is independent of dz. Conversely, in order that $u + iv$ may admit a continuous differential quotient $f'(z)$ at a point z, within whose neighbourhood u, v are continuous, the real functions $u(x, y)$, $v(x, y)$ must give rise to continuous partial differential quotients of the first order. For, by supposition

$$\Delta(u + iv)/\Delta z = \Delta(u + iv)/\Delta x,$$

$$\Delta(u + iv)/\Delta z = \Delta(u + iv)/i\Delta y;$$

and, when Δz tends to zero, the left-hand side tends to $f'(z)$; hence the right-hand sides must tend to

$$\delta u/\delta x + i\delta v/\delta x \text{ and } -i\delta u/\delta y + \delta v/\delta y;$$

i.e.

$$\delta u/\delta x = \delta v/\delta y, \quad \delta u/\delta y = -\delta v/\delta x \quad \cdot \quad \cdot \quad \cdot \quad \cdot \quad \cdot \quad \cdot \quad (1).$$

But $f'(z)$ is continuous by supposition, therefore the components of the right-hand sides must be continuous; hence the condition imposed upon $f(z)$—that it is to have a continuous differential quotient—implies the continuity of $\delta u/\delta x$, $\delta u/\delta y$, $\delta v/\delta x$, $\delta v/\delta y$ in the neighbourhood of z.

The differential equations (1) might have been selected as a starting-point for a theory of functions of a complex variable.* From them it is possible to deduce Cauchy's theorem $\int_C f(z)dz = 0$, where C is a closed curve contained within a region Γ, throughout which u, v, and the first partial differential quotients are one-valued

* This is the plan adopted by Picard in his Traité d'Analyse.

and continuous. As a consequence of this theorem it follows that all the higher differential quotients of u, v exist within Γ, and are finite and continuous; also that $f(z)$ at a point z_1 of the region can be expressed as an integral series $P(z|z_1)$, the circle of convergence extending at least to the rim of Γ.

Since the higher partial differential quotients of u, v exist and are one-valued and continuous in Γ, it follows that

$$\delta^2u/\delta x^2 = \delta^2v/\delta y\delta x,\ \delta^2u/\delta y^2 = -\delta^2v/\delta x\delta y,\ \delta^2u/\delta x^2 + \delta^2u/\delta y^2 = 0 \quad . \quad (2).$$

The discussion of the properties of an analytic function might be carried out, *ab initio*, with the initial assumptions that in the neighbourhood of a non-singular point (x, y) the functions u, $\delta u/\delta x$, $\delta u/\delta y$, $\delta^2u/\delta x^2$, $\delta^2u/\delta y^2$ are continuous, and that $\delta^2u/\delta x^2$, $\delta^2u/\delta y^2$ satisfy the equation

$$\delta^2u/\delta x^2 + \delta^2u/\delta y^2 = 0.$$

The equation (2) is a special case of Laplace's equation

$$\delta^2u/\delta x^2 + \delta^2u/\delta y^2 + \delta^2u/\delta z^2 = 0 \quad . \quad . \quad . \quad . \quad . \quad . \quad . \quad (3),$$

which plays so important a part in Mathematical Physics.

In many branches of Mathematical Physics a problem which is constantly arising is to determine u throughout a region of space, given the values of u on the boundary of the region, and given that u satisfies the equation (3) throughout the region in question. The problem which we propose to discuss in this chapter belongs to this class. On the general subject of the partial differential equations which occur in Mathematical Physics we refer the reader to an important memoir by Poincaré, Amer. Jour. of Math., t. xii., p. 211; this memoir contains among other things a most original discussion of Dirichlet's problem for three dimensions.

Riemann's special mode of treatment of functions, with Laplace's equation as a basis for the theory, was undoubtedly suggested by physical considerations immediately connected with the Potential theory; although his theorems are couched in the language of pure analysis.* The most novel feature of Riemann's work was precisely his attempt to define a function $u + iv$ of $x + iy$ within a region by its discontinuities in that region and by certain boundary conditions, in preference to the use of an arithmetic expression as a basis. The difficulties presented by such a method must of necessity

* Attention is called to the physical aspects of Riemann's work by Klein in his valuable pamphlet Ueber Riemann's Theorie der Algebraischen Functionen und ihrer Integrale.

be formidable, for little is known as yet of the nature of the properties of functions in general; so that when certain properties are postulated of a function it is essential that some proof be given of the existence of a function with these properties. Despite the difficulties associated with it, Riemann's method has proved of immense service in the study of Abelian integrals and automorphic functions.

In view of the yearly increasing importance of the mathematical investigations which have been carried out on Riemann's lines, rigorous proofs of existence theorems are of the highest value, even though the analysis be too indirect and laborious to serve as a method for the calculation of the functions whose existence it establishes. To a student of applied mathematics these proofs may appear unnecessarily long and tedious; the more so as in certain applications the existence of the potential function is indisputable. For his benefit we may be permitted to quote a passage from the memoir of Poincaré's, to which reference was made above: "Néanmoins toutes les fois que je le pourrai, je viserai à la rigueur absolue et cela pour deux raisons; en premier lieu, il est toujours dur pour un géomètre d'aborder un problème sans le résoudre complètement; en second lieu, les équations que j'étudierai sont susceptibles, non seulement d'applications physiques, mais encore d'applications analytiques. C'est sur la possibilité du problème de Dirichlet que Riemann a fondé sa magnifique théorie des fonctions abéliennes. Depuis d'autres géomètres ont fait d'importantes applications de ce même principe aux parties les plus fondamentales de l'Analyse pure. Est il encore permis de se contenter d'une demi-rigueur? Et qui nous dit que les autres problèmes de la Physique Mathématique ne seront pas un jour, comme l'a déjà été le plus simple d'entre eux, appelés à jouer en Analyse un rôle considérable?"

The basis for Riemann's work is a famous proposition known among continental mathematicians as Dirichlet's Principle, or Problem.

§ 253. *Dirichlet's problem for two dimensions.* To find a function $u(x, y)$ which, together with its differential quotients of the first two orders, shall be one-valued and continuous in a region Γ, which shall satisfy Laplace's equation and shall take assigned values upon the boundary of the region.

In the problem just enunciated we must pay attention to the order of connexion of the region, to the character of the rim or rims, and to the manner in which the values are assigned along the rim or rims. In the simplest form of the problem, the region Γ is simply connected and bounded by a simple contour such as the circle or ellipse; also the values on the rim are continuous. In the most general form of the problem, the region Γ is n-ply connected and bounded by an exterior simple contour C_0 and by $n-1$ interior simple contours $C_1, C_2, \cdots, C_{n-1}$, which are exterior to one another.

We shall denote the collective boundary by C, and shall speak of points within the region as points *in* Γ, and of points on the boundary as points *on* C. The curves which constitute C are supposed to have tangents whose directions vary continuously, except at a finite number of points at which there is an abrupt change of direction; a case which would arise, for instance, were the rim a polygon composed of a finite number of circular arcs. The values assigned to the rim will be continuous except at a finite number of places. Any real function $u(x, y)$ which, together with its differential quotients of the first two orders, is one-valued and continuous in Γ, and satisfies Laplace's equation for two dimensions, is called *harmonic* (Neumann, Abel'sche Integrale, 2d ed., p. 390). Dirichlet's problem can be re-stated as follows: to find a function $u(x, y)$ which shall be harmonic in Γ, and which shall take assigned values on C, these values being supposed to be continuous along C except possibly at a finite number of points. It should be remarked that the boundary value at a point s on C is to be identical with the limit to which $u(x, y)$ tends along a path in Γ which leads to s, this limit being assumed to be the same, *however the path may be chosen.**

Green was the first mathematician who enunciated the general problem ("An Essay on Electricity and Magnetism," 1828, § 5), but his proof was based on purely physical considerations. For a treatment of the problem from the physical point of view, the student is referred to Gauss, Ges. Werke, t. v., Allgemeine Lehrsätze in Beziehung auf die im verkehrten Verhältnisse des Quadrats der Entfernung wirkenden Anziehungs- und Abstossungskräfte, Art. 31 to 34; Lord Kelvin (Thomson), Reprint of Papers on Electrostatics and Magnetism, Art. xxviii., and Natural Philosophy, t. i., Appendix A (*d*); Riemann, Schwere, Elektricität und Magnetismus, p. 144; Lamb's Hydrodynamics, pp. 38 *et seq.* Weierstrass has pointed out that Riemann's proof, and in fact all those which make use of the Calculation of Variations, are open to grave objections. On this point the reader should consult Picard, Traité d'Analyse, t. ii., fasc. i., p. 38.

§ 254. If the problem be solvable, the assigned values on C must be, in general, continuous. The following two theorems which relate to the continuity of the rim values were published by M. Painlevé in his thesis, Sur les Lignes Singulières des Fonctions Analytiques, Paris, 1887. They involve the conception of *a passage*

---

* Schwarz has discussed in a very lucid and thorough manner the suppositions which must be made in solving Dirichlet's problem, and the consequences which follow from them. See his memoir Zur Integration der partiellen Differentialgleichung $\Delta u = 0$, Ges. Werke, t. ii. For instance, he points out that an assumption of the continuity of $\delta u/\delta x$, $\delta u/\delta y$ on C is not legitimate, since such an assumption would imply that the rim values are not merely continuous, but also have a differential quotient with respect to the arc.

with uniform continuity to the values on the rim. By this it is implied that to an arbitrarily small number ϵ, given in advance, there corresponds, for *every* point s on C, a region of finite dimensions which penetrates into Γ, and is bounded partly by an arc of C which contains s, partly by a line interior to Γ; the distinctive property of this region being that the difference of any two values of u, in the interior or on the rim of this region, is equal to or less than ϵ in absolute value.

In what follows, we shall use the letter s for a point on C, and also for the variable which determines the position of the point on C.

Theorem I. Let a point s describe a part D of C, and let there be associated with s *a path cs for the function* $u(x, y)$, cs varying continuously with s and leading from the interior point c to s. Then if, along D, u tend uniformly to the value $U(s)$ which is assigned to s, the function $U(s)$ must be a continuous function of s, and $u(x, y)$ tends to $U(s)$, *whatever be the interior path* cs which leads to s.

Theorem II. On the other hand, if $u(x, y)$ tend to $U(s)$, whatever be the path cs, $U(s)$ must be a continuous function of s, and $u(x, y)$ must tend uniformly to $U(s)$ along D.

The following remarks may help to show the *raison d'être* of these theorems. To fix ideas let the region Γ be a circle of centre 0 and radius R, and let the approach of a variable interior point (x, y) to a point s of the rim C take place along the radius from 0 to s. Given ϵ in advance, a positive quantity δ_s can be assigned such that, for all points (x, y) *on the radius from* 0 *to* s at a distance from s less than δ_s,

$$|u(x, y) - U(s)| < \epsilon.$$

Let δ_s be the largest value consistent with this inequality. As s moves round the circle C, δ_s varies; suppose that the lower limit of the quantities δ_s is δ, where δ is different from zero. Then, *for the assigned method of approach* to the points of C, the values $u(x, y)$ tend uniformly to the values $U(s)$. A question which suggests itself is whether the values $u(x, y)$ would tend to the same limiting values $U(s)$ when there is a change in the curves of approach to the points s; and whether, if there be no change in the limiting values, the passage to the rim values is still effected with uniform continuity. There are, so to speak, two degrees of freedom in the question; the infinitely many paths to an assigned point of D, and the infinitely many points of D. That a variety of paths to an assigned point of D may lead to one and the same limit can be illustrated, by analogy, from Chapter III. Suppose that in § 90 the approach to the point 1 is along a straight line $x_0 1$ which does not coincide with a radius; then the limiting value is the same for this path as for the radial path 0 to 1. (Stolz, Schlömilch's Zeitschrift, t. xxix., p. 127.)

I. To an arbitrarily small positive number ϵ, given in advance, there corresponds, by the conditions of Theorem I., a length η such that when an arc sc' of length η is measured off on sc, we have

$$|u(x, y) - U(s)| < \epsilon,$$

for all positions of (x, y) on sc'. Let s_1 be the point $s + \delta s$; and let the path at s_1 which corresponds to the path sc', be s_1c_1', the length of this arc being η. The points c', c_1', ... trace out a continuous line L. Since $u(x, y)$ is continuous along L, and since the path sc varies continuously as s moves along D, there must exist an interval ($s - h$ to $s + h$) such that for all points $s + \delta s$ in this interval,

$$|u_{c'} - u_{c_1'}| < \epsilon.$$

But, by the conditions of Theorem I.,

$$|U(s) - u_{c'}| < \epsilon,$$

$$|U(s + \delta s) - u_{c_1'}| < \epsilon.$$

Hence $$|U(s + \delta s) - U(s)| < 3\epsilon.$$

This proves the continuity of $U(s)$.

Next, let (x, y) be any point interior to the curvilinear quadrilateral bounded by D, L, and those two curves of the set sc which pass through $s - h$, $s + h$. By the conditions imposed initially, (x, y) must lie on some curve of the system sc; suppose that it lies on the curve through $s + \delta s$. Hence,

$$|u(x, y) - U(s + \delta s)| < \epsilon.$$

But

$$|u(x, y) - U(s)| < |u(x, y) - U(s + \delta s)| + |U(s + \delta s) - U(s)|,$$

i.e. $$|u(x, y) - U(s)| < 4\epsilon.$$

This shows that a circle can be drawn with s as centre, and with a radius sufficiently small to ensure that for every point (x, y) interior to the circle, the above inequality shall be satisfied. Hence $u(x, y)$ tends to $U(s)$ at s, whatever be the path of approach to that point.

II. In the second theorem the data show that each point s of the arc D has a domain, say of radius ρ, such that for those points (x, y) of Γ which lie in the domain of s,

$$|u(x, y) - U(s)| \leqq \epsilon;$$

for otherwise there would be paths sc' such that $|u(x, y) - U(s)| > \epsilon$ for points of sc' arbitrarily close to s, contrary to supposition. Measure off two arcs ss', ss'' of lengths ρ, and let a point (x, y) of the domain of s tend to a point $s + \delta s$ of $s's''$. By supposition $u(x, y)$ tends to $U(s + \delta s)$. But throughout the domain of s,

$$|u(x, y) - U(s)| \leqq \epsilon.$$

Hence
$$|U(s + \delta s) - U(s)| \leqq \epsilon.$$

This establishes the continuity of $U(s)$.

If s_1', s_1'' denote the middle points of ss', ss'', there corresponds to every point $s + \delta s$ of the arc $s_1's_1''$ a circular region Δ, of centre $s + \delta s$ and radius $\geqq \rho/2$, such that for every point (x, y) common to Γ and Δ,

$$|u(x, y) - U(s + \delta s)| \leqq 2\epsilon.$$

By reasoning about s_1', s_1'' in the same way as about s, and by a continuation of the process, we must arrive finally at the extremities of D, unless the radius ρ tends to a limit 0, when the centres s tend to some point s_0. To see that this is an impossible case, observe that there corresponds to s_0 a circular region Δ_0, say of radius ρ_0, such that when (x, y) lies inside Δ_0,

$$|u(x, y) - U(s)| \not> \epsilon/2,$$

and therefore to the points s near s_0 there correspond domains of radii at least equal to $\rho_0/2$. These considerations show that we can find a length ζ such that for points inside a circle of radius ζ and centre s,

$$|u(x, y) - U(s)| \not> 2\epsilon,$$

whatever be the position of s on D. Thus $u(x, y)$ passes with uniform continuity to the rim values $U(s)$.

We shall prove, in the next paragraph, some theorems which will be of use at a later stage of the discussion. These theorems embody well-known results in the ordinary theory of the potential.

§ 255. *Green's theorem for functions of two variables.*

Let u_1, u_2 be two functions which are harmonic within Γ', and let Γ be a region which is completely enclosed by Γ'. The functions u_1, u_2, together with their partial differential quotients of the first two orders, are continuous in Γ and on its boundary C. From the remark of Schwarz quoted in the foot-note to p. 377 it appears that it is not permissible to assume in all cases the existence of partial differential quotients on the boundary of Γ'; this is why Γ is chosen. Green's theorem is

$$\iint\left(\frac{\delta u_1}{\delta x}\frac{\delta u_2}{\delta x}+\frac{\delta u_1}{\delta y}\frac{\delta u_2}{\delta y}\right)dxdy=-\iint u_2\left(\frac{\delta^2 u_1}{\delta x^2}+\frac{\delta^2 u_1}{\delta y^2}\right)dxdy-\int u_2\frac{\delta u_1}{\delta n}ds$$

$$=-\iint u_1\left(\frac{\delta^2 u_2}{\delta x^2}+\frac{\delta^2 u_2}{\delta y^2}\right)dxdy-\int u_1\frac{\delta u_2}{\delta n}ds \quad . \quad . \quad (4).$$

Here the differentiation is with regard to a normal drawn *into* Γ and the direction of integration over a rim is positive when that rim is regarded as an independent curve. Since $dx/dn=-dy/ds$, $dy/dn=dx/ds$, we have

$$\frac{\delta u_1}{\delta n}=-\frac{\delta u_1}{\delta x}\frac{dy}{ds}+\frac{\delta u_1}{\delta y}\frac{dx}{ds};$$

and there is a similar formula for u_2. Integration by parts shows that

$$\iint_\Gamma\frac{\delta u_1}{\delta x}\frac{\delta u_2}{\delta x}dxdy=\int_C u_1\frac{\delta u_2}{\delta x}dy-\iint_\Gamma u_1\frac{\delta^2 u_2}{\delta x^2}dxdy,$$

where the first integral is a curvilinear integral taken positively over C. Similarly,

$$\iint_\Gamma\frac{\delta u_1}{\delta y}\frac{\delta u_2}{\delta y}\,dxdy=-\int_C u_1\frac{\delta u_2}{\delta y}\,dx-\iint_\Gamma u_1\frac{\delta^2 u_2}{\delta y^2}\,dxdy.$$

By addition we have

$$\iint_\Gamma\left\{\frac{\delta u_1}{\delta x}\frac{\delta u_2}{\delta x}+\frac{\delta u_1}{\delta y}\frac{\delta u_2}{\delta y}\right\}dxdy$$

$$=\int_C u_1\left(\frac{\delta u_2}{\delta x}dy-\frac{\delta u_2}{\delta y}dx\right)-\iint_\Gamma u_1\left(\frac{\delta^2 u_2}{\delta x^2}+\frac{\delta^2 u_2}{\delta y^2}\right)dxdy,$$

but $$-\frac{\delta u_2}{\delta x}dy+\frac{\delta u_2}{\delta y}dx=\frac{\delta u_2}{\delta n}ds \text{ on } C, \text{ and } \frac{\delta^2 u_2}{\delta x^2}+\frac{\delta^2 u_2}{\delta y^2}=0 \text{ in } \Gamma.$$

Hence $$\iint_\Gamma\left\{\frac{\delta u_1}{\delta x}\frac{\delta u_2}{\delta x}+\frac{\delta u_1}{\delta y}\frac{\delta u_2}{\delta y}\right\}dxdy=-\int_C u_1\frac{\delta u_2}{\delta n}ds \quad . \quad . \quad (5).$$

Similarly

$$\iint_\Gamma\left\{\frac{\delta u_1}{\delta x}\frac{\delta u_2}{\delta x}+\frac{\delta u_1}{\delta y}\frac{\delta u_2}{\delta y}\right\}dxdy=-\int_C u_2\frac{\delta u_1}{\delta n}ds \quad . \quad . \quad (6).$$

An important result can be derived from (6) by writing $u_1 = u_2 = u$;

namely, $$\iint\left\{\left(\frac{\delta u}{\delta x}\right)^2 + \left(\frac{\delta u}{\delta y}\right)^2\right\} dxdy = -\int_C u\frac{\delta u}{\delta n} ds \quad . \quad . \quad (7).$$

This formula shows that there cannot be two harmonic functions which take an assigned system of values on C. For suppose that u_1', u_2' are two such functions. Replacing $u_1' - u_2'$ by u, we have $u = 0$ on C. Hence, if we assume $\frac{\delta u}{\delta n}$ to be finite along C, the formula (7) becomes

$$\iint\left\{\left(\frac{\delta u}{\delta x}\right)^2 + \left(\frac{\delta u}{\delta y}\right)^2\right\} dxdy = 0.$$

That is, a sum whose elements cannot be negative vanishes. This can only happen when every element vanishes. Hence

$$\delta u/\delta x = 0, \quad \delta u/\delta y = 0,$$

and $u_1 - u_2$ = a constant throughout Γ. This constant must be zero; for otherwise there would be a discontinuity in passing to the rim. Thus, *if the problem of Dirichlet admit a solution*, that solution is unique.

Picard has pointed out the limitations to this proof. Besides the necessity of assuming, in general, that $\frac{\delta u}{\delta n}$ does not become infinite, there is the further limitation that the integrals, as also their differential quotients of the first order, are to remain continuous in Γ and on C. (Traité d'Analyse, t. ii., fasc. i., p. 23.)

Theorem. $\int_C \frac{\delta u}{\delta n} ds = 0$, when u is harmonic within a region Γ' which encloses Γ.

The formula which results from the combination of (5) and (6) is

$$\int_C\left(u_1\frac{\delta u_2}{\delta n} - u_2\frac{\delta u_1}{\delta n}\right)ds = 0 \quad . \quad . \quad . \quad . \quad . \quad . \quad . \quad . \quad . \quad . \quad (8).$$

Replacing u_1, u_2 by u, 1, we have

$$\int_C \frac{\delta u}{\delta n} ds = 0 \quad . \quad . \quad . \quad . \quad . \quad . \quad . \quad . \quad . \quad . \quad . \quad . \quad . \quad (9).$$

Theorem. Let Γ be enclosed as before in Γ', and let r be the distance of a point (x, y) from a fixed internal point (a, b), then

$$u(a, b) = \frac{1}{2\pi}\int_C\left(\log r\frac{\delta u}{\delta n} - u\frac{\delta \log r}{\delta n}\right)ds \quad . \quad . \quad . \quad (10).$$

With centre (a, b) describe a small circle C_1 of radius ρ. Let $u_2 = \log r$, $u_1 = u$: then by (8), if the normal to C_1 be drawn towards (a, b),

$$\int_C \left\{ \log r \frac{\delta u}{\delta n} - u \frac{\delta \log r}{\delta n} \right\} ds = \int_{C_1} \left\{ \log r \frac{\delta u}{\delta n} - u \frac{\delta \log r}{\delta n} \right\} ds$$

$$= \log \rho \int_{C_1} \frac{\delta u}{\delta n} ds + \frac{1}{\rho} \int_{C_1} u ds$$

$$= \frac{1}{\rho} \int_{C_1} u ds \text{ by (9).}$$

Since the left-hand side is independent of ρ, the expression $\frac{1}{\rho}\int_{C_1} uds$ must be independent of ρ.

When ρ tends to zero, we know that an approximate value of $\frac{1}{\rho}\int_{C_1} uds$ is $u(a, b) \cdot \frac{1}{\rho}\int_{C_1} ds$, *i.e.*, $2\pi u(a, b)$. Since this value is independent of ρ, it must be the exact value, and the theorem is established.

The result thus obtained is of high importance, because it expresses the value of u at an internal point in term of values along a certain boundary. In the course of the proof, we have established incidentally a theorem which has important consequences in the theory of the potential. This theorem will now be stated explicitly.

Theorem. A harmonic function u cannot have a maximum or minimum in Γ.

If ρ be the radius of a circle C with centre (a, b),

$$2\pi u(a, b) = \frac{1}{\rho}\int_C uds = \int_0^{2\pi} u(\rho, \theta) d\theta \quad . \quad . \quad . \quad (11),$$

in polar co-ordinates. Let ρ be made sufficiently small; then if $u(a, b)$ be greater than all the values within, or on the circumference of the circle (ρ),

$$2\pi u(a, b) < \frac{1}{\rho}\int_C u(a, b) ds,$$

i.e.,

$$2\pi u(a, b) < 2\pi u(a, b).$$

This shows that $u(a, b)$ cannot be a maximum. Similarly it can be shown that it cannot be a minimum. The same result might be deduced from (9).

This theorem extended to three dimensions amounts to the statement that the gravitation potential of an attracting mass has neither a maximum nor a minimum value in empty space. It was discovered by Gauss.

Theorem. If $u(x, y)$ be a function which is harmonic in any region Γ, and continuous on the boundary C, the value of u at every point in Γ is intermediate in value between the greatest and least values on C. Here we assume that there are no points on C at which the function u is discontinuous, and that u is not a constant.

The values of u on C have a finite upper limit G, and a finite lower limit K; and the values in Γ and on C have a finite upper limit G' and a finite lower limit K'. At no points of Γ can the limits G', K' be attained. Let us assume, if possible, that u attains the value G' at an interior point (a, b). The formula (11) shows that on the rim of every circle interior to C, with (a, b) as centre, the value of $u = G'$; for if some of the values be less than G', others must be greater than this quantity, and this is contrary to supposition. Hence at all points of a circular region of centre (a, b), which lies wholly inside Γ and extends to C, u is constant. By choosing a new point (a, b) in this circular region and drawing a new circle, it follows that u is constant throughout the new region. In this way it can be proved that the value of u in Γ is everywhere equal to G'. This is contrary to hypothesis; hence there is no point of Γ at which u attains the value G'. But we know that the continuous function u attains its upper limit at some point (§ 63). This point must lie on C. As the upper limit on C is G, G' must be equal to G; and similar reasoning shows that K' must be equal to K.

Corollary. A harmonic function which is constant on C has the same constant value in Γ.

§ 256. *Schwarz's existence-theorem for a circular region* Γ *of radius* 1. [Schwarz, Werke, t. ii., p. 185.]

Let $f(\alpha)$ be a real function of α, which is continuous and one-valued upon the circumference, and which resumes its value when α increases by 2π; no other restrictions are imposed upon $f(\alpha)$. There exists always one and only one function u which possesses the following properties: —

(1) u is continuous in Γ and on C;

(2) $\delta u/\delta x$, $\delta u/\delta y$, $\delta^2 u/\delta x^2$, $\delta^2 u/\delta y^2$ are one-valued and continuous in Γ and on C, and $\delta^2 u/\delta x^2 + \delta^2 u/\delta y^2 = 0$;

(3) the values on the rim are equal to $f(\alpha)$ at all points of C.

The proof is based upon a consideration of the integral

$$\zeta = \frac{1}{2\pi}\int_0^{2\pi} f(\alpha)\frac{1-r^2}{1-2r\cos(\alpha-\theta)+r^2}d\alpha,$$

where r, θ are the co-ordinates of a point on the rim of a circle whose centre is 0 and radius $r(<1)$. This integral has a perfectly determinate meaning for all points within the circle, and equals the series

$$\frac{1}{2\pi}\int_0^{2\pi} f(\theta)d\theta + \frac{1}{\pi}\sum_{n=1}^{\infty}\int_0^{2\pi} f(\theta)\cos n(\alpha-\theta)d\theta\cdot r^n.$$

This latter series is convergent in Γ. On C it assumes the form of Fourier's series. Schwarz proves that

(i.) ζ is harmonic in Γ;

(ii.) ζ takes the contour value $f(\theta)$.

The proof of (i.) is effected in a very natural and simple manner. We know by the properties of monogenic functions that when $\phi(z)$ is holomorphic in Γ, the real part of $\phi(z)$ is harmonic in Γ. Hence the real part of

$$\phi(z) = \frac{1}{2\pi}\int_0^{2\pi} f(\alpha)\cdot\frac{e^{\alpha i}+z}{e^{\alpha i}-z}d\alpha,$$

is harmonic in Γ. But this real part is precisely Poisson's integral

$$u(x, y) = \zeta = \frac{1}{2\pi}\int_0^{2\pi} f(\alpha)\frac{1-r^2}{1-2r\cos(\alpha-\theta)+r^2}d\alpha.$$

To prove (ii.), Schwarz examines the difference $\zeta - f(\theta)$, where θ is arbitrary, and shows that as (x, y) tends along radii to the various points $(1, \theta)$ of C, $u(x, y)$ tends *uniformly* to the limit $f(\theta)$. To simplify the denominator of the integral write $\alpha+\theta$ for α; this involves no change of limits, since $f(\alpha)$ is periodic. At any stage of the following proof the limits can be replaced by new limits γ, $2\pi+\gamma$, where γ is any positive or negative quantity. The new form of the integral is

$$\frac{1}{2\pi}\int_0^{2\pi}\frac{(1-r^2)f(\alpha+\theta)d\alpha}{1-2r\cos\alpha+r^2},$$

or

$$\frac{1}{2\pi}\int_{-\pi}^{+\pi}\frac{(1-r^2)f(\alpha+\theta)d\alpha}{1-2r\cos\alpha+r^2}.$$

This may be written

$$\frac{1}{2\pi}\int_{-\pi}^{+\pi}\frac{(1-r^2)\{f(\alpha+\theta)-f(\theta)\}d\alpha}{1-2r\cos\alpha+r^2} + \frac{f(\theta)}{2\pi}\int_{-\pi}^{+\pi}\frac{(1-r^2)d\alpha}{1-2r\cos\alpha+r^2}.$$

For all values of r less than 1,

$$\frac{1}{2\pi}\int_{-\pi}^{+\pi}\frac{(1-r^2)\,d\alpha}{1-2r\cos\alpha+r^2}=1.$$

Hence $$\zeta=f(\theta)+\frac{1}{2\pi}\int_{-\pi}^{+\pi}\frac{(1-r^2)\{f(\alpha+\theta)-f(\theta)\}\,d\alpha}{1-2r\cos\alpha+r^2}.$$

We have now to show that when ϵ is an arbitrarily small quantity given in advance, it is possible to find a number η such that for all values of $1-r\begin{matrix}<\eta\\>0\end{matrix}\Big\}$, and for all values of θ,

$$\left|\int_{-\pi}^{+\pi}\frac{(1-r^2)\{f(\alpha+\theta)-f(\theta)\}\,d\alpha}{1-2r\cos\alpha+r^2}\right|<\epsilon.$$

This inequality is required in order that the passage to the contour values may be effected with uniform continuity. It will be observed that the paths of approach to the contour are along radii, but this limitation may be removed by means of Painlevé's theorems. After replacing the limits by $-\delta$, $2\pi-\delta$, let the integral be separated into the two parts

$$\frac{1}{2\pi}\int_{-\delta}^{+\delta}+\frac{1}{2\pi}\int_{+\delta}^{2\pi-\delta}\frac{(1-r^2)\{f(\alpha+\theta)-f(\theta)\}\,d\alpha}{1-2r\cos\alpha+r^2},$$

where δ is a small positive quantity. Let g be the upper limit of $f(\alpha+\theta)-f(\theta)$ in the interval $(-\delta,\delta)$. The absolute value of the former integral is

$$\left|\frac{1}{2\pi}\int_{-\delta}^{+\delta}\frac{(1-r^2)\{f(\alpha+\theta)-f(\theta)\}\,d\alpha}{1-2r\cos\alpha+r^2}\right|<\frac{g}{2\pi}\left|\int_{-\delta}^{+\delta}\frac{(1-r^2)\,d\alpha}{1-2r\cos\alpha+r^2}\right|<g.$$

The absolute value of the latter

$$<\frac{G}{\pi}\int_{\delta}^{2\pi-\delta}\frac{(1-r^2)\,d\alpha}{1-2r\cos\alpha+r^2},$$

where G is the upper limit of the absolute values of $f(\alpha)$ in the interval $(\delta, 2\pi-\delta)$. But

$$\frac{1}{\pi}\int_{\delta}^{2\pi-\delta}\frac{(1-r^2)\,d\alpha}{1-2r\cos\alpha+r^2}<\frac{1-r^2}{r(1-\cos\delta)};$$

for since the expression $\cos\alpha-\cos\delta$ is negative in the interval in question, the denominator $1-2r\cos\alpha+r^2>2r(1-\cos\delta)$. Whatever be the value of the small quantity $\delta(\neq 0)$, it is possible to choose $1-r$ so small that

$$G\frac{1-r^2}{r(1-\cos\delta)}<\epsilon.$$

Hence, when r tends to 1, the integral

$$\frac{1}{2\pi}\int_{-\pi}^{+\pi}\frac{(1-r^2)\{f(\alpha+\theta)-f(\theta)\}d\alpha}{1-2r\cos\alpha+r^2}$$

tends to the limit 0, and ζ tends to limiting values identical with the contour values $f(\theta)$.

§ 257. The solution of the problem for a circle of radius R is very similar to that for the circle of radius 1. It is

$$u(x, y)=\frac{1}{2\pi}\int_0^{2\pi}\frac{(R^2-\rho^2)f(\alpha)d\alpha}{R^2-2R\rho\cos(\alpha-\theta)+\rho^2},$$

where ρ is the distance from the origin of a point c, or (x, y), situated in Γ. The formula can be written

$$u(x, y)=\int_C U(s)\frac{R^2-|c|^2}{2\pi R\cdot|c-s|^2}ds,$$

where $|c-s|$ is the distance from c to a point s of C, and $U(s)$ is the value of u at s. When c coincides with 0, this formula becomes

$$u(0)=\int_0^{2\pi}\frac{U(s)d\theta}{2\pi},$$

or the mean of the values $U(s)$ on the rim of C is equal to the value of u at the centre.

By a known formula in Trigonometry

$$\frac{R^2-\rho^2}{R^2-2R\rho\cos(\alpha-\theta)+\rho^2}=1+2\sum_{n=1}^{\infty}\left(\frac{\rho}{R}\right)^n\cos n(\alpha-\theta),\ (\rho<R).$$

Hence when $x=\rho\cos\theta,\ y=\rho\sin\theta$,

$$u(x, y),=f(\rho,\theta),=\frac{1}{2\pi}\int_0^{2\pi}f(\alpha)d\alpha+\frac{1}{\pi}\sum_{n=1}^{\infty}\left\{\int_0^{2\pi}f(\alpha)\cos n\alpha d\alpha\cdot\cos n\theta\right.$$

$$\left.+\int_0^{2\pi}f(\alpha)\sin n\alpha d\alpha\cdot\sin n\theta\right\}\left(\frac{\rho}{R}\right)^n$$

$$=\frac{a_0}{2}+\sum_{n=1}^{\infty}(a_n\cos n\theta+b_n\sin n\theta)\left(\frac{\rho}{R}\right)^n.$$

This expansion has been proved on the supposition $\rho < R$. Suppose that $u(x, y)$ is continuous, together with its partial differential quotients of the first two orders, in a circle of radius $R_1 > R$. Is the expansion true when $\rho = R$? We know that the limit to which Poisson's integral tends, when ρ tends to R, is $u(x, y)$, or $f(R, \theta)$, where $x = R\cos\theta$, $y = R\sin\theta$; but it is not evident that the integral series in $\frac{\rho}{R}$ tends to the limit

$$\frac{a_0}{2} + \sum_{n=1}^{\infty} (a_n \cos n\theta + b_n \sin n\theta).$$

That this is what actually happens has been proved by Paraf. As soon as it can be shown that the series just written down is convergent, Abel's theorem on integral series (§ 90) enables us to say that this series is the limit of the series in ρ/R.

Paraf's proof of convergence. [Thesis, sur le Problème de Dirichlet et son extension au cas de l'équation linéaire générale du second ordre, p. 56.]

The coefficients a_n, b_n are continuous functions of R, which satisfy the relations $|a_n|$, $|b_n| < h/n^2$, where h is a constant which depends only on the values of $f(R, \alpha)$ and not on R_1. For, after a double integration by parts,

$$a_n = \frac{1}{\pi}\int_0^{2\pi} f(R, \alpha)\cos n\alpha d\alpha,\quad b_n = \frac{1}{\pi}\int_0^{2\pi} f(R, \alpha)\sin n\alpha d\alpha,$$

give

$$a_n = -\frac{1}{\pi}\cdot\frac{1}{n^2}\int_0^{2\pi} f''(R, \alpha)\cos n\alpha d\alpha,\quad b_n = -\frac{1}{\pi}\cdot\frac{1}{n^2}\int_0^{2\pi} f''(R, \alpha)\sin n\alpha d\alpha.$$

Hence $|a_n|$ and $|b_n| < h/n^2$, where h is twice the upper limit of the values $|f''(R, \alpha)|$. This shows that the series

$$\frac{a_0}{2} + \sum_{n=1}^{\infty} (a_n \cos n\theta + b_n \sin n\theta)$$

is absolutely and uniformly convergent. Hence when ρ tends to the limit R,

$$f(\rho, \theta) = \frac{a_0}{2} + \sum_{n=1}^{\infty}\left(\frac{\rho}{R}\right)^n (a_n \cos n\theta + b_n \sin n\theta)$$

gives
$$u(x, y) = f(R, \theta) = \frac{a_0}{2} + \sum_{n=1}^{\infty} (a_n \cos n\theta + b_n \sin n\theta).$$

§ 258. We have arrived at a series for $u(x, y)$, namely

$$u(x, y) = \frac{a_0}{2} + \sum_{n=1}^{\infty} (a_n \cos n\theta + b_n \sin n\theta),$$

in which the coefficients are definite integrals. Let us now make a fresh start with the equations

$$\frac{\delta u}{\delta x} = \frac{\delta v}{\delta y}, \quad \frac{\delta u}{\delta y} = -\frac{\delta v}{\delta x},$$

or with the corresponding equations in polar co-ordinates

$$\frac{\delta u}{\delta r} = \frac{1}{r}\frac{\delta v}{\delta \theta}, \quad \frac{\delta v}{\delta r} = -\frac{1}{r}\frac{\delta u}{\delta \theta},$$

where the pole is at a point (a, b), and (r, θ) are the co-ordinates of an arbitrary point of that circular region of centre (a, b) and radius R which extends to the boundary of Γ. Here Γ is the region of the z-plane for which the functions u, v and their first differential quotients are one-valued and continuous, these differential quotients being subject to the above equations.

Since $\dfrac{\delta(u+iv)}{\delta x}$ and $\dfrac{\delta(u+iv)}{\delta y}$ are finite when $r=0$, we must have for all values of θ,

$$\lim_{r=0}\left[\frac{\delta u}{\delta \theta}\right] = 0, \quad \lim_{r=0}\left[\frac{\delta v}{\delta \theta}\right] = 0.$$

On a circle of radius r ($< R$) we have by Fourier's series

$$u(r, \theta) = \frac{1}{2}a_0 + \sum_{n=1}^{\infty} (a_n \cos n\theta + b_n \sin n\theta),$$

$$v(r, \theta) = \frac{1}{2}c_0 + \sum_{n=1}^{\infty} (c_n \cos n\theta + d_n \sin n\theta),$$

where $$a_0 = \frac{1}{\pi}\int_0^{2\pi} u(r, \theta)\,d\theta, \quad c_0 = \frac{1}{\pi}\int_0^{2\pi} v(r, \theta)\,d\theta,$$

$$a_n = \frac{1}{\pi}\int_0^{2\pi} u(r, \theta) \cos n\theta\,d\theta, \; b_n = \frac{1}{\pi}\int_0^{2\pi} u(r, \theta) \sin n\theta\,d\theta,$$

$$c_n = \frac{1}{\pi}\int_0^{2\pi} v(r, \theta) \cos n\theta\,d\theta, \; d_n = \frac{1}{\pi}\int_0^{2\pi} v(r, \theta) \sin n\theta\,d\theta.$$

From the differential equations

$$\frac{\delta u}{\delta r} = \frac{1}{r}\frac{\delta v}{\delta \theta} \quad \text{(i)},$$

$$\frac{\delta v}{\delta r} = -\frac{1}{r}\frac{\delta u}{\delta \theta} \quad \text{(ii)},$$

we can derive at once equations such as

$$r\int_0^{2\pi} \frac{\delta u}{\delta r} \sin n\theta d\theta - \int_0^{2\pi} \frac{\delta v}{\delta \theta} \sin n\theta d\theta = 0 \quad \text{(iii)}.$$

The second integral, when integrated by parts, gives

$$\int_0^{2\pi} \frac{\delta v}{\delta \theta} \sin n\theta d\theta = -n\int_0^{2\pi} v \cos n\theta d\theta = 0.$$

Since $\frac{\delta u}{\delta r}$ is a continuous function of both r and θ,

$$\int_0^{2\pi} \frac{\delta u}{\delta r} \sin n\theta d\theta = \frac{\delta}{\delta r}\int_0^{2\pi} u \sin n\theta d\theta.$$

Hence, when $n \geqq 1$,

$$r\frac{\delta b_n}{\delta r} + nc_n = 0.$$

Similarly, multiplication by $\cos n\theta$ gives

$$r\frac{\delta a_n}{\delta r} - nd_n = 0.$$

From equation (ii) we can derive, in a similar manner, the two differential equations

$$r\frac{\delta(d_n)}{\delta r} - na_n = 0,$$

$$r\frac{\delta(c_n)}{\delta r} + nb_n = 0.$$

The equations

$$\int_0^{2\pi} \frac{\delta u}{\delta r} d\theta - \frac{1}{r}\int_0^{2\pi} \frac{\delta v}{\delta \theta} d\theta = 0,$$

$$\int_0^{2\pi} \frac{\delta v}{\delta r} d\theta + \frac{1}{r}\int_0^{2\pi} \frac{u}{\delta \theta} d\theta = 0,$$

give $$\frac{\delta a_0}{\delta r} = 0, \quad \frac{\delta c_0}{\delta r} = 0;$$

therefore a_0, c_0 are constants. To find b_n we must integrate the equation

$$r^2\frac{\delta^2 b_n}{\delta r^2} + r\frac{\delta b_n}{\delta r} - n^2 b_n = 0.$$

Hence $\quad b_n = \lambda_n r^n + \lambda_n' r^{-n};$

but knowing that $\delta u/\delta r$ is to be finite for $r = 0$, it is evident that λ_n' must vanish. Corresponding to $b_n = \lambda_n r^n$ we have the three equations

$$c_n = -\lambda_n r^n, \quad a_n = \mu_n r^n, \quad d_n = \mu_n r^n.$$

Hence

$$\begin{aligned} u + iv &= \tfrac{1}{2}(a_0 + ic_0) + \sum_{n=1}^{\infty}(\mu_n - i\lambda_n)(\cos n\theta + i\sin n\theta) r^n \\ &= P(z - z_0), \end{aligned}$$

where $z_0 = a + ib$, $z - z_0 = re^{\theta i}$. This is the Cauchy-Taylor theorem. [Harnack, Fundamentalsätze der Functionentheorie, Math. Ann., t. xxi., p. 305.]

§ 259. In the *generalized problem* for a simply connected region Γ the values $U(s)$ on C are in general continuous, but suffer discontinuities at a finite number of points $a_1, a_2, \cdots, a_h, \cdots, a_m$. The discontinuity at a point a is of a special kind. One condition is that for the region common to Γ and to a small circle of centre a all the values of u are to be finite; but further the values $U(s)$ are to tend to limits $U(a_h + 0)$ or $U(a_h - 0)$, when s tends to a_h in the positive or negative sense, where

$$U(a_h + 0) - U(a_h - 0) = \mu_h, \quad (h = 1, 2, \cdots, m)$$

(see Chapter II., § 61). Let the angles made by the tangents at the points a_h be α_h, where α_h differs from π when there is a sudden change in the direction of the tangent. For simplicity assume that the α's are different from zero. The coefficient of i in

$$\sum_{h=1}^{m}\frac{\mu_h}{\alpha_h}\cdot\log(z - a_h),$$

where the branch of each logarithm is fixed by assigning its value for some point in Γ, is the harmonic function

$$\sum_{h=1}^{m}\frac{\mu_h}{\alpha_h}\tan^{-1}\frac{y - y_h}{x - x_h},$$

x_h, y_h being the co-ordinates of a_h. This sum of inverse tangents suffers a discontinuity μ_h in passing through a point a_h. Hence, by the employment of a method analogous to that used in the proof of Mittag-Leffler's theorem (Chapter V., § 150), the function u can be deprived of its discontinuities by the subtraction of $\Sigma \frac{\mu_h}{\alpha_h} \tan^{-1} \frac{y-y_h}{x-x_h}$. The resulting function $w(x, y)$ is one with assigned values along C, and when (x, y) tends along C to a point a,

$$w_{a-0} - w_{a+0} = 0.$$

At the point a itself let the value 0 be assigned to w. The generalized problem has thus been resolved into the simpler problem of finding a function $w(x, y)$ which is continuous on C, continuous and harmonic in Γ. Thus whenever a solution is known for the ordinary Dirichlet problem in the case of a simply connected region Γ, a solution can be found for the generalized problem.

The solution found in this way is unique. [See M. Jules Riemann's memoir Sur le Problème de Dirichlet, Annales de l'Éc., Norm. Sup., Ser. 3, t. v., 1888.]

We omit the proof of the uniqueness of the solution. The plan to be followed in a proof of this theorem is to establish the fact primarily for a circle and afterwards for every region which can be represented conformly on a circle.

Suppose that a point (x, y) in Γ tends to a point a of C along a path interior to Γ and such that the tangent to the path at a drawn inwards makes angles ϕ_1, ϕ_2 with the two tangents at_1, at_2 to ap_1, ap_2, where p_2ap_1 is a part of C described in the direct sense. The limit to which u tends is

$$U_{a+0} + \frac{\phi_1}{\alpha}\mu$$

$$= U_{a+0} + \frac{\phi_1}{\alpha}(U_{a-0} - U_{a+0})$$

$$= U_{a-0}\frac{\phi_1}{\alpha} + U_{a+0}\frac{\phi_2}{\alpha}.$$

In the generalized problem, as in the ordinary problem, the values of u in Γ lie between the upper limit G and the lower limit K of the values $U(s)$ on C.

M. Jules Riemann has treated the case of a multiply connected region in a very similar manner. By the use of the properties of logarithms he shows that it is possible to solve the generalized problem whenever the ordinary problem can be solved. If, in the generalized problem for a simply connected surface, the restriction be removed that the function is to have a finite absolute value in the neighbourhood of its discontinuities, the theorem of the uniqueness of solution ceases to hold. Picard has given a very simple example of a harmonic function which vanishes at all points save one of a circle C, is continuous within Γ, and yet is not identically zero. The circle C is $x + a(x^2 + y^2) = 0$, where a is a constant, and the function is $a + x/(x^2+y^2)$. The explanation is to be sought in the fact that the function is infinite at the origin. [Picard, Traité d'Analyse, t. ii., fasc. i., p. 46.]

Conform Representation.

§ 260. A problem which is very closely allied with that of Dirichlet is the problem of the conform representation of an assigned surface Γ_1 upon another assigned surface Γ_2. We shall confine our attention to simply connected regions. A simplification can be introduced, for if it be possible to map a region Γ_1 on Γ_3 by means of $z_3 = \phi_1(z_1)$, and Γ_3 upon Γ_2 by $z_2 = \phi_2(z_3)$, then $z_2 = \phi_2\phi_1(z_1)$ will map Γ_1 upon Γ_2. Hence if it be possible to represent two regions conformly on one and the same circle (or half-plane), it is also possible to represent them conformly on each other. In this way the problem reduces itself to the discovery of that function $z' = f(z)$ which maps a simply connected region Γ on a circle, or on the positive half-plane of z. This problem is related to Dirichlet's by reason of the circumstance that the latter problem can be solved for a simply connected region Γ_2, whenever it can be solved for a simply connected region Γ_1 upon which Γ_2 can be represented conformly. Now a solution of Dirichlet's problem can be found for a circle, or the positive half-plane. Hence a proof that every simply connected region can be mapped on a circle is also a proof that Dirichlet's problem can be solved for every simply connected region, and conversely. Since the transformation $\left(z, \dfrac{az+b}{cz+d}\right)$, where a, b, c, d are real and $ad - bc$ positive, transforms a point with a positive ordinate into another point with a positive ordinate, there are at least ∞^3 ways of representing the positive half-plane conformly upon itself. It will be proved (§ 269) that the most general problem of the conform representation of one simply connected region Γ_1 upon another Γ_2 becomes perfectly definite as soon as a point in Γ_1 and a point on C_1 are made to correspond to a point in Γ_2 and a point on C_2. Thus the transformation $\left(z, \dfrac{az+b}{cz+d}\right)$ is the most general one for the conform representation of the positive half-plane upon itself. The problem of the conform representation of a simply connected region Γ upon the positive half-plane becomes perfectly definite as soon as three assigned points on C are made to pass into three assigned points on the real axis of the half-plane; or, what will do equally well, as soon as an assigned point in Γ is made to correspond to an assigned point in the positive half-plane, and an assigned point on C to an assigned point on the real axis.

Statement of the problem. Given two simply connected regions Γ_1 and Γ_2 traced out by z and z', to find a one-valued function $z'=f(z)$ which effects the conform representation of one of the regions upon the other, so that the points interior to Γ_1, Γ_2 correspond one-to-one, as also the points on C_1, C_2.

Let z'_0, z_0 be any two corresponding points in Γ_1, Γ_2; then in the neighbourhoods of z_0, z_0' we must have

$$z'-z_0'=P_1(z-z_0),$$

and, by reversion of series,

$$z-z_0=P_1(z'-z_0');$$

for these relations make dz'/dz, dz/dz' different from 0 in the neighbourhoods of z_0 and z_0', and therefore permit the conform representation of the neighbourhoods of z, z' upon each other. Since the one-valued function $f(z)$ may take the same value at several different points of a region, there is no necessity to restrict the problem of conform representation to a region which covers the plane simply.

Simple examples of conform representation. [Schwarz, Werke, t. ii., p. 148.]

(1) A region bounded by two arcs of circles through the points z_1, z_2 in the z-plane, can be mapped upon half of the z'-plane by the function

$$z'=\{(z-z_1)/(z-z_2)\}^{1/\lambda},$$

where $\lambda\pi$ is the angle at which the arcs intersect.

(2) A region bounded by three arcs of circles which intersect at angles $\pi/2$, $\pi/2$, $\lambda\pi$, is transformed by

$$z'=\{(z-z_1)/(z-z_2)\}^{1/\lambda}$$

into a semicircle, when z_1, z_2 are the two points of intersection of those two arcs which include the angle $\lambda\pi$, and $\lambda\neq 0$. A special case of the above region is the sector of a circle. Here $z_2=\infty$, and the transformation is replaced by

$$z'=(z-z_1)^{1/\lambda}.$$

(3) If in the preceding example λ be made to vanish, the transformations which convert the curvilinear triangle into a sector are

of a different character. Let z_1 be the point at which the two arcs touch; the remaining arc will pass, when produced, through 0. Denoting by $\lambda\pi$ the angle which the real axis makes with the tangent at z_1, the transformation

$$z' = e^{\lambda\pi i}/(z - z_1)$$

is equivalent to a turn of the tangent through an angle $\lambda\pi$ (so as to become parallel to the axis) followed by a quasi-inversion with regard to z_1. The resulting region is bounded by two lines parallel to the real axis in the z'-plane, and by part of a line parallel to the axis of imaginaries. The further transformation

$$z'' = e^{-z'}$$

changes the two parallel straight lines into two straight lines which pass through a point, and the remaining straight line into an arc of a circle with this point as centre. The resulting region is a circular sector.

(4) The transformation $z' = (z - i)/(z + i)$ represents the positive half of the z-plane conformly on a circle in the z'-plane whose centre is at the origin and whose radius is unity.

We have indicated above the existence of a connexion between Dirichlet's Problem and that of conform representation. We shall now prove that this is the case. By supposition there is some transformation $z_1 = \phi(z_2)$ which maps Γ_1 upon Γ_2. Let $z_1 = x_1 + iy_1$, $\phi(z_2) = \chi_1(x_2, y_2) + i\chi_2(x_2, y_2)$. Let $u(x_1, y_1)$ be a harmonic function with assigned rim values U_1 on C_1; let the values U_2 assigned to the rim C_2 be equal to the values U_1 at the corresponding points of C_1.

The function $$u(\chi_1, \chi_2), = u'(x_2, y_2),$$

satisfies the equation $$\frac{\delta^2 u'}{\delta x_2^2} + \frac{\delta^2 u'}{\delta y_2^2} = 0;$$

for $$\frac{\delta^2 u'}{\delta x_2^2} + \frac{\delta^2 u'}{\delta y_2^2} = \left\{\left(\frac{\delta\chi_1}{\delta x_2}\right)^2 + \left(\frac{\delta\chi_1}{\delta y_2}\right)^2\right\}\left(\frac{\delta^2 u}{\delta x_1^2} + \frac{\delta^2 u}{\delta y_1^2}\right) = 0.$$

Also when (x_2, y_2) tends to a point s_2 of C_2, u', $= u$, tends to the assigned rim value $U_1(s_1), = U_2(s_2)$, where s_1 corresponds to s_2.

§ 261. *The conform representation of a simply connected region Γ, bounded by an n-sided rectilinear polygon, upon the positive half-plane.*

Arrange $n + 1$ real numbers $x_0, a_1, a_2, \cdots, a_n$ in ascending order of magnitude, and let z' be defined by

$$z' = \int_{x_0}^{z} (a_1 - z)^{\lambda_1 - 1}(a_2 - z)^{\lambda_2 - 1} \cdots (a_n - z)^{\lambda_n - 1} dz,$$

where the λ's are positive, and $\sum_{\kappa=1}^{n} \lambda_\kappa = n-2$. The function $z'(z)$ is holomorphic throughout the positive half-plane, since the critical points $a_1, a_2, \cdots, a_n$ are all situated on the real axis. Let z move in the positive half-plane, and let $b_1, b_2, \cdots, b_n$ be the points in the z'-plane which correspond to $a_1, a_2, \cdots, a_n$.

The differential equation

$$dz'/dz = (a_1 - z)^{\lambda_1 - 1}(a_2 - z)^{\lambda_2 - 1} \cdots (a_n - z)^{\lambda_n - 1}$$

defines a holomorphic function $z(z')$ at all points in whose neighbourhoods the expression on the right-hand side is holomorphic (Chapter V., § 156).

Let z describe the real axis except near the points $a_1, a_2, \cdots, a_n$, round which it describes infinitely small semicircles situated in the positive half-plane, and let it start from a point a to the left of a_1. As z moves towards a_1, z' is real; near $z = a_1$,

$$dz'/dz \propto (a_1 - z)^{\lambda_1 - 1},$$

so that after the negative description of the semicircle round a_1, the amplitude of z' becomes $\pi(\lambda_1 - 1)$; and as z continues its path from a_1 to a_2, the point z' moves along a straight line $b_1 b_2$, which makes with the direction $b_1 b$ an angle $\pi\lambda_1$. When z describes the small semicircle round a_2 and moves along the straight line $a_2 a_3$, the z'-path changes from the straight line $b_1 b_2$ to the straight line $b_2 b_3$, where $b_2 b_3$ makes with $b_2 b_1$ an angle $\pi\lambda_2$; and so on for the remaining sides. Since $\Sigma\lambda_\kappa = n - 2$, the point z' describes the closed polygon $b_1 b_2 \cdots b_n b_1$, when z describes the complete axis of x. Furthermore, as z moves upwards from the real axis, z' moves into the interior of the polygonal region; and when z' is once inside, it remains inside as long as z remains above the real axis, since an exit from the region would mean a passage of z' over one of the sides, and therefore a passage of z over the real axis into the negative half-plane.

Conversely, while z' remains in the polygonal region, z must remain in the positive half-plane. By the transformation $z'' = \alpha z' + \beta$, one of the corners b_κ can be made to coincide with an angular point c of an assigned polygon $c_1 c_2 \cdots c_n c_1$, so that one of the sides shall coincide in direction with $b_\kappa b_{\kappa+1}$; and the constants a, λ can be chosen so as to make all the sides of the two polygons coincide. The determination of these constants is a matter of considerable difficulty.

The form of this solution is due to Christoffel and Schwarz. Schläfli has elaborated a method for solving the equations which involve the constants in a memoir Zur Theorie der conformen Abbildung, Crelle, t. lxxviii. Objections have been urged by J. Riemann against Schläfli's proof of the possibility of the conform representation of any rectilinear polygon whatsoever upon a circle, while the validity of Schläfli's work has been upheld by Phragmén (Acta Mathematica, t. xiv.). However this may be, there can be no question as to the fact that every rectilinear polygon can be mapped on a circle; and when this is granted, Schläfli's method provides a definite process for the determination of the constants. On the problem of the conform representation, the reader should consult Riemann's Werke, The Inaugural Dissertation; the many important memoirs on this subject contained in the second volume of Schwarz's Werke; Darboux's Leçons sur la Théorie Générale des Surfaces, t. i., chapter iv.; Harnack, Die Grundlagen der Theorie des logarithmischen Potentiales, 1887; Picard's Traité d'Analyse, t. ii., Fascicule 1; Laurent's Traité d'Analyse, t. vi., chapter vii. He should also consult M. Jules Riemann's valuable commentary on Schwarz's memoirs, Sur le Problème de Dirichlet, Ann. Sc. de l'École Norm. Sup., Ser. 3, t. v., 1888; Schottky, Ueber die conforme Abbildung mehrfach zusammenhängender ebener Flächen, Crelle, t. lxxxiii.; Christoffel, Sul problema delle temperature stazionarie e la rappresentazione di una data superficie, Annali di Matematica, Ser. 2, t. i., Sopra un problema proposte da Dirichlet, ib., t. iv., Ueber die Integration von zwei partiellen Differentialgleichungen, Göttingen Nachrichten, nr. 18, 1871.

By the use of inversion (a special case of orthomorphic transformation), Dirichlet's problem for a region which encloses the point ∞ can be made to depend upon the simpler problem for a region in the finite part of the plane. Suppose that two regions Γ, Γ_1 can be represented conformly on each other by $x + iy = \phi(x_1 + iy_1)$; then the harmonic function $u(x, y)$ in Γ transforms into the harmonic function $u_1(x_1, y_1)$ in Γ_1. In particular suppose that $x + iy$ is inverted with respect to a point c of Γ; then the region bounded by the curves $C_0, C_1, \cdots, C_{n-1}$ of § 253 changes into a region exterior to the n curves C' which are the inverses of the C's, and the point c passes to ∞. The value assumed by u_1 at ∞ must be finite and determinate, and this value is determined by the rim values of $u_1(x_1, y_1)$. Dirichlet's problem for a region which extends to ∞ can be stated in the following manner:—

The exterior problem. Given a region Γ which is formed by the part of the plane exterior to n curves $C_0, C_1, C_2, \cdots, C_{n-1}$, which are themselves exterior to one another, to find a function $u(x, y)$ which shall be harmonic in Γ, which shall take assigned values on C, and finally which shall tend to a perfectly definite value u_∞ when x, y tend to ∞, this value ∞ being determined by the rim values $U(s)$.

Some theorems with regard to the solution of $\frac{\delta^2 u}{\delta x^2} + \frac{\delta^2 u}{\delta y^2} = 0$ are useful. In the neighbourhood of an ordinary point it can be proved by Fourier's series that

$$u = a_0 + A_0 \log r + \sum_{\kappa=1}^{\infty} \{\cos \kappa\theta(a_\kappa r^\kappa + b_\kappa r^{-\kappa}) + \sin \kappa\theta(c_\kappa r^\kappa + d_\kappa r^{-\kappa})\}.$$

Suppose that the function u is zero at ∞, then $A_0 = 0$, $a_\kappa = 0$, $c_\kappa = 0$; thus, for the region Γ exterior to a circle C with arbitrarily great radius,

$$u = \sum_{\kappa=1}^{\infty} \left(\frac{1}{r}\right)^\kappa \{b_\kappa \cos \kappa\theta + d_\kappa \sin \kappa\theta,\}$$

and $\int_C u d\theta = 0$. This result implies one or other of two things. Either $u = 0$ on C and throughout the exterior plane, or some values of u on C are negative. In the latter case the value 0 at ∞ is neither a maximum nor a minimum, and the extreme values for the region lie on the boundary C. [See Harnack, Grundlagen, p. 48. Further theorems relating to the point ∞ are to be found in § 27 of the same work.]

§ 262. Schwarz's method for solving Dirichlet's Problem employs theorems from the theory of conform representation relative to regions bounded by regular arcs of analytic lines. After showing that a simply connected region bounded by regular arcs of analytic lines can be mapped upon a circle of assigned radius, Schwarz employs his alternating process and proves the existence theorem of the harmonic function for simply and multiply connected regions with rims of highly arbitrary shapes.*

Regular arcs of analytic lines. Let $P(t - t_0)$, $Q(t - t_0)$, where t_0 is real, be two integral series with real coefficients. These series can be continued along the real axis, by the method explained in Chapter III., until further progress is debarred by the presence of singular points. In this way two functions $f_1(t)$, $f_2(t)$ are defined for two portions of the real axis which contain t_0. Let these portions overlap along a stroke $t_1't_2'$, and let this stroke be shortened by making t_1', t_2', move towards t_0 until they attain positions t_1, t_2 such that, within the interval t_1t_2, inclusive of the terminal points, $f_1'(t)$ and $f_2'(t)$ are nowhere simultaneously zero. As t moves along the real axis from t_1 to t_2, the point (x, y) defined by

$$x = f_1(t), \ y = f_2(t),$$

traces out a regular arc of an analytic line.

An important property of a regular arc of an analytic line is that it always maps into an arc of the same character. For let $z' = \phi(z)$, where $z = x + iy = f_1(t) + if_2(t) = f(t)$. Suppose that z' traces out an arc $a'b'$ when z traces out the regular arc ab. For each point of ab, $\phi(f(t))$ is expressible as an integral series in t; and since

* Although the methods of Schwarz and Neumann leave some outstanding cases unconsidered, their proofs are of so general a character that they leave no doubt as to the truth of the existence theorem for harmonic functions u in a continuous region of the plane with assigned continuous rim values.

$\dfrac{dz'}{dt} = \phi'(f(t)) \cdot f'(t)$, it follows that $\dfrac{dz'}{dt}$ does not vanish on $a'b'$; *i.e.* $\dfrac{dx'}{dt}, \dfrac{dy'}{dt}$ are not simultaneously zero. Hence z' traces out a regular arc of an analytic line. We are now in a position to prove an important theorem due to Schwarz: —

If a harmonic function be an analytic function of t along a regular arc ab of an analytic line, it can be continued beyond the boundary ab.

The problem can be simplified by reducing the general case of a curvilinear arc ab to the easier case of a straight line. We have to show that it is possible to enclose the arc ab in a simply connected region Γ_1, which can be represented conformly on another simply connected region Γ_2, so that ab passes into a straight line within Γ_2. The simplest plan is to use the equations

$$x = x_0 + a_1(t - t_0) + a_2(t - t_0)^2 + \cdots,$$

$$y = y_0 + b_1(t - t_0) + b_2(t - t_0)^2 + \cdots,$$

which exist in the neighbourhood of a point (x_0, y_0) of ab, and thence form the equation

$$z - (x_0 + iy_0) = (a_1 + ib_1)(t - t_0) + (a_2 + ib_2)(t - t_0)^2 + \cdots.$$

Now let t receive complex values. Because $a_1 + ib_1 \neq 0$, this transformation establishes a conform representation of a region of the z-plane, which includes $z_0 (= x_0 + iy_0)$, upon the circle of convergence of the integral series in $t - t_0$.

By giving t_0 various positions between t_1 and t_2 it is possible to represent a region which contains the arc ab conformly upon a composite region constructed out of a system of circles of convergence with centres at the various points t_0, so that the interior arc ab becomes the interior straight line t_1t_2 (§ 264). It is sufficient therefore if we prove Schwarz's theorem for the case when ab forms part of the real axis of x.

(1) Let a function u be harmonic in a simply connected region Γ whose contour C consists partly of a curvilinear arc D drawn from a to b in the positive half-plane, and partly of the straight line ab; and let the values $U(s)$ along ab be zero. Let $\bar{D}$, $\bar{z}$ be the reflexions in the real axis of the curve D and the point z. On the axis ab take any point x_0 and describe a semicircle C' with centre x_0 so as to lie wholly within the region Γ. Suppose that u takes values $U(s)$ along C', and assign to the remaining semicircle, re-

quired to complete the circle, the values $U(\bar{s}) = -U(s)$. Use Poisson's formula

$$I = \frac{1}{2\pi}\int_0^{2\pi} \frac{(R^2-\rho^2)f(\alpha)\,d\alpha}{R^2 - 2R\rho\cos(\alpha-\theta)+\rho^2}$$

for the function which is harmonic in the complete circle and has the rim values $U(s)$, $U(\bar{s})$; the equation $U(\bar{s}) = -U(s)$ shows that $f(\alpha) = -f(2\pi - \alpha)$, and that I takes, consequently, the values 0 on that part of the real axis which lies within the circle. Since u and I are harmonic functions with the same rim values, they must be identically equal. But I has a meaning for the complete circle; therefore we have proved that u can be continued beyond the real axis into the negative half-plane.

(2) Next let the values along ab of the harmonic function u be given by an analytic function of x. Near x_0 let the value of u be

$$a_0 + a_1(x - x_0) + a_2(x - x_0)^2 + \cdots.$$

Replacing the real variable x by a complex variable z, the resulting series is an integral series in $z - x_0$ whose real part $u'(x, y)$ is harmonic and takes the same values as $u(x, y)$ upon the real axis in the neighbourhood of x_0. Hence $u(x, y) - u'(x, y)$ is a function which is harmonic in the positive half of the circular region and takes zero values on that diameter which lies along the real axis. The function is of precisely the same kind as that considered in (1). Hence $u(x, y)$ can be continued into the negative half-plane. [Schwarz, Werke, t. ii., p. 144, Ueber die Integration der partiellen Differentialgleichung $\partial^2 u/\partial x^2 + \partial^2 u/\partial y^2 = 0$, unter vorgeschriebenen Grenz- und Unstetigkeitsbedingungen.]

§ 263. Suppose that it is known that a series $u_1 + u_2 + u_3 + \cdots$, whose individual terms are harmonic in Γ and continuous on C, converges uniformly throughout Γ; it will follow that the series converges uniformly on C. For, by supposition,

$$|u_{n+1} + u_{n+2} + \cdots + u_{n+p}| < \epsilon \quad (n > \mu),$$

for every interior point (x, y); hence, when the point (x, y) tends to a point s of C, $|u_{n+1} + u_{n+2} + \cdots + u_{n+p}|$ tends to a limit which cannot be greater than ϵ. We proceed to the proof of two important theorems due to Harnack.

Theorem I. Let the component terms of the series $\sum_{\kappa=1}^{\infty} u_\kappa$ be harmonic in Γ and continuous on C, and let $u_\kappa(x, y)$ tend to $U_\kappa(s)$ when the point (x, y) in Γ tends to a point s of C by any route whatsoever in Γ. Then if the series $u = \sum_{\kappa=1}^{\infty} u_\kappa$ be uniformly convergent for the points of C, it must also be uniformly convergent throughout Γ, and $u(x, y)$ tends to $U(s) = \sum_{\kappa=1}^{\infty} U_\kappa(s)$, whatever be the route in Γ from (x, y) to s.

This theorem is the converse of that just established. To prove it we must show that

(i.) $\sum_{\kappa=1}^{\infty} u_\kappa$ is continuous in Γ,

(ii.) $\sum_{\kappa=1}^{\infty} u_\kappa$ is harmonic in Γ.

(i.) Let

$$\begin{aligned}
S_n &= U_1 + U_2 + \cdots + U_n, \\
{}_pR_n &= U_{n+1} + U_{n+2} + \cdots + U_{n+p}, \\
R_n &= U_{n+1} + U_{n+2} + \cdots \text{ to } \infty, \\
\sigma_n &= u_1 + u_2 + \cdots + u_n, \\
{}_p\rho_n &= u_{n+1} + u_{n+2} + \cdots + u_{n+p}, \\
\rho_n &= u_{n+1} + u_{n+2} + \cdots \text{ to } \infty.
\end{aligned}$$

If ϵ be given in advance, we can find a number μ such that

$$|{}_pR_n| < \epsilon \quad (n > \mu),$$

for all points of C. But ${}_p\rho_n$ is *harmonic* in Γ, and tends to ${}_pR_n$, when the interior point (x, y) tends to a point s of C. Hence $|{}_p\rho_n| < \epsilon$ throughout Γ. Thus the series $\sum_1^{\infty} u_\kappa$ is uniformly convergent in Γ; and the sum u of this series must consequently be continuous in Γ.

(ii.) Next we have to prove that u is harmonic in Γ. To prove this, select any point c of Γ, and describe round c as centre a circle C_1 of radius R, such that the enclosed region Γ_1 lies wholly within Γ. Let a function u' be constructed which has the same values as u upon C_1 and is harmonic throughout Γ_1. For the points s_1 of C_1 the series

$$u'(s_1) = u_1(s_1) + u_2(s_1) + u_3(s_1) + \cdots$$

is uniformly convergent; let it be integrated term by term after multiplication by $(R^2 - r^2)d\alpha/(R^2 - 2Rr\cos(\alpha - \theta) + r^2)$. Hence

$$\frac{1}{2\pi}\int_0^{2\pi} u'(s_1)\frac{R^2 - r^2}{R^2 - 2Rr\cos(\alpha - \theta) + r^2}d\alpha = u_1(x, y) + u_2(x, y) + \cdots$$

$$+ u_n(x, y) + \frac{1}{2\pi}\int_0^{2\pi} \rho_n(R, \alpha)\cdot\frac{R^2 - r^2}{R^2 - 2Rr\cos(\alpha - \theta) + r^2}d\alpha,$$

where r, θ are the polar co-ordinates of a point (x, y) in Γ_1, and $\rho_n(R, \alpha)$ is the value of $\rho_n(x, y)$ at (R, α).

Since $|\rho_n| < \epsilon$ upon C_1, it follows that

$$\lim_{n=\infty} \sigma_n(x, y) = \frac{1}{2\pi}\int_0^{2\pi} \frac{R^2 - r^2}{R^2 - 2Rr\cos(\alpha - \theta) + r^2} u'(s_1)\,d\alpha\,;$$

but the expression on the right-hand side defines a function which is harmonic in Γ_1. Therefore the expression on the left-hand side, namely u, is harmonic in Γ_1. By varying the point c it can be shown that u is harmonic in Γ.

Let s be any point of C. From s as centre describe a small circle of radius δ. A value of δ can be found such that, for all points (x, y) of Γ which lie inside this circle,

$$|S_n(s) - \sigma_n(x, y)| < \epsilon,$$

for, since $\sigma_n(x, y)$ is a continuous function of (x, y) in Γ, σ_n tends to the rim value S_n, when (x, y) tends to s on C. Hence

$$\begin{aligned} |u(x, y) - U(s)| &= |\sigma_n(x, y) + \rho_n(x, y) - S_n(s) - R_n(s)| \\ &< |\sigma_n(x, y) - S_n(s)| + |\rho_n(x, y)| + |R_n(s)| \\ &< 3\epsilon. \end{aligned}$$

Theorem II. Let the terms of $\sum_1^\infty u_\kappa$ be harmonic, *and all positive,* throughout a region Γ. If this series be convergent at any point of Γ, it is uniformly convergent at all points of Γ.

(i.) Let C be a circle of radius R and centre 0; and let $f(x, y)$ be a function which is harmonic in Γ and everywhere positive. Then at a point c, or (x, y), inside this circle,

$$f(x, y) = \int F(s) \frac{R^2 - |c|^2}{2\pi R \cdot |c - s|^2} ds,$$

where $F(s)$ is the rim value of f at s, and $|c - s|$ denotes the distance from c to s. Now $F(s)$ maintains one sign on C, and $\dfrac{R^2 - |c|^2}{|c - s|^2}$ is intermediate between $\dfrac{R - |c|}{R + |c|}$, $\dfrac{R + |c|}{R - |c|}$. Therefore,

$$\frac{R - |c|}{R + |c|}\int F(s)\frac{ds}{2\pi R} < f(x, y) < \frac{R + |c|}{R - |c|}\int F(s)\frac{ds}{2\pi R}.$$

Let a circle of radius R', where $R' < R$, be constructed with the centre 0. Then if c, c', or (x, y), (x', y'), be two points interior to Γ', and if U_κ' be the values on C' of u_κ, we have

$$\frac{R'-|c|}{R'+|c|}\int_{C'}\frac{U_\kappa' ds'}{2\pi R'} < u_\kappa(x, y) < \frac{R'+|c|}{R'-|c|}\int_{C'}\frac{U_\kappa' ds'}{2\pi R'},$$

$$\frac{R'-|c'|}{R'+|c'|}\int_{C'}\frac{U_\kappa' ds'}{2\pi R'} < u_\kappa(x', y') < \frac{R'+|c'|}{R'-|c'|}\int_{C'}\frac{U_\kappa' ds'}{2\pi R'}.$$

Hence,

$$u_\kappa(x, y) < \frac{R'+|c|}{R'-|c|}\cdot\frac{R'+|c'|}{R'-|c'|}\,u_\kappa(x', y').$$

Therefore, if c remain within a circle of radius $R''(<R')$,

$$u_\kappa(x, y) < \frac{R'+R''}{R'-R''}\cdot\frac{R'+|c'|}{R'-|c'|}\,u_\kappa(x', y').$$

Hence

$$\sum_1^\infty u_\kappa(x, y) < A\sum_1^\infty u_\kappa(x', y'),$$

where A is a finite quantity whose value is independent of the position of (x, y). Suppose that $\sum_1^\infty u_\kappa(x, y)$ is convergent at (x', y'). Then throughout the circle (R''), $\sum_1^\infty u_\kappa(x, y)$ is uniformly convergent; and R'' can be made to differ from R by an arbitrarily small quantity.

(ii.) Let Γ be any connected region; and let the series $\sum_1^\infty u_\kappa$ be convergent at an interior point c'. Let a finite number of circles $C_1, C_2, \cdots, C_\lambda$ be constructed in Γ, such that C_2 cuts C_1, C_3 cuts $C_2, \cdots, C_\lambda$ cuts $C_{\lambda-1}$, the circles being so selected that c', c lie in C_1, C_λ. Since $\sum_1^\infty u_\kappa$ is convergent in C_1, it must be convergent in $C_2, C_3, \cdots, C_\lambda$; also the convergence is uniform. By Harnack's first theorem, the function defined by the series is harmonic.

§ 264. Schwarz's solution of the problem of Dirichlet is effected by the employment partly of the method of conform representation, partly of a process known as the *alternating process.* The object of the latter process is to effect the solution for a composite region $\Gamma_1+\Gamma_2-\Gamma'$, supposing that the solution is known for *two component parts* Γ_1, Γ_2, which overlap along a region Γ'.

Schwarz first establishes an important lemma. Suppose that the contours of Γ_1, Γ_2 (which are taken to be simple) intersect at two points a; and that Γ_1 is divided into arcs C_1, C_3 and Γ_2 into arcs C_2, C_4; the arcs C_3, C_4 bounding Γ'.

(i.) Let u_1 be a function which is harmonic in Γ_1 and which takes values 1, 0 on C_1, C_3. The values of u_1 in Γ_1 must be all positive and less than 1. Upon C_4 there must be an upper limit for the values of u_1, say θ. This number cannot be greater than 1, for all the values of u_1 in Γ_1 lie in the interval $(0, 1)$. It cannot be 1 inside the region Γ_1; nor can it be 1 at a point a where the line C_4 meets the rim of Γ_1, for this would imply that the line C_4 touches C_1. Hence θ is a proper fraction.

(ii.) Let values with an upper limit G be assigned along C_1, such that there exists in Γ_1 a harmonic function u_2 with these values on C_1 and the value 0 on C_3. Then $Gu_1 + u_2$ is 0 upon C_3, is nowhere negative on C_1, and therefore nowhere negative inside Γ_1. Its values along C_4 are consequently positive. Writing

$$Gu_1 + u_2 = G\theta + u_2 + G(u_1 - \theta),$$

and noticing that the second term is nowhere positive on C_4, it follows that $G\theta + u_2$ is nowhere negative on C_4. Similarly, from the function $Gu_1 - u_2$, it follows that $G\theta - u_2$ is nowhere negative on C_4. Hence, at each point of C_4,

$$|u_2| \not> G\theta.$$

With the aid of this preliminary proposition, Schwarz establishes his method of combination. Let values U be assigned for the lines C_1, C_2, and let G, K be the upper and lower limits of these rim values. For the surface Γ_1 there can be constructed a function u_1 which takes the prescribed values U on C_1 and the value K on C_3. This can be done since the generalized problem can always be reduced to the ordinary problem. As the line C_4 lies within Γ_1, the values $u_1^{(4)}$ along C_4 must be $\geqq K$, and $\leqq G$. For the surface Γ_2 a harmonic function u_2 is constructed which satisfies analogous conditions; namely, it takes along C_2 the values U, along C_4 the values $u_1^{(4)}$. Since the line C_3 lies in Γ_2, the values $u_2^{(3)}$ along C_3 must be $\geqq K$. Hence $u_2 - u_1$ must be positive in Γ'; for on C_4 it vanishes, and on C_3 it equals $u_2^{(3)} - u_1^{(3)} =$ (a positive quantity which is equal to or greater than K) $- K$. Thus $u_2 - u_1$ is positive throughout Γ'.

Next, two new functions u_3, u_4 are constructed for Γ_1, Γ_2. Along C_1, u_3 takes the values U; along C_3 it takes the values $u_2^{(3)}$. By reasoning similar to that already used, $u_3 \geqq u_1$. Let the values of u_3 along C_4 be $u_3^{(4)}$. Along C_2, u_4 takes the values U; along C_4, u_4 takes the values $u_3^{(4)}$; therefore $u_4 \geqq u_2$. Along C_3 call the values of u_4, $u_4^{(3)}$. By proceeding in this way, two series arise,

$$u_1 + (u_3 - u_1) + (u_5 - u_3) + \cdots + (u_{2n+1} - u_{2n-1}) + \cdot \cdot \quad (A),$$

$$u_2 + (u_4 - u_2) + (u_6 - u_4) + \cdots + (u_{2n} - u_{2n-2}) + \quad (B),$$

whose terms are all positive. Along C_4 two terms of the same rank in (A), (B) are equal, while along C_3 a term of rank κ in (A) is equal to the term of rank $\kappa-1$ in (B). The convergence of the series can be established by proving the existence of $\lim u_{2n+1}$, $\lim u_{2n}$. The existence of the former limit for Γ_1 follows from the fact that $u_1, u_3, u_5, \cdots$ are in ascending order of magnitude and all less than G; hence, if $\lim u_{2n+1} = u'$, the sum of the series (A) is u'. Similarly, the sum of (B) is u'', where $u'' = \lim u_{2n}$. From the fact that all the terms of (A) and (B) are positive, it follows that u', u'' are harmonic (§ 263).

The series under consideration are uniformly convergent. For consider, in the first place, the term $u_3 - u_4$. As this term vanishes on C_1 and is less than $G - K (= D)$ on C_3, it must be less than D throughout Γ_1. In particular

$$u_3^{(4)} - u_1^{(4)} < \theta_1 D,$$

where θ_1 is a proper fraction. Hence in Γ_2,

$$u_4 - u_2 < \theta_1 D.$$

In particular $u_4^{(3)} - u_2^{(3)} < \theta_1\theta_2 D$, where θ_2 is a proper fraction.

The term $u_5 - u_3$ vanishes on C_1, and takes along C_3 the value $u_4^{(3)} - u_2^{(3)}$. Hence in Γ_1,

$$u_5 - u_3 < \theta_1\theta_2 D.$$

A continuation of this process of reasoning leads to the results

$$u_{2n+1} - u_{2n-1} < D(\theta_1\theta_2)^{n-1} \text{ in } \Gamma_1,$$

$$u_{2n+2} - u_{2n} < D\vartheta_1(\theta_1\theta_2)^{n-1} \text{ in } \Gamma_2.$$

These results prove that the series u', u'' are uniformly convergent. Since the above inequalities apply also to rim values, it follows that the two series

$$U_1 + (U_3 - U_1) + (U_5 - U_3) + \cdots,$$

$$U_2 + (U_4 - U_2) + (U_6 - U_4) + \cdots,$$

converge to sums U', U'', where $U' = \lim U_{2n+1}$ and $U'' = \lim U_{2n}$. The functions u', u'' are equal throughout Γ' and are both harmonic. The function U' takes the values U along C_1; and the function U'' the values U along C_2. Along C_3 and C_4 the values of U', U'' are equal, since at a point in these lines $\lim U_{2n+2} = \lim U_{2n+3}$, by hypothesis.

The following reasoning shows that the limiting values to which u', u'' tend, when they approach a point a at which C_1, C_3 meet by some path interior to Γ', are equal. Let the path in question make angles ϕ_1, ϕ_3 with C_1, C_3, and angles ϕ_2, ϕ_4 with C_2, C_4. Then the limiting values to which u', u'' tend at a are

$$V' = (U_1)_a\left(1 - \frac{\phi_1}{\pi}\right) + (U_3)_a\left(1 - \frac{\phi_3}{\pi}\right),$$

$$V' = (U_2)_a\left(1 - \frac{\phi_2}{\pi}\right) + (U_4)_a\left(1 - \frac{\phi_4}{\pi}\right).$$

But
$$(U_4)_a = (U_1)_a\left(1 - \frac{\phi_1 - \phi_4}{\pi}\right) + (U_3)_a\left(1 - \frac{\phi_3 + \phi_4}{\pi}\right),$$

$$(U_3)_a = (U_2)_a\left(1 - \frac{\phi_2 - \phi_3}{\pi}\right) + (U_4)_a\left(1 - \frac{\phi_3 + \phi_4}{\pi}\right).$$

Hence, using $\phi_1 + \phi_3 = \pi$, $\phi_2 + \phi_4 = \pi$,

$$V' = (U_3)_a\left(1 - \frac{\phi_3}{\pi}\right) + \left(1 - \frac{\phi_1}{\pi}\right)\frac{\pi}{\pi - \phi_1 + \phi_4}\left\{(U_4)_a - (U_3)_a \cdot \frac{\pi - \phi_3 - \phi_4}{\pi}\right\}$$

$$= (U_3)_a \frac{\phi_4}{\phi_3 + \phi_4} + (U_4)_a \frac{\phi_3}{\phi_3 + \phi_4},$$

$V'' =$ the same expression.

Thus the functions u', u'' are harmonic in Γ', take the same values on C_3, C_4, and must be supposed equal even at the corners a, since the limits to which u'. u'' tend along any path to a interior to Γ' are equal. Thus u', u'' are identical in Γ'. A function u, which is equal to u' in Γ_1 and to u'' in Γ_2, solves Dirichlet's problem for the combined region $\Gamma_1 + \Gamma_2 - \Gamma'$. [See M. Jules Riemann's memoir, Sur le problème de Dirichlet, Ann. Sc. d. l'Éc. Norm. Sup., 1888, p. 334; Harnack, Die Grundlagen der Theorie des Logarithmischen Potentiales, pp. 109–113. The original proof is to be found in Schwarz's Werke, t. ii., pp. 156–160.]

§ 265. *Solution of Dirichlet's problem for a simply connected region Γ bounded by a polygon whose sides are regular arcs of analytic lines.* All that portion of Γ which does not lie in immediate proximity to the rim can be covered by circles; the part which *is* in proximity to the rim can be enclosed within regions for which solutions of the problem are known. By employing Schwarz's alternating process the solution can be obtained for Γ. We have, then, to study the parts close to the rim C. Let ab, a regular arc of an analytic line, form part of the boundary C, and suppose that the consecutive arcs do not touch ab. We know that ab can be enclosed within a region Γ_1, which can be mapped on a region Γ_2 so that the arc ab becomes a straight line $a'b'$ in Γ_2. Join a', b' by an arc D' of a circle, and let D be that regular arc (from a to b) of an analytic line which maps into D' when $a'b'$ maps into ab. The region contained between D and ab can be mapped on a circle, for the segment made by D' and $a'b'$ can be mapped on a circle. Thus, when there are no points at which two consecutive arcs of C touch, there exists a harmonic function for those parts of Γ which are close to C.

In this way Schwarz has solved Dirichlet's problem for all simply connected surfaces bounded by regular arcs of analytic lines which intersect at angles different from zero; he has also solved the problem for certain cases in which consecutive arcs touch each other. But his solution (as given in the memoir Ueber die Integration der partiellen Differentialgleichung, loc. cit., pp. 144–171) does not cover the case in which the arcs are not regular analytic arcs. With the help of the alternating process the problem may be regarded as solved for any composite simply connected region which can be formed by the combination of simply connected regions with boundaries composed of regular analytic arcs.

§ 266. In a letter to Klein, written in February, 1882, and published in the Mathematische Annalen, t. xxi., Schwarz has shown that his alternating process can be used, in the case of a multiply connected surface, to prove the existence of a potential function u which takes assigned values on the boundary, and suffers constant discontinuities in passing over cross-cuts. The proof bears a close resemblance to that already used for simply connected regions.

Suppose that Γ is a doubly connected region, and let Q_1, Q_2 be two cross-cuts each of which would separately reduce the region to simple connexion, and which do not intersect. For the sake of definiteness suppose that Γ is bounded by an exterior oval curve $C_1 + C_1'$ and an interior oval curve $C_2 + C_2'$; where C_1, C_2 are contained between Q_2^- (the negative side of Q_2) and Q_1^+ (the positive side of Q_1), and C_1', C_2' are the remaining parts of the two rims. Let Γ_1, Γ_2 be the two simply connected regions created by Q_1, Q_2. A harmonic function which vanishes on the complete rim of Γ_1, and is equal to 1 on the two banks of Q_1, must be less than 1 throughout the interior of Γ_1. As the line Q_2 lies within Γ_1, the values along Q_2 must have a maximum value $\theta_2 (< 1)$. Similarly, a harmonic function which vanishes on the rim of Γ_2, and which is equal to 1 on the two banks of Q_2, must be less than 1 along Q_1. Let $\theta_1 (< 1)$ be its maximum value along Q_1. The general problem reduces in the present case to the proof of the existence of a harmonic function u with given values on the rims $C_1 + C_1'$, $C_2 + C_2'$ of Γ, and such that $u^+ - u^- = A$ along Q_1.

A series of harmonic functions $u_1, u_3, u_5, \cdots,$

$$u_2, u_4, u_6, \cdots,$$

can be found for Γ_1, Γ_2 with the following properties:—

(i.) u_1 has on the rim of Γ *prescribed* values $U(s)$, and the values on Q_1^+ are greater than those on Q_1^- by A;

(ii.) u_2 has on the rims C_1', C_2' the values $U(s)$, on the rims C_1, C_2 the values $U(s) - A$, and on Q_1^+, Q_1^-, the same values as u_1, $u_1 - A$;

(iii.) u_3 has on the rim of Γ the values $U(s)$, along Q_1^- the values of u_3, u_2 are equal, and along Q_1^+ the values of u_3 are equal to those of $u_2 + A$; and so on for $u_4, u_5, \cdots$. Consider the two series

$$u_1 + (u_3 - u_1) + (u_5 - u_3) + \cdots,$$

$$u_2 + (u_4 - u_2) + (u_6 - u_4) + \cdots.$$

By the use of Harnack's theorem these series can be shown to be uniformly convergent. For if the maximum value of $| u_1 - u_3 |$ on Q_1 be G, the maximum value of the same expression on $Q_2 < G\theta_2$. On both banks of Q_2, $u_1 - u_3 = u_2 - u_4$. Therefore $| u_2 - u_4 |$ is a function which equals 0 on the rim of Γ, and $\leqq G\theta_2$ on Q_2. Its value on Q_1, *a line interior to* Γ_2, must therefore $\leqq G\theta_1\theta_2$. Similarly, $| u_3 - u_5 | \leqq G\theta_1\theta_2$, $| u_4 - u_6 | \leqq G\theta_1\theta_2^2$, etc. The uniform convergence of the series can be established by exactly the same method as before, and the sums be proved equal to $\lim_{n=\infty} u_{2n+1}$, $\lim_{n=\infty} u_{2n}$. On the rim $Q_1^- + C_1 + Q_2^+ + C_2$ of Γ', $u' = u'' + A$; hence in the interior of Γ' the same equation holds. On the rim of $\Gamma - \Gamma'$, $u' = u''$; therefore in $\Gamma - \Gamma'$, $u' = u''$.

The function u' is precisely the solution required by the conditions of the problem for Γ_1. For let $u = u'$ throughout Γ_1; the analytic continuation of u over the line Q_1 is u'', which equals $u' - A$. The function u defined by u' and its continuation u'' is the required solution.

§ 267. *Green's function.* Let (x, y) be a point in a region Γ, or on the contour C, whose distance from an interior point (a, b), or c, is r. We shall prove presently that there exists a function $g(x, y)$ which is equal to $\log 1/r$ for all points (x, y) on C, and which is harmonic in Γ. This function is often called Green's function, but Klein has pointed out that this name is more suitably given to that one-valued potential function which is infinite at (a, b) like $\log r$ for $r=0$, and is continuous elsewhere in Γ, the rim values being zero. *Green's function* $G(x, y)$ is therefore $g(x, y) - \log 1/r$. To avoid confusion we shall speak of g as the g-function; also for the sake of simplicity we shall suppose the region Γ to be simply connected.

Assuming the existence of $g(x, y)$ for the region Γ, we proceed to the proof of some of the properties which make it specially suitable as an auxiliary potential function in the investigation of Dirichlet's problem. Since $G(x, y)$ is 0 upon C, and a very large negative quantity near c, the function must be negative throughout Γ. Formally the proof runs as follows: Describe a small circle C' round c; then the function G is zero on C, negative on C', and harmonic throughout the region between C and C'. The function must therefore be constantly negative within this region. Changing from Green's function to the g-function, the theorem states that *throughout* Γ *the value of* g *is less than that of* $\log 1/r$.

Theorem. Let c_1, c_2 be two points in Γ, and let g_1, g_2 be the corresponding g-functions; then the value of g_1 at c_2 is equal to the value of g_2 at c_1.

Let C' be a curve, interior to C and arbitrarily close to C, upon which Green's function G_1 (relative to c_1) takes the constant negative value a. Ultimately, a will be made to tend to zero. Apply Green's theorem

$$\iint \left\{ \frac{\delta(G_1 - a)}{\delta x} \frac{\delta G_2}{\delta x} + \frac{\delta(G_1 - a)}{\delta y} \frac{\delta G_2}{\delta y} \right\} dxdy = -\int (G_1 - a) \frac{\delta G_2}{\delta n} ds,$$

where G_2 is Green's function relative to c_2, to that region of Γ which is bounded by C', and two small circles of radii ρ_1, ρ_2 round c_1, c_2 as centres. The simple integral is taken positively over the outer rim and negatively over the inner rims with respect to the region, and the normal is drawn inwards. The part relative to C' vanishes; the remaining two parts contribute

$$-\int (G_1 - a) \frac{\delta G_2}{\delta n} \rho_1 d\theta - \int (G_1 - a) \frac{\delta G_2}{\delta n} \rho_2 d\theta$$

$$= -\int (G_1 - a) \frac{\delta G_2}{\delta n} \rho_1 d\theta - \int (G_1 - a) \frac{\delta g_2}{\delta n} \rho_2 d\theta - \int (G_1 - a) d\theta.$$

Now let ρ_1, ρ_2 tend to 0; the expression reduces to

$$-\int(G_1-\alpha)d\theta, \text{ or } -2\pi(G_1-\alpha)_{c_2}.$$

Hence, when α becomes zero,

$$\int\int\left\{\frac{\delta G_1}{\delta x}\frac{\delta G_2}{\delta x}+\frac{\delta G_1}{\delta y}\frac{\delta G_2}{\delta y}\right\}dxdy=-2\pi(G_1)_{c_2}.$$

Symmetry shows that the right-hand side can also be written $2\pi(G_2)_{c_1}$.

Hence, $$(G_1)_{c_2}=(G_2)_{c_1},$$

and $$(g_1)_{c_2}=(g_2)_{c_1},$$

a result of great importance.

The next property to be proved is that when c_2 tends to a point s on the rim, g_2 at c_1 tends to $\log 1/r$, where r is the distance of a point c_1 of Γ from s. Let c_2 lie on a curve for which $g_1-\log 1/r_1=\alpha$, where r_1 is the distance of a point of the curve from c_1, and α is a negative constant which is arbitrarily small. Since the value of g_2 at c_1 is equal to the value of g_1 at c_2, it follows that the value of g_2 at c_1 is $\log 1/|c_1-c_2|+\alpha$, where $|c_1-c_2|$ represents the distance apart of c_1 and c_2. When c_2 tends to s, α tends to 0 and the function g_2 tends to $\log 1/|c_1-s|$, *i.e.* $\log 1/r$. Let Γ' be the region which remains after the removal of a small region which contains s and c_2; the passage of g_2 to the rim values $\log 1/r$ for the region Γ' is uniformly continuous. For $g_2-\log 1/r_2$, where r_2 is the distance from c_2 of any point in Γ', or on its rim, is equal to the small negative quantity α at c_1, is harmonic in Γ', and negative or zero on the rim of Γ'. Hence, at any interior point $g_2-\log 1/r_2$ is equal to the product of α by a finite quantity (see § 263). When c_2 tends to s, α tends to 0 and the function g_2 in Γ' tends with uniform continuity to $\log 1/r$.

Further, all the partial derivatives at c_1 of the harmonic function $G=g_2-\log 1/r_2$ tend to 0 when c_2 tends to s. For at each point ρ, θ of a circle with c_1 as centre and R as radius, we have (§ 258)

$$\frac{\delta G}{\delta\rho}=\frac{1}{\pi R}\sum_{n=1}^{\infty}n\left(\frac{\rho}{R}\right)^{n-1}\left[\cos n\theta\int_{-\pi}^{+\pi}G'\cos n\theta d\theta+\sin n\theta\int_{-\pi}^{+\pi}G'\sin n\theta d\theta\right],$$

$$\frac{1}{\rho}\frac{\delta G}{\delta\theta}=\frac{1}{\pi R}\sum_{n=1}^{\infty}n\left(\frac{\rho}{R}\right)^{n-1}\left[-\sin n\theta\int_{-\pi}^{+\pi}G'\cos n\theta d\theta+\cos n\theta\int_{-\pi}^{+\pi}G'\sin n\theta d\theta\right],$$

where G' stands for the values on the rim of the circle. [Harnack, Existenzbeweise zur Theorie des Potentiales, Math. Ann., t. XXXV.]

The absolute value of each integral in the brackets is increased by the suppression of the factors $\cos n\theta$, $\sin n\theta$ in the integrands. But the integrals then become $\int_{-\pi}^{+\pi} G'd\theta = 2\pi G_0$, where G_0 is the value of G at the centre of the circle; and the new series converge. Hence, throughout the circle (R), $|\delta G/\delta\rho|$, $|\delta G/\delta\theta|$ are equal to G_0 multiplied by finite quantities. Therefore, when G_0 tends to 0, the expressions $\delta G/\delta\rho$ and $\delta G/\rho\delta\theta$ tend with uniform continuity to 0 for all points of the circle (R). By selecting a point in (R) as centre for a new circle, it can be shown that the same theorem holds for the new circle, and so on until the whole region Γ has been accounted for. Similarly, every partial derivative of G tends to 0 under the above conditions.

Harnack's Method.

§ 268. Suppose that the solution can be found for Dirichlet's problem in the case of any simply connected region bounded by a rectilinear polygon. Harnack has shown that the g-function can then be found for any region whatsoever in the plane of x, y, and thence that Dirichlet's problem can be solved with complete generality. The theorem may be proved for simply connected regions and the passage to multiply connected regions effected by the alternating process in the manner described above.*

Regard the region Γ as the limit of a series of regions $\Gamma_1, \Gamma_2, \cdots$, bounded by rectilinear polygons which are subject to the following conditions:—

(i.) Each polygon is supposed to be interior to Γ;

(ii.) The rim of Γ_{n-1} lies wholly within the rim of Γ_n, whatever be the value of n;

(iii.) As n tends to ∞, the rim of Γ_n tends to coincidence with C.

One method for satisfying (iii.) is to select a sequence of descending real positive numbers $(\alpha_1, \alpha_2, \cdots)$ which defines 0 (see § 50), and then let C_n lie entirely within that region of Γ which is covered by a circle of radius α_n, when its centre describes the complete boundary C. Construct for each of these polygons the g-function with respect to a point c of Γ; and let these g-functions be

* Harnack defines a simply connected continuum in the finite part of the (x, y) plane by means of masses of points. See Die Grundlagen der Theorie des Logarithmischen Potentiales, § 39.

$g_1, g_2, \cdots$. Upon the rim of Γ_1, $g_1 = \log 1/r$ and $g_2 < \log 1/r$, where r is the distance of a point of the rim from c. Therefore throughout Γ_1

$$g_1 > g_2 > g_3 > \cdots.$$

The same reasoning shows that within the larger region Γ_p,

$$g_p > g_{p+1} > g_{p+2} > \cdots,$$

whatever be the value of the positive integer p. Harnack's theorems with regard to series of harmonic functions show that

$$g = \lim_{n=\infty} g_n = g_1 + (g_2 - g_1) + (g_3 - g_2) + \cdots,$$

is convergent. For each of the quantities g_n is greater than $\log 1/\mu$, where μ is the maximum distance of c from C, therefore $\lim g_n$ exists and is finite; and further the terms in brackets are harmonic and all negative. Thus by Theorem II. of § 263, the function g is harmonic in Γ. It must be proved that g takes on C the proper rim values, viz. $\log 1/r$.

To simplify the proof Harnack introduced in the place of g the new function ϕ defined by $\phi = re^g$. The harmonic function ϕ is equal to 1 on C and has the value 0 at c; its values at points interior to Γ and other than c are all included between 0 and 1. Consider the curve $\phi = k$, where k is a proper fraction. This curve cannot be a closed curve which excludes c, for it would follow that $\log \phi$ is constant on the perimeter of a closed curve and harmonic inside, and therefore, also, constant throughout Γ. That the curve must be a closed curve can be proved from the properties of harmonic functions; for a point at which the curve stops abruptly would be a point at which $\log \phi$ is a maximum or minimum. As the value k changes from 0 to 1, the curves of level expand from an infinitely small curve round c to the complete boundary. Let Γ be contained wholly within a region Γ' bounded by a rectilinear polygon C', and let one of the sides of C' pass through a point of C. Suppose that g' is the g-function for Γ' relative to the point c, and that $\psi = re^{g'}$. We have to prove that $\phi = 1$ at each point s of C. Let Γ' be so selected that s is the point common to C' and C. Then, when (x, y) approaches s along any path (interior to Γ) which ends at s, the values of ψ, ϕ are always subject to the inequalities

$$\psi \leqq \phi < 1.$$

For, throughout Γ,' $\psi < \phi_n$, for all orders of n however great.

But when the point (x, y) tends to s, the function ψ tends, by the definition of the g-function, to 1. Hence the limit of ϕ is also 1.*

* This modification of Harnack's proof is due to M. Jules Riemann.

The existence of the g-function has been established for a simply connected region whose rim is subject to the sole restrictions that it is not to intersect itself and that it is to be a continuous line. The rim may suffer infinitely many abrupt changes of direction; it is not even in fact necessary that it should have a determinate direction at any of its points, or an element of length. The existence of the g-function for a multiply connected region in the plane of x, y, follows by the use of Schwarz's method for creating a multiply connected surface by combinations of simply connected surfaces.

§ 269. With the help of Green's function it is possible to represent any simply connected region Γ in the finite part of the plane, conformly upon a circle of radius unity. Let Green's function $G(x, y) = g(x, y) - \log 1/r$ be constructed with respect to an interior point (a, b) of Γ. Draw a line from (a, b) to the rim C, and let the surface Γ be called Γ' when this line is treated as a barrier. Let $H(x, y)$ be conjugate to $G(x, y)$, so that

$$H(x, y) = \int_{x_0, y_0}^{x, y} \left(\frac{\delta G}{\delta x} dy - \frac{\delta G}{\delta y} dx \right).$$

The values of $H(x, y)$ are determined (to a constant term) in Γ' and differ on opposite banks of the barrier by 2π. When (x, y) describes a line $G =$ constant, the equation $\delta H/\delta s = -\delta G/\delta n$ shows that H increases constantly; therefore $G + iH$ takes an assigned value only once in Γ. Consider now the function

$$z' = e^{G+iH} = re^{g+iH}.$$

The values of this function in Γ are all less than 1, and on C the absolute value is precisely 1. The point (a, b) passes into the origin $0'$ of the z'-plane and the points of Γ correspond one-to-one with the points of a region Γ' in the z'-plane, bounded by a circle C' of centre $0'$ and radius unity. It will be remembered that H contained an undetermined constant; hence, *assuming* that the points of C, C' are related one-to-one by the above transformation, the simply connected region Γ can be mapped upon the circular region Γ' so that the arbitrarily selected interior point (a, b) becomes $0'$, and an arbitrarily selected point of C becomes an assigned point of C'. Harnack has called attention to this assumption. Writing

$$z' = ze^{g+iH},$$

when z tends to a point z_0 of C, z' tends to C', but it is not evident that it tends to a definite point of C'. Painlevé has removed this obstacle to a complete proof for the case which we have been considering, namely, that of a rim with continuous curvature except at a finite number of points at which there is an abrupt change of direction. [Sur la théorie de la réprésentation conforme, Comptes Rendus, t. cxii., p. 653, 1891.]

In this proof we have had occasion to make use of the function $H(x, y)$ conjugate to $G(x, y)$. It is worth while to call the reader's attention to the fact that when u is harmonic in an n-ply connected surface, the conjugate function

$$v = \int_{x_0, y_0}^{x, y} \left(\frac{\delta u}{\delta x} dy - \frac{\delta u}{\delta y} dx \right)$$

has periods at some of $n-1$ barriers (§ 168).

§ 270. It is now possible to complete the solution of Dirichlet's problem for a simply connected region Γ, whose rim C has an integrable element ds and suffers only a finite number of abrupt changes of direction. Here, as elsewhere, we assume that at a point s where there is such a change of direction the two parts of the curve C which meet at s do not touch. Harnack assumes the existence of a function u which is harmonic in Γ and has assigned rim values $U(s)$, and arrives at a formula which gives the value at any interior point c whatsoever. The values u in Γ are to pass with uniform continuity into the values $U(s)$ on the rim.

First, then, we assume the existence of a function u with the given values $U(s)$ on the rim. In order to arrive at the formula referred to above, let us suppose that Γ is contained in a slightly larger region Γ'. This supposition is made in order to avoid difficulties as to integration over a natural boundary (§ 104). At a point c of Γ whose co-ordinates are x, y, u is given by

$$u(x, y) = \frac{1}{2\pi} \int_C U \frac{\delta \log 1/r}{\delta n} ds - \frac{1}{2\pi} \int_C \frac{\delta U}{\delta n} \log 1/r \cdot ds \quad \cdot \quad (1),$$

where r is the distance of a point s of C from c. Now, although we have proved the existence in Γ of the g-function relative to c, we know nothing about the nature of $\delta g/\delta n$ along C. For this reason let us consider, in place of C, a curve C' which is interior and very close to C; and let us form, relatively to C', the equation

$$\frac{1}{2\pi} \int \left(U' \frac{\delta g'}{\delta n'} - g' \frac{\delta U'}{\delta n'} \right) ds' = 0 \quad \cdot \quad \cdot \quad \cdot \quad \cdot \quad (2),$$

where accents denote that the values are those taken by $U, \delta g/\delta n, \cdots$, on C', and where $\delta g'/\delta n'$ is constructed with respect to an inner normal. When C' approaches indefinitely near to C, g' tends to $\log 1/r$ and $\delta U'/\delta n'$ to $\delta U/\delta n$ (for u is supposed to exist in a region which extends beyond Γ). Hence $\lim\limits_{C'=C}\int g'\,\delta U'/\delta n'\cdot ds'$ exists and equals $\int g\delta U/\delta n\cdot ds$; and therefore also $\lim\limits_{C'=C}\int U'\delta g'/\delta n'\cdot ds'$ exists and equals the same quantity. Subtracting (2) from (1), we have

$$u=\frac{1}{2\pi}\int U\frac{\delta \log 1/r}{\delta n}ds-\frac{1}{2\pi}\lim_{C'=C}\int U'\frac{\delta g'}{\delta n'}ds' \quad . \quad . \quad (3).$$

This is the formula to which reference was made above. We shall now consider the expression

$$\frac{1}{2\pi}\int U\frac{\delta \log 1/r}{\delta n}ds-\frac{1}{2\pi}\lim_{C'=C}\int U'\frac{\delta g'}{\delta n'}ds'$$

on its own merits, without reference to what precedes. The problem is to show that *this expression defines a function u which is harmonic in Γ and takes the rim values $U(s)$.* Consider separately the two parts of which it is composed.

I. The integral $\int U\dfrac{\delta \log 1/r}{\delta n}ds$.

The form of this integral shows that some restriction must be placed upon the rim C, since an integration is to be effected with regard to an arc ds. The rim *must* possess an integrable arc ds. Harnack, to simplify the work as far as possible, imposes the further restrictions that the rim C is to have, in general, a definite tangent whose direction alters continuously along C; and that $\cos\psi$, where ψ is the inclination of the tangent to the axis of x (*an arbitrary line*), is not to take the same value at infinitely many points of C, unless it be along a rectilinear portion of C. The geometric significance of these restrictions can be found without difficulty. The curve C is assumed to have only a finite number of points at which the tangent alters abruptly. Suppose, if possible, that the axis of x meets the curve in infinitely many points; between two consecutive points there must be a point at which the tangent is parallel to the axis of x, or else a point at which the tangent changes abruptly. But the latter points are finite in number; hence the number of points at which $\cos\psi=0$ is infinite, contrary to hypothesis. Thus no straight line (not included in C) can intersect C in infinitely many points.

A simple meaning can be assigned to the expression $\dfrac{\delta \log 1/r}{\delta n}ds$. Whatever be the position of the interior point c, let θ stand for the angle between the normal dn (drawn inwards) and the line from ds to c. Then

$$\frac{\delta \log 1/r}{\delta n} = \cos\theta/r,$$

and the expression

$$\frac{\delta \log 1/r}{\delta n} ds = \text{the angle subtended by } ds \text{ at } c$$
$$= (ds)_c, \text{ in Neumann's notation.}$$

(Neumann, Abel'sche Integrale, 2d ed., p. 403.) The element $(ds)_c$ (*i.e.* the apparent magnitude of ds as viewed from c) is to be treated as positive or negative according as the inner or outer side of ds is turned towards c.

Suppose that the point c tends to a point t of the rim C. The rim C can be separated into two parts C_1, C_2, the former of which contains t; we suppose that C_1 is small. Call u_1, u_2 the two parts of the integral which correspond to C_1, C_2 respectively. When c moves to t, the integrand of u_2 remains finite and continuous. In the limit, when the extremities of C_2 tend to t, let u_2 tend to $\int U(ds)_c$, whatever be the position of the interior point c. Since the length of C_1 is at our disposal, and since the values $U(s)$ are continuous near t, C_1 can be chosen so short that for every point of its arc we shall have, in virtue of the continuity of U,

$$| U - U_t | < \epsilon,$$

where ϵ is arbitrarily small and U is any value of the function $U(s)$ upon C_1. Then

$$| u_1 - U_t \int (ds_1)_c | < \epsilon \int | (ds_1)_c |.$$

When c moves up to t, $\int (ds_1)_c$ becomes that angle between the two lines from t to the extremities of C_1 which does not contain c. Let C_1 shrink indefinitely; then $\int (ds_1)_c$ tends to a limit $2\pi - \alpha$, where α is the angle between the two tangents at t measured *within* Γ, this angle being ordinarily equal to π. In the limit when C_1 shrinks up to t and c moves to t,

$$\lim u_1 = \pi U_t, \text{ or } (2\pi - \alpha) U_t,$$

and

$$\frac{1}{2\pi} \lim (u_1 + u_2), \text{ or } \frac{1}{2\pi} \lim \int U \frac{\delta \log 1/r}{\delta n} ds,$$
$$= \frac{1}{2} U_t + \frac{1}{2\pi} \int U(ds)_t,$$

or

$$= \frac{2\pi - \alpha}{2\pi} U_t + \frac{1}{2\pi} \int U(ds)_t,$$

according as the point t is, or is not, a point at which the tangent changes its direction abruptly.

II. *The integral* $\frac{1}{2\pi}\lim_{C'=C}\int U'\frac{\delta g'}{\delta n'}ds'$.

We have to prove that the value of this integral is

$$-\frac{1}{2}U_t+\frac{1}{2\pi}\int U(ds)_t,$$

or

$$-\frac{\alpha}{2\pi}U_t+\frac{1}{2\pi}\int U(ds)_t.$$

In the first place the value of the integral must be defined more precisely, for the values U' at the points of C' are not known. Let the values U' assigned to the points of C' be equal to the values U at corresponding points of C, the correspondence between the points of C and C' being in general one-to-one. The method by which Harnack effects this is to separate the curve C into parts which are concave and into parts which are convex to the axis of x. The former parts are translated through an arbitrarily small distance δ parallel to the axis of y downwards, and the other parts through the same distance upwards. It is clear that a point s' of the new curve C' will correspond to that point s of the original curve C, which is at a distance δ from s'. As the curve C' is to be wholly comprised within C, the above method has to be modified near points of C, at which the tangent is vertical or suffers an abrupt change of direction. Consider the case in which a tangent, parallel to the axis of y, touches C externally; near the point of contact there must be a vertical chord which passes through two points of C at a distance 2δ apart. The translation-process brings these two points together, at a point p say; these two points and all the points between the chord and the tangent are to be represented by p. It is true that there is a discontinuity in the values of U at p, but this discontinuity can be made arbitrarily small by making δ tend to 0. The case in which a vertical tangent touches C internally differs very slightly from the preceding. The point of contact changes into two points, and there is a gap in the curve C'. This gap can be filled up by connecting the two extremities of the separate arcs by an arbitrarily short arc interior to C. Let all the points of this arc correspond to the single point of contact on C; thus the value of U' on this short arc is constant.

§ 271. Having assigned in this way a definite meaning to the U' of $\frac{1}{2\pi}\int U'\frac{\delta g'}{n'}ds'$, let us now consider the function

$$v=\frac{1}{2\pi}\int U\frac{\delta g'}{\delta n'}ds',$$

where g' is the value on C' of the g-function for Γ relative to a point c, and $\frac{\delta g'}{\delta n'}$ is the derivative with regard to a normal of C' drawn inwards, U replacing U'.

Let (x', y') be a point on C', and let (a, b) be the point c. Construct the functions $g(x, y; a, b)$ and $g(x, y; x', y')$ relative to (a, b) and (x', y').

Then $\qquad g(x', y'; a, b)=g(a, b; x', y').$

But in the neighbourhood of (a, b) the function on the right-hand side can be expressed by a Fourier series, when (a, b) is chosen as the origin for polar co-ordinates. The coefficients in this expansion can be expressed, in the neighbourhood of (a, b), as definite integrals, which can in turn be expressed as integral series in x', y' throughout the neighbourhood of (x', y'). In this way it can be proved that all the composite derivatives of g with regard to a, b, x', y' are harmonic in the neighbourhoods of (a, b), (x', y').

Therefore $\qquad \frac{\delta^2 v}{\delta a^2}+\frac{\delta^2 v}{\delta b^2}=\frac{1}{2\pi}\int U\frac{\delta}{\delta n'}\left(\frac{\delta^2 g'}{\delta a^2}+\frac{\delta^2 g'}{\delta b^2}\right)ds'=0.$

This proves that v is harmonic in Γ. Now when c tends to a point t of C, g' and $\frac{\delta g'}{\delta n'}$ pass with uniform continuity into $\log 1/r$ and $\frac{\delta}{\delta n'}\log 1/r$, where r stands for the distance of t from the points of C'. Hence when c tends to the point t on C, v tends to a value V' which is the same as the value of $\frac{1}{2\pi}\int U\frac{\delta}{\delta n'}\log 1/r\cdot ds'$. As C' tends to C, the values of the harmonic function v and the rim values V' alter. We have to determine the exact value of the limit V to which V' tends. As before, separate C into two parts, one being an arbitrarily short arc s_1 on which t is situated. On s_1 the values of the continuous function U differ by less than ϵ. Through the extremities of s_1 draw vertical lines to cut C' in an arc s_1' parallel to s_1; denote the remainders of C, C' by s_2, s_2'. Now let the parts of the integral which are associated with s_1' and s_2' be

$$A=\frac{1}{2\pi}\int U\frac{\delta}{\delta n'}\log 1/r\cdot ds_1',$$

and $$B=\frac{1}{2\pi}\int U\frac{\delta}{\delta n'}\log 1/r\cdot ds_2'.$$

When δ tends to 0, the curve C' tends to C and the values of r in B pass with uniform continuity into the lengths of the radii from t to points of s_2. Therefore

$$\lim B=\frac{1}{2\pi}\int U\frac{\delta}{\delta n}\log 1/r\cdot ds_2=\frac{1}{2\pi}\int U(ds_2)_t,$$

the passage to the limit being accomplished with uniform continuity for all points t of the rim.

To find $\lim A$, denote as before the value of U at t by U_t. The values of U on the arc s_1' can be made to differ from U_t by an arbitrarily small amount by diminishing the length of the arc. Thus the equation

$$A=\frac{1}{2\pi}\int U(ds_1')_t$$

yields the inequality

$$\left|A-\frac{1}{2\pi}U_t\int(ds_1')_t\right|<\epsilon\cdot\frac{1}{2\pi}\int\left|(ds_1')_t\right|.$$

But $\int(ds_1')_t=-$ (the angle which s_1' subtends at t) $=-\beta'$,

where the negative sign is used because it is the outer side of ds_1' which is turned towards t. As C' moves up to its limiting position C, β' tends to the value β, where β is the angle between the two lines from t to the extremities of s_1. When the arc s_1 is made to shrink indefinitely, β tends to the limit π or α, where α has the same meaning as before, and we have

$$\lim A=-\frac{1}{2}U_t \text{ or } -\frac{\alpha}{2\pi}U_t,\ \lim B=\frac{1}{2\pi}\lim_{s_1=0}\int U(ds_2)_t=\frac{1}{2\pi}\int U(ds)_t;$$

and $$V_t=\lim A+\lim B=-\frac{1}{2}U_t+\frac{1}{2\pi}\int U(ds)_t,$$

or $$V_t=-\frac{\alpha}{2\pi}U_t+\frac{1}{2\pi}\int U(ds)_t.$$

It remains to be shown that the passage of V_t' to the limit V_t is effected with uniform continuity for all positions of t; in other words, that there is always a position of C' such that the values of v

relative to C' differ from V_t by less than an arbitrarily small positive number η. To prove this, observe that

$$V_t' = -\frac{\beta'}{2\pi}U_t + \frac{1}{2\pi}\int_{s'_2} U'\frac{\delta}{\delta n'}\log 1/r'ds_2' + \epsilon_1,$$

$$V_t = -\frac{\beta}{2\pi}U_t + \frac{1}{2\pi}\int_{s'_2} U\frac{\delta}{\delta n}\log 1/rds_2 + \epsilon_2.$$

Here r, r' are the distances of t from corresponding points of C, C'. In considering $V_t' - V_t$, we have to examine $\epsilon_1 - \epsilon_2$, $\beta' - \beta$, and

$$\frac{1}{2\pi}\int_{s'_2} U'\frac{\delta}{\delta n'}\log 1/r'ds_2' - \frac{1}{2\pi}\int_{s'_2} U\frac{\delta}{\delta n}\log 1/rds_2.$$

The absolute value of the first of these can be made less than $\eta/3$ by choosing s_1 sufficiently small; the difference between β and β' can be made less than $\eta/3$ by making C' approach sufficiently near to C, *i.e.* by making δ less than some number δ_1; finally the third expression can be made less than $\eta/3$ by making δ less than some number δ_2, for r' can be made to differ arbitrarily little from r by making δ sufficiently small. Here δ_1, δ_2 are determined for all points t of C.

Now let $(\delta', \delta'', \delta''', \cdots)$ be a regular sequence which defines 0, the members of the sequence being positive numbers arranged in descending order of magnitude. Let the harmonic functions v which correspond to the values $\delta', \delta'', \delta''', \cdots,$ of δ be $v', v'', v''', \cdots,$ and let the values of these functions at t be $V_{1,t}, V_{2,t}, V_{3,t}, \cdots$. We have just seen that the sequence $(V_{1,t}, V_{2,t}, V_{3,t}, \cdots)$ defines a value V_t; that is, $V_{1,t} + (V_{2,t} - V_{1,t}) + (V_{3,t} - V_{2,t}) + \cdots = V_t$. By § 263 the series

$$v' + (v'' - v') + (v''' - v'') + \cdots$$

defines a function v which is harmonic in Γ and continuous on C; and the value of this function at the point t is V_t.

The combination of the two integrals

$$\frac{1}{2\pi}\int U\frac{\delta \log 1/r}{\delta n}ds - \frac{1}{2\pi}\lim_{C'=C}\int U\frac{\delta g'}{\delta n'}ds'$$

is a function u which is harmonic in Γ and passes with uniform continuity into the assigned rim values U. Thus Dirichlet's problem has been solved and solved uniquely.

§ 272. Before passing on to the discussion of potential functions with discontinuities we shall establish a proposition which relates to Dirichlet's problem for the circle. Suppose that u is the harmonic function which takes assigned values $U(s)$ on a circle of radius R, and let a concentric circle be described whose radius ρ is less than R. The difference between a value u on the circle (ρ) and the value u_0 at the centre is

$$\frac{1}{\pi}\int_0^{2\pi} f(\alpha+\theta)\frac{-\rho(\rho-R\cos\alpha)d\alpha}{R^2-2R\rho\cos\alpha+\rho^2},$$

$$=\frac{1}{\pi}\int_{\alpha=0}^{\alpha=2\pi} f(\alpha+\theta)d\left(\sin^{-1}\frac{\rho\sin\alpha}{\sqrt{R^2-2R\rho\cos\alpha+\rho^2}}\right).$$

The expression $\sin^{-1}\left(\frac{\rho\sin\alpha}{\sqrt{R^2-2R\rho\cos\alpha+\rho^2}}\right)$ has for its maximum value $\sin^{-1}\rho/R$.

Suppose that as α increases from 0 to 2π this maximum is first attained when $\alpha=\phi$. Then, between the limits $\alpha=0$, $\alpha=\pi$, the expression under consideration increases, while α increases up to ϕ, and then decreases to 0. Hence the absolute value of each of the partial integrals

$$\int_{\alpha=0}^{\alpha=\phi},\int_{\alpha=\phi}^{\alpha=\pi},\int_{\alpha=\pi}^{\alpha=2\pi-\phi},\int_{\alpha=2\pi-\phi}^{\alpha=2\pi} f(\alpha+\theta)d\left(\sin^{-1}\frac{\rho\sin\alpha}{\sqrt{R^2-2R\rho\cos\alpha+\rho^2}}\right),$$

is less than $G\sin^{-1}\rho/R$, and

$$|u-u_0|<\frac{4G}{\pi}\sin^{-1}\frac{\rho}{R},$$

where G is the upper limit of the values u on the circle (R). Suppose that K is the lower limit of the same values. Then the function $u'=u-(K+G)/2$ has, for its upper limit on R, the value $(G-K)/2$, or $D/2$, where D is the oscillation on the circle (R).

Hence

$$|u'-u_0'|<D\cdot\frac{2}{\pi}\sin^{-1}\rho/R,$$

and the absolute value of the difference of any two values of u' on the circle (ρ) is less than

$$D\cdot\frac{4}{\pi}\sin^{-1}\rho/R.$$

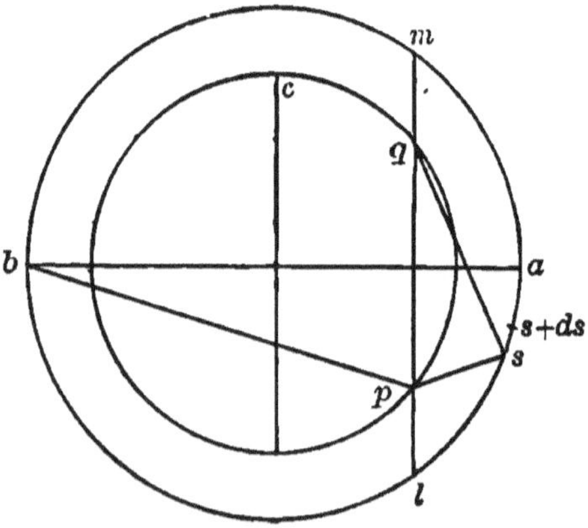

Fig. 82

[Schwarz, Werke, t. ii., p. 359; Harnack, Grundlagen der Theorie des Logarithmischen Potentiales, § 19.]

A closer limit has been found by Neumann, Abel'sche Integrale, p. 412. Let p, q (Fig. 82) be points on the circle of radius ρ, the radius of the outer circle being R. The values u_p, u_q are (§ 257)

$$\frac{1}{\pi}\int\left\{\frac{\delta}{\delta n}\log 1/r_1 - 1/2R\right\} U ds,$$

and

$$\frac{1}{\pi}\int\left\{\frac{\delta}{\delta n}\log 1/r_2 - 1/2R\right\} U ds;$$

that is,

$$u_p = \frac{1}{\pi}\int U(ds)_p - \frac{1}{2\pi R}\int U ds,$$

$$u_q = \frac{1}{\pi}\int U(ds)_q - \frac{1}{2\pi R}\int U ds.$$

Thus

$$u_p - u_q = \frac{1}{\pi}\int U\{(ds)_p - (ds)_q\} = \frac{1}{\pi}\int U\{d\psi - d\psi'\}$$

$$= \frac{1}{\pi}\int U d\chi,$$

where $\psi =$ the angle lps, $\psi' =$ the angle lqs, $\chi =$ the angle qsp.

Similarly, the function $u - (K+G)/2$ gives

$$u_p - u_q = \frac{1}{\pi}\int\{U - (K+G)/2\} d\chi.$$

Let the angles qap, pbq be γ_1, γ_2 respectively, and let the integral be taken over the four parts l to a, a to m, m to b, b to l. The absolute value of each of the first two parts is less than

$$\frac{1}{\pi}\{G - (K+G)/2\}\gamma_1;$$

that is, less than $D\gamma_1/2\pi$; and the absolute value of each of the remaining two parts is less than $D\gamma_2/2\pi$. Thus

$$|u_p - u_q| < D(\gamma_1 + \gamma_2)/\pi < 4D\,(\text{angle } cab)/\pi.$$

But

$$\tan cab = \rho/R;$$

therefore $|u_p - u_q| < \frac{4}{\pi} D \tan^{-1}\rho/R$. Here the multiplier $\frac{4}{\pi}\tan^{-1}\rho/R$, or θ, is a proper fraction whose value depends solely on ρ, R.

§ 273. *Discontinuities.* Hitherto the solution u of

$$\delta^2 u/\delta x^2 + \delta^2 u/\delta y^2 = 0$$

has been supposed to be continuous in Γ. Let us now extend the treatment of the potential function for a multiply connected region Γ of the plane so as to include partially the discontinuities of the kind met with in the discussion of Abelian integrals on a Riemann surface. These two kinds of discontinuity are of the following species:—

(1) In the region Γ the function $u(x, y)$ is infinite at a point (x_0, y_0) like the real part $f(x, y)$ of

$$\sum_{\kappa=1}^{\lambda}(\alpha_\kappa + i\beta_\kappa)/(z - z_0)^\kappa + (\alpha + i\beta) \log (z - z_0),$$

while $u - f(x, y)$ is one-valued, continuous, and satisfies Laplace's equation in the neighbourhood of (x_0, y_0), that point included;

(2) In the simply connected region Γ', which arises from Γ by the drawing of cross-cuts, the function $u(x, y)$ is one-valued and continuous, and its values along the cross-cuts differ by constant quantities (the real parts of the periods associated with these cross-cuts).

The boundary values are supposed to be assigned arbitrarily, except that they are subject to the usual restrictions as to continuity.

Suppose that a potential function u takes assigned values $U(s)$ on the rim of Γ, and that in Γ it is continuous except at a point c, at which it is infinite like $\log r$. The existence theorem for this case follows at once from the existence theorem for the g-function. For let u' be that harmonic function which takes the values $U(s)$ on C, and is everywhere one-valued and continuous in Γ, and let g_c be the g-function relative to Γ and c. Then $u' + g - \log 1/r$ is the required function. For $g - \log (1/r)$ is zero on C and infinite at c in the required manner, and $u' + g - \log 1/r$ has the values U on C and is infinite in the required manner.

The case of a purely algebraic discontinuity can be dealt with equally easily. For let $f(x, y)$ be the real part of

$$\sum_{\kappa=1}^{\lambda}(\alpha_\kappa + i\beta_\kappa)/(z - z_0)^\kappa,$$

and let the values of f on C be F. Construct the harmonic function u', which takes the values $U(s) - F$ on C, and is everywhere one-valued and continuous in Γ. Then $u = u' + f$ is the potential function with the assigned discontinuity and the assigned values on the rim.

Next, suppose that there is a solution f of the differential equation which is one-valued and continuous in a region Γ, except along lines or at points such that the integrals $\frac{1}{2\pi}\int_0^{2\pi}\frac{\delta f}{\delta n}ds$, taken over curves round the separate lines or points, all vanish. Harnack, following Schwarz and Neumann, has proved that there exists a potential function ϕ, which is one-valued and continuous at all points (∞ inclusive) of the plane other than the places of discontinuity of f, and which is discontinuous like f at the above-mentioned lines and points (*i.e.* $\phi - f$ is harmonic throughout Γ). [Harnack, Grundlagen der Theorie des Logarithmischen Potentiales, § 45; Neumann, Abel'sche Integrale, 2d ed., pp. 445, 461; Schwarz, Werke, t. ii., pp. 163, 166.]

§ 274. Take a circular region; call this region T_1, and let its rim be C_1. Let C_2 be a concentric circle of smaller radius, such that all the places of discontinuity of f are situated within the circular region bounded by C_2. Let the region exterior to C_2 be called T_2. Along the rim of any circle with centre 0 and radius intermediate to the radii of C_1, C_2,

$$\frac{1}{2\pi}\int_0^{2\pi} f d\theta = \text{a constant } A;$$

or, replacing $f - A$ by f,

$$\frac{1}{2\pi}\int_0^{2\pi} f d\theta = 0.$$

Apply Schwarz's alternating process to the two regions T_1, T_2. Construct for T_2 a harmonic function u_1 with the values F_2 on C_2, where F_2 stands for the values of f on C_2; for T_1 a harmonic function u_2 with the values $u_1^{(1)} - F_1$ on C_1, where $u_1^{(1)}$ stands for the values of u_1 on C_1; for T_2 a harmonic function u_3 with the values $u_2^{(2)} + F_2$ on C_2; for T_1 a harmonic function with the values $u_3^{(1)} - F_1$ on C_1; and so on indefinitely. Since the upper limit of the values F_2 on C_2 is finite, the absolute value of $u_1^{(1)} - F_1$ is everywhere finite on C_1. Also we know that on C_1,

$$\int_0^{2\pi} u_1^{(1)} d\theta = 0, \qquad \int_0^{2\pi} F_1 d\theta = 0,$$

therefore, from the equation $u_2^{(1)} = u_1^{(1)} - F_1$,

$$\int_0^{2\pi} u_2^{(1)} d\theta = 0, \quad \text{and} \quad \int_0^{2\pi} u_2^{(2)} d\theta = 0.$$

The latter equation shows that the value of u_2 at the centre of C_2 is 0. Applying the theorem of § 272, the absolute value* and the oscillation of u_2 on C_2 are less than θD, where D is the greatest oscillation of the rim values $u_1^{(1)} - F_1$, and θ is a proper fraction which depends solely upon the radii of C_1, C_2.

Since u_3 takes on C_2 the values $u_2^{(2)} + F_2$, we have

$$\int_0^{2\pi} u_3^{(2)} d\theta = \int_0^{2\pi} u_3^{(1)} d\theta = 0.$$

Since $u_3^{(2)} - u_1^{(2)} = u_2^{(2)}$, the absolute value and the oscillation of $u_3 - u_1$ on C_2 are less than θD. Again, the equations

$$\int_0^{2\pi} u_4^{(1)} d\theta = \int_0^{2\pi} u_3^{(1)} d\theta = 0,$$

show that the absolute value and the oscillation of $u_4^{(1)} - u_2^{(1)}$ are less than θD, and therefore that the absolute value and the oscillation of $u_4^{(2)} - u_2^{(2)}$ are less than $\theta^2 D$. The two systems

$$\begin{cases} u_1, u_3, u_5, \cdots, \\ u_2, u_4, u_6, \cdots, \end{cases}$$

are such that in T_2

$$u_3 - u_1 < \theta D, \; u_5 - u_3 < \theta^2 D, \cdots,$$

and in T_1 $\quad u_4 - u_2 < \theta D, \; u_6 - u_4 < \theta^2 D, \cdots.$

Hence, the two series

$$u_1 + (u_3 - u_1) + (u_5 - u_3) + \cdots,$$

$$u_2 + (u_4 - u_2) + (u_6 - u_4) + \cdots,$$

define harmonic functions u', u'', which are the limits of u_{2n+1}, u_{2n}, when n tends to ∞. On the two rims of the region T', common to T_1 and T_2, we have

$$u_{2n}^{(1)} = u_{2n-1}^{(1)} - F_1, \; u_{2n}^{(2)} = u_{2n+1}^{(2)} - F_2;$$

therefore $u'' + f$ takes the same values as u' throughout T', and $u'' + f$ serves for the prolongation of u' over the boundary of T_1. It should be noticed that the region formed by the combination of T_1, T_2 is a closed region, a result of importance in view of the fact that a Riemann surface may be closed.

---

* When the value at the centre is 0, the lower limit is 0 or negative. Therefore the absolute value on the rim must be equal to or less than the oscillation.

References. Neumann and Schwarz share the honour of being the first mathematicians to give a satisfactory proof of Dirichlet's Principle. Those memoirs of the latter which relate to this problem are to be found in his Werke, t. ii.; while Neumann's researches have been incorporated largely in his treatise on the Abelian Functions. Riemann's original investigations were published in his Inaugural Dissertation. Picard has given an admirably lucid account of the methods of Schwarz, Neumann, and Poincaré in his Traité d'Analyse, t. ii., fasc. 1. Harnack's method is expounded in his work, Grundlagen der Theorie des Logarithmischen Potentiales, a contribution of high value to the Potential-theory. Two important memoirs on the methods of Schwarz and Poincaré have appeared within recent years; one by M. Jules Riemann, Sur le Problème de Dirichlet, Ann. de l'Éc. Norm. Sup., ser. 3, t. v., the other by M. Paraf, Sur le Problème de Dirichlet, Paris, 1892. The reader should consult Klein's memoir, Beiträge zur Riemann'schen Functionentheorie, Math. Ann., t. xxi., Klein's Schrift, and Klein-Fricke, Modulfunctionen, t. i., pp. 504 et seq. For the history of the subject he is referred to the notes at the end of Schwarz's collected works. The discussion of the potentials which are associated with Abelian integrals will be taken up again at the end of the next chapter.

CHAPTER X.

Abelian Integrals.

§ 275. Riemann's classical memoir on the Abelian Functions (Theorie der Abelschen Functionen) was published in the 54th volume of Crelle's Journal in 1857. His treatment of the subject was based on the brilliant conception that a theory could be constructed from the definition of Abelian integrals by certain characteristic properties on a Riemann surface. To establish the existence of functions with the assigned properties he used Dirichlet's principle; but it has been pointed out (in Chapter IX.) that Dirichlet's problem has not been solved satisfactorily by methods drawn from the Calculus of Variations. To secure solid foundations for Riemann's superstructure, recourse must be had to the existence-theorems of Neumann and Schwarz. In the chapter on Riemann surfaces everything was referred back to a basis-equation $F(w, z) = 0$, the number of sheets of T in the z-plane and the distribution of the branch-places being determined by the equation; but it was indicated (§ 190) that the process might be reversed. One of the distinguishing features of Riemann's method is that he starts with a surface given independently of an equation $F = 0$ (§ 162), and passes thence to algebraic functions of the surface. In this way the surface becomes associated with a whole system of equations which pass into one another by birational transformations, and no one of these equations can claim precedence over the others. The surface being spread over the z-plane, it is often convenient to single out from the algebraic functions of the surface the particular function z, and to express the other algebraic functions in terms of z (§ 154). Kronecker's theory of the discriminant shows that amongst these functions w there is one which gives rise to an equation $F(w, z) = 0$ with no higher singularities than nodes; in this equation the branch-points arise solely from vertical tangents. (See Nöther, Rationale Ausführung der Operationen in der Theorie der Algebraischen Functionen, Math. Ann., t. xxiii.) There is no loss of generality in selecting for a basis-curve one with no singularities higher than nodes.*

* Using the language of Higher Plane Curves, the Cramer-Nöther transformations of Chapter IV. are one-to-one transformations of the plane of co-ordinates x_1, x_2, x_3, while the transforma-

The plan of this chapter is (1) to find what discontinuities can arise when there is a basis equation $F(w^n, z^m)=0$; (2) to find what theorems follow from an assumption of the truth of the existence-theorems for integrals of the three kinds on a surface T given independently of a basis-equation; (3) to work out results, using the Clebsch-Gordan form of equation $f_n(x_1, x_2, x_3)=0$, or $F(w, z,)=0$, in which the branch-points are simple and situated in the finite part of the plane, and the multiple points are not higher than nodes; (4) to show that the existence-theorems for Abelian integrals on the surface T follow from the theorems of Chapter IX. In justification of this plan we may observe that the theorems of Chapter IX. indicate the existence of the Abelian integrals on T; also that when these existence-theorems are known to hold for the most general case, there is no advantage to be gained by selecting a basis-equation with complicated expansions, since the character of Riemann's results can be appreciated adequately by using one which gives simple branch-points.

§ 276. Let us start, in the first place, with an irreducible equation $F(w^n, z^m)=0$ of deficiency p; and let T, T′ have their ordinary meanings. An examination of the nature of the discontinuities of the Abelian integral $\int_{z_0, w_0}^{z, w} R(w, z)dz$ will lead to a classification of Abelian integrals.

Near each place s of the surface T, R can be expanded in a series of ascending powers of $(z-c)^{1/r}$, where r is a positive integer and c is the value of z at s. The first exponent q_1/r may be negative, but cannot be infinite. If $q_1<0$, the place s is an infinity, and if $r>1$, it is also a branch-place at which r sheets hang together. Since, near s,

$$R=(z-c)^{q_1/r}P_0(z-c)^{1/r},$$

the value of $\int Rdz$ is

$$\int (z-c)^{q_1/r}\{a_0+a_1(z-c)^{1/r}+a_2(z-c)^{2/r}+\cdots\}dz,$$
$$=b_0(z-c)^{(q_1+r)/r}+b_1(z-c)^{(q_1+r+1)/r}+\cdots.$$

Case I. If $q_1+r>0$, $\int Rdz=P_{q_1+r}(z-c)^{1/r}$. Here s is a branch-place, or an ordinary place, according as $r \gtreqless 1$.

tions of Kronecker are one-to-one transformations of a curve in the plane. The former transformations separate coincident tangents, but do not resolve a multiple point with distinct tangents into its component nodes.

Case II. If $q_1 + r \leqq 0$, let $q_1 + r = -t$; then

$$\int Rdz = b_0(z-c)^{-t/r} + b_1(z-c)^{(1-t)/r} + \cdots + b_t \log(z-c) + P(z-c)^{1/r}.$$

Here s is both a branch-place and an infinity. The nature of the infinity depends on the values of the coefficients. When

$$b_0 = b_1 = b_2 = \cdots = b_{t-1} = 0,$$

the infinity is purely logarithmic. On the other hand, if $b_t = 0$, the infinity is purely algebraic; but if $b_0 = b_1 = b_2 = \cdots = b_t = 0$, we fall back on Case I. For every other system of values the infinity is logarithmic-algebraic.

Case II. shows that the integral is infinite at an infinity of R, except when that infinity is a branch-place on T such that $q_1 + r > 0$. We have established the fact that the integral has only a finite number of infinities which are algebraic, logarithmic, or logarithmic-algebraic; and that no other infinities can occur. An integral is said to be of the first kind when it has no infinity on T, of the second kind when its infinities are algebraic solely, and of the third kind when it has logarithmic or logarithmic-algebraic infinities. The simplest integral of the second kind is that which has only one infinity on T, that infinity being a pole s of the first order; the integral of the second kind is then said to be *elementary.* The simplest integral of the third kind is that which is infinite at two places of T like $+\log(z-c_1)$ and $-\log(z-c_2)$, and is elsewhere finite; the integral of the third kind is then said to be *elementary.* We shall denote elementary integrals of the second and third kinds by Z and Q, and shall indicate by suffixes the places at which the integrals are infinite. Thus in the present notation Z_s is an elementary integral of the second kind, $Q_{s_1 s_2}$ one of the third.* We have had instances in § 219.

The integrals of the three kinds are continuous functions of the place on T, but are not one-valued. When the upper limit of an Abelian integral describes a closed path on T, the final value of the integral may differ from the initial value by multiples of periods † due to the cross-cuts or to the logarithmic discontinuities.

It should therefore be borne in mind that the integral of the first kind may become infinite by the addition of infinitely many periods.

---

* In cases where no ambiguity will arise we shall say that an integral is ∞ at a place s like $(z-s)^{-1}$ or $\log(z-s)$, where what is meant is that it is ∞ like $(z-z(s))^{-1}$ or $\log(z-z(s))$, $z(s)$ being the *point* of the z-plane at which s lies.

† The word period is used for shortness. It would be more correct to use modulus of periodicity and reserve period for the Abelian functions which arise from the inversion problem.

In Case II., writing $z-c=z_1^r$, we have

$$\int Rdz = b_0 z_1^{-t} + b_1 z_1^{1-t} + \cdots + rb_t \log z_1 + P(z_1).$$

If the path of integration be a small *closed* curve round the place s, this gives

$$\int_{(s)} Rdz = 2\pi i r b_t.$$

The coefficient rb_t is called the residue at the point in question (§ 180). Round an ordinary point, and round a critical point such that in the expansion of R no term $(z-c)^{-1}$ enters,

$$\int_{(s)} Rdz = 0.$$

Let $s_1, s_2, \cdots, s_\kappa$ be the logarithmic infinities of $\int Rdz$ on the surface T. The surface T can be reduced by the ordinary cross-cuts A, B, C to T′. As in § 180, all the logarithmic infinities must be cut out; the surface is now $(\kappa+1)$-ply connected. Let it be reduced to a simply connected surface T″ by κ further cuts. On T″ the integral has no logarithmic infinities; hence taken positively round the complete boundary of T″ it vanishes. The parts of this boundary which are traversed twice *in opposite directions* contribute nothing to the sum-total, since each element Rdz is cancelled by an element $-Rdz$. The only parts which are not of this category are the small closed curves round the places s; these contribute $2\pi i$ (the sum of the residues). Hence the important theorem that the sum of the residues of $\int Rdz$ on T is zero.

In the proof of this theorem we have integrated round the boundary of a dissected surface. This method is used constantly in Riemann's theory. It seems desirable therefore to point out explicitly that elements on opposite banks of a cross-cut are combined with each other,* and that the cross-cuts C play a subsidiary part, the period across a C being zero (see § 180). The theorem with regard to the residues shows why an integral with *two* logarithmic infinities is chosen as the elementary integral; for an integral with only one logarithmic infinity is non-existent.

In practice use is made of a special equation, and the results are open to objections on the score of incomplete generality. To meet these objections it is necessary to place the Riemann surface T in the fore-front of the theory, and to prove that Abelian

*Integration round the rim of a dissected surface is analogous to integration round the rim of a parallelogram of periods in the case of a doubly periodic function. See Chapter VII., § 197.

integrals of the three kinds exist upon it. It is assumed that this surface is spread over the z-plane in a finite number of sheets (say n), that its order of connexion is $2p+1$, and that the sum of the orders of its branchings is $2p+2n-2$. To avoid interrupting the discussion of properties, we shall assume the truth of certain existence theorems, and give the proofs at a later stage. Strictly we ought not to speak of Abelian integrals of the first three kinds for such a surface, but rather of functions of the place of the first three kinds, since there is, at first, no basis-equation. But no confusion will be caused inasmuch as the functions prove to be of the form $\int R(w, z)dz$, where w is an algebraic function of T which is connected with z by an algebraic equation.

§ 277. The existence theorem for integrals of the first kind establishes the fact that there exists upon T a function W with the following properties:

(1) W is continuous on T;

(2) W has constant periods $\alpha_\kappa + i\beta_\kappa$, $\alpha_{\kappa+p} + i\beta_{\kappa+p}$, at the cross-cuts A_κ, B_κ, where the α's are assigned arbitrarily.

That the name Abelian integral is bestowed appropriately on the function can be shown by the following considerations. On opposite banks of a cross-cut the difference between W_+ and W_- is constant. Hence

$$dW_+/dz = dW_-/dz,$$

and the function dW/dz is continuous except at the branch-points, where it may have algebraic discontinuities. Such a function is necessarily a rational function $R(w, z)$ of w and z, where $F(w^n, z)=0$ is an irreducible equation associated with the surface. Thus W is an Abelian integral. The property (1) shows that it is an Abelian integral of the first kind.

Suppose, to take a more general case, that there exists on T a function I which has the following three properties:

(1) It is continuous on T except at isolated places;

(2) Its discontinuity at each place of discontinuity is algebraic, logarithmic, or logarithmic-algebraic;

(3) The value at any place of T is altered merely by an additive constant after the description of a closed path on T.

The derivative dI/dz is continuous except at certain places where the discontinuities are algebraic; for the logarithmic infinities are converted by differentiation into algebraic infinities, and dI/dz

has no periods at the cross-cuts which change T into T″. Thus the function dI/dz is of the form $R(w, z)$, and I is an Abelian integral.

A special case of Green's theorem. Let W be an integral of the first kind; let $W = U + iV$, where U, V are real. The only places on T′ at which the first derivatives of U, V can become infinite are the branch-places of T. Let these be cut out from T′ by small closed curves. By an application of Green's theorem

$$\int U dV = \int\left(-U\frac{\delta U}{\delta y}dx + U\frac{\delta U}{\delta x}dy\right)$$
$$= \int\int\left\{\left(\frac{\delta U}{\delta x}\right)^2 + \left(\frac{\delta U}{\delta y}\right)^2\right\}dxdy,$$

since $\delta^2 U/\delta x^2 + \delta^2 U/\delta y^2 = 0$. The single integral is taken round the boundary of T′ and round the small curves which enclose the branch-points. The integrals round the branch-points become infinitely small when the curves become infinitely small; therefore we have the theorem that

$$\int_{T'} U dV = \int\int\left\{\left(\frac{\delta U}{\delta x}\right)^2 + \left(\frac{\delta U}{\delta y}\right)^2\right\}dxdy \quad . \quad . \quad (1),$$

where the single integral is taken round the rim of T′. [See Ch. IX., § 255, and Ch. VI., § 176.]

By integrating round the lines A, B, C, we get, since $U_+ - U_-$ is equal to α_κ along A_κ and to $\alpha_{\kappa+p}$ along B_κ,

$$\int_{T'} U dV = \sum_{\kappa=1}^{p}\left\{\alpha_\kappa \int_{A_\kappa} dV + \alpha_{\kappa+p}\int_{B_\kappa} dV\right\};$$

for each cross-cut is traversed twice in opposite directions and $dV_+ = dV_-$. The integrals on the right-hand side are equal to $\beta_{\kappa+p}$, $-\beta_\kappa$; hence the expression

$$\sum_{\kappa=1}^{p}\begin{vmatrix}\alpha_\kappa & \alpha_{\kappa+p}\\ \beta_\kappa & \beta_{\kappa+l}\end{vmatrix} = \text{a positive quantity, or } 0 \quad . \quad (2).$$

In particular, W reduces to a constant when all the α's or all the β's vanish. That is, *an integral of the first kind with periods all real, or an integral with periods all purely imaginary, is a constant.* Referring to what was said at the beginning of this paragraph, we can now see that the conditions imposed upon W determine it to a constant term. For let W' be an integral of the first kind whose α's are the same as those of W. The function $W - W'$ is an integral of the first kind which has purely imaginary periods at the cross-cuts; therefore, by the theorem which we have just proved, $W - W'$ is *a constant*. If the ratios of the periods be all real, the periods themselves are of the form $k a_\kappa$, $k a_{\kappa+p}$, where k may be complex; W/k is now a function whose periods are all real, and therefore W/k and W are constant.

Example. An integral whose periods at the curves A_κ (or B_κ) are all zero, is a constant.

§ 278. It is of great importance to determine how many integrals W can be linearly independent. Let there be q linearly independent integrals W_λ, $(\lambda = 1, 2, \cdots, q)$; and consider the expression

$$k_0 + \sum_{\lambda=1}^{q} k_\lambda W_\lambda,$$

where the coefficients k_λ are constants. The expression is continuous on T and has constant differences $\sum_{\lambda=1}^{q} k_\lambda \omega_{\lambda\kappa}$, $\sum_{\lambda=1}^{q} k_\lambda \omega_{\lambda, \kappa+p}$ at the cross-cuts A_κ, B_κ, where $\omega_{\lambda\kappa}$, $\omega_{\lambda, \kappa+p}$ are the periods of W_λ at these cross-cuts. It is therefore an integral of the first kind. Denote it by W and its periods by ω_κ, $\omega_{\kappa+p}$; also let ω_κ, $\omega_{\lambda\kappa}$, k_λ be resolved into their real and imaginary parts $a_\kappa + i\beta_\kappa$, $a_{\lambda\kappa} + i\beta_{\lambda\kappa}$, $\mu_\lambda + i\nu_\lambda$. We have the system of $2p$ equations

$$\left.\begin{aligned} &\sum_{\lambda=1}^{q} (\mu_\lambda + i\nu_\lambda)(a_{\lambda\kappa} + i\beta_{\lambda\kappa}) = a_\kappa + i\beta_\kappa, \\ &\sum_{\lambda=1}^{q} (\mu_\lambda + i\nu_\lambda)(a_{\lambda, \kappa+p} + i\beta_{\lambda, \kappa+p}) = a_{\kappa+p} + i\beta_{\kappa+p}, \end{aligned}\right\} \quad (3),$$

$\kappa = 1, 2, \cdots, p$. By equating the real parts, we get $2p$ equations, from which the $2q$ quantities μ_λ, ν_λ can be eliminated, provided $q < p$; and there remain relations between the quantities $a_{\lambda\kappa}$, $\beta_{\lambda\kappa}$, $a_{\lambda, \kappa+p}$, $\beta_{\lambda, \kappa+p}$, a_κ, $a_{\kappa+p}$. Choosing a set of values for a_κ which do not obey these relations, we have an integral of the first kind which is not linearly related to the W_λ's. If $q + 1 < p$, we can by the same argument obtain $q + 2$ linearly independent W's, and so on until p is reached. Hence there are not less than p linearly independent integrals of the first kind. Suppose that W_1, W_2, $\cdots$, W_p are p linearly independent integrals of the first kind; then *every other integral W of the first kind is expressible in the form*

$$W = k_0 + \sum_{\lambda=1}^{p} k_\lambda W_\lambda \quad \cdots\cdots\cdots\cdots \quad (4),$$

a theorem of fundamental importance. In proof of this theorem observe that whatever be the real parts of the periods of W it is always possible to assign such values to μ_λ, ν_λ as will make the real parts of the periods of the linear combination $\sum_{\lambda=1}^{p} k_\lambda W_\lambda$ identical with the real parts of the periods of W. And when this has been done, $W - \sum_{\lambda=1}^{p} k_\lambda W_\lambda$ is an integral of the first kind whose periods are all purely imaginary, and is therefore a constant.

The theorem which has just been enunciated cannot be regarded as proved until it is shown that the determinant formed by the coefficients of the quantities $a_{\lambda\kappa}$, etc., in the equations for a_κ and $a_{\kappa+p}$, does not vanish. Neumann shows by considerations drawn from determinants that the vanishing of this determinant is inconsistent with the linear independence of $W_1, W_2, \cdots, W_p$. Neumann, Abel'sche Integrale, 2d ed., p. 243.

Corollary. When $p = 0$, there is no integral of the first kind.

The theorems of this and the preceding paragraph are the ones referred to at the end of § 220.

§ 279. *The bilinear relations for integrals of the first kind.* The integral $\int W_\lambda dW_{\lambda'}$, taken over the complete boundary of T′, is zero. The cross-cut A_κ is described twice in opposite directions, and

$$\begin{aligned}\int_{A_\kappa} W_\lambda dW_{\lambda'} &= \int_{A_\kappa} (W_\lambda^+ dW_{\lambda'}^+ - W_\lambda^- dW_{\lambda'}^-)\\ &= \int_{A_\kappa} (W_\lambda^+ - W_\lambda^-) dW_{\lambda'}^+,\\ &= \omega_{\lambda\kappa} \int_{A_\kappa} dW_{\lambda'},\\ &= \omega_{\lambda\kappa}\omega_{\lambda', p+\kappa};\end{aligned}$$

see Fig. 74, § 180.

Similarly,

$$\int_{B_\kappa} W_\lambda dW_{\lambda'} = -\omega_{\lambda'\kappa}\omega_{\lambda, p+\kappa}.$$

Also, $\int_{C_\kappa} W_\lambda dW_{\lambda'} = 0.$

Therefore, after addition,

$$\sum_{\kappa=1}^{p} (\omega_{\lambda\kappa}\omega_{\lambda', p+\kappa} - \omega_{\lambda'\kappa}\omega_{\lambda, p+\kappa}) = 0,$$

r

$$\sum_{\kappa=1}^{p} \begin{vmatrix} \omega_{\lambda\kappa} & \omega_{\lambda, p+\kappa} \\ \omega'_{\lambda'\kappa} & \omega'_{\lambda', p+\kappa} \end{vmatrix} = 0 \quad \cdots\cdots\cdots \quad (5).$$

In the most general case when I, I' are two Abelian integrals on T,

$$\int_{T'} IdI' = \sum_{1}^{p} \begin{vmatrix} \omega_\kappa & \omega_{p+\kappa} \\ \omega'_\kappa & \omega'_{p+\kappa} \end{vmatrix} \quad \cdots\cdots\cdots \quad (6),$$

where ω, ω' stand for periods of I, I'; and each side of the equation is equal to the integral round the infinities of both I and I'. This may be called the general bilinear relation.*

Theorem. The determinant

$$\Delta \equiv \begin{vmatrix} \omega_{11} & \omega_{12} \cdots & \omega_{1p} \\ \omega_{21} & \omega_{22} \cdots & \omega_{2p} \\ \cdot & \cdot \quad \cdot \quad \cdot & \cdot \\ \cdot & \cdot \quad \cdot \quad \cdot & \cdot \\ \omega_{p1} & \omega_{p2} \cdots & \omega_{pp} \end{vmatrix},$$

formed by the p^2 periods of the p linearly independent integrals W_λ at the cross-cuts A_κ, cannot vanish.

For if $\Delta = 0$, the theory of determinants enables us to assign constants k_λ, not all zeros, such that

$$\sum_{\lambda=1}^{p} k_\lambda \omega_{\lambda\kappa} = 0, \ (\kappa = 1, 2, \cdots, p).$$

The integral $\Sigma k_\lambda W_\lambda$ has now no periods at the cross-cuts, and is therefore a constant. Hence the W_λ's are not linearly independent, contrary to hypothesis.

§ 280. *Normal integrals of the first kind.* An integral W may have the period 1 at the cross-cut A_ι, and the period 0 at all the other cross-cuts A_κ. For the equations

$$\sum_{\lambda=1}^{p} k_\lambda \omega_{\lambda\kappa} = 0, \quad (\kappa \neq \iota),$$

$$\sum_{\lambda=1}^{p} k_\lambda \omega_{\lambda\kappa} = 1, \quad (\kappa = \iota),$$

determine the p quantities k_λ without ambiguity. Hence W is completely determinate, save as to an added constant.

Such an integral is called *a normal integral of the first kind.* We shall denote it by u_ι and its period at B_κ by $\tau_{\iota\kappa}$. Corresponding to the p cuts A_κ there are p such integrals, whose periods form the scheme

	A_1	$A_2 \cdots$	A_p	B_1	$B_2 \cdots$	B_p
u_1	1	$0 \cdots$	0	τ_{11}	$\tau_{12} \cdots$	τ_{1p}
u_2	0	$1 \cdots$	0	τ_{21}	$\tau_{22} \cdots$	τ_{2p}
.	.	. . .	.	.	. . .	.
.	.	. . .	.	.	. . .	.
u_p	0	$0 \cdots$	1	τ_{p1}	$\tau_{p2} \cdots$	τ_{pp}

* Other relations besides the bilinear one connect the periods of two Abelian integrals of the first kind. See Clebsch, t. iii., p. 185.

The p normal integrals are linearly independent. For, if possible, let

$$k_0 + \Sigma k_\iota u_\iota = 0.$$

Then at A_κ

$$k_\kappa \cdot 1 = 0,$$

and the vanishing of the coefficients implies that there is no linear relation.

The bilinear relation becomes, in the case of the normal integrals, u_ι, u_κ,

$$\begin{vmatrix} 1 & \tau_{\iota\iota} \\ 0 & \tau_{\kappa\iota} \end{vmatrix} + \begin{vmatrix} 0 & \tau_{\iota\kappa} \\ 1 & \tau_{\kappa\kappa} \end{vmatrix} = 0,$$

or

$$\tau_{\kappa\iota} = \tau_{\iota\kappa}.$$

In virtue of this relation the array formed by the τ's is symmetric. The simple character of the properties of the p normal integrals $u_1, u_2, \cdots, u_p$ makes them specially suitable for use as the elements in terms of which all integrals W are expressed linearly.

In § 277 it was proved that

$$\sum_{\kappa=1}^{p} \begin{vmatrix} \alpha_\kappa & \alpha_{p+\kappa} \\ \beta_\kappa & \beta_{p+\kappa} \end{vmatrix} \geqq 0.$$

If we apply this result to the integral

$$n_1 u_1 + n_2 u_2 + \cdots + n_p u_p,$$

where $n_1, n_2, \cdots, n_p$ are real, we have, if $\tau_{\iota\kappa} = \alpha_{\iota\kappa} + \iota\beta_{\iota\kappa}$,

$$\alpha_\kappa = n_\kappa,$$

$$\beta_\kappa = 0,$$

$$\alpha_{p+\kappa} = \sum_{\iota=1}^{p} n_\iota \alpha_{\iota\kappa},$$

$$\beta_{p+\kappa} = \sum_{\iota=1}^{p} n_\iota \beta_{\iota\kappa},$$

and therefore

$$\sum_{\kappa=1}^{p} \sum_{\iota=1}^{p} n_\kappa n_\iota \beta_{\iota\kappa} \geqq 0;$$

that is,

$$\beta_{11} n_1^2 + 2\beta_{12} n_1 n_2 + \cdots + \beta_{pp} n_p^2 \geqq 0.$$

The sign of equality obtains only when the integral $\sum_{\iota=1}^{p} n_\iota u_\iota$ is constant, a case which implies the vanishing of $n_1, n_2, \cdots, n_p$.

§ 281. *The functions Φ.* The derivative, dW/dz, of an integral of the first kind belongs to a system of algebraic functions on the surface which is of paramount importance. We shall denote such a function by Φ. Thus $\int \Phi dz$ is always an integral of the first kind.

These functions are $(2p+2n-2)$-placed. For near a place s at which $z=c$ let

$$W = P(z-c)^{1/r}$$
$$= W(s) + a_1(z-c)^{1/r} + a_2(z-c)^{2/r} + \cdots.$$

If $a_1 \neq 0$, $\quad \Phi = b_1(z-c)^{(1-r)/r} + \cdots,$

and Φ is ∞^{r-1} at s.

Therefore Φ is β-placed, where

$$\beta = \Sigma(n-1) = 2p + 2(n-1), \text{ (§ 170)}.$$

Of the $2p+2(n-1)$ zeros of Φ we can at once account for $2n$. For by supposition there is no branching at ∞, and accordingly, when $c=\infty$,

$$W = W(s) + P_1(1/z),$$
$$\Phi = P_2(1/z).$$

Therefore each of the n places at ∞ is a zero of the second order. There remain $2p-2$ zeros, which are said to be *moveable*, inasmuch as they are different for different Φ's. That this is the case will appear in § 288.

It has been assumed in the proof that a_1, and therefore b_1, does not vanish. When $r=1$, the vanishing of a_1 evidently means that the place s is itself a zero of Φ, and in this way also we account for the vanishing of a_1 when $r>1$, by saying that then a zero of the Φ in question happens to be at a branch-point. And similarly when in the expansions at ∞ the coefficient of $1/z$ in the series for W is missing, one of the moveable zeros happens to coincide with a fixed zero.

This proof is from Klein-Fricke, t. i., p. 544.

The ratio of two functions Φ is evidently, at most, $(2p-2)$-placed.

There can be chosen p functions Φ between which there is no homogeneous linear relation; but between these and every other Φ there must be a homogeneous linear relation. This is proved at once by differentiating (4). From this again it follows that half the moveable zeros of a Φ can be assigned arbitrarily. For, if s_1, $s_2, \cdots, s_{p-1}$ be the assigned zeros, the constants k can be chosen so as to satisfy the $p-1$ equations.

$$\sum_{\lambda=1}^{p} k_\lambda \Phi_\lambda(s_\kappa) = 0,$$

where $\qquad \kappa = 1, 2, \cdots, p-1;$

that is, we can make $\sum_{\lambda=1}^{p} k_\lambda \Phi_\lambda$, which is the general expression of a Φ, vanish at the $p-1$ places.

It appears at first sight as if the $p-1$ assigned zeros determine the other $p-1$ zeros; but this is not necessarily the case, for the above $p-1$ equations need not be independent. For example, when the surface is two-sheeted the basis-equation can be written

$$w^2 = \prod_{\kappa=1}^{2p+2} (z - a_\kappa),$$

we can take (§ 182) for the p independent Φ's z^λ/w, $\lambda = 0, 1, \cdots, p-1$, and for any Φ, $\Sigma k_\lambda z^\lambda/w$. The $2p-2$ moveable zeros form $p-1$ *pairs* of places, say $s\bar{s}$, each pair having the same z. Now provided that among the assigned zeros there are no pairs, the remaining zeros are all determined, for they are $\bar{s}_1, \bar{s}_2, \cdots, \bar{s}_{p-1}$; but when among the assigned zeros there are ι pairs, and therefore $p-1-2\iota$ unpaired places, of the remaining zeros $p-1-2\iota$ are determined, and 2ι form pairs unspecified in position.

The case of a two-sheeted Riemann surface, or surface which can be rendered two-sheeted, is called the *hyperelliptic* case.

The surface depends on the $2p+2$ branch-points, which we can regard as arbitrary points of the z-plane. By a bilinear transformation, 3 of these can be made to take assigned positions. Therefore the number of independent co-ordinates of such a surface is $2p-1$.

An important property of the functions Φ is that the ratio of two such functions is an invariant with regard to birational transformations. This arises from the fact that in such transformations an integral of the first kind remains an integral of the first kind. See Weber, Abelsche Functionen vom Geschlecht 3; and Klein-Fricke, t. i., p. 546.

§ 282. *Integrals of the second kind.* It is a consequence of the existence-theorems that on the arbitrary Riemann surface T there is a function Z_s, which has the following properties:—

(1) It is continuous on T except at a single place s;

(2) It is infinite at s like $1/(z-c)$;*

(3) It has constant periods at the cuts A_κ, B_κ, whose real parts are assigned arbitrarily.

Such a function must be an Abelian integral. For dZ_s/dz is everywhere continuous except at s, and possibly certain branch-places. At the places at which it is discontinuous, it is algebraically

* If s be a branch-place at which r sheets hang together, Z_s is infinite like $1/(z-c)^{1/r}$.

infinite. It is therefore an algebraic function of the surface, and is of the form $R(w, z)$. The conditions (1), (2), (3) determine the integral to a constant term. For if Z_s, Z_s' be two functions with these conditions, $Z_s - Z_s'$ is nowhere infinite, and has at the cross-cuts purely imaginary periods. Accordingly $Z_s - Z_s'$ is a constant. We shall denote the period of Z_s at A_κ by η_κ, and that at B_κ by $\eta_{p+\kappa}$. The expression

$$Z_s^{(0)} = Z_s + k_0 + \sum_{\lambda=1}^{p} k_\lambda u_\lambda$$

satisfies the conditions (1), (2), (3); and since the p constants k_λ can be so chosen as to make the real parts of the periods of $Z_s^{(0)}$ take assigned real values, while c is at our disposal, $Z_s^{(0)}$ is the most general elementary integral of the second kind.

The normal integrals of the second kind. Let $-k_\lambda =$ the period of $Z_s^{(0)}$ at the cross-cut A_λ, $(\lambda = 1, 2, \cdots, p)$. This choice of the constants makes all the periods of $Z_s^{(0)}$ at the curves A_κ equal to 0. Denote the integral so defined by ζ_s, and call it *the normal integral of the second kind with an infinity at s.* To obtain the periods of ζ_s at the cross-cut B_λ, observe that the integral $\int_{T'} \zeta_s du_\lambda$ must be equal to $\int_{(s)} \zeta_s du_\lambda$ (§ 178).

Now
$$\int_{T'} \zeta_s du_\lambda = \sum_{\kappa=1}^{p} \int_{A_\kappa} \zeta_s du_\lambda + \sum_{\kappa=1}^{p} \int_{B_\kappa} \zeta_s du_\lambda$$
$$= 0 + \sum_{\kappa=1}^{p} \int_{B_\kappa} \eta_{p+\kappa} du_\lambda,$$

the integral being taken once along B_κ and in the positive direction,

$$= \sum_{\kappa=1}^{p} \eta_{p+\kappa} \int_{B_\kappa} du_\lambda.$$

But
$$\int_{B_\kappa} du_\lambda = 0 \text{ or } -1 \text{ according as } \kappa \neq \lambda \text{ or } = \lambda.$$

Hence
$$\int_{T'} \zeta_s du_\lambda = -\eta_{p+\kappa}.$$

On the other hand, we have in the neighbourhood of s (which we assume to be an ordinary place)

$$\zeta_s = 1/(z-c) + P(z-c),$$

and therefore,

$$\int_{(s)} \zeta_s du_\lambda = \int_{(s)} \zeta_s \frac{du_\lambda}{dz} dz = \left(\frac{du_\lambda}{dz}\right)_s \int_{(s)} \frac{dz}{z-c}$$
$$= 2\pi i \left(\frac{du_\lambda}{dz}\right)_s.$$

Hence $$\eta_{p+\kappa} = -2\pi i\left(\frac{du_\lambda}{dz}\right)_s.$$

Let du_λ/dz be denoted by Φ_λ. Then the scheme of periods for ζ_s is

$$\begin{array}{cccc|cccc} A_1 & A_2 & \cdots & A_p & B_1 & B_2 & \cdots & B_p \\ 0 & 0 & \cdots & 0 & -2\pi i\Phi_1(s) & -2\pi i\Phi_2(s) & \cdots & -2\pi i\Phi_p(s). \end{array}$$

§ 283. *Algebraic functions on a Riemann surface.* We are now in a position to prove that upon an arbitrary Riemann surface T there exist algebraic functions which have simple poles at μ distinct but otherwise arbitrary places, whenever $\mu > p$.

Let $s_1, s_2, \cdots, s_\mu$ be any μ places on T which are subject to the restrictions that no two coincide in the same sheet or lie in the same vertical; and assume that there is an algebraic function w of z, which is infinite like $k_\lambda/(z - c_\lambda)$ at the place s_λ, $(\lambda = 1, 2, \cdots, \mu)$. Construct the integrals ζ_{s_λ} and call them, for shortness, ζ_λ. The expression

$$w - \sum_{\lambda=1}^{\mu} k_\lambda \zeta_\lambda$$

is a function of z which is continuous on T. It is therefore an integral of the first kind. As w and the ζ_λ's experience no discontinuities at the cross-cuts A_κ, such a function, if existent, must reduce to a constant. Hence, the form of w is

$$w = k_0 + \sum_{\lambda=1}^{\mu} k_\lambda \zeta_\lambda.$$

It remains for us to find out when the last expression is an algebraic function on the surface. The necessary and sufficient condition is that it has no periods at the cross-cuts B_κ. This condition is expressed by the p equations

$$E_\kappa \equiv -2\pi i \sum_{\lambda=1}^{\mu} k_\lambda \Phi_\kappa(s_\lambda) = 0.$$

These equations may or may not be linearly independent. When the former is the case, p of the constants k_λ are determined as linear functions of the other $\mu - p$. Thus there remain, including k_0, $\mu - p + 1$ arbitrary constants in w. But if the equations be not all linearly independent, there will be a certain number τ of linearly independent relations

$$\sum_{\kappa=1}^{p} a_\kappa^{(h)} E_\kappa = 0, \quad (h = 1, 2, \cdots, \tau);$$

in this case the places $s_1, s_2, \cdots, s_\mu$ are said to be τ-ply *tied*, and there remain $\mu - p + \tau + 1$ arbitrary constants which enter linearly. The τ equations in E show that τ is the number of linearly independent combinations $\sum_{\kappa=1}^{p} b_\kappa \Phi_\kappa$ which vanish at the places s. This theorem was given by Roch (Crelle, t. lxiv.) as the extension of a result of Riemann's, and is known as *the Riemann-Roch theorem.*

It is clear that if $\mu > p$, there is a solution of the p equations which is distinct from $k_1 = k_2 = \cdots = k_\mu = 0$; that is, there is an algebraic function of z which is infinite only at the μ places or (if some of the k_λ's be zero) at some of these places. When w is infinite at μ_1 of the μ places, it is a μ_1-placed function of z and satisfies an algebraic equation $F(w^n, z^{\mu_1}) = 0$. In this way the existence of an algebraic function of the arbitrary Riemann surface T follows from the existence theorems for the Abelian integrals. It is especially worthy of note that in Riemann's theory the integrals come first and algebraic functions second.

From the Riemann-Roch theorem we can infer that the poles of algebraic functions are necessarily tied when $\mu < p + 1$. For when $\mu < p + 1$, τ is necessarily greater than 1, unless the function is itself a constant. A function whose poles are tied is called a *special* function, and thus any function for which $\mu < p + 1$ is a special function. Further, the theorem is the basis of our knowledge of what values of μ are permissible. Riemann proved in his memoir (Werke, p. 101) that in general $\mu \not< \frac{1}{2} p + 1$, but for particular surfaces smaller values of μ may occur; for instance, the hyper-elliptic case is characterized by the presence of two-placed functions. Weierstrass has proved that there is no function which is ∞^μ at a single *arbitrary* place when $\mu < p + 1$. At a finite number of special places there may be exception.

See Klein-Fricke, t. i., p. 554; Brill and Nöther, Math. Ann., t. vii.; Nöther, Crelle, t. xcii., p. 301.

§ 284. *The class-moduli.*

When $\mu > 2p - 2$, the function cannot be special, as there are only $2p - 2$ moveable zeros of a Φ (§ 281). Let the μ poles be moveable; then there are on the whole $\mu + \mu - p + 1$ independent constants, or $\infty^{2\mu - p + 1}$ functions with μ poles. Each of these maps the surface T into a μ-sheeted surface T_μ, so that the correspondence of T and T_μ is (1, 1). Let there be ∞^ρ transformations of T (and therefore of T_μ) into itself. Then the number of surfaces T_μ is $\infty^{2\mu - p + 1 - \rho}$. These are all in a (1, 1) correspondence. Let $\beta = 2p + 2\mu - 2$, then (§ 128) β is the number of branch-places (supposed simple). When the *branch-points* are assigned, there is only a finite number of possible surfaces, for the number of ways in which the sheets can hang together at these points is finite. Thus there are ∞^{β} μ-sheeted surfaces; and on each of them algebraic functions exist. These considerations show that the number of essentially distinct surfaces is $\infty^{\beta-2\mu+p-1+\rho}$ or $\infty^{3p-3+\rho}$.

Hence the number of characteristic constants of a surface is $3p-3+\rho$. These must all be equal for two surfaces which can be mapped on each other.

These characteristic constants are called by Riemann the *class-moduli*.

When $p=0$, we can take for the surface the z-plane itself, and the birational transformation is bilinear, say $w=(az+b)/(cz+d)$. There are therefore three quantities at our disposal when we seek to transform the plane into itself, and $\rho=3$. Hence there is no modulus.

It can be proved that when $p=1$, $\rho=1$; and that when $p>1$, $\rho=0$. Hence when $p=1$, there is one modulus, and this can be taken as the absolute invariant g_2^3/g_3^2 of the quartic which occurs when the surface is reduced to the standard form (§ 218). When $p>1$, there are $3p-3$ moduli. In the hyperelliptic case we found that there are $2p-1$ characteristic constants (§ 281). Now $3p-3>2p-1$ when $p>2$. Hence when $p>2$, there must be relations between the moduli when the surface can be brought to the two-sheeted form. In this paragraph we have followed Klein's tract Ueber Riemann's Theorie, to which (pp. 64–72) the reader is referred for more information. The proof is similar to the one given by Riemann (Werke, p. 113). Proofs drawn from the theory of curves are given in Clebsch, t. iii., ch. 1, and by Cayley, Math. Ann., t. viii. On the relation of the hyperelliptic case to the general one, when $p>2$, see Klein-Fricke, t. i., p. 571.

§ 285. *The integrals of the third kind.* The most general Abelian integral has logarithmic infinities which may be $c_1, c_2, \cdots, c_\kappa$. For simplicity, assume $c_1, c_2, \cdots, c_\kappa$ to be ordinary places of the surface. After cutting out these points we must render the surface simply connected by a further cut D which extends from c_1 to c_2, from c_2 to c_3, $\cdots$, and finally from c_κ to some point of the boundary, as for instance the point where A_1, B_1 meet. When this cut has been made, the surface may be named T″. Let a_λ be the residue at c_λ $(\lambda=1, 2, \cdots, \kappa)$. The part D_κ which extends from c_κ to the boundary furnishes no period, for the integral along a path on T″

which encloses all the points $c_1, c_2, \cdots, c_\kappa$, and leads from one side of D_κ to the other, is equal to $2\pi i \Sigma a_\lambda, = 0$. Hence there is no discontinuity at D_κ, which behaves in this respect like the cuts C_κ. That part $D_{\kappa-1}$ of D which leads from $c_{\kappa-1}$ to c_κ furnishes a period $2\pi i a_\kappa$, the part $D_{\kappa-2}$ a period $2\pi i(a_\kappa + a_{\kappa-1})$, and so on. This may be illustrated by $\int d \log R$, where R is a μ-placed function of the surface. If c_λ be the zeros, c_λ' the poles of R, where $\lambda = 1, 2, \cdots, \mu$, the cut D may be drawn from c_1' to c_1, from c_1 to c_2', from c_2' to c_2, and so on. Across a cut from c_λ' to c_λ there is a period $2\pi i$, and across a cut from c_λ to $c'_{\lambda+1}$ there is no period.

Let us assume that there exists on the arbitrary Riemann surface T a function which

(1) is continuous on T except at two arbitrary places c_1, c_2;

(2) is infinite at c_1 like $\log(z - c_1)$ and at c_2 like $-\log(z - c_2)$;

(3) has constant periods at the cross-cuts of T″, with arbitrary real parts.

By the same reasoning as before we can prove that the function in question is an Abelian integral.

Denote it by $Q_{c_1c_2}$. If there be a second function with the same properties and whose periods have the same real parts as those of Q, reasoning similar to that already used shows that the two functions differ merely by a constant. As an elementary integral of the third kind remains of the same kind after the addition of any integral of the first kind, the most general elementary integral of the third kind is

$$Q_{c_1c_2} + k_0 + \sum_{\lambda=1}^{p} k_\lambda u_\lambda;$$

for the k_λ's can be so chosen as to ensure that the real parts of the periods of this expression take values which are assigned arbitrarily. By a proper choice of the k_λ's we can make the periods at the cross-cuts A_κ all vanish. The resulting integral is *the normal integral of the third kind* with the prescribed logarithmic infinities, and may be denoted by $\Pi_{c_1c_2}$.

The remaining periods of $\Pi_{c_1c_2}$. In the general bilinear relation write

$$I = W, \quad I' = Q_{c_1c_2},$$

and denote the periods by the scheme

	A_1	$A_2 \cdots A_p$	B_1	$B_2 \cdots B_p$
W	ω_1	$\omega_2 \cdots \omega_p$	ω_{p+1}	$\omega_{p+2} \cdots \omega_{2p}$
$Q_{c_1c_2}$	ϵ_1	$\epsilon_2 \cdots \epsilon_p$	ϵ_{p+1}	$\epsilon_{p+2} \cdots \epsilon_{2p}$.

We have

$$\sum_{\kappa=1}^{p}\begin{vmatrix}\omega_\kappa & \omega_{p+\kappa}\\ \epsilon_\kappa & \epsilon_{p+\kappa}\end{vmatrix}=\int_{(c_1)+D+(c_2)} W dQ_{c_1c_2},$$

where the path, as in Fig. 77, describes both sides of a line D which runs on T′ from a place near c_1 to a place near c_2, and describes small circles (c_1), (c_2) round the places c_1, c_2. Now the integral $\int_D W dQ_{c_1c_2}$, taken along both sides of D, is zero, since W has no discontinuity across D. Near c_1,

$$Q_{c_1c_2}=\log(z-c_1)+P(z-c_1),$$

$$\frac{dQ_{c_1c_2}}{dz}=\frac{1}{z-c_1}+P(z-c_1),$$

$$W=W(c_1)+P_1(z-c_1);$$

and therefore $$\int_{(c_1)} W dQ_{c_1c_2}=+2\pi i W(c_1).$$

Similarly, $$\int_{(c_2)} W dQ_{c_1c_2}=-2\pi i W(c_2).$$

Therefore $$\sum_{\kappa=1}^{p}\begin{vmatrix}\omega_\kappa & \omega_{p+\kappa}\\ \epsilon_\kappa & \epsilon_{p+\kappa}\end{vmatrix}=2\pi i\{W(c_1)-W(c_2)\}.$$

It is now an easy matter to find the periods of the normal integral $\Pi_{c_1c_2}$ at the cross-cuts B_κ, $(\kappa=1, 2, \cdots, p)$. In the bilinear relation just established, let W and $Q_{c_1c_2}$ be replaced by the normal integrals u_λ and $\Pi_{c_1c_2}$. Then $\omega_\kappa=0$ when $\kappa\neq\lambda$, and $\omega_\lambda=1$; also $\epsilon_\kappa=0$, $(\kappa=1, 2, \cdots, p)$. The bilinear relation becomes

$$\begin{vmatrix}1 & \omega_{p+\lambda}\\ 0 & \epsilon_{p+\lambda}\end{vmatrix}=2\pi i\{u_\lambda(c_1)-u_\lambda(c_2)\},$$

or $$\epsilon_{p+\lambda}=2\pi i\int_{c_2}^{c_1} du_\lambda.$$

Hence the scheme of periods of $\Pi_{c_1c_2}$ is

D	A_1	$A_2 \cdots$	A_p	B_1	$B_2 \cdots$	B_p
$-2\pi i$	0	$0 \cdots$	0	$2\pi i\int_{c_2}^{c_1} du_1$	$2\pi i\int_{c_2}^{c_1} du_2 \cdots$	$2\pi i\int_{c_2}^{c_1} du_p$

where the path of integration from c_1 to c_2 lies on T′.

For two elementary integrals of the third kind $Q_{c_1c_2}$ and $Q_{c_1'c_2'}$, the bilinear relation takes the form

$$\sum_{\kappa=1}^{p}\begin{vmatrix}\epsilon_\kappa & \epsilon_{p+\kappa}\\ \epsilon'_\kappa & \epsilon'_{p+\kappa}\end{vmatrix}=\int Q_{c_1c_2}dQ_{c_1'c_2'},$$

the integral being taken along a closed curve which surrounds the infinities c_1, c_2, c_1', c_2'. By contraction the path of integration can be replaced by two double loops $(c_1)+D+(c_2)$, $(c_1')+D'+(c_2')$, and the integral becomes

$$\int_{D'}+\int_{(c_1')}+\int_{(c_2')}+\int_{D}+\int_{(c_1)}+\int_{(c_2)}Q_{c_1c_2}dQ_{c_1'c_2'}.$$

The first of these is zero since $Q^+_{c_1c_2}=Q^-_{c_1c_2}$; the second is $+2\pi i Q_{c_1c_2}(c_1')$; and the third is $-2\pi i Q_{c_1c_2}(c_2')$. The integrand in the remaining terms may be changed to $-Q_{c_1'c_2'}dQ_{c_1c_2}$, since for a closed path on T″ the product $Q_{c_1'c_2'}Q_{c_1c_2}$ is one-valued. The values of the last three integrals are now obtained in the same way as those of the former three, and we have

$$-2\pi i\int_{c_1'}^{c_2'}dQ_{c_1c_2}+2\pi i\int_{c_1}^{c_2}dQ_{c_1'c_2'}=\sum_{\kappa=1}^{p}\begin{vmatrix}\epsilon_\kappa & \epsilon_{p+\kappa}\\ \epsilon'_\kappa & \epsilon'_{p+\kappa}\end{vmatrix}.$$

When the relation

$$\sum_{\kappa=1}^{p}\begin{vmatrix}\epsilon_\kappa & \epsilon_{p+\kappa}\\ \epsilon'_\kappa & \epsilon'_{p+\kappa}\end{vmatrix}=0$$

is satisfied, the equation becomes

$$\int_{c_1'}^{c_2'}dQ_{c_1c_2}=\int_{c_1}^{c_2}dQ_{c_1'c_2'};$$

or, in words, the *interchange of limits and parameters is permissible.* In particular this is the case for normal integrals for which

$$\epsilon_1=\epsilon_2=\cdots=\epsilon_p=0,\text{ and }\epsilon_1'=\epsilon_2'=\cdots=\epsilon_p'=0.$$

In the case of $\Pi_{c_1c_2}$ the theorem may be stated in the form

$$\int_{c_1'}^{c_2'}d\Pi_{c_1c}=\int_{c_1}^{c_2}d\Pi_{c_1'c_2'},$$

or, in a notation often convenient,

$$\Pi_{c_1c_2}^{c_1'c_2'}=\Pi_{c_1'c_2'}^{c_1c_2}.$$

The difference of two elementary integrals of the third kind with the same infinities is an integral of the first kind. Hence in particular

$$Q^{zy}_{c_1c_2} = \Pi^{zy}_{c_1c_2} + \Sigma k_\lambda u_\lambda^{zy}.$$

The differential quotient of $Q_{c_1c_2}$ with regard to c_2 is infinite at c_2 like $1/(z-c_2)$, and is elsewhere finite. It is therefore an elementary integral of the second kind. When Q is replaced by Π, the integral of the second kind is normal.

§ 286. *The most general Abelian integral.* By § 276 the most general Abelian integral I derived from the basis-equation $F(w, z)=0$ has infinities which are logarithmic or algebraic, and has periods due to the cross-cuts on T and to the logarithmic infinities. It can be found *by addition* from

(1) Integrals $u_1, u_2, \cdots, u_p$, which are everywhere finite;

(2) Integrals $Q_{\xi\eta}$ with two logarithmic infinities;

(3) Integrals Z_s, dZ_s/dz_0, $d^2Z_s/dz_0^2, \cdots$, which are ∞ at s like $(z-z_0)^{-1}$, $(z-z_0)^{-2}$, $2!(z-z_0)^{-3}, \cdots$, where $z_0 = z(s)$.

For suppose that I is infinite like

$$a_1 \log(z-c_1),\ a_2 \log(z-c_2),\ \cdots,\ a_\kappa \log(z-c_\kappa),$$

where $a_1 + a_2 + \cdots + a_\kappa = 0$ (§ 276).

Then $I_1 = I - [a_1 Q_{c_1c} + a_2 Q_{c_2c} + \cdots + a_\kappa Q_{c_\kappa c}]$,

where c is any place whatsoever on the surface, is an integral which is nowhere logarithmically infinite. The integral I has been robbed of its logarithmic infinities at $c_1, c_2, \cdots, c_\kappa$, by the subtraction of the expression in brackets, and no new infinity has been introduced at c because $\sum_{\iota=1}^{\kappa} a_\iota = 0$. The algebraic infinities can be removed by the subtraction of integrals of the second kind and of their differential quotients. Suppose, for example, that I_1 is infinite at a place s at which $z = z_0$ like

$$(z-z_0)^{-\lambda}[b_0 + b_1(z-z_0) + b_2(z-z_0)^2 + \cdots + b_{\lambda-1}(z-z_0)^{\lambda-1}],$$

and that accents denote differentiations with regard to z_0; then

$$I_1 - \left[b_{\lambda-1}Z_s + b_{\lambda-2}Z'_s + \cdots + \frac{1}{(\lambda-1)!}b_0 Z_s^{(\lambda-1)}\right]$$

is an integral which has no algebraic infinity at s. By proceeding in this way we can remove from I every infinity; what is left must be an integral which is everywhere finite on T, and is therefore a linear combination of the u's, plus a constant.

§ 287. In Chapters IV. and VI. the general equation $F(w, z) = 0$ was simplified by various processes of transformation. To avoid difficulties connected with the presence of higher singularities, Clebsch and Gordan use throughout their treatise one of these simplified forms for their basis-equation; their results are therefore lacking in complete generality. Let $f(x_1, x_2, x_3) = 0$, or shortly $f = 0$, be a curve of order n derived from $F = 0$ by considering w, z as parameters of two pencils of rays (Chapter VI., § 189); and let $f = 0$ be a curve *which has no singularities higher than nodes.**

We shall assume (as in § 189) that the centres of the two pencils have no specialty of position with regard to the curve $f = 0$. Having indicated the consequences of the existence-theorems, we shall simplify our work by using a curve $f = 0$ with no singularities other than nodes, and an equation $F = 0$, all of whose branch-points are simple and situated in the finite part of the plane; F or f will be used as occasion demands.

Since $f = 0$ is homogeneous, we have by Euler's theorem

$$x_1 f_1 + x_2 f_2 + x_3 f_3 = nf = 0,$$

where f_κ denotes the partial derivative with regard to x_κ; and also we have

$$f_1 dx_1 + f_2 dx_2 + f_3 dx_3 = 0.$$

Hence

$$(x_2 dx_3 - x_3 dx_2)/f_1 = (x_3 dx_1 - x_1 dx_3)/f_2 = (x_1 dx_2 - x_2 dx_1)/f_3,$$

$$= \frac{|\, c_1 x_2 dx_3 \,|}{c_1 f_1 + c_2 f_2 + c_3 f_3},$$

where the numerator stands for the determinant

$$\begin{vmatrix} c_1 & c_2 & c_3 \\ x_1 & x_2 & x_3 \\ dx_1 & dx_2 & dx_3 \end{vmatrix},$$

c_1, c_2, c_3 being arbitrary constants which can be regarded as the co-ordinates of an arbitrary point. It is customary to replace the expression

$$\frac{|\, c_1 x_2 dx_3 \,|}{c_1 f_1 + c_2 f_2 + c_3 f_3}$$

by $d\omega$.†

---

* Clebsch and Gordan allow $f = 0$ to have cusps, but this is an unnecessary complication (see § 189). It may be added that the work of Chapter IV. shows that multiple points with separated tangents cause no difficulties as regards algebraic expansions.

† This differential expression $d\omega$ is called by Cayley the *displacement of the current point* on the curve $f = 0$. See Cayley's memoir on the Abelian and Theta Functions, Am. Jour., t. v., p. 138.

The general Abelian integral is now $\int \chi d\omega$, where χ is a homogeneous form in x_1, x_2, x_3, the degree of whose numerator must exceed that of its denominator by $n-3$, in order that the integral may not be altered when x_1, x_2, x_3 become kx_1, kx_2, kx_3.

Clebsch and Gordan assume that w and z in $F(w, z)=0$ are the parameters of two pencils

$$w \mid c_1b_2x_3 \mid + \mid a_1b_2x_3 \mid = 0, \quad z \mid b_1c_2x_3 \mid + \mid a_1c_2x_3 \mid = 0,$$

where the constants are the coefficients in the homographic transformation defined by the equations

$$\left.\begin{aligned} \lambda x_1 &= a_1t + b_1z + c_1w \\ \lambda x_2 &= a_2t + b_2z + c_2w \\ \lambda x_3 &= a_3t + b_3z + c_3w \end{aligned}\right\}$$

which connect x_1, x_2, x_3, with t, z, w, and which change $F(w, z, t)$ into $\lambda^n f(x_1, x_2, x_3)$. To pass to the non-homogeneous form of an Abelian integral, use the relations

$$\lambda^n f(x_1, x_2, x_3) = F(w, z, t),$$

$$\lambda^{n-3}(x_1, x_2, x_3)_{n-3} = (w, z, t)_{n-3},$$

$$c_1f_1 + c_2f_2 + c_3f_3 = \lambda\delta f/\delta w = \frac{1}{\lambda^{n-1}}\delta F/\delta w,$$

$$\lambda^2 \mid c_1x_2dx_3 \mid = \mid a_1b_2c_3 \mid (tdz - zdt),$$

then the Abelian integral

$$\int (x_1, x_2, x_3)_{n-3} \frac{\mid c_1x_2dx_3 \mid}{c_1f_1 + c_2f_2 + c_3f_3}$$

passes, when $t=1$, into

$$\int (w, z, 1)_{n-3} \frac{dz}{\delta F/\delta w}.$$

Here it should be observed that $F=0$ is of order n in the quantities w, z combined. This is Clebsch and Gordan's F, and not Riemann's.

The process used by Clebsch and Gordan for the classification of Abelian integrals is purely algebraic. (See their treatise, pp. 5 to 8.) The result at which they arrive is that every Abelian integral is expressible linearly in terms of the following kinds of integrals:—

(1) Integrals of rational functions of z;

(2) Integrals of the form $\int \frac{Hdz}{\delta F/\delta w}$, where H is an integral polynomial of order $n-3$ in w, z;

(3) Integrals of the form $\int \frac{Hdz}{(z-a)\delta F/\delta w}$, where H is an integral polynomial of order $n-2$ in w, z;

(4) Integrals which are derived from (3) by differentiations with regard to a.

They then proceed to an examination of the nature of the discontinuities of these integrals, and to a discussion of the integrals of the first, second, and third kinds.

§ 288. *The adjoint curves* ϕ_{n-3}. The integral $\int \frac{Hdz}{\delta F/\delta w}$, where H is a polynomial of order $n-3$ in w, z, can be finite for all values of z, w. To see that this is the case, observe that the integral can become infinite (if at all) only when $\delta F/\delta w = 0$; that is, at the branch-points and nodes of $F=0$.

(1) Near a branch-point (a, b)

$$w - b = P_1(z-a)^{1/2},$$

and the integrand is

$$\frac{Hdz}{P_1(z-a)^{1/2}}, = \frac{Hdz}{(z-a)^{1/2}} P_0(z-a)^{1/2};$$

hence the integral is $P_\kappa(z-a)^{1/2}$, where $\kappa \geqq 1$. This proves that the integral is finite at a branch-point.

(2) Near a node (a, b),

$$w - b = P_1(z-a),$$

and the integrand is

$$\frac{Hdz \cdot P_0(z-a)}{z-a};$$

hence the integral is infinite, like $\log(z-a)$, except when $H=0$ at the node.

It appears that the integral $\int \frac{Hdz}{\delta F/\delta w}$ cannot be everywhere finite unless $H=0$ passes through all the δ double points. When this happens, the number of arbitrary constants in H is

$$\tfrac{1}{2}(n-1)(n-2) - \delta, = p.$$

That is, the number of linearly independent integrals which are finite for all values of z is equal to the deficiency of the basis-curve. These p integrals are Abelian integrals of the first kind. As H is of order $n-3$, n is at least equal to 3. If $p=0$, there is no curve $H=0$ which passes through all the double points. But in this case w and z can be expressed as rational functions of a parameter t; and every Abelian integral becomes the integral of a rational function of t.

In the homogeneous form the integral becomes

$$\int \phi_{n-3} d\omega,$$

where $\phi_{n-3}=0$ is an adjoint curve of order $n-3$, *i.e.* a curve of order $n-3$ which passes through all the double points of $f=0$. If $\phi_{n-3}=0$ were not an adjoint curve, the nodes would make f_1, f_2, f_3 vanish, but not ϕ_{n-3}; they would therefore be infinities of the integral. The only other points of $f=0$ at which the integral could possibly become infinite are the points at which tangents from (c_1, c_2, c_3) touch $f=0$. These points are points of intersections of $f=0$ with the first polar of c. The integral must be finite at these points, for its value does not depend on c_1, c_2, c_3, and consequently the position of its infinities cannot be dependent in any way on the position of the point c.

The curve $\phi_{n-3}=0$ is to pass through the δ double points of $f=0$; the number of the remaining intersections with $f=0$ is $n(n-3)-2\delta$, or $2p-2$; *i.e.* the curve $\phi=0$ has $2p-2$ moveable points of intersection with $f=0$. Since there are p arbitrary constants in a curve $\phi_{n-3}=0$, it can be made to pass through $p-1$ arbitrary points; in general $p-1$ of the intersections of a curve $\phi=0$ with $f=0$ determine the remaining points of intersection.

A function $H/\frac{\delta F}{\delta w}$ whose integral is always finite is evidently identical with a function Φ of § 281. And the $2p-2$ moveable zeros of a Φ are, on the curve, the intersections of an adjoint curve ϕ_{n-3} with $f=0$. More generally the points at which Φ/Φ' takes a given value λ are, on the curve, the moveable points at which the adjoint curve $\phi-\lambda\phi'=0$ cuts $f=0$. There are $2p-2$ such points unless $\phi=0$ and $\phi'=0$ intersect on the curve.

Let any rational function $R(w, z)$ become, in homogeneous co-ordinates, $M(x_1, x_2, x_3)/N(x_1, x_2, x_3)$. The points at which the function takes the value λ are the points in which the curve $M-\lambda N=0$ cuts the curve $f=0$, exclusive of the common intersections of $M=0$, $N=0$, $f=0$. At such a point the value of M/N is easily seen to be λ', if $M-\lambda' N=0$ be that curve of the pencil which touches $f=0$ at the point (§ 217).

Of the moveable points how many are arbitrary? The question is answered by the theorem of Riemann and Roch. Namely, if there be μ moveable points, and if through them τ independent adjoint curves ϕ_{n-3} can be drawn, there are $\mu-p+\tau+1$ arbitrary constants in the expression of a function which is nowhere infinite except at these μ points. If M_κ/N be such a function, the μ points are those intersections of $N=0$ with $f=0$ which are not double points or basis-points of the pencil; and the most general function is

$$(a_0N+a_1M_1+a_2M_2+\cdots+a_{\mu-p+\tau}M_{\mu-p+\tau})/N,$$

where the a's are arbitrary constants.

For example, let $p=4$, and let the curve be a quintic with two nodes; the curves ϕ are conics through the nodes. Let $\mu=2$. Through the two points and the two nodes we can draw ∞^1 conics, of which two are linearly independent. Therefore $\tau=2$; $\mu-p+\tau+1=1$; and the function reduces to a constant, which need not be ∞ at the two points. This does not stultify the theorem, which refers to functions which are not infinite at other than the μ points without asserting that they *are* infinite at the μ points. Next let $\mu=3$. The 3 points and the 2 nodes determine a conic; therefore $\tau=1$, $\mu-p+\tau+1=1$. There is no algebraic function which is ∞ at the 3 points. But if the 3 points be in a line L with a node, the conic becomes this line and any line L' through the other node. Now $\tau=2$, $\mu-p+\tau+1=2$; and there exists a function of the form $(a_0N+a_1M_1)/N$, which *is* ∞ at the three points, and there only. We can take as N the line L itself, and as M_1 any other line through the same node. Lastly, let $\mu=4$. Here $\tau=0$, $\mu-p+\tau+1=2$, and the function required is still of the form $(a_0N+a_1M_1)/N$. Let N be an adjoint cubic through the 4 points; the cubic cuts $f=0$ again at 7 points, and we take for M_1 an adjoint cubic through the 7 points.

The reader is referred for an exhaustive account of Riemann and Roch's theorem to Clebsch, t. iii., Chapter 1, § 2.

The system of points on the curve or on the Riemann surface, at which an algebraic function takes a given value, is said to be coresidual * or equivalent with the system of points at which the function takes any other given value (§ 217).

Before leaving the subject of the adjoint curves ϕ_{n-3}, it should be mentioned that they are employed by Clebsch (t. iii., ch. 1, § 3) in the problem of selecting, from the class of curves which arise from a given curve by birational transformations, that of the lowest order. The result is that when $p>2$ a curve of deficiency p can be transformed into a *normal* curve of order $p-\pi+2$, where $p=3\pi$, $3\pi+1$, or $3\pi+2$. For $p=0$, 1, 2 the normal curves are respectively a

* The word coresidual has the same meaning as in the theory of residuation, explained in Salmon's Higher Plane Curves, Chapter V. The theorem of Riemann and Roch is fundamental in this theory.

line, a non-singular cubic, and a nodal quartic. [Clebsch and Gordan, § 3; Salmon's Higher Plane Curves, § 365; Brill and Nöther, Math. Ann., t. vii., p. 287; Poincaré, Acta Math., t. vii.]

§ 289. *Homogeneous form of Q.** Consider the integral

$$\int \frac{\psi_{n-2}}{a_1x_1 + a_2x_2 + a_3x_3}\, d\omega.$$

Its possible infinities are those points of $f = 0$ at which

(1) f_1, f_2, f_3 vanish simultaneously. These are double points of $f=0$, and make the integral infinite unless $\psi=0$ passes through them;

(2) $c_1f_1 + c_2f_2 + c_3f_3 = 0$, without f_1, f_2, f_3 being all separately equal to 0. These are branch-points and do not contribute infinities to the integral;

(3) $a_1x_1 + a_2x_2 + a_3x_3 = 0$. The number of intersections of this straight line with $f = 0$ is n; at each of these n points the integral will be infinite unless $\psi = 0$ happens to pass through the point.

To get rid of the infinities due to double points, make $\psi_{n-2} = 0$ pass through the double points. The curve $\psi_{n-2} = 0$ cannot be made to pass through more than $n - 2$ points of intersection of

$$a_1x_1 + a_2x_2 + a_3x_3 = 0$$

with $f = 0$, except when $\psi_{n-2} = 0$ degenerates into the straight line and an adjoint curve $\phi_{n-3}=0$ of order $n-3$. In this exceptional case

$$\int \frac{\psi_{n-2}}{a_1x_1 + a_2x_2 + a_3x_3}\, d\omega = \int \phi_{n-3}\, d\omega;$$

that is, the integral is of the first kind. Ruling out this case, the number of infinities of $\int \frac{\psi_{n-2}}{a_1x_1 + a_2x_2 + a_3x_3}\, d\omega$ is two, and these are due to those two points of intersections of $a_1x_1 + a_2x_2 + a_3x_3 = 0$ with $f = 0$, through which $\psi_{n-2} = 0$ does not pass.

To determine shortly the nature of these two infinities, we have in the non-homogeneous form

$$\int \frac{\psi_{n-2}}{aw + \beta z + \gamma} \frac{dz}{\delta F/\delta w}.$$

Let z_0, w_0 be one of the infinities; then

$$aw_0 + \beta z_0 + \gamma = 0,$$

and $$w - w_0 = P_1(z - z_0),$$

the point being assumed to be an ordinary point.

* Clebsch and Gordan, p. 17.

Hence $\qquad \alpha w + \beta z + \gamma = \alpha(w - w_0) + \beta(z - z_0) = P_1(z - z_0),$

while ψ_{n-2} and $\delta F/\delta w$ are both finite.

Therefore the integral is of the form

$$\int \frac{dz}{P_1(z - z_0)}$$

and is ∞ like $k \log(z - z_0)$, where k is a constant.

Hence the integral is an elementary integral of the third kind, multiplied by a constant factor.

§ 290. *The adjoint curves ψ_{n-2}.* The elementary integral of the second kind can be derived from the integral of the third kind by letting the infinities coincide, so that the line $\alpha_1 x_1 + \alpha_2 x_2 + \alpha_3 x_3$ is now the tangent of the curve f at the one infinity. The nature of the infinity can be seen from the fact that now dw/dz at the place z_0, w_0 is the same for the line $\alpha w + \beta z + \gamma = 0$ and the curve $f = 0$; whence it follows that

$$\alpha(w - w_0) + \beta(z - z_0) = P_2(z - z_0);$$

and the integral is, near the point, of the form

$$\int \frac{dz}{P_2(z - z_0)};$$

that is, it is ∞ like $\dfrac{k}{z - z_0}$.

The adjoint curve ψ_{n-2} which appears in the numerator in Clebsch and Gordan's form of an integral of the second kind meets the curve at the nodes and at the points where the tangent at the infinity meets the curve again.* It therefore contains

$$(n-2)(n+1)/2 + 1 - (n-2) - \delta, \text{ or } p + 1$$

arbitrary constants. It meets the curve f again at

$$n(n-2) - (n-2) - 2\delta, \text{ or } 2p$$

moveable points.

The p ratios of constants of the adjoint curve ψ_{n-2} are sufficient to enable the $2p$ moveable points in which it meets f to coincide in pairs. In point of fact the problem admits of 2^{2p} solutions. The curve ψ has then p-tuple contact with f, and is spoken of as a

* These points will be meant when mention is made of the residuals of the infinity.

contact curve. For the complete discussion of this problem, which is of great importance in the sequel (see § 299), we refer to Clebsch and Gordan, and to Clebsch's Géométrie. The points of contact will be denoted by $a_1, a_2, \cdots, a_p$.

§ 291. *Abel's theorem for integrals of the first kind.* Let $R(w, z)$ be a μ-placed function of the surface T whose zeros are x_s and poles y_s $(s=1, 2, \cdots, \mu)$. Let us apply the extension of Cauchy's theorem to the integral

$$\int_{T'} WdR/R,$$

where W is any integral of the first kind. We have

$$\int_{T'} WdR/R = \sum_{s=1}^{\mu}\int_{(x_s)} WdR/R + \sum_{s=1}^{\mu}\int_{(y_s)} WdR/R.$$

Near x_s,

$$R = P_1(z-z_s),$$

$$dR/R = dz/(z-z_s) + dzP_0(z-z_s),$$

and therefore

$$\int_{(x_s)} WdR/R = 2\pi i\, W(x_s),$$

where $W(x_s)$ is an abbreviated expression for an integral $\int_{z_0}^{z} Wdz$ whose path of integration lies wholly on T′. Similarly,

$$\int_{(y_s)} WdR/R = -2\pi i\, W(y_s).$$

On the other hand, precisely as in § 279,

$$\int_{A_\kappa} WdR/R = \omega_\kappa \int_{A_\kappa} dR/R = 2\pi i\nu_\kappa\omega_\kappa,$$

where $2\pi i\nu_\kappa$ is the increment of $\log R$ due to the description of A_κ. Here ν_κ must be an integer. In the same way it can be shown that

$$\int_{B_\kappa} WdR/R = 2\pi i\nu_{\kappa+p}\omega_{\kappa+p},$$

where $\nu_{\kappa+p}$ is some integer. Collecting these facts, we have

$$\sum_{s=1}^{\mu} \{W(x_s) - W(y_s)\} = \sum_{\kappa=1}^{p} (\nu_\kappa\omega_\kappa + \nu_{\kappa+p}\omega_{\kappa+p}).$$

The places x, y have been defined as zeros and poles of a function $R(w, z)$; it might therefore appear, at first sight, that a distinction is to be made between 0, ∞, and ordinary constant values k, k'. That no such distinction is inherent in the nature of the case will be apparent from the consideration that we have, if

$$R=(R'-k)/(R'-k'),$$

$$R'=k,\ k' \text{ when } R=0,\ \infty.$$

The places x_s, y_s are characterized adequately by affirming that they are the points of T at which the μ-placed function R' takes given values k, k'. Two such systems of places have been called equivalent systems (§ 217). As R' changes continuously from k' to k, we may regard the system y_s as changing continuously into the system x_s, and may say that the μ paths are *equivalent*. The paths are supposed subject to the restrictions that no path passes through a branch-place, and that no two paths intersect. Suppose that in the above equation the system x_s shifts into the equivalent system y_s+dy_s. On the supposition that none of the places concerned lie near a cross-cut, the equation holds equally for the new system; and therefore, by subtraction,

$$\sum_{s=1}^{\mu}\{W(y_s+dy_s)-W(y_s)\}=0,$$

or

$$\sum_{s=1}^{\mu} d\,W(y_s)=0.$$

Integrating from the system y_s to the system x_s, we have

$$\sum_{s=1}^{\mu}\int_{y_s}^{x_s} d\,W(x_s)=\text{a constant},$$

where the paths of integration are equivalent paths. So long as no path crosses a cross-cut, the constant is zero; but each time that the path crosses a cross-cut, the constant must include the period at that cross-cut. Generally the constant may be written

$$\sum_{\kappa=1}^{p}(\nu_\kappa\omega_\kappa+\nu_{\kappa+p}\omega_{\kappa+p}),$$

where ν_κ, $\nu_{\kappa+p}$ are the numbers of positive crossings of A_κ, B_κ, less the number of negative crossings. Comparing this result with the equation of § 285, we infer that the value of $\log R$ taken round a cross-cut is to be obtained by considering the points at which that cross-cut is crossed by a system of equivalent paths which lead from the poles to the zeros of R; in fact, $\frac{1}{2\pi i}\int d\log R$ is the excess of the number of these positive crossings over the number of negative crossings. [Neumann, ch. xi.]

§ 292. *Abel's theorem for normal integrals of the third kind.* Let $\Pi_{\xi\eta}$ be a normal integral of the third kind. Cut out from the surface a dumb-bell-shaped double loop formed by two small circles (ξ), (η) round ξ and η and by a connecting line D whose direction is from ξ to η; to reduce the resulting surface to simple connexion a second line D' must be drawn from c_2 to the boundary of T'. Let R have the same zeros and infinities as before and let these 2μ places be cut out from the surface by 2μ small circles and by connecting lines $L_{12}, L_{23}, \cdots, L_{2\mu-1,\,2\mu}$; the last circle can be connected with the boundary of T' by a line D''. On the simply connected surface T'' which results from these various cuts, D' and D'' play a part similar to that played by the C's; they may therefore be neglected.* By applying the extension of Cauchy's theorem to the integral

$$\int_{\mathrm{T}''}\Pi_{\xi\eta}dR/R,$$

we find that

$$\int_{\mathrm{T}''}\Pi_{\xi\eta}dR/R=\sum_{s=1}^{\mu}\int_{(x_s)}\Pi_{\xi\eta}dR/R+\sum_{s=1}\int_{(y_s)}\Pi_{\xi\eta}dR/R+\int_{(\xi)+D+(\eta)}\Pi_{\xi\eta}dR/R.$$

The last integral can be transformed into

$$-\int_{(\xi)+D+(\eta)}\log R\cdot d\Pi_{\xi\eta},$$

because $\Pi_{\xi\eta}\log R$ is one-valued on the surface T''; the value of the integral is $-2\pi i\{\log R(\xi)-\log R(\eta)\}$, since $d\Pi^{+}=d\Pi^{-}$ and $dR^{+}=dR^{-}$ along the cut D. The reduction of the other integrals can be effected as in the preceding paragraph; and there results

$$\sum_{s=1}^{\mu}\{\Pi_{\xi\eta}(y_s)-\Pi_{\xi\eta}(x_s)\}+\log\frac{R(\xi)}{R(\eta)}=\sum_{\kappa=1}^{p}(\nu_\kappa\epsilon_\kappa+\nu_{\kappa+p}\epsilon_{\kappa+p}),$$

where ν_κ, $\nu_{\kappa+p}$ have the same meanings as before and ϵ_κ, $\epsilon_{\kappa+p}$ are the periods of $Q_{\xi\eta}$ at A_κ, B_κ. If we substitute for R the function $\dfrac{R-k}{R-k'}$ which takes the value k' at the points y_λ and the value k at the points x_λ, the formula becomes

$$\sum_{s=1}^{\mu}\{\Pi_{\xi\eta}(x_s)-\Pi_{\xi\eta}(y_s)\}=\log\frac{R(\eta)-k'}{R(\eta)-k}\cdot\frac{R(\xi)-k}{R(\xi)-k'}-\sum_{\kappa=1}^{p}(\nu_\kappa\epsilon_\kappa+\nu_{\kappa+p}\epsilon_{\kappa+p}).$$

* The lines D, L are supposed not to cross themselves, each other, or the cross-cuts A, B, C It must be borne in mind that the cross-cuts A, B, C admit of deformation.

§ 293. These proofs of Abel's theorems for integrals of the first and third kinds are taken from Neumann's Abel'sche Integrale, Chapter XI. We shall next give Riemann's proof.

Let $x_1, x_2, \cdots, x_\mu, y_1, y_2, \cdots, y_\mu$ have the same meanings as before; namely, let them be places at which a rational function R takes values $0, \infty$. By means of the μ-placed function R, map the surface $\mathbf{T}$ which is spread in n sheets over the z-plane, upon a surface $\mathbf{T}_\mu$ which is spread in μ sheets over the R-plane. The μ places (z, R) on $\mathbf{T}$ at which R takes an assigned value map into μ places x *in the same vertical line* on $\mathbf{T}_\mu$. We have to connect the μ places x with the μ places y in a determinate manner. To do this, draw in the surface $\mathbf{T}_\mu$ a path for which the initial value of R is 0 and the final value ∞, this path being supposed not to pass through any branch-place of $\mathbf{T}_\mu$. Call the initial place x_1 and the final place y_1. Through this path draw a vertical section of the surface $\mathbf{T}_\mu$ and thus cut out $\mu - 1$ other paths; let the initial and final places for these paths be $(x_2, y_2), (x_3, y_3), \cdots, (x_\mu, y_\mu)$. Returning to the surface $\mathbf{T}$ on the z-plane, these μ paths map into paths on $\mathbf{T}$ which connect the two systems of places $x_1, x_2, \cdots, x_\mu$ and $y_1, y_2, \cdots, y_\mu$.

Suppose that I is an Abelian integral on $\mathbf{T}$, we have to form the sum

$$\int_{y_1}^{x_1} dI + \int_{y_2}^{x_2} dI + \cdots + \int_{y_\mu}^{x_\mu} dI,$$

where the paths from y_s to x_s, $(s = 1, 2, \cdots, \mu)$, are those just described. Change to the new variable R; hence we have to find the value of

$$\sum_{s=1}^{\mu} \int_{R_0}^{R_1} \left(\frac{dI}{dR}\right)_s dR = \int_{R_0}^{R_1} \left\{ \sum_{s=1}^{\mu} \frac{dI}{dR} \right\} dR,$$

where $\left(\frac{dI}{dR}\right)_s$ is the value of $\frac{dI}{dR}$ on that path which starts from the place x_s in $\mathbf{T}_\mu$. The sum of the μ values obtained by making s pass from 1 to μ is equal to the sum of the μ values $\frac{dI}{dR}$ which lie in a vertical line on $\mathbf{T}_\mu$, and is therefore a rational function of R. Hence

$$\sum_{s=1}^{\mu} \int_{y_s}^{x_s} dI = \int_{R_0}^{1} \{\text{a rational function of } R\} dR$$

$$= \text{a combination of logarithmic and algebraic terms.}$$

For supposing that the indefinite integral on the right-hand side becomes

$$r(R) + \log r'(R) + \text{a constant},$$

where r and r' denote rational functions, the above equation becomes

$$\sum_{s=1}^{\mu} \int_{s}^{x_s} dI = r(R_1) - r(R_0) + \log \frac{r'(R_1)}{r'(R_0)}.$$

This is *the general form of Abel's theorem.* It is evident that if we keep the limits fixed but vary the paths from the lower to the upper limits, periods may be added to the right-hand side of the equation; again, if $x_1, x_2, \cdots, x_\mu$ be associated with $y_1, y_2, \cdots, y_\mu$ in a new order, periods may be added.

Suppose that I is an integral W of the first kind; when lower limits y_s are assigned, it is impossible to find an equivalent system x_s which will make the sum of the μ integrals infinite. Therefore the expression in R_1 on the right-hand side cannot become infinite for any value of R_1. This requires that $r(R)$, $r'(R)$ shall reduce to constants. In the simplified equation

$$\sum_{s=1}^{\mu} \int_{y_s}^{x_s} d\,W = \text{a constant},$$

the constant must be zero, for when the equivalent systems tend to coincidence, the paths which connect x_s, y_s vanish. This proof was given originally by Riemann in his memoir on the Abelian functions as already pointed out, but the mode of presentation is that of Klein-Fricke (Modulfunctionen, t. ii., pp. 481 to 484). We have not given a separate statement of Abel's theorem for normal integrals of the second kind; remembering that ζ_ξ is derived from $\Pi_{\xi\eta}$ by diffentiation with respect to ξ, the theorem for the normal integrals of the second kind can be derived at once from the theorem for normal integrals of the third kind.

The following remarks may assist the reader in arriving at a clear understanding of what is involved in these statements of Abel's theorem; for simplicity we shall take a normal integral u_λ of the first kind and consider the theorem

$$\sum_{s=1}^{\mu} \int_{y_s}^{x_s} du_\lambda = 0 \quad \cdot \quad \cdot \quad \cdot \quad \cdot \quad \cdot \quad \cdot \quad (A).$$

(1) Attention must be paid to the number μ, to the composition of the integrand, and to the situation of the places on T, or points on $f=0$, which are associated with the limits.

(2) By giving to λ the values $1, 2, \cdots, p$ we can derive p equations; the equation for an integral W is derivable from these by linear composition.

(3) In order that the right-hand side of (A) may be 0, the systems of places (x), (y) must be equivalent; also attention must be paid to the paths.

(4) For arbitrary paths between the limits the sign of equality in (A) must be replaced by a sign of congruence.

(5) The number μ is determined by a μ-placed function $R(w, z)$ and not by the integrand.

On the subject of Abel's theorem see Abel's Works, p. 145, Mémoire sur une propriété générale d'une classe étendue de fonctions transcendantes; Riemann, § 14 of the memoir on Abelian Functions; Clebsch and Gordan, Abelsche Functionen, Chapter II., also pp. 124 to 127; Cayley, a memoir on the Abelian and Theta Functions, Amer. Journ. of Math., tt. v., vii.; Rowe, a memoir on Abel's Theorem, Phil. Trans. R. Soc., 1881.

§ 294. *Properties of special p-tuple theta-functions.* Let

$$\theta(v_1, v_2, \cdots, v_p)$$

be the theta-function

$$\sum_{n_1=-\infty}^{n_1=+\infty}\sum\cdots\sum_{n_p=-\infty}^{n_p=\infty} \exp \pi i[(*)(n_1, n_2, \cdots, n_p)^2+2(n_1v_1+n_2v_2+\cdots+n_pv_p)],$$

with the mark $\begin{bmatrix}0 & 0\cdots 0\\ 0 & 0\cdots 0\end{bmatrix}$ and the moduli $\tau_{11}, \tau_{12}, \cdots, \tau_{pp}$. Write

$$v_\lambda = u_\lambda - e_\lambda, \quad \lambda = 1, 2, \cdots, p,$$

where u_λ is a normal integral (§ 280), and e_λ is constant; and let the coefficients of $(*)(n_1, n_2, \cdots, n_p)^2$ be the τ's of the u's. The functions are *special* theta-functions, since the number of moduli τ is greater than the number of class-moduli of the surface T. The τ's are admissible as moduli of the theta-function, since they satisfy the conditions for convergence of the theta-series (§ 280). The theorems which will be proved in the next few paragraphs will be needed in the discussion of the Inversion Problem.

Theorem I. $\theta(v_1, v_2, \cdots, v_p)$ is a one-valued function which admits p zeros on T′.

The integral $\dfrac{1}{2\pi i}\displaystyle\int d \log\theta$, taken over the boundary of T′ is equal to the number of zeros of θ upon T′. The value is found by considering each line A, B, C in turn, and by noticing that each of these lines is traversed twice in opposite directions. Along a cross-cut A_κ, $u_\lambda^+ - u_\lambda^- = 0$ or 1, according as κ is or is not equal to λ, and along a cross-cut B_κ we have $u_\lambda^+ - u_\lambda^- = \tau_{\kappa\lambda}$. Hence, along A_κ, $\theta(v_\lambda^+) = \theta(v_\lambda^-)$, and, along B_κ, $\theta(v_\lambda^+) = \theta(v_\lambda^- + \tau_{\kappa\lambda})$; or more shortly,

$$\theta_+ = \theta_- \text{ along } A_\kappa;\ \theta_+ = \exp - 2\pi i(v_\kappa^- + \tau_{\kappa\kappa}/2)\cdot\theta_- \text{ along } B_\kappa.\text{*}$$

Hence $$d\log\theta_+ = d\log\theta_- \text{ along } A_\kappa,$$

$$d\log\theta_+ = -2\pi i du_\kappa + d\log\theta_- \text{ along } B_\kappa.$$

Also $$d\log\theta_+ = d\log\theta_- \text{ along } C_\kappa.$$

In the second formula du_κ means, indifferently, du_κ^+, or du_κ^-. These relations show that the lines A, C contribute nothing to the integral $\frac{1}{2\pi i}\int d\log\theta$; there remain the lines B. The part of the integral which is contributed by B_κ is

$$\frac{1}{2\pi i}\int_{B_\kappa} -2\pi i du_\kappa;$$

this expression = the period across A_κ, namely 1.

Thus on the whole

$$\frac{1}{2\pi i}\int d\log\theta = p.$$

Now θ cannot become infinite on T′; therefore $\frac{1}{2\pi i}\int d\log\theta =$ the number of zeros of θ in T′, and we have the theorem that θ has p zeros on T′.

Let the p places on T′ at which $\theta(u_1 - e_1, u_2 - e_2, \cdots, u_p - e_p)$ vanishes be named $x_1, x_2, \cdots, x_p$. In what follows the determination of suitable lower limits for the integrals which make their appearance is of great importance. Clebsch and Gordan allow different lower limits for the same integral; for instance,

$$\int_{c_1}^{x_1} du_\kappa,\ \int_{c_2}^{x_2} du_\kappa,\ \cdots,\ \int_{c_p}^{x_p} du_\kappa.$$

Riemann uses, for a definite u_κ, one and the same lower limit whatever be the upper limit; while, for different values of κ, the lower limits may be different. This difference of notation should be borne in mind in comparing the results of Riemann with those

* Here use is made of a theorem for the addition of periods to the arguments of a p-tuple theta-function. This theorem is stated in formula (5), § 232, when r is allowed to mean any of the integers 1, 2, ..., p; and the proof is precisely analogous to the one there given.

of Clebsch and Gordan. When the lower limit is not specified for $\int^x du_\kappa$, where κ has a definite value, it is to be understood that this lower limit is the same however x may vary.

§ 295. *Theorem II.* The p zeros $x_1, x_2, \cdots, x_p$ satisfy the relations

$$\sum_{\kappa=1}^{p} u_\lambda(x_\kappa) - e_\lambda = k_\lambda + h_\lambda + g_1\tau_{1\lambda} + g_2\tau_{2\lambda} + \cdots + g_p\tau_{\lambda p}(\lambda = 1, 2, \cdots, p),$$

where the quantities g, h are integers and where k_λ stands for a quantity independent of the g's, h's, x's, and e's.

When the lower limits of the p normal integrals of the first kind are left undetermined, these p quantities are determined on T′ save as to additive constants. By making a suitable choice of these lower limits it is possible to simplify the work. Let definite lower limits be assigned to $u_1, u_2, \cdots, u_p$, and, in accordance with Riemann's notation, let $\int_\mu^x du_\lambda$ have the *same* lower limit, for a definite λ, whatever be the upper limit x.

The theorem which has been enunciated above can be proved by means of the integral

$$\frac{1}{2\pi i}\int u_\lambda d\log\theta \quad (\lambda = 1, 2, \cdots, p),$$

taken over the boundary of T′. Knowing that u_λ is everywhere finite on T′ and that $\log\theta$ is finite except at the p places $x_1, x_2, \cdots, x_p$ at which it is simply infinite with residues 1, we derive at once the equation

$$\frac{1}{2\pi i}\int u_\lambda d\log\theta = \text{the sum of the residues of the integrand,}$$
$$= u_\lambda(x_1) + u_\lambda(x_2) + \cdots + u_\lambda(x_p).$$

The integral on the left-hand side of the equation can be treated as the sum of $2p$ integrals over the $2p$ lines A_κ, B_κ, each line being described twice in opposite directions. Along A_κ, B_κ we have, as before, $d\log(\theta_+/\theta_-) = 0$, $u_\lambda^+ - u_\lambda^- = 0$ or 1, according as $\lambda \neq \kappa$ or $= \kappa$; $d\log(\theta_+/\theta_-) = -2\pi i du_\lambda$, $u_\lambda^+ = u_\lambda^- + \tau_{\kappa\lambda}$.

A line A_κ contributes the part $\frac{1}{2\pi i}\int(u_\lambda^+ - u_\lambda^-)d\log\theta_+$; the value of this part is 0 when $\lambda \neq \kappa$, and $\frac{1}{2\pi i}\int_{A_\kappa} d\log\theta$ when $\lambda = \kappa$. But $\int_{A_\lambda} d\log\theta = \log(\theta_+/\theta_-)$, where the signs refer to B_λ, and we know from the values of θ along B_λ that

$$\log(\theta_+/\theta_-) = 2\pi i\{u_\lambda(c_1)_- - e_\lambda + \tau_{\lambda\lambda}/2 + h_\lambda\},$$

$u_\lambda(c_1)_-$ being the value of u_λ at the point c_1 of T′ (see Fig. 74, p. 254). Thus the p lines A contribute a part

$$e_\lambda - u_\lambda(c_1)_- + h_\lambda - \tau_{\lambda\lambda}/2.$$

The sum of the p integrals over the lines B_κ

$$= \frac{1}{2\pi i}\sum_{\kappa=1}^{p}\int_{B_\kappa}(u_\lambda^+ d\log\theta_+ - u_\lambda^- d\log\theta_-)$$

$$= \frac{1}{2\pi i}\sum_{\kappa=1}^{p}\int_{B_\kappa}[(u_\lambda^- + \tau_{\kappa\lambda})(d\log\theta_- + 2\pi i du_\kappa) - u_\lambda^- d\log\theta_-]$$

$$= \frac{1}{2\pi i}\sum_{\kappa=1}^{p}\int_{B_\kappa}[\tau_{\kappa\lambda} d\log\theta_- + 2\pi i\tau_{\kappa\lambda} du_\kappa + 2\pi i u_\lambda^- du_\kappa].$$

To find the value of this sum, consider separately the integrals

$$\int_{B_\kappa} d\log\theta_-, \int_{B_\kappa} du_\kappa, \int_{B_\kappa} u_\lambda^- du_\kappa.$$

The first of these is equal to $\log(\theta_-/\theta_+)$ along A_κ; that is, its value is $g_\kappa \cdot 2\pi i$. The second integral is equal to $u_\kappa^- - u_\kappa^+$ along A_κ; that is, its value is -1. The third integral may be written

$$\tfrac{1}{2}\int_{B_\kappa}[(u_\lambda^+ + u_\lambda^-) - (u_\lambda^+ - u_\lambda^-)]du_\kappa$$
$$= \tfrac{1}{2}\int_{B_\kappa}(u_\lambda^+ + u_\lambda^-)du_\kappa - \frac{\tau_{\kappa\lambda}}{2}\int_{B_\kappa} du_\kappa.$$

Thus the sum of the $2p$ integrals over the lines A_κ, B_κ is

$$e_\lambda - u_\lambda(c_1)_- + h_\lambda - \tau_{\lambda\lambda}/2 + \frac{1}{2\pi i}\sum_{\kappa=1}^{p}[2\pi i g_\kappa\tau_{\kappa\lambda} - 2\pi i\tau_{\kappa\lambda} + \pi i\tau_{\kappa\lambda}$$
$$+ \pi i\int_{B_\kappa}(u_\lambda^+ + u_\lambda^-)du_\kappa].$$

Hence,

$$\frac{1}{2\pi i}\int_{T'} u_\lambda d\log\theta = e_\lambda - u_\lambda(c_1)_- + h_\lambda - \tau_{\lambda\lambda}/2 + \sum_{\kappa=1}^{p}[g_\kappa\tau_{\kappa\lambda} - \tau_{\kappa\lambda}/2$$
$$+ \tfrac{1}{2}\int_{B_\kappa}(u_\lambda^+ + u_\lambda^-)du_\kappa],$$

and

$$\sum_{\kappa=1}^{p} u_\lambda(x_\kappa) = e_\lambda + k_\lambda + h_\lambda + g_1\tau_{1\lambda} + g_2\tau_{2\lambda} + \cdots + g_p\tau_{\lambda p} \quad \cdot \quad (A),$$

where

$$k_\lambda = -u_\lambda(c_1)_- - \tau_{\lambda\lambda}/2 + \sum_{\kappa=1}^{p}\left[-\tau_{\kappa\lambda}/2 + \tfrac{1}{2}\int_{B_\kappa}(u_\lambda^+ + u_\lambda^-)du_\kappa\right].$$

Theorem III. The lower limits of $u_1, u_2, \cdots, u_p$ can be so chosen as to make all the quantities k_λ vanish, and then

$$u_\lambda(x_1)+u_\lambda(x_2)+\cdots+u_\lambda(x_p)=e_\lambda+h_\lambda+g_1\tau_{1\lambda}+g_2\tau_{2\lambda}+\cdots+g_p\tau_{\lambda p}.$$

Riemann proves this by the following argument:—

The fact that u_λ possesses at A_λ the period 1 and at the other cross-cuts A the periods 0, determines u_λ only to an additive constant. Let definite values be assigned to the quantities $u_1, u_2, \cdots, u_p$, and let us still use Riemann's lower limits for these integrals. A simultaneous increase of u_λ, $e_\lambda (\lambda=1, 2, \cdots, p)$, by a_λ leaves the function $\theta(u_1-e_1, u_2-e_2, \cdots, u_p-e_p)$ unaffected and produces, accordingly, no change in the positions of $x_1, x_2, \cdots, x_p$, or in the values of g_λ, h_λ. The left-hand side of the equation (A) increases by pa_λ, the right-hand side by a_λ + the increment of k_λ. Hence the increment of k_λ is $(p-1)a_\lambda$. Choose a_λ so as to make the new $k_\lambda=0$. This will be effected if the increment of k_λ be $-k_\lambda$, that is, if

$$a_\lambda=-k_\lambda/(p-1).$$

When the lower limits of the integrals are changed by the inclusion of the constants a_λ just defined, the quantities $e_1, e_2, \cdots, e_p$ must be congruent to

$$u_1(x_1)+u_1(x_2)+\cdots+u_1(x_p),\ u_2(x_1)+u_2(x_2)+\cdots+u_p(x_p), \cdots,$$
$$u_p(x_1)+u_p(x_2)+\cdots+u_p(x_p)$$

with respect to the array of periods

$$\begin{array}{cccc|cccc} 1 & 0 & \cdots & 0 & \tau_{11} & \tau_{12} & \cdots & \tau_{1p} \\ 0 & 1 & \cdots & 0 & \tau_{12} & \tau_{22} & \cdots & \tau_{2p} \\ \cdot & \cdot & \cdot & \cdot & \cdot & \cdot & \cdot & \cdot \\ \cdot & \cdot & \cdot & \cdot & \cdot & \cdot & \cdot & \cdot \\ 0 & 0 & \cdots & 1 & \tau_{1p} & \tau_{2p} & \cdots & \tau_{pp}. \end{array}$$

§ 296. The function $\theta(u_1-e_1, u_2-e_2, \cdots, u_p-e_p)$ differs from

$$\theta(\cdots, u_\lambda(x)-u_\lambda(x_1)-u_\lambda(x_2)-\cdots-u_\lambda(x_p), \cdots)$$

by an exponential factor. Owing to the length of the p arguments it is usual to print only one of them, and in order to show more clearly the exact character of these arguments and the position of the p zeros, Clebsch and Gordan, Weber, and others write the theta-function in the form

$$\theta\left(\int_e^x du_\lambda-\int_{e_1}^{x_1} du_\lambda-\int_{e_2}^{x_2} du_\lambda-\cdots-\int_{e_p}^{x_p} du_\lambda+k_\lambda\right),$$

where $c, c_1, c_2, \cdots, c_p$ are fixed places which are assigned *arbitrarily* on T′. Owing to the arbitrariness of the lower limits, an expression k_λ must be reintroduced. Denoting by $x_1, x_2, \cdots, x_p$ arbitrary places on T′, it is possible to determine the quantities $k_1, k_2, \cdots, k_p$, *independently of* $x, x_1, x_2, \cdots, x_p$, so that the theta-function shall vanish at the p places $x_1, x_2, \cdots, x_p$. It should be observed expressly that we have assumed that the theta-function, considered as a function of the co-ordinates z, w of x does not vanish identically. See § 300.

§ 297. Having established those properties of the theta-function which will be needed in the sequel, we shall now prove the converse of Abel's theorem for integrals of the first kind. Suppose that $x_1, x_2, \cdots, x_\mu$ and $y_1, y_2, \cdots, y_\mu$ are two systems of places on T at which a μ-placed function $R(w, z)$ takes two assigned values; then by Abel's theorem

$$\int_{y_1}^{x_1} du_\lambda + \int_{y_2}^{x_2} du_\lambda + \cdots + \int_{y_\mu}^{x_\mu} du_\lambda \equiv 0 \quad (\lambda = 1, 2, \cdots, p)$$

where the sign of congruence means that the right-hand sides are of the form

$$h_\lambda + g_1\tau_{1\lambda} + g_2\tau_{2\lambda} + \cdots + g_p\tau_{\lambda p},$$

the g's and h's being integers.

The converse of Abel's theorem. Suppose that the p congruences

$$u_\lambda^{x_1 y_1} + u_\lambda^{x_2 y_2} + \cdots + u_\lambda^{x_\mu y_\mu} \equiv 0$$

are satisfied, and that a function $R(w, z)$ takes one and the same value R_0 at the μ places $y_1, y_2, \cdots, y_\mu$. Then R must take one and the same value R_1 at the μ places $x_1, x_2, \cdots, x_\mu$.

To prove the theorem we shall construct a function $R(w, z)$ which is ∞^1 at the places $y_1, y_2, \cdots, y_\mu$ and 0^1 at the places $x_1, x_2, \cdots, x_\mu$. The method for constructing such a function has been described by Weber, Zur Theorie der Umkehrung der Abelschen Integrale, Crelle, t. lxx.

The building up of algebraic functions on T, with assigned zeros and assigned infinities, can be effected, whenever the problem admits of solution, by the use of methods closely analogous to those employed in the construction of the general elliptic function with the σ-function as element. In the latter case σu is one-valued and continuous throughout the u-plane, has one zero and no infinity within the parallelogram of periods, and is quasi-periodic (see Ch. VII., § 206). When the quotient of suitably chosen σ-products was multiplied by a suitably chosen exponential, the one-valuedness remained, the quasi-periodicity was converted into the periodicity of the whole function, and the zeros and infinities of the function were the zeros of the elements in the numerator and denominator respectively. In the present case the p-tuple theta-function and the surface T′ take the place of the σ-function and the parallelogram of periods, while the single zero of a σ-function is replaced by the p zeros of a theta-function. These p zeros are denoted by the upper limits $x_1, x_2, \cdots, x_p$ of the arguments

$$\int_c^x du_\lambda - \int_{c_1}^{x_1} du_\lambda - \int_{c_2}^{x_2} du_\lambda - \cdots - \int_{c_p}^{x_p} du_\lambda + k_\lambda \ (\lambda = 1, 2, \cdots, p),$$

or, (as they are often written for shortness)

$$\int_c^x - \int_{c_1}^{x_1} - \int_{c_2}^{x_2} - \cdots - \int_{c_p}^{x_p} du_\lambda + k_\lambda;$$

here it is to be understood that k_λ is supposed to be related to the lower limits in such a way that the zeros shall be $x_1, x_2, \cdots, x_p$, and also that these p zeros are not so situated as to make the theta-function vanish identically.

Suppose that μ is an exact multiple of p, say $\mu = sp$. If the number of integrals be not an exact multiple of p, add on the required number of integrals with coincident limits. It is required to construct a quotient Q of theta-functions which shall be 0^1 at sp places $x_1, x_2, \cdots, x_{sp}$, and ∞^1 at sp places $y_1, y_2, \cdots, y_{sp}$. Separate the sp places $x_1, x_2, \cdots, x_{sp}$ into s sets of p, $(x_1, x_2, \cdots, x_p)$, $(x_{p+1}, x_{p+2}, \cdots, x_{2p})$, $\cdots$, $(x_{(s-1)p+1}, x_{(s-1)p+2}, \cdots, x_{sp})$; and the sp places $y_1, y_2, \cdots, y_{sp}$ into s sets of p, $(y_1, y_2, \cdots, y_p)$, $(y_{p+1}, y_{p+2}, \cdots, y_{2p})$, $\cdots$, $(y_{(s-1)p+1}, y_{(s-1)p+2}, \cdots, y_{sp})$.

The factors of the numerator of the quotient Q are

$$\theta\left(\int_c^x - \int_{c_1}^{x_{(r-1)p+1}} - \int_{c_2}^{x_{(r-1)p+2}} - \cdots - \int_{c_p}^{x_{rp}} du_\lambda + k_\lambda\right) \quad (r = 1, 2, \cdots, s),$$

where the places $c, c_1, c_2, \cdots, c_p$ are arbitrary but fixed. The factors of the denominator are obtained from those of the numerator by replacing x's by y's.

The theta-functions in question are discontinuous at the cross-cuts B and continuous at the cross-cuts A. At B_κ,

$$\theta(u_\lambda^+ - e_\lambda) = \theta(u_\lambda^- - e_\lambda) \cdot \exp - 2\pi i(u_\kappa^- - e_\kappa + \tau_{\kappa\kappa}/2),$$

hence
$$Q_+ = Q_- \cdot \exp 2\pi i \left[\sum_{r=1}^{sp} \int_{c_{r'}}^{x_r} du_\kappa - \sum_{r=1}^{sp} \int_{c_{r'}}^{y_r} du_\kappa\right],$$

where r' is the remainder after r is divided by p.

Thus $Q_+ = Q_- \cdot \exp 2\pi i \sum_{r=1}^{sp} \int_{y_r}^{x_r} du_\kappa.$

Now, by supposition,

$$\sum_{r=1}^{sp} \int_{y_r}^{x_r} du_\kappa = m_\kappa + n_1\tau_{1\kappa} + n_2\tau_{2\kappa} + \cdots + n_p\tau_{\kappa p};$$

therefore $Q_+ = Q_- \cdot \exp 2\pi i\, (n_1\tau_{1\kappa} + n_2\tau_{2\kappa} + \cdots + n_p\tau_{\kappa p}).$

Let $P = \exp 2\pi i \left(n_1 \int_c^x du_1 + n_2 \int_c^x du_2 + \cdots + n_p \int_c^x du_p \right);$

then along a cross-cut A_κ we have $P_+ = P_-$; while along B_κ

$$P_+ = P_- \cdot \exp 2\pi i\, (n_1\tau_{1\kappa} + n_2\tau_{2\kappa} + \cdots + n_p\tau_{\kappa p}).$$

Hence Q/P is a function of the place x on T, which is not affected with a discontinuity by a passage over any of the lines A_κ, B_κ, and which possesses the same zeros and infinities as Q. This proves that Q/P is one-valued throughout T and continuous at the cross-cuts, therefore Q/P is a function $R(w, z)$. Thus we have proved the converse of Abel's theorem by constructing synthetically the algebraic function $R(w, z)$, which is 0^1 at the upper limits and ∞^1 at the lower limits. It may appear, at first sight, as if too much had been proved; for the theorem appears to imply that the number and position of the infinities y being assigned arbitrarily, there exists an algebraic function $R(w, z)$ which is infinite at these places and nowhere else. The proof, however, ceases to be valid if some of the theta-functions in Q vanish identically. This is a case which always arises when $\mu \leqq p$. [Weber, loc. cit., p. 317.]

The identical vanishing of a theta-function. We have seen that

$$\theta\left(\int_c^x - \int_{c_1}^{x_1} - \cdots - \int_{c_p}^{x_p} du_\lambda + k_\lambda \right)$$

vanishes at the places $x_1, x_2, \cdots, x_p$, when the constants k are properly chosen. Make the place x coincide with x_p; then, noticing that the theta-function is even,

$$\theta\left(\int_{c_p}^{c} + \int_{c_1}^{x_1} + \cdots + \int_{c_{p-1}}^{x_{p-1}} du_\lambda - k_\lambda \right)$$

must vanish identically whatever be the positions of the places $x_1, x_2, \cdots, x_{p-1}$. We shall return later on to this question of the identical vanishing of a theta-function; but for full information we must refer the reader to the memoirs of Riemann, and the treatise of Clebsch and Gordan.

§ 298. Weber has shown how to deal with the case $\mu \leq p$, *when the congruence*

$$u_\lambda^{x_1 y_1} + u_\lambda^{x_2 y_2} + \cdots + u_\lambda^{x_\mu y_\mu} \equiv 0 \quad (\lambda = 1, 2, \cdots, p)$$

is satisfied by the two systems x, y.

It is always possible to give such a position to x_1 that

$$\theta\left(\int_c^x - \int_{c_1}^{x_1} - \int_{c_2}^{t_2} - \cdots - \int_{c_p}^{t_p} du_\lambda + k_\lambda\right)$$

shall not vanish for all values of x, t_2, t_3, $\cdots$, t_p; for the p arguments of this theta-function can be made congruent to an arbitrarily assigned system of periods by a proper choice of x and the t's. In proof of this statement, observe that the p normal integrals of the first kind are linearly independent, and that therefore the p arguments can be made to receive arbitrarily assigned increments when the p places x, t_2, t_3, $\cdots$, t_p are subjected to independent translations on T. In the fraction

$$Q = \prod_{r=1}^{\mu} \frac{\theta\left(\int_c^x - \int_{c_1}^{x_r} - \int_{c_2}^{t_2^{(r)}} - \cdots - \int_{c_p}^{t_p^{(r)}} du_\lambda + k_\lambda\right)}{\theta\left(\int_c^x - \int_{c_1}^{y_r} - \int_{c_2}^{t_2^{(r)}} - \cdots - \int_{c_p}^{t_p^{(r)}} du_\lambda + k_\lambda\right)},$$

it is possible to arrange the quantities t so that the theta-functions in the numerator and denominator shall not vanish identically. This being done, the t's in the numerator neutralize those in the denominator; there are left μ zeros x_1, x_2, $\cdots$, x_μ, and μ infinities y_1, y_2, $\cdots$, y_μ. Multiplying Q, as before, by an exponential factor P, *and assuming the existence of the p congruences*, we have an algebraic function Q/P with the upper and lower limits of the μ integrals as zeros and infinities.

§ 299. When we replace the lower limits c_1, c_2, $\cdots$, c_p, which have not been specialized as yet, by the places a_1, a_2, $\cdots$, a_p, which made their appearance in Clebsch and Gordan's theory as the points of contact of a contact-curve, we can determine the values of the added constants k_λ very conveniently. When we pass to the non-homogeneous form, let the integral of the second kind become

$$\int \frac{\psi dz}{(aw + bz + c)\,\delta f/\delta w}.$$

Here $\psi/(aw + bz + c)$ is ∞^2 at one place a of the surface, at which $aw + bz + c = 0^2$, and 0^1 at the p places a_κ. Let any integral of the first kind be $\int \frac{\phi dz}{\delta f/\delta w}$, where $\phi = 0$ is what the equation of an adjoint curve ϕ_{n-3} becomes in the non-homogeneous form. Then ϕ is 0^1 at $2p-2$ places. Hence $R, = \frac{(aw+bz+c)\phi}{\psi}$, has $2p$ zeros, among which a is counted twice, and $2p$ poles, namely, the places a_κ each counted twice.

Given that $\theta\left(\int_a^x - \sum_{\kappa=1}^{p}\int_{a_\kappa}^{x_\kappa} du_\lambda + k_\lambda\right)$ . . . . (A)

vanishes at $x_1, x_2, \cdots, x_p$, we have, making x coincide with x_1,

$$\theta\left(\int_{a_1}^{a} + \sum_{\kappa=2}^{p}\int_{a_\kappa}^{x_\kappa} du_\lambda - k_\lambda\right) = 0$$

for all positions of $x_2, x_3, \cdots, x_p$. Now let $\phi = 0^1$ at $x_2, x_3, \cdots, x_p$; and let the remaining zeros be $x'_2, x'_3, \cdots, x'_p$. Abel's theorem gives, when applied to the equivalent systems $a, a, x_2, \cdots, x_p, x'_2, \cdots, x'_p$ and $a_1, a_1, a_2, \cdots, a_p, a_2, \cdots, a_p$,

$$2\int_{a_1}^{a} du_\lambda + \sum_{\kappa=2}^{p}\int_{a_\kappa}^{x_\kappa} du_\lambda + \sum_{\kappa=2}^{p}\int_{a_\kappa}^{x'_\kappa} du_\lambda \equiv 0.$$

Thus $$\int_{a_1}^{a} du_\lambda + \sum_{\kappa=2}^{p}\int_{a_\kappa}^{x_\kappa} du_\lambda \equiv -\left\{\int_{a_1}^{a} du_\lambda + \sum_{\kappa=2}^{p}\int_{a_\kappa}^{x'_\kappa} du_\lambda\right\}.$$

Hence, on inserting this value in the expression (A), we find that

$$\theta\left(\int_{a_1}^{a} + \int_{a_2}^{x'_2} + \cdots + \int_{a_p}^{x'_p} du_\lambda + k_\lambda\right) \quad . \quad . \quad . \quad . \quad \text{(B)}$$

vanishes identically. Since the $p-1$ points $x_2, x_3, \cdots, x_{p-1}$ were chosen perfectly arbitrarily, the $p-1$ residuals $x'_2, x'_3, \cdots, x'_{p-1}$ must be perfectly arbitrary. We may therefore suppress the accents in (B). When this has been done, we have two theta-functions

$$\theta\left(\int_a^x - \sum_{\kappa=1}^{p}\int_{a_\kappa}^{x_\kappa} du_\lambda + k_\lambda\right),\quad \theta\left(\int_a^x - \sum_{\kappa=1}^{p}\int_{a_\kappa}^{x_\kappa} du_\lambda - k_\lambda\right),$$

which vanish at the same places $x_1, x_2, \cdots, x_p$. But this means that the arguments are congruent. Hence the differences of the arguments must be congruent to 0, or

$$2k_1,\ 2k_2, \cdots, 2k_p \equiv 0.$$

This gives
$$k_1 = h_1/2 + g_1\tau_{11}/2 + g_2\tau_{12}/2 + \cdots + g_p\tau_{1p}/2,$$
$$k_2 = h_2/2 + g_1\tau_{12}/2 + g_2\tau_{22}/2 + \cdots + g_p\tau_{2p}/2,$$
$$\cdots\cdots\cdots\cdots$$
$$k_p = h_p/2 + g_1\tau_{1p}/2 + g_2\tau_{2p}/2 + \cdots + g_p\tau_{pp}/2,$$

where g_κ and h_κ are integers. That is, the constants k_λ are a system of half-periods of the theta-function. Allowing $x_1, x_2, \cdots, x_p$ to coincide with $a_1, a_2, \cdots, a_p$, we find that

$$\theta\left(\int_a^x du_\lambda - k_\lambda\right)$$

vanishes in general at the p places $a_1, a_2, \cdots, a_p$, and only at these places. If the function vanish when x is at a, then

$$\theta(k_\lambda) = 0,$$

and the mark $\begin{vmatrix} g_\kappa \\ h_\kappa \end{vmatrix}$ is *odd*. But this occurs only when one of the places $a_1, a_2, \cdots, a_p$ coincides with a. Now recurring to the theory of § 290, a contact curve ψ_{n-2} passes through the $n-2$ residuals of the point a, and if, in addition, it touch the basis-curve at a, it has more than $n-2$ intersections with the tangent at a. Therefore it breaks up into this tangent, and an adjoint curve ϕ_{n-3} which touches the curve at $p-1$ points. Thus, of the 2^{2p} non-congruent marks, we are led to associate the $2^{p-1}(2^p-1)$ odd marks (§ 229) with *improper* contact curves, formed of the tangent at a and an adjoint curve ϕ_{n-3}; and the $2^{p-1}(2^p+1)$ even marks with *proper* contact curves ψ_{n-2}. When $p=3$, the normal curve is a non-singular quartic, and the improper contact curves are an arbitrary tangent and the 28 double tangents. [See Clebsch, Géométrie, t. iii., Chapter I., § 9.]

The method followed is that of Weber in his memoir on the Inversion Problem to which reference has already been made. In a different form, it is to be found in Clebsch and Gordan. To these authors, and to Fuchs, Crelle, t. lxxiii., the reader is referred. In Weber's memoir will be found the proof that to each system of half-periods there corresponds a system of lower limits $a_1, a_2, \cdots, a_p$ (points of contact of a contact curve), when the theta-function has the prescribed zeros $x_1, x_2, \cdots, x_p$.

We shall suppose, for the rest of the chapter, that such a canonical dissection of the surface has been selected that all the constants k_λ are zero.

§ 300. *The identical vanishing of a theta-function.*

The theta-function whose arguments are

$$\int_a^x du_\lambda - \sum_{\kappa=1}^{p} \int_{a_\kappa}^{x_\kappa} du_\lambda \quad . \quad . \quad . \quad . \qquad . \quad . \quad . \quad (1)$$

vanishes at all places x when the function

$$\theta\left(-\int_{a_p}^{a} du_\lambda - \sum_{\kappa=1}^{p-1}\int_{a_\kappa}^{x_\kappa} du_\lambda\right) \quad . \quad . \quad . \quad . \quad . \quad . \quad . \quad (2),$$

whose arguments are obtained by allowing x to coincide with x_p, vanishes for an arbitrary choice of the $p-1$ places $x_1, x_2, \cdots, x_{p-1}$.

Let the p places $x_1, x_2, \cdots, x_p$ be tied by a function ϕ, and let the remaining moveable zeros of ϕ be $x_{p+1}, x_{p+2}, \cdots, x_{2p-2}$. Then Abel's theorem when applied to the function R of § 299 gives

$$\sum_{\kappa=1}^{p}\int_{a_\kappa}^{x_\kappa} du_\lambda + \sum_{\kappa=1}^{p-2}\int_{a_\kappa}^{x_{p+\kappa}} du_\lambda + \int_{a_{p-1}}^{a} du_\lambda + \int_{a_p}^{a} du_\lambda \equiv 0.$$

Therefore the arguments (1) are congruent to

$$\int_{a_{p-1}}^{x} du_\lambda + \sum_{\kappa=1}^{p-2}\int_{a_\kappa}^{x_{p+\kappa}} du_\lambda + \int_{a_p}^{a} du_\lambda.$$

As θ is an even function, the sign of each argument can be changed, and we then have arguments of the form in (2), in which x and $x_{p+1}, x_{p+2}, \cdots, x_{2p-2}$ can be chosen arbitrarily.

Hence we have the important theorem: when the p zeros, which θ must have, are tied by a function ϕ, θ vanishes identically. [Clebsch and Gordan, p. 211; Clebsch, Géométrie, t. iii., ch. i., § 8; Riemann, Crelle, t. lxvi., or Werke, p. 198.]

With the aid of § 283, this may be stated as follows: all the zeros of an algebraic function of the surface cannot be included among the zeros of a theta-function unless the theta-function vanishes identically.

Jacobi's Inversion Problem.

§ 301. *Introductory remarks.* The importance of the inversion problem in such simple cases as

$$v = \int_0^x \frac{dx}{\sqrt{1-x^2}} \text{ and } v = \int_0^x \frac{dx}{\sqrt{(1-x^2)(1-k^2x^2)}}$$

is well known. It permits the passage from many-valued integrals to one-valued functions. In § 193 we indicated that the problem is altered as soon as the quantity under the radical is assumed to be of higher order than 4; for when we pass from elliptic to hyper-elliptic integrals x becomes a function of v with more than two independent periods, and therefore cannot be a one-valued or a finitely many-valued function of v. The essential distinction between the function of one variable with two distinct periods and the function of one variable with more than two distinct periods is to be sought in the theorem that $m_1\omega_1 + m_2\omega_2$ cannot, and $m_1\omega_1 + m_2\omega_2 + \cdots + m_q\omega_q$ can, be made less than any assigned quantity. Jacobi pointed out in his memoirs (Crelle, tt. ix., xiii.) that one-valuedness could be secured by introducing functions of p variables in the place of functions of one variable.

A theta-function affords an illustration of a one-valued function of p variables. Expressions involving θ can be constructed, on a plan analogous to that of § 297, which reproduce themselves exactly when the variables v are altered as in the scheme (B) below; they are then $2p$-tuply periodic in the strict sense, whereas θ itself is quasi-periodic. [It should be recalled that $2p$ is the maximum number of independent systems of periods of a one-valued function of p variables (§ 193).] The expressions in question arise in the solution of Jacobi's inversion problem. This problem is always soluble, and in general the solution is unique. It consists in finding the system of p places $x_1, x_2, \cdots, x_p$ from the p equations

$$\left.\begin{array}{l}
\int_{c_1}^{x_1} du_1 + \int_{c_2}^{x_2} du_1 + \cdots + \int_{c_p}^{x_p} du_1 = v_1, \\
\int_{c_1}^{x_1} du_2 + \int_{c_2}^{x_2} du_2 + \cdots + \int_{c_p}^{x_p} du_2 = v_2, \\
\cdots\cdots\cdots\cdots \\
\int_{c_1}^{x_1} du_p + \int_{c_2}^{x_2} du_p + \cdots + \int_{c_p}^{x_p} du_p = v_p,
\end{array}\right\} \quad \cdots \quad \text{(A)},$$

where the lower limits are any fixed places, and the v's are arbitrary variables.

The paths in any one column are supposed to be the same. We may, if we like, suppose initially that all the paths are drawn on T′; then, removing the restriction that the paths are not to pass over cross-cuts, we have for the most general values of the right-hand sides in (A), the system

$$\left.\begin{array}{l}
v'_1 = v_1 + h_1 \cdot 1 + h_2 \cdot 0 + \cdots + h_p \cdot 0 + g_1\tau_{11} + g_2\tau_{12} + \cdots + g_p\tau_{1p}, \\
v'_2 = v_2 + h_1 \cdot 0 + h_2 \cdot 1 + \cdots + h_p \cdot 0 + g_1\tau_{12} + g_2\tau_{22} + \cdots + g_p\tau_{2p}, \\
\cdots\cdots\cdots\cdots \\
\cdots\cdots\cdots\cdots \\
v'_p = v_p + h_1 \cdot 0 + h_2 \cdot 0 + \cdots + h_p \cdot 1 + g_1\tau_{1p} + g_2\tau_{2p} + \cdots + g_p\tau_{pp},
\end{array}\right\} \text{(B)}.$$

The difficulty which arises in the case of functions of one variable with more than two periods does not present itself when we consider a function of p variables, $v_1, v_2, \cdots, v_p$ with $2p$ independent periods. It can be proved by algebraic work that, when the p differences $v'_\kappa - v_\kappa$ are resolved into their real and imaginary parts, $2p-1$ of the $2p$ resulting expressions can be made arbitrarily small by a proper choice of the integers g, h; but when this has been done, the remaining expression is not infinitely small. The algebraic lemma upon which this theorem is based, will be found in Clebsch and Gordan (pp. 130 *et seq.*).

In passing from one-valued functions of one variable to one-valued functions of many variables, a new phenomenon presents itself. The one-valued function $\frac{x^2-y^2}{x^2+y^2}$ is completely indeterminate when $x=0$, $y=0$, its value depending on the mode of approach of x, y to 0, 0. More generally the one-valuedness of a function of p variables which belong to a p-dimensional space is not necessarily affected by the circumstance that it may take infinitely many values in a space of $p-2$ dimensions (Clebsch and Gordan, p. 186).

Definition of Abelian functions. Let R be any algebraic function of T. Construct the symmetric products of $R(x_1)$, $R(x_2)$, $\cdots$, $R(x_p)$ r at a time, and let it equal $(-1)^r M_r$. Then the functions M_1, M_2, $\cdots$, M_p, when considered in virtue of the equations (A) as functions of the v's, are said to be *Abelian functions of* v_1, v_2, $\cdots$, v_p.

This definition of Abelian functions is based on that in Clebsch and Gordan (p. 139). Riemann's definition is altogether different. See Riemann, Werke, 1st ed., p. 457; Clebsch, Géométrie, t. iii., p. 222.

Inasmuch as for the same system of upper limits, the right-hand sides of the equations (A) are capable of the values given by the scheme (B), the Abelian functions are unaltered when v_1, v_2, $\cdots$, v_p are changed, as in (B), by simultaneous periods compounded from the scheme

$$\left.\begin{array}{c|cccc|cccc} v_1 & 1 & 0 & 0 & \cdots 0 & \tau_{11} & \tau_{12} & \tau_{13} & \cdots \tau_{1p}, \\ v_2 & 0 & 1 & 0 & \cdots 0 & \tau_{12} & \tau_{22} & \tau_{23} & \cdots \tau_{2p}, \\ \vdots & \vdots & \vdots & \vdots & \vdots & \vdots & \vdots & \vdots & \vdots \\ v_p & 0 & 0 & 0 & \cdots 1 & \tau_{1p} & \tau_{2p} & \tau_{3p} & \cdots \tau_{pp}, \end{array}\right\} \quad . \quad . \quad . \quad . \quad \text{(C)},$$

and are therefore $2p$-tuply periodic.

Among the many memoirs dealing with Jacobi's inversion problem we shall only mention those of Weierstrass (Crelle, t. xlvii.), H. Stahl (Crelle, t. lxxxix.), and Nöther (Math. Ann., t. xxviii.).* For the inversion problem when the number of integrals in each of the

* The history of the problem is given in the introductory part of Casorati's valuable work, Teorica delle Funzioni di Variabili Complesse, Pavia, 1868.

equations (A) is other than p, as well as for Jacobi's problem, see Klein-Fricke, t. ii.

§ 302. Clebsch and Gordan solve Jacobi's Inversion Problem with the help of integrals of the third kind. We shall state their solution in terms of the Riemann surface as is done by Weber (Crelle, t. lxx.) and Neumann.

Let the equations be

$$u_\kappa^{x_1c_1} + u_\kappa^{x_2c_2} + \cdots + u_\kappa^{x_pc_p} = v_\kappa \quad \cdot \cdot \cdot \cdot \cdot \cdot \cdot \quad (1),$$

$$\kappa = 1, 2, \cdots, p.$$

The quantities v_κ are assigned arbitrarily, and therefore the lower limits $c_1, c_2, \cdots, c_p$ can be taken and will be taken at the places $a_1, a_2, \cdots, a_p$ of § 290. It is assumed that the places $x_1, x_2, \cdots, x_p$ are not tied by a function ϕ; and the problem is to express these places, or symmetric combinations of them, as functions of $v_1, v_2, \cdots, v_p$.

Take any algebraic function of the surface, say R. Let it be μ-placed, and let ξ_ι and η_ι, where $\iota = 1, 2, \cdots, \mu$, be equivalent systems of places. Let them be associated in pairs so that when η_1 moves into the position ξ_1, η_2 moves to ξ_2, and so on; and further let none of the μ equivalent paths, which begin at η_ι and end at ξ_ι, pass over a cut B_λ.

By Abel's theorem for integrals of the third kind we have

$$\sum_{\iota=1}^{\mu} \int_{\eta_\iota}^{\xi_\iota} d\Pi_{x_\lambda a_\lambda} = \log \frac{R(x_\lambda) - R(\xi)}{R(x_\lambda) - R(\eta)} \cdot \frac{R(a_\lambda) - R(\eta)}{R(a_\lambda) - R(\xi)}.$$

For simplicity let the places η_κ be poles of R; then the right-hand expression is

$$\log \{R(x_\lambda) - R(\xi)\} - \log \{R(a_\lambda) - R(\xi)\}.$$

Using the law of interchange of parameters and arguments, we have

$$\log \{R(x_\lambda) - R(\xi)\} - \log \{R(a_\lambda) - R(\xi)\} = \sum_{\iota=1}^{\mu} \int_{a_\lambda}^{x_\lambda} d\Pi_{\xi_\iota \eta_\iota} \quad \cdot \quad (2).$$

The paths of integration in this equation are assigned to suit Abel's theorem (§ 291). The path of the integral $\int_{a_\lambda}^{x_\lambda} du_\lambda$ which occurs in (1) has not been assigned, and will be taken to be the same as in (2). This may require the addition of periods to the arbitrary quantities v_κ in (1), but it will be proved later that the periods so added will disappear from the final formula.

In equation (2) we let $\lambda = 1, 2, \cdots, p$; there results, if we add the p equations and take the exponentials of both sides,

$$\prod_{\lambda=1}^{p} \frac{R(x_\lambda) - R(\xi)}{R(a_\lambda) - R(\xi)} = \exp \sum_{\iota=1}^{\mu} \sum_{\lambda=1}^{p} \int_{a_\lambda}^{x_\lambda} d\Pi_{\xi_\iota \eta_\iota} \quad . \quad . \quad . \quad (3).$$

The problem is reduced to the finding of an expression for the right-hand side, in terms of $v_1, v_2, \cdots, v_p$.

Let
$$H(\xi) = \frac{\theta\left(\int_a^\xi du_\lambda - \sum_{\kappa=1}^{p} \int_{a_\kappa}^{x_\kappa} du_\lambda\right)}{\theta\left(\int_a^\xi du_\lambda\right)},$$

$$K(\xi) = \exp \sum_{\lambda=1}^{p} \int_{a_\lambda}^{x_\lambda} d\Pi_{\xi\eta_\iota}.$$

Remembering (§ 285) that $\int_{a_\lambda}^{x_\lambda} d\Pi_{\xi\eta}$, or $\int_\eta^\xi d\Pi_{x_\lambda a_\lambda}$, regarded as a function of ξ, is infinite at x_λ like $+\log(\xi - x_\lambda)$ and at a_λ like $-\log(\xi - a_\lambda)$, it follows that the exponential of this integral is 0^1 at x_λ and is ∞^1 at a_λ. Therefore $K(\xi)$ is 0^1 at $x_1, x_2, \cdots, x_p$ and ∞^1 at $a_1, a_2, \cdots, a_p$. The function $H(\xi)$ has the same zeros and poles. The functions have no discontinuities at a cut A_λ, while at opposite places of a cut B_λ each of the ratios $H(\xi_+)/H(\xi_-)$ and $K(\xi_+)/K(\xi_-)$ is equal to $\exp 2\pi i \Sigma \int_{a_\kappa}^{x_\kappa} du_\lambda$. (See § 297.)

Hence the ratio $H(\xi)/K(\xi)$ is one-valued and continuous everywhere on **T**; that is, it is a constant.

Thus,

$$H(\xi_\iota)/H(\eta_\iota) = K(\xi_\iota)/K(\eta_\iota) = \exp \sum_{\lambda=1}^{p} \int_{a_\lambda}^{x_\lambda} d\Pi_{\xi_\iota \eta_\iota} \quad . \quad . \quad . \quad . \quad . \quad (4).$$

From (1),
$$H(\xi) = \frac{\theta\left(\int_a^\xi du_\lambda - v_\lambda\right)}{\theta\left(\int_a^\xi du_\lambda\right)};$$

therefore,
$$\exp \sum_{\lambda=1}^{p} \int_{a_\lambda}^{x_\lambda} d\Pi_{\xi_\iota \eta_\iota} = \frac{\theta\left(\int_a^{\xi_\iota} du_\lambda - v_\lambda\right)\theta\left(\int_a^{\eta_\iota} du_\lambda\right)}{\theta\left(\int_a^{\eta_\iota} du_\lambda - v_\lambda\right)\theta\left(\int_a^{\xi_\iota} du_\lambda\right)};$$

and from (3),

$$\prod_{\lambda=1}^{p} \frac{R(x_\lambda) - R(\xi)}{R(c_\lambda) - R(\xi)} = \prod_{\iota=1}^{\mu} \frac{\theta\left(\int_a^{\xi_\iota} du_\lambda - v_\lambda\right)\theta\left(\int_a^{\eta_\iota} du_\lambda\right)}{\theta\left(\int_a^{\eta_\iota} du_\lambda - v_\lambda\right)\theta\left(\int_a^{\xi_\iota} du_\lambda\right)} \quad . \quad (5).$$

When the product of the numerators on the left is multiplied out, the coefficients of the powers of $R(\xi)$ are Abelian functions of $x_1, x_2, \cdots, x_p$; and by giving p different values to $R(\xi)$ we get sufficient equations to determine these Abelian functions in terms of the $p\mu$ places ξ_ι, and the quantities v_λ.

The algebraic function R is at our disposal. Neumann points out that the simplest supposition, namely $R = z$, is sufficient and convenient. The equivalent system ξ, is then of course a system of places in the same vertical, and μ is the number of sheets.

It remains to show that the expression on the right hand of (5) is not altered by the addition of periods to v_λ.

Let the periods added be given by (B), § 301.

Then the expression in question acquires the factor

$$\prod_{\iota=1}^{\mu} \exp 2\pi i \sum_{\lambda=1}^{p} h_\lambda \int_{\eta_\iota}^{\xi_\iota} du_\lambda,$$

or

$$\prod_{\lambda=1}^{p} \exp 2\pi i h_\lambda \sum_{\iota=1}^{\mu} \int_{\eta_\iota}^{\xi_\iota} du_\lambda.$$

Inasmuch as no path of integration has been allowed to cross a cut B_κ, Abel's theorem shows that the factor is 1.

§ 303. *The case* $p = 2$. It will be well to indicate how some of the preceding general theory can be brought to bear in the case $p = 2$, which is next in order to the elliptic case already considered. Following Clebsch and Gordan's method, we take as normal curve a nodal quartic f. An adjoint curve ϕ_{n-3} is a line through the node; so that to places x, $\bar{x}$ in the same vertical on the surface T (§ 281) there correspond points x, x collinear with the node on the curve f. From the node 6 tangents can be drawn; their points of contact correspond to the coincident tied places; that is, to the branch-points. We call these points $a_1, a_2, \cdots, a_6$.

An adjoint curve ψ_{n-2} is one of a system of conics through the node and the points where the tangent at a meets f again. The remaining 4 points in which a conic of the system meets f are coresidual with a, a, x, $\bar{x}$; and the points a_1, a_2 of § 290 are the points of contact of a conic, of the system, which touches f twice. Clebsch and Gordan's theory shows that there are 2^4 such bitangent conics, corresponding to the 16 non-congruent pairs of half-periods. Of these 6 are improper, and consist of the tangent at a and one of the 6 tangents from the node; these can be shown to correspond to the 6 odd pairs of half-periods.

For the sake of simplicity, let a coincide with a branch-point a_1. Then the bitangent conic passes twice through the node, and becomes a pair of tangents from the node. Hence a_1, a_2 are also branch-points, say a_ι and a_κ. The arguments of θ_{00} are now

$$\int_{a_1}^{s} du_\lambda - \int_{a_\iota}^{x_1} du_\lambda - \int_{a_\kappa}^{x_2} du_\lambda,$$

and the zeros are x_1 and x_2. Let $s = x_1$; then

$$\theta\left(\int_{a_1}^{a_\iota} - \int_{a_\kappa}^{x_2}\right) = 0$$

for all values of x_2, and in particular, letting x_2 coincide in turn with a_ι and a_κ,

$$\theta\left(\int_{a_1}^{a\kappa}\right) = \theta\left(\int_{a_1}^{a\iota}\right) = 0.$$

We can determine ι and κ as follows:

In the canonically dissected surface (Fig. 68), let a_2 be that branch-point which lies within A_1 and B_1, and let $a_2, a_3, a_4, a_5, a_6, a_1$ lie in *negative* order, so that for example a_5 and a_6 lie within A_2, a_1 and a_2 within A_1.*

The periods across the cross-cuts are given by the scheme

	A_1	A_2	B_1	B_2
u_1	1	0	τ_{11}	τ_{12}
u_2	0	1	τ_{12}	τ_{22},

and the integrals from each branch-point to the next are found, as in § 182, to be

	$\int_{a_2}^{a_1}$	$\int_{a_1}^{a_6}$	$\int_{a_6}^{a_5}$	$\int_{a_5}^{a_4}$	$\int_{a_4}^{a_3}$	$\int_{a_3}^{a_2}$
u_1	$-\tau_{11}/2$	$1/2$	$-\tau_{12}/2$	0	$(\tau_{11}+\tau_{12})/2$	$-1/2$
u_2	$-\tau_{12}/2$	$-1/2$	$-\tau_{22}/2$	$1/2$	$(\tau_{12}+\tau_{22})/2$	$0.$

On the other hand, the pairs of half-periods which make θ_{00} vanish are given in § 236, and, on comparison with the preceding scheme, prove to be congruent with

$$\int_{a_1}^{a_3},\ \int_{a_3}^{a_5},\ \int_{a_5}^{a_1},\ \int_{a_2}^{a_4},\ \int_{a_4}^{a_6},\ \int_{a_6}^{a_2}.$$

---

* Any other arrangement of names for the branch-points would suit our immediate purpose, but the selected arrangement is in harmony with the notation of Chapter VIII.

Among these the arguments $\int_{a_1}^{a_\iota}$ and $\int_{a_1}^{a_\kappa}$ are to appear. Therefore $\iota, \kappa = 3, 5$; and finally the arguments of θ_{00} can be written

$$\int_{a_1}^{x} du_\lambda - \int_{a_3}^{x_1} du_\lambda - \int_{a_5}^{x_2} du_\lambda.$$

We see that, corresponding to the selected canonical system of cuts, θ_{00} is associated with the separation of the six branch-points into the two triads (a_1, a_3, a_5) and (a_2, a_4, a_6).

The arguments of the other theta-functions are now readily determined by the addition of half-periods (§ 232). For example, the half-periods $\int_{a_3}^{a_5} du_\lambda$ are $(\tau_{11}+\tau_{12})/2$, $(1+\tau_{12}+\tau_{22})/2$; from § 232 the mark of the new function is $\begin{vmatrix} 1 & 1 \\ 0 & 1 \end{vmatrix}$, or (35); hence the arguments of θ_{35} are obtained by adding $\int_{a_3}^{a_5}$ to the arguments of θ_{00}, and are

$$\int_{a_1}^{x} - \int_{a_1}^{x_1} - \int_{a_1}^{x_2}.$$

In this way the duad notation, which in Chapter VIII. appeared merely as an algebraic artifice, begins to acquire a vital significance in its connexion with the surface.

Each of the ten even theta-functions is associated with one of the ten pairs of triads into which the branch-points can be separated; each of the six odd functions is associated with a branch-point.

§ 304. We shall proceed to show in the case $p = 2$ that a theta-function, with arguments $\int_y^x du_\lambda$, where x and y are arbitrary, can be expressed in terms of algebraic functions of x and y, and of a single integral of the third kind.

Let
$$H(\xi) = \frac{\theta\left(\int_{a_1}^{\xi} - \int_{a_3}^{x_1} - \int_{a_5}^{x_2}\right)}{\theta\left(\int_{a_1}^{\xi}\right)};$$

then equation (4) of § 302 gives

$$H(\xi)/H(\eta) = \exp\left(\int_{a_3}^{x_1} d\Pi_{\xi\eta} + \int_{a_5}^{x_2} d\Pi_{\xi\eta}\right).$$

Here ξ, η, x_1, x_2 are arbitrary places. If we replace x_1, x_2 by y_1, y_2 and divide, we have

$$\frac{\theta\left(\int_{a_1}^{\xi}-\int_{a_3}^{x_1}-\int_{a_5}^{x_2}\right)\theta\left(\int_{a_1}^{\eta}-\int_{a_3}^{y_1}-\int_{a_5}^{y_2}\right)}{\theta\left(\int_{a_1}^{\eta}-\int_{a_3}^{x_1}-\int_{a_5}^{x_2}\right)\theta\left(\int_{a_1}^{\xi}-\int_{a_3}^{y_1}-\int_{a_5}^{y_2}\right)}=\exp\Sigma\int_{y_\lambda}^{x_\lambda}d\Pi_{\xi\eta}\cdot \quad . \quad . \quad . \quad (6).$$

Let
$$\int_{s_1}^{\xi}du_\lambda=\int_{a_3}^{x_1}du_\lambda+\int_{a_5}^{x_2}du_\lambda,$$

$$\int_{a_1}^{\eta}du_\lambda=\int_{a_3}^{y_1}du_\lambda+\int_{a_5}^{y_2}du_\lambda,$$

where
$$\lambda=1,\ 2.$$

Let $\bar{s}$ denote as before the place in the same vertical with s; recalling that $\int_{a_1}^{a}+\int_{a_1}^{\bar{s}}=0$, we see that the conditions just imposed make $\bar{\xi}$, x_1, x_2 and $\bar{\eta}$, y_1, y_2 coresidual with a_1, a_3, a_5.

Now let $R(s)$ be $\dfrac{(z-a_1)(z-a_3)(z-a_5)}{w}$; the fundamental equation being

$$w^2=\prod_{\kappa=1}^{6}(z-a_\kappa),$$

$R(s)$ is 0^1 at a_1, a_3, a_5 and ∞^1 at a_2, a_4, a_6; and (§ 291) must take the same values at $\bar{\xi}$, x_1, x_2 (or at $\bar{\eta}$, y_1, y_2). Therefore by Abel's theorem

$$\Sigma\int_{y_\lambda}^{x_\lambda}d\Pi_{\xi\eta}+\int_{\bar{\eta}}^{\bar{\xi}}d\Pi_{\xi\eta}=\log\frac{R(\bar{\xi})-R(\xi)}{R(\bar{\xi})-R(\eta)}\cdot\frac{R(\bar{\eta})-R(\eta)}{R(\bar{\eta})-R(\xi)},$$

$$=\log\frac{4R(\xi)R(\eta)}{\{R(\xi)+R(\eta)\}^2},$$

since
$$R(\bar{s})=-R(s).$$

Thus while the left side of (6) has become $\dfrac{\theta^2(0)}{\theta^2\left(\int_\eta^\xi\right)}$, the right side has become
$$\frac{4R(\xi)R(\eta)}{\{R(\xi)+R(\eta)\}^2}\exp-\Pi_{\xi\eta}^{\bar{\xi}\bar{\eta}}.$$

Therefore, finally,

$$\frac{\theta_{00}\left(\int_\eta^\xi du_\lambda\right)}{c_{00}}=\frac{R(\xi)+R(\eta)}{2\sqrt{R(\xi)}\sqrt{R(\eta)}}\exp\tfrac{1}{2}\Pi_{\xi\eta}^{\bar{\xi}\bar{\eta}}.$$

If f_1 and f_2 be the two cubics

$$z-a_1\cdot z-a_3\cdot z-a_5,\quad z-a_2\cdot z-a_4\cdot z-a_6,$$

the algebraic factor in the result is

$$\frac{\sqrt{f_1(\xi)}\sqrt{f_2(\eta)}+\sqrt{f_1(\eta)}\sqrt{f_2(\xi)}}{2\sqrt{w(\xi)\cdot w(\eta)}}.$$

For other even thetas we have merely to substitute the appropriate cubics into which the sextic divides (§ 303).

The important theorem of this paragraph is taken from Klein's first memoir on Hyperelliptic Sigma-Functions (Math. Ann., t. xxvii.).

§ 305. For an odd θ, with which is associated a single branch-point a, or a determinate division of the sextic into a linear factor $z-a$ and a quintic, we replace each of the lower limits of the arguments of θ on the left of equation by a.

Let
$$\int_a^x du_\lambda=\int_a^{x_1}du_\lambda+\int_a^{x_2}du_\lambda,$$
$$\int_a^y du_\lambda=\int_a^{y_1}du_\lambda+\int_a^{y_2}du_\lambda,$$
$$\lambda=1,\ 2,$$

x being very near to ξ, and y to η.

Then $\bar{x}$, x_1, x_2 and $\bar{y}$, y_1, y_2 are coresidual to a, a, a. There is no algebraic function of T which is 0^3 at a and not zero at all other places; but we can satisfy the above equations by letting $x_2=y_2=a$, $x_1=x$, $y_1=y$; the algebraic function $R(s)$ which is 0^2 at a is simply $z-a$, a function which takes the same value at $\bar{x}$, x (or at $\bar{y}$, y). We have then by Abel's theorem

$$\int_y^x d\Pi_{\xi\eta}+\int_{\bar{y}}^{\bar{x}} d\Pi_{\xi\eta}=\log\frac{R(\bar{x})-R(\xi)}{R(\bar{x})-R(\eta)}\,\frac{R(\bar{y})-R(\eta)}{R(\bar{y})-R(\xi)},$$

or if the values of z at ξ, η, $\bar{x}$, $\bar{y}$ be z, z', $z+dz$, $z'+dz'$,

$$=\log\frac{dzdz'}{(z+dz-z')(z'+dz'-z)}.$$

Therefore

$$\frac{\theta\left(\int_x^\xi\right)\theta\left(\int_y^\eta\right)}{\theta\left(\int_x^\eta\right)\theta\left(\int_y^\xi\right)}=\frac{dzdz'}{(z+dz-z')(z'+dz'-z)}\exp-\int_{\bar{y}}^{\bar{x}}d\Pi_{\xi\eta};$$

whence $$\theta^2\left(\int_\eta^\xi\right) = (z-z')^2 \exp \Pi_{\xi\eta}^{\bar{\xi}\bar{\eta}} \cdot \lim \frac{\theta\left(\int_{z+dz}^{z}\right)\theta\left(\int_{z'+dz'}^{z'}\right)}{dz \cdot dz'}.$$

Let the value of $\dfrac{\delta\theta(v_1, v_2)}{\delta v_\lambda}$, when $v_1 = v_2 = 0$, be denoted by $c^{(\lambda)}$.

Then $$\theta\left(\int_{z+dz}^{z} du_\lambda\right) = -\Sigma c^{(\lambda)} \frac{du_\lambda}{dz} dz,$$

and therefore the required limit is

$$\Sigma c^{(\lambda)} \frac{du_\lambda}{dz} \cdot \Sigma c^{(\lambda)} \frac{du_\lambda}{dz'}.$$

Now the normal integrals u_1, u_2 are linear combinations of the two integrals $\int z dz/w$, $\int dz/w$ (§ 281), and therefore the required limit can be written

$$\frac{cz + c'}{w(z)} \cdot \frac{cz' + c'}{w(z')},$$

where c and c' are constants.

Lastly, as we know that $\theta\left(\int_a^\xi\right)$ has its two zeros at the point a, we obtain $ca + c' = 0$. Therefore

$$\frac{\theta\left(\int_\eta^\xi du_\lambda\right)}{2c} = \frac{(z-z')\sqrt{z-a \cdot z'-a}}{2\sqrt{w(\xi) \cdot w(\eta)}} \exp \tfrac{1}{2} \Pi_{\xi\eta}^{\bar{\xi}\bar{\eta}} \quad . \quad . \quad (7),$$

a factor 2 being introduced to make the denominator on the right agree with that in (6).

The connexion of the thetas with the surface is not completed until the constants which appear on the left of equations (6) and (7) are expressed in terms of the sextic. This has been done by Thomæ (Schlömilch's Zeitschrift, t. xi., p. 427) for the present case.*

But, while it is beyond our purpose to go further into the matter, we can recognise the determinability of these constants; and it is

* The general problem, as stated by Riemann (Werke, p. 231), is to determine $\log \theta(0, 0, \cdots, 0)$ in terms of the $3p-3$ *class-moduli* (§ 284). The problem is reduced to one of integration by Thomæ in Crelle, t. lxvi., and the integration is effected, in the general hyperelliptic case, by the same writer in Crelle, t. lxxi. Fuchs (Crelle, t. lxxiii.) obtains the same results as Thomæ more simply by the use of the system of lower limits introduced by Clebsch and Gordan. Klein (Abel'sche Functionen, Math. Ann., t. xxxvi., p. 68) considers the case $p=3$, using the coefficients of the normal curve (the plane non-singular quartic) as class-moduli; whereas previous writers on the problem have assumed a knowledge of the branch-places.

found that when they are incorporated with the thetas, there results the same advantage as was gained when Jacobi's thetas (Chapter VII.) were replaced by Weierstrass's sigma-functions. In fact, the left sides of equations (6) and (7) are sigma-functions in the case $p=2$; but they are not yet in their final form, for which the reader is referred to Klein's memoirs, already cited.

The introduction of the higher sigma-functions was due to Weierstrass. See Schottky's Abriss einer Theorie der Abel'schen Functionen dreier Variabeln; Staude, Math. Annalen, t. xxiv. The above definition of a sigma-function agrees with that of Staude.

§ 306. *Existence theorems for the Riemann surface* T. It is assumed that the surface is spread in a finite number of sheets (say n) over the z-plane, that the sum of the orders of its branchings is $2p+2n-2$, and that its order of connexion is $2p+1$ (see § 275). A *potential function of the first kind* is everywhere continuous on T and its values at any point differ merely by periods due to the description of periodic paths on the surface T. We have stated in § 277 the properties of Abelian integrals on T; let each integral be expressed in the form $\xi+i\eta$, where ξ, η are real functions of x, y. The functions ξ, η are called potential functions of the first, second, or third kinds, according as they arise from integrals of the first, second, or third kinds. The potentials ξ, η are conjugate; when ξ is given, η is determined save as to an additive constant which arises from the description of periodic paths.

The following simple example shows how ξ, η behave at a discontinuity. Regarding ξ as a velocity-potential for a fluid motion in the plane of x, y, the equations $\xi=$ a constant, $\eta=$ a constant give lines of level and lines of flow. When the combination $\xi+i\eta=A\log(z-c)$, where A is real and $z-c=\rho\exp i\theta$, we have

$$\xi=A\log\rho,\ \eta=A\theta;$$

the lines of level are circles with centre c, and the lines of flow are straight lines which radiate from c. The motion is due to a *source* of strength $2\pi A$. The function η is many-valued. When A is purely imaginary and equal to iB,

$$\xi=-B\theta,\ \eta=B\log\rho,$$

and the motion is that of a liquid whirling round a point c in concentric circles. For a general discussion of the potentials which arise from logarithmic and algebraic discontinuities, see Klein's Schrift, Section I.

A *potential of the second kind* is continuous on T, except at a finite number r of places z_κ at which it is infinite like the real part of a function

$$(z-z_\kappa)^{-\lambda_\kappa}\{a_0+a_1(z-z_\kappa)+\cdots+a_{\lambda_\kappa-1}(z-z_\kappa)^{\lambda_\kappa-1}\},$$

and possesses the usual periodic properties at the cross-cuts. In the case of an *elementary potential of the second kind*, r and λ equal 1 [Klein-Fricke, p. 506].

A similar definition applies to *potentials* and to *elementary potentials of the third kind.**

The materials for establishing the existence on T of potentials of the first kind, and of elementary potentials of the second and third kinds, lie ready to hand. For instance, we have explained in Chapter IX. Schwarz's solution for a doubly connected region when the discontinuity along a cross-cut was an arbitrarily assigned real period. By an extension of this method there exists a function on a $(2p+1)$-ply connected Riemann surface with a puncture at a point, which has arbitrarily assigned periods at the $2p$ cross-cuts which reduce T to simple connexion, and has an assigned value at the puncture. In this way Schwarz's solution of § 266 establishes the existence-theorem for potentials of the first kind. The proofs in the text are those given in Klein-Fricke, t. i., p. 504 *et seq.*

Proof of the existence on T *of one-valued elementary potentials of the second kind.* Let Γ be a circular region on any sheet of T, and let it contain no branch-place of T in its interior or on its rim. The investigation of § 271 shows that there exists on Γ a function which is harmonic in the interior, and which takes an assigned system of values on the rim; and by § 273 there exists on Γ a potential function ξ_c which has all the usual properties of the harmonic function except that at one place of the interior it is infinite like the real part of $a_0/(z-c)$. The exclusion of a branch-place is not necessary; for a branch-place at which r sheets hang together can be enclosed by a region bounded by an r-fold circle, without affecting the solvability of Dirichlet's problem (§ 260). Recalling the fact that the method of combination leads to the solution of Dirichlet's problem for composite regions, it is easy to prove the existence on T of ξ_c. For the surface T can be covered, without leaving any gaps, by a *finite* number of overlapping regions with boundaries which are 1-fold or r-fold circles. In the process of construction of the composite region there are two stages. During the first stage the initial region is continued over T without including c, and the corresponding potential function is continuous throughout the interior of the composite region. The second stage begins with the inclusion of c, and the potential function has now the added property that it is discontinuous at c in the assigned manner.

* Attention must of course be paid to the additional periods due to the logarithmic discontinuities of the integral of the third kind.

To derive general existence-theorems for potentials of the kind met with in considering the integrals of algebraic functions, we can use two principles: the principle of superposition and the principle of juxtaposition. The latter of these is due to Klein (Math. Ann., t. xxi., pp. 161, 162, and Klein-Fricke, pp. 518–522). The principle of superposition can be explained immediately. Let ξ_c be infinite at c like the real part of $1/(z-c)$; then $d\xi_c/dc$, $d^2\xi_c/dc^2$, $\cdots$, are ∞^2, ∞^3, $\cdots$ at c. The potential function

$$\lambda_1\xi_c + \lambda_2\cdot\frac{d\xi_c}{dc} + \lambda_3\frac{d^2\xi_c}{dc^2} + \cdots + \lambda_r\frac{d^{r-1}\xi_c}{dc^{r-1}},$$

where the λ's are constants, is a potential function with an algebraic infinity of order r at the place c. We proceed to show how Klein derives potentials of the first and third kinds from ξ_c.

Let ξ be a one-valued elementary potential of the second kind which is infinite at z_0 like the real part of $c_0/(z-z_0)$. Multiply c_0 by an arbitrarily small *real* number dc, and let the result be the stroke dz_0. There exists on T a one-valued elementary potential of the second kind ξ_{z_0} which is infinite at z_0 like $dz_0/(z-z_0)$; to this Klein applies a process which amounts to integration. Let z_0 move from a place a to a place b along a chain of strokes dz_0, and to each stroke let there be attached the corresponding potential ξ_{z_0}; the potential which arises from the juxtaposition of these potentials is

$$\int_{z_0=a}^{z_0=b}\xi_{z_0}, = \xi_{a,b} \text{ (say)}.$$

As regards its discontinuities $\xi_{a,b}$ must behave like the real part of

$$\int_{z_0=a}^{z_0=b}\frac{dz_0}{z-z_0}, = \log\left(\frac{z-a}{z-b}\right).$$

Preserving the chain of strokes from a to b, construct for each stroke dz an elementary potential ξ'_{z_0} which is one-valued and continuous and which behaves at z_0 like the real part of $i dz_0/(z-z_0)$. Write

$$\xi'_{a,b} = \frac{-1}{2\pi}\int_{z_0=a}^{z_0=b}\xi'_{z_0}.$$

Regarding the chain of strokes as a barrier on T, the function ξ'_{ab} is one-valued and continuous on the surface which results when a cut is made along the barrier. The values on opposite sides of the barrier differ by 1, for ξ'_{ab} behaves along the barrier like the real part of

$$\frac{-1}{2\pi}\int_{z_0=a}^{z_0=b}\frac{i dz_0}{z-z_0}, = \frac{-i}{2\pi}\log\frac{z-a}{z-b}.$$

When the restriction is removed that the barrier is not to be crossed, ξ'_{ab} ranges itself as one branch of the system $\xi'_{ab} \pm m$, where $m=1, 2, 3, \cdots$.

Suppose that the chain of strokes begins and ends at $z_0 = a$, and that the corresponding cut does not sever the surface.

The potential ab loses its discontinuities on **T** and has no periods; therefore it degenerates into a constant; but ξ'_{ab} passes into a potential which is one-valued and continuous on the surface **T** after a retrosection has been drawn along the chain of strokes, and which has a period 1 along this retrosection.

Let us now consider the surface **T** and construct with reference to the cross-cuts $A_1, A_2, \cdots, A_p, B_1, B_2, \cdots, B_p$, $2p$ normal potentials of the first kind, $U_1, U_2, \cdots, U_{2p}$ which have the property that U_κ ($U_{\kappa+p}$) has the period 1 at A_κ (B_κ) where $\kappa = 1, 2, \cdots, p$, and the period 0 at the remaining $2p-1$ cross-cuts. These potentials are determined completely save as to additive constants; since the co-existence of two distinct potentials of the first kind U_κ, U'_κ with the period 1 at the κth cross-cut and the periods 0 at the remaining cross-cuts involves the existence of a potential U of the first kind which is one-valued and continuous throughout **T**, *i.e.* the difference $U_\kappa - U'_\kappa$ is a constant. From the $2p$ potentials U_κ it is possible to construct the most general real potential of the first kind. For the expression

$$U = \sum_{\kappa=1}^{p} (\lambda_\kappa U_\kappa + \lambda_{\kappa+p} U_{\kappa+p}) + \lambda \quad . \quad . \quad . \quad . \quad . \quad . \quad . \quad \text{(i.)}$$

in which the λ's are supposed real, is one-valued and continuous on **T**$'$ and has the *real* periods λ_κ, $\lambda_{\kappa+p}$ at A_κ, B_κ which can be assigned perfectly arbitrarily; and by the preceding reasoning the periods determine the function, a potential of the first kind, save as to an additive constant. To see that the potentials $U_1, U_2, \cdots, U_{2p}$ must be themselves linearly independent, assume a relation

$$\sum_{\kappa=1}^{p} (a_\kappa U_\kappa + a_{\kappa+p} U_{\kappa+p}) = 0 \quad . \quad . \quad . \quad . \quad . \quad . \quad . \quad . \quad . \quad \text{(ii.)}.$$

Since the non-vanishing periods of $a_\kappa U_\kappa$, $a_{\kappa+p} U_{\kappa+p}$ are a_κ, $a_{\kappa+p}$, it follows that the expression on the left-hand side of (ii.) has periods a_κ, $a_{\kappa+p}$; therefore the equation (ii.) requires that all the coefficients a_κ, $a_{\kappa+p}$ shall vanish, and consequently the potentials $U_1, U_2, \cdots, U_{2p}$ cannot be connected by a linear relation.

By associating with each of the potentials $U_1, U_2, \cdots, U_{2p}$ its conjugate potential, we can construct a system of $2p$ integrals of the first kind

$$W_1 = U_1 + iV_1, \;\; W_2 = U_2 + iV_2, \cdots, \;\; W_{2p} = U_{2p} + iV_{2p};$$

and the preceding work shows that every integral of the first kind can be expressed in one way, and in one way only, as a linear combination

$$(\lambda_\kappa W_\kappa + \lambda_{\kappa+p} W_{\kappa+p}) + \alpha + i\beta,$$

with real coefficients λ. [Klein-Fricke, t. i., p. 526.]

When combinations with complex coefficients are admitted it is possible to express all integrals W of the first kind as linear combinations of p linearly independent integrals. That it is impossible to express all integrals $W = U + iV$, in terms of fewer than p integrals, follows from the consideration that this would involve the theorem, that all potentials U of the first kind can be expressed as linear combinations of fewer than $2p$ selected potentials of the first kind. It is possible to express all integrals W in terms of more than p integrals of the first kind, but the representation will then be no longer unique; for assuming such a representation, each integral of the first kind can be resolved into its real and imaginary parts, and as a necessary consequence there results the erroneous theorem that every potential U can be expressed *uniquely* in terms of more than $2p$ potentials. [Klein-Fricke, t. i., p. 527.]

Enough has been said to indicate Klein's mode of passage from the theorems of Chapter IX. to the Riemann surface.

§ 307. The existence-theorem for the general Abelian integral on the Riemann surface T can be proved by synthesis from the potentials whose existence on T has been established, use being made of conjugate potentials and of differentiation with regard to parameters. It can also be established on Schwarz's lines (Schwarz, Ges. Werke, t. ii., p. 163). Suppose that the integral is $\xi + i\eta$ and that ξ is to be infinite, like the real part ξ' of a sum of algebraic functions

$$\sum_{\kappa=1}^{r} \{ (z-c_\kappa)^{-\lambda_\kappa} (a_0 + a_1(z-c_\kappa) + \cdots + a_{\lambda_\kappa - 1}(z-c_\kappa)^{\lambda_\kappa - 1}) + b_\kappa \log (z-c_\kappa) \};$$

suppose further that ξ has assigned periods at the cross-cuts A_κ, B_κ, and at the further cross-cuts made necessary by the excision of the places with logarithmic discontinuities. The function $\xi - \xi'$ is free from discontinuities on T″ other than the periods due to the cross-cuts, and is therefore one-valued and continuous on T″ (for the meaning of T″ see § 285). The unique existence of a function with these properties has been established; hence also the existence of ξ. By combining ξ with its conjugate η, we get the Abelian integral $\xi + i\eta$.

This brief sketch of the existence-theorems of Schwarz and Neumann is, we hope, sufficient to show the bearing of the theorems of Chapter IX. on Riemann's theory of the Abelian integrals. It must not be supposed that this is the sole application which can be made of existence-theorems.

In verification of this remark the reader may consult a memoir by Ritter on automorphic functions (Math. Ann., t. xli.). The discussion of Abelian integrals on an n-sheeted $(2p+1)$-ply connected closed surface spread over a plane forms but a small part of the programme drawn up by Riemann and elaborated by Klein, Poincaré, and others.

Mention may be made of two directions in which extensions have been sought. Without altering the surface T it is possible to establish the existence upon it of functions which have properties different from those of Abelian integrals or algebraic functions; examples of this kind are Appell's *Fonctions à multiplicateurs*, Acta Mathematica, t. xiii.* The other extension consists in removing the restriction that a Riemann surface is to be a surface *spread over a plane;* passing to the sphere with p handles, Klein has discussed the properties of the associated potentials and has indicated under what conditions a doubly extended manifoldness in hyper-space can be used as a Riemann surface.

References. — The materials for this chapter have been drawn mainly from the memoirs of Riemann, and from the treatises of Clebsch and Gordan, Fricke, and Neumann.

The importance of Riemann's work in the field of Abelian functions cannot readily be overestimated, but unfortunately it is expressed in too concise a form to be easily intelligible without the use of commentaries. The works of Neumann and Prym (Neue Theorie d. ult. Funct.), mentioned below, will be found useful helps to the acquisition of a working knowledge of Riemann's methods. The standard treatise on Abelian functions is that of Clebsch and Gordan.

Books on hyperelliptic and Abelian functions. — Abel's Works, p. 145, Mémoire sur une propriété générale d'une classe très-étendue de fonctions transcendantes; Briot, Fonctions abéliennes; Clebsch and Gordan, Abel'sche Functionen; Jacobi's Works; Klein, Ueber Riemann's Theorie der Algebraischen Functionen; Klein-Fricke, Modulfunctionen; Königsberger, Ueber die Theorie der Hyperelliptischen Integrale; Neumann, Abel'sche Integrale; Prym, Neue Theorie der ultraelliptischen Functionen, 2d ed., and Zur Theorie der Functionen in einer zweiblättrigen Fläche; Schottky, Abriss einer Theorie der Abel'schen Functionen vom Geschlecht 3; Weber, Theorie der Abel'schen Functionen vom Geschlecht 3.

Memoirs. — The memoir-literature on Abelian integrals and functions is very extensive. The following short list is added solely with the view of directing the reader's attention to a few of the numerous developments of the subject: —

(1) Memoirs based on Weierstrass's work.

Weierstrass, Zur Theorie der Abel'schen Functionen, Crelle, tt. xlvii., lii.

Henoch, De Abelianarum functionum periodis (Berlin dissertation, 1867).

Forsyth, On Abel's Theorem and Abelian Functions. Phil. Trans., 1882.

* Prym stated in his memoir (Crelle, t. lxx.) the existence of functions of the class discussed by Appell; the values of these functions on opposite banks of a cross-cut are connected by lineo-linear relations.

(2) Transformation.

Hermite, Sur la théorie de la transformation des fonctions abéliennes, Comptes Rendus, t. xl. Paris, 1855.

Königsberger, Ueber die Transformation der Abel'schen Functionen erster Ordnung, Crelle, t. lxiv. This memoir is written on Weierstrass's lines.

(3) Memoirs based on Klein's work.

Klein, Ueber hyperelliptische Sigmafunctionen, Math. Ann., tt. xxvii., xxxii.

Burkhardt, Beiträgen zur Theorie der hyperelliptischen Sigmafunctionen, Math. Ann., t. xxxii.

Klein, Abel'sche Functionen, Math. Ann., t. xxxvi.

Thompson (H. D.), Hyperelliptische Schnittsysteme und Zusammenordnung der algebraischen und transcendentalen Thetacharacteristiken.

(4) Memoirs on special systems of points on a basis-curve, class-moduli, exceptional cases in the theory of Abelian Functions, reality considerations, etc.

Brill and Nöther, Math. Ann., t. vii.

Schwarz, Ueber diejenigen algebraischen Gleichungen zwischen zwei veränderlichen Grössen, welche eine Schaar rationaler eindeutig umkehrbarer Transformationen in sich selbst zulassen, Crelle, t. lxxxvii., p. 140.

Poincaré, Sur un théorème de M. Fuchs, Acta Math., t. vii.

Painlevé, Sur les équations différentielles du premier ordre, Ann. Sc. de l'Éc. Norm. Sup., Ser. 3, t. viii., p. 103.

Weber, Ueber gewisse in der Theorie der Abel'schen Functionen auftretende Ausnahmefälle, Math. Ann., t. xiii.

Nöther, Ueber die invariante Darstellung algebraischer Functionen, Math. Ann., t. xvii.

Pick, Zur Theorie der Abel'schen Functionen, Math. Ann., t. xxix.

Hurwitz, Ueber Riemann'sche Flächen mit gegebenen Verzweigungspunkten, Math. Ann., t. xxxix.

Klein, Ueber Realitätsverhältnisse bei der einem beliebigen Geschlechte zugehörigen Normalcurve der ϕ, Math. Ann., t. xlii.

EXAMPLES.

(1) Prove that $|z+\sqrt{z^2-c^2}|+|z-\sqrt{z^2-c^2}|=|z+c|+|z-c|$.

(2) The condition of similarity of two triangles *abc*, *xyz* is

$$\begin{vmatrix} 1 & 1 & 1 \\ a & b & c \\ x & y & z \end{vmatrix}=0. \quad [\S\ 10.]$$

(3) Given a triangle *abc*, points *x*, *y* can be found such that the triangles *axy*, *ybx*, *xyc*, *abc* are similar.

(4) In the preceding question, the triangles *abc*, $xy\infty$ have the same Hessian points. [§ 33.]

(5) In § 25, show that when *a* and *b* are complex the question is reduced to the one considered by proper rotations of the two real axes.

(6) If $w=2z+z^2$, prove that the circle $|z|=1$ corresponds to a cardioid in the *w*-plane, and that an equilateral triangle in the circle corresponds to the points of contact of parallel tangents of the cardioid.

(7) The three points of contact of parallel tangents of a fixed cardioid have one fixed Hessian point.

(8) Prove that the Brocard points of a triangle $z_1z_2z_3$ are the Hessian points of the triangle h_+h_-k. [§§ 31, 33.]

(9) The middle points of the strokes $z_\kappa j_\kappa (\kappa=1, 2, 3)$ of § 31 have the same Hessian points as the triangles z_1, z_2, z_3 and j_1, j_2, j_3.

(10) Prove that the polar pair of the point ∞ with regard to a triangle are the foci of the maximum inscribed ellipse. [§ 35. See Quarterly Journal, June, 1891.]

(11) Show how to invert a triangle and its centroid into a triangle and its centroid. [Ib.]

(12) A convenient way of introducing homogeneity is to name any point x by the ratio x_1/x_2 of the strokes to it from two fixed points of the plane. Show that the point $(x_1+\lambda y_1)/(x_2+\lambda y_2)$ divides the stroke from x to y in the ratio $\lambda/1$. Here λ may be complex. [§ 35.]

(13) The points z_1, z_2, z_3, z_4 are concyclic if, α, β, γ being real,

$$\begin{vmatrix} 1 & 1 & 1 \\ \alpha & \beta & \gamma \\ z_2z_3+z_1z_4 & z_3z_1+z_2z_4 & z_1z_2+z_3z_4 \end{vmatrix} = 0. \quad \text{(Beltrami.)}$$

(14) Through any three of four points z_1, z_2, z_3, z_4 a circle is drawn. Prove that an anharmonic ratio of the inverses of any fifth point with regard to the four circles is equal to the corresponding anharmonic ratio of z_1, z_2, z_3, z_4.

(15) Let a_κ, b_κ ($\kappa = 1$ to 4) be quadrangles having a common Jacobian. Prove that, when the points are suitably paired,

$$\sum_\kappa 1/(a_\kappa - b_\kappa) = 0. \quad [\S\ 39.]$$

(16) The four points whose first polars with regard to a quartic U have a double point are the zeros of

$$g_3U - g_2H,$$

where H is the Hessian of U. [§ 36. See Clebsch, Géométrie, t. i., ch. 3, sec. 5.]

(17) In the preceding question, the first polar points which do not coincide are given by

$$3g_3U + g_2H = 0.$$

The quartic having the same Hessian as U is

$$3g_3U - 2g_2H.$$

(18) The intersections of three mutually orthogonal circles are the Jacobian points of a system of quartics. [§ 40.]

(19) Show how to invert a harmonic quadrangle into a square. [Laisant, Théorie des Équipollences.]

(20) If $u + iv$ be a one-valued function of z, and if normals be drawn to the z-plane at each point z, of lengths u and v, the two surfaces so formed have the same curvature at points on the same normal. [§ 17. See Briot et Bouquet, Fonctions Elliptiques, § 10.]

(21) A convex polygon which includes all the zeros of a rational polynomial $f(z)$ must also include the zeros of the derived polynomial $f'(z)$. [This generalization of Rolle's Theorem in the Theory of Equations was given by F. Lucas, Journal de l'École Polytechnique, 1879.]

(22) Discuss the amount of truth in Siebeck's paradox, that when $y=f(x)$, to an arbitrary curve in the x-plane corresponds a curve in the y-plane with fixed foci, namely the finite branch-points. [Siebeck, Crelle, t. lv., p. 236.]

(23) The series $\sum\limits_0^\infty a^n z^{n^2}$, where a is positive and < 1, is convergent in and on the circle $|z|=1$; and the same is true of the series formed by the rth derivatives of the terms. But the function defined by the series cannot be extended beyond the circle. [§ 104. See Freedholm, Comptes Rendus, 1890.]

(24) The deficiency of the equation $w^5+z^5-5wz^2=0$ is $p=2$. [Raffy. In his thesis, Recherches algébriques sur les intégrales abéliennes, ch. 3, M. Raffy shows how to find the number p by means of divisions and eliminations, without the solution of algebra equations.]

(25) The terms of a series $\Sigma f_n(z), = F(z)$, are holomorphic in any region Γ and are continuous on its boundary C, and the series $F(z)$ converges uniformly on C. Prove that

(i) The series $F(z)$ converges in every region Γ' which lies wholly within Γ (*i.e.* which has no point in common with C);

(ii) The series formed by the successive derivatives of the terms $f_n(z)$ converge uniformly in Γ' and represent the successive derivatives of $F(z)$. [See Painlevé, sur les lignes singulières des fonctions analytiques, p. 11.]

(26) The integral $\dfrac{1}{2\pi i}\displaystyle\int_C \dfrac{g(z)\,dz}{z-t}$ is equal to $g(t)$ or 0 according as t is within or without the contour C. Hence, write an expression which is equal to $g_1(t)$ within C_1, to $g_2(t)$ within $C_2, \cdots$, and to $g_\kappa(t)$ within C_κ, where the contours $C_1, C_2, \cdots, C_\kappa$ lie exterior to one another. [§ 139. Compare § 103. An important theory, based on this method of introducing lines of discontinuity is given in Hermite's Cours.]

(27) Let $\quad ax^2+2hxy+by^2+2gx+2fy+c=0.$

If x describe an ellipse whose foci are the branch-points in the x-plane, the two points y describe ellipses whose foci are the branch-points in the y-plane.

(28) Let $F(x, y)=0$ be an algebraic equation, and let the planes of x and y be covered with the Riemann surfaces which belong to the equation. On either surface a non-critical place, at which $d^2y/dx^2=0$, is such that to a line through it corresponds on the other surface a curve with an inflexion at the corresponding place.

Also a non-critical place at which the Schwarzian Derivative vanishes (that is,

$$\frac{y'''}{y'}-\frac{3}{2}\left(\frac{y''}{y'}\right)^2=0,$$

where accents denote differentiations for x) is such that to a line or circle through it corresponds a curve which has maximum or minimum curvature at the corresponding place.

(29) The necessary and sufficient condition that the region bounded by two confocal ellipses can be mapped on a circular ring is that the sums of the axes of the ellipses are proportional to the diameters of the circles. [Schottky, Crelle, t. lxxxiii.]

(30) Let $F(x, y)$ become $\{\phi(x, y)\}^\kappa$ by means of a transformation $x=R_1(x, y)$, $y=R_2(x, y)$. When the deficiency of $F=$ the deficiency of $\phi, >1$, then $\kappa=1$. [§ 183. See Weber, Crelle, t. lxxvi.]

(31) The expression $dx/\sqrt{U}$ is transformed by means of $z=-H/U$ into $\frac{1}{2}dz/\sqrt{4z^3-g_2z-g_3}$, where H is the Hessian, g_2 and g_3 the invariants, of the quartic U. [§ 203. This theorem, which brings the elliptic integral into Weierstrass's form, is due to Hermite. See Halphen, t. ii., p. 360.]

(32) If
$$u+v+w=0,$$
then
$$\sigma u\sigma_3 v\sigma_2 w+\sigma_3 u\sigma v\sigma_1 w+\sigma_2 u\sigma_1 v\sigma w=0. \quad [\S\ 211.]$$

(33) If
$$a+b+c+d=0,$$
then
$$(e_2-e_3)\sigma_1 a\sigma_1 b\sigma_1 c\sigma_1 d+(e_3-e_1)\sigma_2 a\sigma_2 b\sigma_2 c\sigma_2 d+(e_1-e_2)\sigma_3 a\sigma_3 b\sigma_3 c\sigma_3 d$$
$$=-(e_2-e_3)(e_3-e_1)(e_1-e_2)\sigma a\sigma b\sigma c\sigma d. \quad \text{[Cayley.]}$$

(34) With the notation of § 212,

$$\psi_r=\frac{(-1)^{r-1}}{\{1!\,2!\cdots(r-1)!\}^2}\begin{vmatrix}\wp' & \wp'' & \cdots & \wp^{(r-1)}\\ \wp'' & \wp''' & \cdots & \wp^{(r)}\\ \cdot & \cdot & \cdot & \cdot\\ \wp^{(r-1)} & \wp^{(r)} & \cdots & \wp^{(2r-3)}\end{vmatrix}. \quad \text{[Weierstrass.]}$$

(35) If
$$f(u) = \sigma(u-u_1)\sigma(u-u_2)\cdots\sigma(u-u_n),$$
$$g(u) = \sigma(u-v_1)\sigma(u-v_2)\cdots\sigma(u-v_n),$$
and
$$\sum_{\kappa=1}^{n} u_\kappa = \sum_{\kappa=1}^{n} v_\kappa,$$
then
$$\sum_{\kappa=1}^{n} \frac{g(u_\kappa)}{f'(u_\kappa)} = 0.$$

[Frobenius and Stickelberger, Crelle, t. lxxxviii.]

(36) $\sigma(3u)/\sigma^3 u = \frac{\wp'^3 u}{4}\frac{d^2\wp' u}{d\wp^2 u}$,

$\sigma(3u)\{\zeta(3u) - 3\zeta u\} = \sigma^3 u \wp'^3 u$. [Ib.]

(37) Prove that $\omega_1 = \frac{\pi i}{24}\frac{\delta \log \Delta}{\delta \eta_2}$, $\omega_2 = -\frac{\pi i}{24}\frac{\delta \log\Delta}{\delta\eta_1}$. [§ 214.]

(38) The Jacobian of g_2, g_3 with regard to ω_1, ω_2 is

$\frac{\delta(g_2, g_3)}{\delta(\omega_1, \omega_2)} = -\frac{8\Delta}{3\pi i}$. [§ 214. See Klein-Fricke, t. i., p. 120.]

(39) $\frac{\delta(\eta_1, \eta_2)}{\delta(\omega_1, \omega_2)} = -\frac{g_2}{12}$.

(40) $\frac{\delta(g_3, \log\Delta)}{\delta(\omega_1, \omega_2)} = \frac{8 g_2^2}{\pi i}$. [Klein-Fricke, t. i., p. 120.]

(41) $\frac{\delta(g_2, \log\Delta)}{\delta(\omega_1, \omega_2)} = \frac{144 g_3}{\pi i}$. [Ib. p. 119.]

(42) The Hessian of $\log\Delta$ with regard to ω_1, ω_2 is

$48 g_2/\pi^2$. [Ib. p. 123.]

(43) The Hessian of $\log\Delta$ with regard to η_1, η_2 is

$2^8 \cdot 3^3/\pi^2 g_2$.

(44) In § 219, show that

$$\int^{z, w} \frac{z+w}{2(z_0-w_0)}\frac{dz}{w} - \int^{z_0, w_0} \frac{z_0+w_0}{2(z-w)}\frac{dz_0}{w_0} = u\zeta c - c\zeta u + (2n+1)\pi i.$$

[Weierstrass-Schwarz, p. 89.]

(45) In § 221, show that the lines which bisect opposite edges of the rectangle of periods in the u-plane map into circles in the z-plane. [Weierstrass-Schwarz, p. 75.]

(46) Let w be an analytic function of z, such that lines parallel to a w-axis are mapped into algebraic curves in the z-plane. Then either

(1) w is an algebraic function of z;

(2) μw is the logarithm of an algebraic function of z, where μ is either real or pure imaginary; or

(3) $\mu w = \int_0^t \frac{dt}{\sqrt{(1-t^2)(1-k^2t^2)}}$, where k is real, and t is an algebraic function of z. [Schwarz, Ueber ebene algebraische Isothermen, Werke, t. ii., p. 260.]

(47) The orthogonal trajectory of the system of algebraic curves of the preceding example is also algebraic. [Ib.]

(48) With the notation of § 222,

$$\wp' u = \frac{\sqrt{f(y)}\,a_x^3 a_y + \sqrt{f(x)}\,a_x a_y^3}{(xy)^2}.$$

[Klein, Math. Ann., t. xxvii., p. 455.]

(49) When two functions u_1, u_2 are harmonic in regions which have a common two-dimensional simply connected region Γ, and coincide throughout a portion of Γ, however small, they coincide throughout Γ, and their continuations beyond Γ coincide. [Schwarz, Werke, p. 202.]

(50) In the case of the equation $x^4 + y^4 = 1$, we can take as the linearly independent integrals of the first kind

$$\int dx/y^3,\ \int dy/x^3,\ \int (ydx - xdy).$$

These integrals are elliptic. [§ 280. See Amstein, Bulletin de la Société Vaudoise, 1888, where this case is fully worked out as an example of Weber's memoir on Abelian functions of deficiency 3.]

(51) Given the equation

$$F(w, z) = 2w^5 - 5w^2 - z(az^2 + 2bz + c)^2 = 0,$$

where c, $b^2 - ac$ are different from 0, prove that

$$\int \frac{\lambda(az^2 + 2bz + c) + 2Bwz + Cw^2 + 2Ew}{w^4 - w} dz$$

is an integral of the first kind, whatever be the values of λ, B, C, E. [Raffy.]

(52) The number of linearly independent integrals of the first kind may be $> p$ when the basis-equation is reducible. [Christoffel.]

(53) Prove that $$Q^{xy}_{c_2c_3} + Q^{xy}_{c_3c_1} + Q^{xy}_{c_1c_2}$$
is an integral of the first kind. [§ 285. See Clebsch and Gordan, p. 22 *et seq.*]

(54) The number of three-sheeted surfaces with β assigned branch-points is $\frac{1}{2}(3^{\beta-2}-1)$. [§ 284. See Kasten, Zur Theorie der dreiblättrigen Riemann'schen Fläche; Hurwitz, Math. Ann., t. xxxix.]

(55) The comparison drawn in § 297 between the function σ in the case $p=1$, and the function θ in the general case, is not exact, inasmuch as in the former case the surface T′ has been mapped by means of an integral u of the first kind on the u-plane. Show that, in the general case, the surface T′ is mapped by means of an integral of the first kind into p parallelograms which lie one over another, and which hang together at $2p-2$ branch-places. By a proper choice of the cross-cuts of T′, the parallelograms become rectilinear. [Riemann, Werke, 1st ed., p. 114.]

(56) In § 303, when the sextic contains only even powers of z (a form into which a sextic can be inverted when the roots are in involution), prove that $$\tau_{12}^2 - \tau_{11}\tau_{22} = 1.$$

If the sextic be of the form $z^6 + a^6$, then
$$\tau_{11} = \tau_{22} = 2i/\sqrt{3},\ \tau_{12} = -i/\sqrt{3}.$$

[A discussion of the values of $\tau_{\alpha\beta}$, for sextics which can be linearly transformed into themselves, is given in section iii. of Bolza's paper, American Journal, t. x.]

(57) Between four odd double sigma-functions there is the square relation
$$\begin{vmatrix} 1 & 1 & 1 & 1 \\ a_\alpha & a_\beta & a_\gamma & a_\delta \\ a_\alpha^2 & a_\beta^2 & a_\gamma^2 & a_\delta^2 \\ \sigma_\alpha^2 & \sigma_\beta^2 & \sigma_\gamma^2 & \sigma_\delta^2 \end{vmatrix} = 0,$$
a_α being the branch-point associated with σ_α. [§ 305. Compare § 244.]

(58) Prove the four product-relations
$$\left\| \begin{matrix} 1 & 1 & 1 & 1 \\ a_1 & a_3 & a_5 & a_2 \\ \sigma_1\sigma_{12} & \sigma_3\sigma_{32} & \sigma_5\sigma_{52} & \sigma_2\sigma_{00} \end{matrix} \right\| = 0,$$
where $\sigma_{\alpha\beta} = \theta_{\alpha\beta}/c_{\alpha\beta}$ for any even mark. [§ 304. Compare § 247.]

INDEX.

[The numbers refer to the pages.]

GLOSSARY.

Abbildung, conforme, conform representation.
Abweichung, amplitude.
Absolute value, *module, absoluter Betrag*, p. 2.
Absoluter Betrag, absolute value.
Algebraic configuration (or formation), *algebraisches Gebilde*, p. 328.
Algebraisches Gebilde, algebraic configuration.
Amplitude, *argument, Abweichung*, p. 2.
Analysis Situs, topology, geometry of situation (H. J. S. Smith), geometry of position (Clerk Maxwell), p. 232.
Argument, amplitude.
Ausserwesentlicher singulärer Punkt, non-essential singularity, pole.

Bank (of a cross-cut), *Ufer*.
Barrier, *coupure* (Hermite); see cross-cut.
Bereich, region.
Betrag, absoluter, absolute value.
Bilinear transformation, *Kreisverwandschaft*.
Birational transformation, correspondence, *transformation birationelle, correspondance uniforme ou birationelle, eindeutige Transformation*.
Branch, *Zweig*.
Branch-cut, cross-line (Clifford), *Verzweigungschnitt*.
Branch-place, *Verzweigungsstelle*. See branch-point.
Branch-point, spire, spiral point (Smith), *point singulier d'embranchement, point de ramification, Verzweigungspunkt, Windungspunkt*.

Charakteristik, mark.
Circuit, used by Briot and Bouquet in the sense of a closed path which encloses all the loops drawn from some point to the branch-points, *vollständiger Umgang*.

Class-moduli, *Moduln einer Klasse* (Riemann), p. 441.
Conform representation, orthomorphic transformation, *représentation conforme, conforme Abbildung*, p. 14.
Connexion, *Zusammenhang*.
Continuation, *extension analytique, Fortsetzung*.
Contour élémentaire (Puiseux), loop.
Convergenz in gleichem Grade, uniform convergence.
Correspondance uniforme, birational correspondence.
Coupure (Hermite), cross-cut, barrier.
Critical points, p. 12, p. 127.
Cross-cut, *section* (sometimes *coupure*), *Querschnitt*.

Deficiency, rank, genus (H. J. S. Smith), *Geschlecht, Rang*.
Delimited, *vollständig begränzt*.
Discontinuität erster Gattung, non-essential singularity, pole, infinity.
Domain, *domaine, Umgebung*. Sometimes used in other senses.
Doppelfläche, unilateral surface.

Einändrig, one-valued (not in common use).
Eindeutig, one-valued.
Eindeutige Transformation, birational transformation.
Einfach zusammenhängend, simply connected.
Equivalent paths, *simultane Bahnen der Niveaupunkte* (Neumann).
Essential singularity, *point singulier essentiel, wesentlicher singulärer Punkt*.
Extension analytique, continuation.

Facteur primaire, primary factor.
Fonction entière, holomorphic function, integral function.
fractionnaire, meromorphic function, fractional function.

Fonction (*continued*).
holomorphe, holomorphic function.
méromorphe, meromorphic function.
monogène, function, monogenic function.
synectique, holomorphic function.
Fortsetzung, continuation.
Functionenpaar, p. 141.

Ganze algebraische Function, integral algebraic function.
Ganze eindeutige Function, or *ganze Function*, holomorphic throughout finite part of plane.
Gebiet, region.
Genre, deficiency. Used by Laguerre in another sense, p. 187.
Geometry of situation, see Analysis Situs.
Geschlecht, deficiency.
Gleichmässig convergent, uniformly convergent.
Gleichverzweigt, like-branched.
Grenzpunkte, limiting points.
Grundzahl, index.

Harmonic function, *harmonisch*. The term is used in a special sense in the chapter on Dirichlet's problem, p. 377.
Holomorphic function, *fonction holomorphe*, *entière* (Halphen), *synectique* (Cauchy), *ganze eindeutige Function*, *ganze Function*, (holomorphic throughout finite part of plane, p. 161).

Index of a surface (or system of surfaces), order of connexion (for a single surface), *ordre adelphique* (Laurent), *Grundzahl*.
Infinitely slow (convergence), *unendlich verzögerte*.
Infinity, *infini*, *Unstetigkeitspunkt*. See non-essential singularity.
Integral algebraic function of a Riemann surface (p. 263), *ganze algebraische Function*.
Integral function, whether rational (p. 9) or transcendental (p. 112), *fonction entière*, *ganze Function*.
Integral series, power-series, *série entière*, *série de puissances*, *Potenzreihe*.
Isogonal transformation, *isogonale Verwandtschaft*.

Kreisverwandtschaft, bilinear transformation.
Kreuzungspunkt, turn-point.

Lacet, loop.
Like-branched, *gleichverzweigt*, p. 245.
Limiting points, *Grenzpunkte*.
Loop, *lacet*, *contour élémentaire* (Puiseux), *Schleife*.

Many-placed, *mehrwerthig*.
Many-valued, *multiforme*, *plurivoque*, *polytrope*, *mehrdeutig*, sometimes *mehrwerthig*.
Mark, *Characteristik*.
Mass of points, *masse*, *Punktmenge*.
Mehrdeutig, many-valued.
Mehrfach zusammenhängend, multiple connected.
Mehrwerthig, many-placed (sometimes many-valued).
Meromorphic function, Fonction méromorphe, Fonction fractionnaire (Halphen).
Module, absolute value.
Moduln einer Klasse, class-moduli.
Moduli of a theta-function, *Theta-moduln*, p. 343.
Monadelphe, simply connected.
Monodrome, one-valued.
Monogenic function, function, *fonction monogène*, *monogene Function*.
Monotrope, one-valued.
Multiforme, many-valued.
Multiply connected (region), *aire à multiple connection*, *polyadelphe*, *mehrfach zusammenhängend*.

Nähe, neighbourhood.
Neighbourhood, *voisinage*, *Nähe*.
Non-essential singularity, *pôle*, *ausserwesentlicher singulärer Punkt*, *Discontinuität erster Gattung*, *Pol*.

One-valued, *uniforme*, *univoque*, *monodrome*, *monotrope*, *eindeutig*, *einwerthig*, *einändrig*.
Order of connexion, see index.
Ordre adelphique, index.
Orthomorphic transformation (Cayley), see conform representation.

Place, *Stelle*.
Point de ramification, branch-point;
singulier d'embranchement, branch-point;
singulier essentiel, essential singularity.
Pole, see non-essential singularity.
Polyadelphe, multiply connected.
Polytrope, many-valued.
Potenzreihe, integral series.
Power-series, integral series.
Primary factor, *facteur primaire*, *Primfactor*, p. 190.

Prim-factor, primary factor.
Punktmenge, mass of points.

Querschnitt, cross-cut.

Ramification, point de, branch-point.
Rang, deficiency.
Region, *Bereich*, *Gebiet*;
of continuity, *Stetigkeitsbereich*.
Regular sequence, *reguläre Folge*.
Retrosection, *rétrosection*, *Rückkehrschnitt*.
Rückkehrschnitt, retrosection.

Schleife, loop.
Section, cross-cut.
Série de puissances, integral series;
entière, integral series.
Simply connected (region), (*aire*) *à simple connexion*, *monadelphe*, *einfach zusammenhängend*.
Simultane Bahnen der Niveaupunkte, equivalent paths.
Spire, see branch-point.
Stelle, place.
Strecke, stroke.
Stroke, *strecke*.
Synectique, holomorphic.

Tied, *verknüpft*, p. 441.
Transcendental integral function, *transcendente ganze Function*, p. 112.
Turn-point, *Kreuzungspunkt*.

Ufer, bank (of a cross-cut).
Ultraelliptic functions, hyperelliptic functions in the case $p=2$. This term is falling into disuse.
Umgang, circuit.
Umgebung, domain.
Unbedingt convergent, unconditionally convergent.
Unconditionally convergent, *unbedingt convergent*.
Unendlich verzögerte (Convergenz), infinitely slow (convergence).
Uniforme, one-valued.
Uniform (convergence), *gleichmässig*, *in gleichem Grade*.
Unilateral surface, one-sided, *Doppelfläche*.
Unstetigkeitspunkt, discontinuity, often infinity.

Verknüpft, tied.
Verzweigungschnitt, branch-cut.
Verzweigungspunkt, branch-point.
Voisinage, neighbourhood.
Vollständig begränzt, delimited.

Werthigkeit, number of times that a function takes any assigned-value.
[An m-placed function has '*Werthigkeit*' m.]
Wesentlicher singulärer Punkt, essential singularity.
Wesentliche singuläre Stelle, essential singularity.
Windungsfläche, winding-surface; example on p. 207.
Windungspunkt, branch-point.

Zusammenhang, connexion.
Zusammenhängend, *einfach*, simply connected.
mehrfach, multiply connected.
Zweig, branch.

TABLE OF REFERENCES.

[1] Explicit references to Clebsch's Lectures on Geometry are not to Lindemann's edition, but to Benoist's translation of Lindemann's work.

MATHEMATICAL TEXT-BOOKS

PUBLISHED BY

MACMILLAN & CO.

Mechanics.

RIGID DYNAMICS.

An Introductory Treatise. By W. STEADMAN ALDIS, M.A. $1.00.

ELEMENTARY APPLIED MECHANICS.

Part II. Transverse Stress. By T. ALEXANDER and A. W. THOMPSON. 8vo. $2.75.

EXPERIMENTAL MECHANICS.

A Course of Lectures delivered to the Royal College of Science for Ireland. By Sir R. S. BALL, LL.D., F.R.S. Second Edition. With Illustrations. 12mo. $1.50.

Teachers of experimental mechanics can have no better assistant than this book, for the lectures are happy in their illustrations, clear in their demonstrations, and admirable in their arrangement. — *Manchester Guardian.*

LESSONS ON THERMODYNAMICS.

By R. E. BAYNES, M.A. 12mo. $1.90.

WORKS BY W. H. BESANT, D.Sc.,

Lecturer of St. John's College, Cambridge.

A TREATISE ON HYDROMECHANICS.

Fifth Edition, revised. Part I. Hydrostatics. 12mo. $1.25.

A TREATISE ON DYNAMICS. $1.75.

ELEMENTARY HYDROSTATICS. 16mo. $1.00.

ELEMENTS OF DYNAMIC.

An Introduction to the Study of Motion and Rest in Solid and Fluid Bodies. By W. KINGDON CLIFFORD, F.R.S.

Part I. Books I.–III. 12mo. $1.90.

Part II. Book IV. and Appendix. 12mo. $1.75.

APPLIED MECHANICS.

An Elementary General Introduction to the Theory of Structures and Machines. By JAMES H. COTTERILL, F.R.S. 8vo. $5.00.

The book is a valuable one for students. . . . A very admirable feature of the book is the frequent interjecting of practical problems for solution, at the close of the different sections. — *School of Mines Quarterly.*

This work constitutes a manual of applied mechanics which, in its completeness and adaptation to the requirements of the engineer student, is without a rival. — *Academy.*

ELEMENTARY MANUAL OF APPLIED MECHANICS.

By Prof. J. H. COTTERILL, F.R.S., and J. H. SLADE. 12mo. $1.25.

GRAPHICAL STATICS.

By LUIGI CREMONA. Two Treatises on the Graphical Calculus and Reciprocal Figures in Graphical Calculus. Authorized English Translation by T. HUDSON BEARE. 8vo. $2.25.

A TREATISE ON ELEMENTARY DYNAMICS.

For the Use of Colleges and Schools. By WILLIAM GARNETT, M.A., D.C.L. Fifth Edition, revised. $1.50.

ELEMENTARY STATICS.

By H. GOODWIN, D.D., Bishop of Carlisle. Second Edition. 75 cents.

WORKS BY JOHN GREAVES, M.A.

A TREATISE ON ELEMENTARY STATICS.

Second Edition, revised. 12mo. $1.90.

It can be unreservedly recommended. — *Nature.*

STATICS FOR BEGINNERS. 16mo. 90 cents.

HYDROSTATICS.

By ALFRED GEORGE GREENHILL. *In preparation.*

THE APPLICATIONS OF PHYSICAL FORCES.

By A. GUILLEMIN. Translated and Edited by J. NORMAN LOCKYER, F.R.S. With Colored Plates and Illustrations. Royal 8vo. $6.50.

ELEMENTARY DYNAMICS OF PARTICLES AND SOLIDS.

By W. M. HICKS. 12mo. $1.60.

ELEMENTARY MECHANICS.

Stage I. By J. C. HOROBIN, B.A. With Numerous Illustrations. 12mo. Cloth. 50 cents. *Stages II. and III. in preparation.*

THE ELEMENTS OF GRAPHICAL STATICS.

A Text-Book for Students of Engineering. By LEANDER M. HOSKINS, C.E., M.S., Prof. of Applied Mechanics, Leland Stanford Jr. University, Palo Alto, Cal.

A TREATISE ON THE THEORY OF FRICTION.

By J. H. JELLETT, B.D., late Provost of Trinity College, Dublin. $2.25.

THE MECHANICS OF MACHINERY.

By ALEXANDER B. W. KENNEDY, F.R.S. With Illustrations. 12mo. $3.50.

HYDRODYNAMICS.

By H. LAMB. A Treatise on the Mathematical Theory of Fluid Motion. 8vo. $3.00.

WORKS BY THE REV. J. B. LOCK, M.A.

Fellow of Gonville and Caius College, Cambridge.

DYNAMICS FOR BEGINNERS. 16mo. $1.00.

Mr. Lock writes like one whose heart is in his work. He shows rare ingenuity in smoothing the difficulties in the path of the beginner, and we venture to predict that his book will meet with a wide and hearty welcome among practical teachers. — *Athenæum.*

ELEMENTARY STATICS. 16mo. $1.10.

The whole subject is made interesting from beginning to end, and the proofs of the various propositions are very simple and clear. We have no doubt that the book will be appreciated by all who have an opportunity of judging of its merits.—*Nature.*

MECHANICS FOR BEGINNERS.

Part I. Mechanics of Solids. 90 cents.

An introductory book, which, it is hoped, may be found useful in schools and by students preparing for the elementary stage of the Science and Art and other public examinations.

ELEMENTARY HYDROSTATICS. *In preparation.*

WORKS BY S. L. LONEY, M.A.,

Fellow of Sidney Sussex College, Cambridge.

A TREATISE ON ELEMENTARY DYNAMICS.

New and Enlarged Edition. 12mo. $1.90.

Solutions of the Examples contained in the Above. *In the Press.*

THE ELEMENTS OF STATICS AND DYNAMICS.

Part I. Elements of Statics. $1.25. Part II. Elements of Dynamics. $1.00. Complete in one volume. 16mo. $1.90.

AN ELEMENTARY TREATISE ON KINEMATICS AND DYNAMICS.

By JAMES GORDON MACGREGOR, M.A., D.Sc., Munro Professor of Physics, Dalhousie College, Halifax. 12mo. $2.60.

The problems and exercises seem to be well chosen, and the book is provided with a useful Index. Altogether it will repay a careful study. — *Educational Times.*

WORKS BY G. M. MINCHIN, M.A.,

Professor of Mathematics in the Royal Indian Engineering College.

A TREATISE ON STATICS.

Third Edition, corrected and enlarged. Vol. I. Equilibrium of Coplanar Forces. 8vo. $2.25. Vol. II. Statics. 8vo. $4.00.

UNIPLANAR KINEMATICS OF SOLIDS AND FLUIDS. 12mo. $1.90.

A TREATISE ON ELEMENTARY MECHANICS.

For the use of the Junior Classes at the Universities, and the Higher Classes in Schools. With a collection of examples by S. PARKINSON, F.R.S. Sixth Edition. 12mo. $2.25.

LESSONS ON RIGID DYNAMICS.

By the Rev. G. PIRIE, M.A. 12mo. $1.50.

ELEMENTARY STATICS.

By G. RAWLINSON, M.A. Edited by E. STURGES. 8vo. $1.10.

WORKS BY E. J. ROUTH, LL.D., F.R.S.,

Fellow of the Senate of the University of London.

A TREATISE ON THE DYNAMICS OF A SYSTEM OF RIGID BODIES.

With Examples. New Edition Revised and Enlarged. 8vo.

Part I. Elementary. Fifth Edition Revised and Enlarged. $3.75.
Part II. Advanced. $3.75.

STABILITY OF A GIVEN STATE OF MOTION,

Particularly Steady Motion. 8vo. $2.25.

HYDROSTATICS FOR BEGINNERS.

By F. W. SANDERSON, M.A. 16mo. $1.10. *New edition preparing.*

Mr. Sanderson's work has many merits, and is evidently a valuable text-book, being the work of a practical and successful teacher.—*Canada Educational Monthly.*

SYLLABUS OF ELEMENTARY DYNAMICS.

Part I. Linear Dynamics. With an Appendix on the Meanings of the Symbols in Physical Equations. Prepared by the Association for the Improvement of Geometrical Teaching. 4to. 30 cents.

A TREATISE ON DYNAMICS OF A PARTICLE.

By P. G. TAIT, M.A., and W. J. STEELE. Sixth Edition, Revised. $3.00.

WORKS BY ISAAC TODHUNTER, F.R.S.

Late Fellow and Principal Mathematical Lecturer of St. John's College.

MECHANICS FOR BEGINNERS.

With Numerous Examples. New Edition. 18mo. $1.10. KEY. $1.75.

A TREATISE ON ANALYTICAL STATICS.

Fifth Edition. Edited by Prof. J. D. EVERETT, F.R.S. 12mo. $2.60.

A COLLECTION OF PROBLEMS IN ELEMENTARY MECHANICS.

By W. WALTON, M.A. Second Edition. $1.50.

PROBLEMS IN THEORETICAL MECHANICS.

Third Edition, Revised. With the addition of many fresh Problems. By W. WALTON, M.A. 8vo. $4.00.

Higher Pure Mathematics.

WORKS BY SIR G. B. AIRY, K.C.B., formerly Astronomer-Royal.

ELEMENTARY TREATISE ON PARTIAL DIFFERENTIAL EQUATIONS.

With Diagrams. Second Edition. 12mo. $1.50.

ON THE ALGEBRAICAL AND NUMERICAL THEORY OF ERRORS

of Observations and the Combination of Observations. Second Edition, revised. 12mo. $1.75.

NOTES ON ROULETTES AND GLISSETTES.

By W. H. BESANT, D.Sc., F.R.S. Second Edition, Enlarged. $1.25.

A TREATISE ON THE CALCULUS OF FINITE DIFFERENCES.

By the late GEORGE BOOLE. Edited by J. F. MOULTON. Third Edition. 12mo. $2.60.

ELEMENTARY TREATISE ON ELLIPTIC FUNCTIONS.

By ARTHUR CAYLEY, D.Sc., F.R.S. *New Edition preparing.*

DIFFERENTIAL CALCULUS.

With Applications and Numerous Examples. An Elementary Treatise. By JOSEPH EDWARDS, M.A. 12mo. $2.75.

A useful text-book both on account of the arrangement, the illustrations, and applications and the exercises, which contain more easy examples than are usually given. — *Educational Times.*

AN ELEMENTARY TREATISE ON SPHERICAL HARMONICS

And Subjects Connected with Them. By Rev. N. M. FERRERS, D.D., F.R.S. 12mo. $1.90.

WORKS BY ANDREW RUSSELL FORSYTH, M.A.

A TREATISE ON DIFFERENTIAL EQUATIONS. 8vo. $3.75.

As Boole's "Treatise on Different Equations" was to its predecessor, Wymer's, this treatise is to Boole's. It is more methodical and more complete, as far as it goes, than any that have preceded it. — *Educational Times.*

THEORY OF DIFFERENTIAL EQUATIONS.

Part I. Exact Equations and Pfaff's Problems. 8vo. $3.75.

AN ELEMENTARY TREATISE ON CURVE TRACING.

By PERCIVAL FROST, M.A. 8vo. $3.00.

DIFFERENTIAL AND INTEGRAL CALCULUS.

With Applications. By ALFRED GEORGE GREENHILL, M.A. 12mo. $2.00.

APPLICATION OF ELLIPTIC FUNCTIONS.

By ALFRED GEORGE GREENHILL, M.A. 12mo. $3.50.

AN ELEMENTARY TREATISE ON THE DIFFERENTIAL AND INTEGRAL CALCULUS.

By G. W. HEMMING, M.A. 8vo. $2.50.

THEORY OF FUNCTIONS.

By JAMES HARKNESS, M.A., Professor of Mathematics, Bryn Mawr College, and FRANK MORLEY, A.M., Professor of Mathematics, Haverford College, Pa. *In preparation.*

INTRODUCTION TO QUATERNIONS.

With Examples. By P. KELLAND, M.A., and P. G. TAIT, M.A. Second Edition. 12mo. $2.00.

HOW TO DRAW A STRAIGHT LINE.

A Lecture on Linkages. By A. B. KEMPE, B.A. With Illustrations. *Nature Series.* 12mo. 50 cents.

DIFFERENTIAL CALCULUS FOR BEGINNERS.

With Examples. By ALEXANDER KNOX, B.A. 16mo. 90 cents.

MESSENGER OF MATHEMATICS.

Edited by J. W. L. GLAISHER. Published Monthly. 35 cents each number.

WORKS BY THOMAS MUIR,
Mathematical Master in the High School, Glasgow.

A TREATISE ON THE THEORY OF DETERMINANTS.

With Examples. 12mo. *New Edition in Preparation.*

THE THEORY OF DETERMINANTS

In the Historical Order of its Development. Part I. Determinants in General. Leibnitz (1693) to Cayley (1841). 8vo. $2.50.

TREATISE ON INFINITESIMAL CALCULUS.

By Bartholomew Price, M.A., F.R.S., Professor of Natural Philosophy, Oxford.

Vol. I. Differential Calculus. Second Edition. 8vo. $3.75.

Vol. II. Integral Calculus, Calculus of Variations, and Differential Equations. 8vo. *Reprinting.*

Vol. III. Statics, including Attractions; Dynamics of a Material Particle. 8vo. $4.00.

Vol. IV. Dynamics of Material Systems. Together with a Chapter on Theoretical Dynamics, by W. F. Donkin, M.A. 8vo. $4.50.

A TREATISE ON THE THEORY OF DETERMINANTS,

And their Applications in Analysis and Geometry. By Robert Scott, M.A. 8vo. $3.50.

FACTS AND FORMULÆ IN PURE MATHEMATICS

And Natural Philosophy. Containing Facts, Formulæ, Symbols, and Definitions. By the late G. R. Smalley, B.A., F.R.A.S. New Edition by J. M'Dowell, M.A., F.R.A.S. 16mo. 70 cents.

MATHEMATICAL PAPERS OF THE LATE REV. J. S. SMITH.

Savilian Professor of Geometry in the University of Oxford. With Portrait and Memoir. 2 vols. 4to. *In preparation.*

AN ELEMENTARY TREATISE ON QUATERNIONS.

By P. G. Tait, M.A., Professor of Natural Philosophy in the University of Edinburgh. Third Edition, Much Enlarged. 8vo. $5.50.

Works by Isaac Todhunter, F.R.S.,

Late Principal Lecturer on Mathematics in St. John's College, Cambridge.

AN ELEMENTARY TREATISE ON THE THEORY OF EQUATIONS.

12mo. $1.80.

A TREATISE ON THE DIFFERENTIAL CALCULUS.

12mo. $2.60. Key. $2.60.

A TREATISE ON THE INTEGRAL CALCULUS

And its Applications. 12mo. $2.60. Key. $2.60.

AN ELEMENTARY TREATISE ON LAPLACE'S,

Lamé's and Bessel's Functions. 12mo. $2.60.

TRILINEAR CO-ORDINATES,

And Other Methods of Modern Analytical Geometry of Two Dimensions. An Elementary Treatise. By W. Allen Whitworth, M.A. 8vo. $4.00.

Analytical Geometry. (PLANE AND SOLID

SOLID GEOMETRY.

By W. STEADMAN ALDIS. 12mo. $1.50.

ELEMENTS OF PROJECTIVE GEOMETRY.

By LUIGI CREMONA, LL.D. Translated by CHARLES LEUDESDORF, M.A. 8vo. $3.25.

EXERCISES IN ANALYTICAL GEOMETRY.

By J. M. DYER, M.A. With Illustrations. 12mo. $1.25.

A TREATISE ON TRILINEAR CO-ORDINATES,

The Method of Reciprocal Polars, and the Theory of Projections. By the Rev. N. M. FERRERS. Fourth Edition. 12mo. $1.75.

WORKS BY PERCIVAL FROST, D.Sc., F.R.S.

Lecturer in Mathematics, King's College, Cambridge.

AN ELEMENTARY TREATISE ON CURVE TRACING. 8vo. $3.00.

SOLID GEOMETRY.

A New Edition, revised and enlarged, of the Treatise, by FROST and WOLSTENHOLME. Third Edition. 8vo. $6.00.

Hints for the Solution of Problems in the above. 8vo. $3.00.

THE ELEMENTS OF SOLID GEOMETRY.

By R. BALDWIN HAYWARD, M.A., F.R.S. 12mo. 75 cents.

THE GEOMETRY OF THE CIRCLE.

By W. J. MCCLELLAND, M.A., Trinity College, Dublin; Head Master of Santry School. Crown 8vo. Illustrated. $1.60.

AN ELEMENTARY TREATISE ON CONIC SECTIONS

And Algebraic Geometry. With Examples and Hints for their Solution. By G. HALE PUCKLE, M.A. Fifth Edition, Revised and Enlarged. 12mo. $1.90.

WORKS BY CHARLES SMITH, M.A.

Fellow and Tutor of Sidney Sussex College, Cambridge.

AN ELEMENTARY TREATISE ON CONIC SECTIONS.

Seventh Edition. 12mo. $1.60. KEY. $2.60.

The best elementary work on these curves that has come under our notice. — *Academy.*

AN ELEMENTARY TREATISE ON SOLID GEOMETRY. 12mo. $2.60.

WORKS BY ISAAC TODHUNTER, F.R.S.

Late Principal Lecturer on Mathematics in St. John's College.

PLANE CO-ORDINATE GEOMETRY,

As Applied to the Straight Line and the Conic Sections. 12mo. $1.80. KEY. By C. W. BOURNE, M.A. 12mo. $2.60.

EXAMPLES IN ANALYTICAL GEOMETRY

Of Three Dimensions. New Edition, Revised. 12mo. $1.00.

ANALYTICAL GEOMETRY FOR SCHOOLS.

By Rev. T. G. VYVYAN. Fifth Edition, Revised. $1.10.

NEW AND FORTHCOMING BOOKS

ON

PURE AND APPLIED MATHEMATICS

PUBLISHED BY

MACMILLAN & CO.

READY IMMEDIATELY.

A SHORT COURSE IN THE THEORY OF DETERMINANTS.

By LAENAS GIFFORD WELD, M.A., Professor of Mathematics, State University of Iowa.

THE MECHANICS OF HOISTING MACHINERY:
Including Accumulators, Excavators, and Pile Drivers.

A Text-book for Technical Schools and a Guide for Practical Engineers. By Dr. JULIUS WEISBACH and Prof. GUSTAV HERMANN. With 177 illustrations. Authorized translation from the second German edition by KARL P. DAHLSTROM, M.E., Instructor of Mechanical Engineering at the Lehigh University, Pa.

AN ELEMENTARY TREATISE ON THEORETICAL MECHANICS.

By ALEXANDER ZIWET, C.E., Assistant Professor of Mathematics in the University of Michigan. Part I. Kinematics. *Shortly.*

THE APPLICATIONS OF ELLIPTIC FUNCTIONS.

By ALFRED GEORGE GREENHILL, M.A., F.R.S., Author of "Differential and Integral Equations," etc. 8vo, $3.00, *net.*

CO-PLANAR VECTORS AND TRIGONOMETRY.

By R. BALDWIN HAYWARD, M.A., F.R.S., Author of "The Elements of Solid Geometry," etc. $2.00, *net.*

AN ELEMENTARY TREATISE ON PURE GEOMETRY.

With numerous Examples. By JOHN WELLESLEY RUSSELL, M.A. 12mo, $2.60, *net.*

AN ELEMENTARY TREATISE ON MODERN PURE GEOMETRY.

By R. LACHLAN, M.A. 12mo, $2.25, *net.*

DIFFERENTIAL CALCULUS FOR BEGINNERS.

By JOSEPH EDWARDS, M.A., formerly Fellow of Sidney Sussex College, Cambridge. Author of "A Treatise on the Differential Calculus." 16mo, $1.10, *net.*

A TREATISE ON THE FUNCTIONS OF A COMPLEX VARIABLE.

By A. R. FORSYTH, Sc.D., F.R.S., Fellow of Trinity College, Cambridge, Author of "A Treatise on Differential Equations," etc. Royal 8vo, $8.50, *net.*

MACMILLAN & CO., PUBLISHERS, NEW YORK.

www.ingramcontent.com/pod-product-compliance
Lightning Source LLC
Chambersburg PA
CBHW020856210326
41598CB00018B/1681
* 9 7 8 3 7 4 2 8 9 2 4 5 4 *